Methods in Molecular Biology

Series Editor
John M. Walker
School of Life and Medical Sciences
University of Hertfordshire
Hatfield, Hertfordshire, AL10 9AB, UK

For further volumes: http://www.springer.com/series/7651

Induced Pluripotent Stem (iPS) Cells

Methods and Protocols

Edited by

Kursad Turksen

Ottawa Hospital Research Institute
Ottawa, ON, Canada

Andras Nagy

Mount Sinai Hospital Samuel Lunenfeld Research Institue
Toronto, ON, Canada

Humana Press

Editors
Kursad Turksen
Ottawa Hospital Research Institute
Ottawa, ON, Canada

Andras Nagy
Mount Sinai Hospital Samuel Lunenfeld
Research Institue
Toronto, ON, Canada

ISSN 1064-3745 ISSN 1940-6029 (electronic)
Methods in Molecular Biology
ISBN 978-1-4939-3054-8 ISBN 978-1-4939-3055-5 (eBook)
DOI 10.1007/978-1-4939-3055-5

Library of Congress Control Number: 2015960271

Springer New York Heidelberg Dordrecht London

Printed on acid-free paper

Humana Press is a brand of Springer
Springer Science+Business Media LLC New York is part of Springer Science+Business Media (www.springer.com)

Preface

The demonstration of reprogramming of somatic cells with four transcription factors by Yamanaka's group in 2006 provided a molecular basis for the earlier results of J. Gurdon published in 1962. These ground-breaking developments have led to a flurry of studies aimed at better understanding many facets of stem cell biology, lineage commitment, and differentiation. The scope and volume of the ongoing work made it timely to put together a series of protocols that can be used by both experts and novices interested in stem cells. We are grateful to all the contributors who agreed to describe their detailed protocols. Without their generosity and input, this volume would not have been possible.

One of us [K.T.] would like to acknowledge the outstanding sabbatical spent in the Nagy lab and all the stimulating discussions on reprogramming that occurred during that time. A very special thank you goes to Masha for her expert guidance in training in the techniques of reprogramming and differentiation of iPS cells

We are also grateful to John Walker, Editor in Chief of the Methods in Molecular Biology series, for supporting our idea to put together this volume. We acknowledge Patrick Morton, Editor of the Springer Protocol Series, for his continuous encouragement as the volume was put together. Lastly, but not the least, we thank David Casey for his keen eye for missing details and for his availability to answer many questions during the preparation of this volume.

Ottawa, Canada — *Kursad Turksen*
Toronto, ON, Canada — *Andras Nagy*

Contents

Contributors

MARIELLE AFANASSIEFF • *INSERM, U846, Bron, France; Stem Cell and Brain Research Institute, Bron, France; Université de Lyon, Villeurbanne, France*
MARCOS J. ARAÚZO-BRAVO • *Computational Biology and Systems Biomedicine, Biodonostia Health Research Institute, San Sebastián, Spain; IKERBASQUE, Basque Foundation for Science, Bilbao, Spain*
TUGBA BAGCI-ONDER • *Brain Cancer Research and Therapy Laboratory, Koç University School of Medicine, Istanbul, Turkey*
MARK P BARTOLD • *Colgate Australian Clinical Dental Research Centre, School of Dentistry, University of Adelaide, Adelaide, SA, Australia*
PAOLA A. BIGNONE • *Mandala Biosciences LLC, San Diego, CA, USA*
SARA BRIGHTWELL • *MRC Centre for Regenerative Medicine, University of Edinburgh, Edinburgh BioQuarter, Edinburgh, UK*
GAEL CAGNONE • *The Mitochondrial Genetics Group, Centre for Genetic Diseases, MIMR-PHI Institute of Medical Research, Clayton, VIC, Australia; The Department of Molecular and Translational Science, Monash University, Clayton, VIC, Australia*
ISHITA CHATTERJEE • *Department of Pharmacology, University of Illinois at Chicago, Chicago, IL, USA*
POOJA CHAUDHARI • *Department of Oncology, The Sidney Kimmel Comprehensive Cancer Center Cellular and Molecular Medicine Graduate Program, Johns Hopkins University School of Medicine, Baltimore, MD, USA*
SAI H. CHAVALA • *Department of Ophthalmology, University of North Carolina, Chapel Hill, NC, USA*
HYUN WOO CHOI • *Department of Animal Biotechnology, Konkuk University, Seoul, South Korea*
DUNCAN E. CROMBIE • *Centre for Eye Research Australia and Royal Victorian Eye and Ear Hospital, East Melbourne, VIC, Australia; Department of Ophthalmology, University of Melbourne, East Melbourne, VIC, Australia*
CHUN CUI • *Wuxi Medical School, Jiangnan University, Wuxi, People's Republic of China*
JASON DANG • *Department of Bioengineering, Department of Pediatrics, University of California San Diego School of Medicine, La Jolla, CA, USA*
RADBOD DARABI • *Center for Stem Cell and Regenerative Medicine, The Brown Foundation Institute of Molecular Medicine, The University of Texas Health Science Center at Houston, Houston, TX, USA*
LIANI DEVITO • *Stem Cell Laboratories, Assisted Conception Unit, Guy's Hospital, London, UK*
RODNEY J. DILLEY • *Department of Surgery, O'Brien Institute, University of Melbourne, Fitzroy, VIC, Australia; Ear Science Institute Australia and Ear Sciences Centre, School of Surgery, University of Western Australia, Nedlands, WA, Australia*
JEONG TAE DO • *Department of Animal Biotechnology, Konkuk University, Seoul, South Korea; Department of Animal Biotechnology, College of Animal Bioscience and Technology, South Korea*

GREGORY J. DUSTING • *Department of Surgery, O'Brien Institute, University of Melbourne, Fitzroy, VIC, Australia; Centre for Eye Research Australia and Royal Victorian Eye and Ear Hospital, East Melbourne, VIC, Australia*
JAMES R. DUTTON • *Stem Cell Institute, University of Minnesota, Minneapolis, MN, USA*
TAKUMI ERA • *Kumamoto University, Kumamoto, Japan*
STEVEN L. FARNSWORTH • *South Texas Veterans Health Care System, University of Texas Health Science Center, San Antonio, TX, USA*
OLIVIER GOUREAU • *Institut de la Vision, Institut National de la Santé et de la Recherche Médicale, U968, Paris, France; Université Pierre-et-Marie-Curie Paris 6, Sorbonne Universités, UMR S968, Paris, France; Centre National de la Recherche Scientifique, UMR 7210, Paris, France*
THOMAS GRAF • *Gene Regulation, Stem Cells and Cancer Programme, Centre for Genomic Regulation (CRG), Barcelona, Spain; Universitat Pompeu Fabra (UPF), Barcelona, Spain; Institució Catalana de Recerca i Estudis Avançats (ICREA), Barcelona, Spain*
VERONICA KON GRAVERSEN • *Department of Surgery, University of North Carolina at Chapel Hill, Chapel Hill, NC, USA*
LUCAS V. GREDER • *MRC Molecular Haematology Unit, Weatherall Institute of Molecular Medicine, Radcliffe Department of Medicine, John Radcliffe Hospital, University of Oxford, Oxford, UK*
STAN GRONTHOS • *Mesenchymal Stem Cell Laboratory, Cancer Theme, Level 5 South, SAHMRI, Adelaide, SA, Australia; Mesenchymal Stem Cell Laboratory, School of Medical Sciences, University of Adelaide, Adelaide, SA, Australia*
YOSHIHIKO HAYASHI • *Department of Cariology, Nagasaki University Graduate School of Biomedical Sciences, Nagasaki, Japan*
JACOB J. HEMMI • *South Texas Veterans Health Care System, University of Texas Health Science Center, San Antonio, TX, USA*
YEAN JU HONG • *Department of Animal Biotechnology, Konkuk University, Seoul, South Korea*
PETER J. HORNSBY • *South Texas Veterans Health Care System, University of Texas Health Science Center, San Antonio, SA, USA*
KIM HYNES • *Colgate Australian Clinical Dental Research Centre, School of Dentistry, University of Adelaide, Adelaide, SA, Australia; Mesenchymal Stem Cell Laboratory, School of Medical Sciences, University of Adelaide, Adelaide, SA, Australia; Mesenchymal Stem Cell Laboratory, Cancer Theme, Level 5 South, SAHMRI, Adelaide, SA, Australia*
DUSKO ILIC • *Stem Cell Laboratories, Assisted Conception Unit, Guy's Hospital, London, UK*
GUN-IL IM • *Department of Orthopaedics, Dongguk University Ilsan Hospital, Goyang, South Korea*
YOON-YOUNG JANG • *Department of Oncology, The Sidney Kimmel Comprehensive Cancer Center Cellular and Molecular Medicine Graduate Program Institute for Cell Engineering, Johns Hopkins University School of Medicine, Baltimore, MD, USA*
JUSTIN C. ST. JOHN • *The Mitochondrial Genetics Group, Centre for Genetic Diseases, MIMR-PHI Institute of Medical Research, Clayton, VIC, Australia; Department of Molecular and Translational Science, Monash University, Clayton, VIC, Australia*
JACQUELINE JOHNSON • *The Mitochondrial Genetics Group, Centre for Genetic Diseases, MIMR-PHI Institute of Medical Research, Clayton, VIC, Australia; The Department of Molecular and Translational Science, Monash University, Clayton, VIC, Australia*
KEISUKE KAJI • *MRC Centre for Regenerative Medicine, University of Edinburgh, Edinburgh BioQuarter, Edinburgh, UK*

Akihide Kamiya • *Laboratory of Stem Cell Therapy, Institute of Innovative Science and Technology, Tokai University, Isehara, Kanagawa, Japan*
Işil Karanfil • *Brain Cancer Research and Therapy Laboratory, Koç University School of Medicine, Istanbul, Turkey*
Shin-Il Kim • *Center for iPS Cell Research and Application (CiRA), Kyoto University, Kyoto, Japan*
Jong Soo Kim • *Department of Animal Biotechnology, Konkuk University, Seoul, South Korea*
Ji-Yun Ko • *Department of Orthopaedics, Dongguk University, Ilsan Hospital, Goyang, South Korea*
Erin E. Kohler • *Department of Pharmacology, University of Illinois at Chicago, Chicago, IL, USA*
Rachel A. Krupa • *Mandala Biosciences LLC, San Diego, CA, USA*
Shoen Kume • *Graduate School of Bioscience and Biotechnology, Tokyo Institute of Technology, Yokohama, Kanagawa, Japan*
David Larocca • *Mandala Biosciences LLC, San Diego, CA, USA; BioTime Inc., Alameda, CA, USA*
William Lee • *The Mitochondrial Genetics Group, Centre for Genetic Diseases, MIMR-PHI Institute of Medical Research, Clayton, VIC, Australia; The Department of Molecular and Translational Science, Monash University, Clayton, VIC, Australia*
Miao Li • *South Texas Veterans Health Care System, University of Texas Health Science Center, San Antonio, TX, USA*
Fei Li • *Department of Pharmacology, University of Illinois at Chicago, Chicago, IL, USA*
Jing Liao • *Broad Institute of MIT and Harvard, Cambridge, MA, USA; Harvard Stem Cell Institute, Cambridge, MA, USA; Department of Stem Cell and Regenerative Biology, Harvard University, Bauer Laboratories-Administrative Suite, Cambridge, MA, USA*
Shiang Y. Lim • *Department of Surgery, O'Brien Institute, University of Melbourne, Fitzroy, VIC, Australia*
Lin Liu • *State Key Laboratory of Medicinal Chemical Biology, Department of Cell Biology and Genetics, College of Life Sciences, Nankai University, Tianjin, China*
Asrar B. Malik • *Department of Pharmacology, University of Illinois at Chicago, Chicago, IL, USA*
Jian Mao • *State Key Laboratory of Medicinal Chemical Biology, Department of Cell Biology and Genetics, College of Life Sciences, Nankai University, Tianjin, China*
Danijela Menicanin • *Colgate Australian Clinical Dental Research Centre, School of Dentistry, University of Adelaide, Adelaide, SA, Australia; Mesenchymal Stem Cell Laboratory, School of Medical Sciences, University of Adelaide, Adelaide, SA, Australia; Mesenchymal Stem Cell Laboratory, School of Medical Sciences, University of Adelaide, Adelaide, SA, Australia*
Cristian Miere • *Stem Cell Laboratories, Assisted Conception Unit, Guy's Hospital, London, UK*
Anuja Mishra • *South Texas Veterans Health Care System, University of Texas Health Science Center, San Antonio, TX, USA*
Hiroyuki Miwa • *Kumamoto University, Kumamoto, Japan*
Andras Nagy • *Lunenfeld-Tanenbaum Research Institute, Mount Sinai Hospital, Toronto, Canada*

Masato Nakagawa • *Center for iPS Cell Research and Application (CiRA), Kyoto University, Kyoto, Japan*
Hiromitsu Nakauchi • *Division of Stem Cell Therapy, Center for Stem Cell Biology and Regenerative Medicine, Institute of Medical Science, The University of Tokyo, Tokyo, Japan; Institute for Stem Cell Biology and Regenerative Medicine, Stanford University School of Medicine, Stanford, CA, USA*
Amanda Neises • *Division of Regenerative Medicine MC1528B, Department of Medicine, Loma Linda University, Loma Linda, CA, USA*
Fabian Oceguera-Yanez • *Center for iPS Cell Research and Application (CiRA), Kyoto University, Kyoto, Japan*
Alice Pébay • *Centre for Eye Research Australia and Royal Victorian Eye and Ear Hospital, East Melbourne, VIC, Australia; Department of Ophthalmology, University of Melbourne, East Melbourne, VIC, Australia*
Roger A. Pedersen • *Anne McLaren Laboratory for Regenerative Medicine and Department of Surgery, Wellcome Trust – Medical Research Council Cambridge Stem Cell Institute, University of Cambridge, Cambridge, UK*
Rita C. R. Perlingeiro • *Department of Medicine, Lillehei Heart Institute, University of Minnesota, Minneapolis, MN, USA*
Alexander V. Pickering • *South Texas Veterans Health Care System, , University of Texas Health Science Center, San Antonio, TX, USA*
Jason Post • *Stem Cell Institute, University of Minnesota, Minneapolis, MN, USA*
Neha Prasad • *Department of Oncology, The Sidney Kimmel Comprehensive Cancer Center, Johns Hopkins University School of Medicine, Baltimore, MD, USA*
Zhifang Qiu • *South Texas Veterans Health Care System, University of Texas Health Science Center, San Antonio, TX, USA*
Tariq M. Rana • *Department of Pediatrics, University of California San Diego School of Medicine, La Jolla, CA, USA*
Jalees Rehman • *Department of Pharmacology, University of Illinois at Chicago, Chicago, IL, USA*
Sacha Reichman • *Institut de la Vision, Institut National de la Santé et de la Recherche Médicale, U968, Paris, France; Université Pierre-et-Marie-Curie Paris 6, UMR S968, Sorbonne Universités, Paris, France; Centre National de la Recherche Scientifique, UMR 7210, Paris, France*
Chiho Sakurai • *Center for iPS Cell Research and Application (CiRA), Kyoto University, Kyoto, Japan*
Pierre Savatier • *INSERM, U846, Bron, France; Stem Cell and Brain Research Institute, Bron, France; Université de Lyon, Villeurbanne, France*
Nobuhiro Shimozawa • *Tsukuba Primate Research Center (TPRC), National Institute of Biomedical Innovation (NIBIO), Tsukuba, Ibaraki, Japan*
Nobuaki Shiraki • *Department of Stem Cell Biology, Institute of Molecular Embryology and Genetics (IMEG), Kumamoto University, Kumamoto, Japan*
Priyadharshini Sivakumaran • *Department of Surgery, O'Brien Institute, University of Melbourne, Fitzroy, VIC, Australia*
Filipa A.C. Soares • *Anne McLaren Laboratory for Regenerative Medicine and Department of Surgery, , Wellcome Trust – Medical Research Council Cambridge Stem Cell Institute, University of Cambridge, Cambridge, UK*
Bruno Di Stefano • *Gene Regulation, Stem Cells and Cancer Programme, Centre for Genomic Regulation (CRG), Barcelona, Spain*

Ruijun Jeanna Su • *Division of Regenerative Medicine MC1528B, Department of Medicine, Loma Linda University, Loma Linda, CA, USA*
K. Sugimoto • *Department of Cardiology, Graduate School of Biomedical Sciences, Nagasaki University, Nagasaki, Japan*
Claire Sun • *The Mitochondrial Genetics Group, Centre for Genetic Diseases, MIMR-PHI Institute of Medical Research, Clayton, VIC, Australia; The Department of Molecular and Translational Science, Monash University, Clayton, VIC, Australia*
Yann Tapponnier • *INSERM, U846, Bron, France; Stem Cell and Brain Research Institute, Bron, France; Université de Lyon, Villeurbanne, France*
Lipeng Tian • *Department of Oncology, The Sidney Kimmel Comprehensive Cancer Center, Johns Hopkins University School of Medicine, Baltimore, MD, USA*
Vincent Truong • *Stem Cell Institute, University of Minnesota, Minneapolis, MN, USA*
Kahoko Umeda • *Department of Stem Cell Biology, Institute of Molecular Embryology and Genetics (IMEG), Kumamoto University, Kumamoto, Japan; Program for Leading Graduate Schools "HIGO (Health life science; Interdisciplinary and Glocal Oriented) Program", Kumamoto University, Kumamoto, Japan*
Vijesh Vaghjiani • *The Mitochondrial Genetics Group, Centre for Genetic Diseases, MIMR-PHI Institute of Medical Research, Clayton, VIC, Australia; The Department of Molecular and Translational Science, Monash University, Clayton, VIC, Australia*
Ludovic Vallier • *Anne McLaren Laboratory for Regenerative Medicine and Department of Surgery, Wellcome Trust – Medical Research Council Cambridge Stem Cell Institute, University of Cambridge, Cambridge, UK; Wellcome Trust Sanger Institute, Hinxton, UK*
Patrick J. Walsh • *Stem Cell Institute, University of Minnesota, Minneapolis, MN, USA*
Kishore K. Wary • *Department of Pharmacology, University of Illinois at Chicago, Chicago, IL, USA*
Michael D. West • *BioTime Inc., Alameda, CA, USA*
Knut Woltjen • *Center for iPS Cell Research and Application (CiRA), Kyoto University, Kyoto, Japan*
Shinya Yamanaka • *Center for iPS Cell Research and Application (CiRA), Kyoto University, Kyoto, Japan; Gladstone Institute of Cardiovascular Disease, University of California, San Francisco, CA, USA*
Taiji Yamazoe • *Department of Stem Cell Biology, Institute of Molecular Embryology and Genetics (IMEG), Kumamoto University, Kumamoto, Japan*
Ayaka Yanagida • *Division of Stem Cell Therapy, Center for Stem Cell Biology and Regenerative Medicine, Institute of Medical Science, The University of Tokyo, Tokyo, Japan*
Ka-Yu Yeung • *The Mitochondrial Genetics Group, Centre for Genetic Diseases, MIMR-PHI Institute of Medical Research, Clayton, VIC, Australia; The Department of Molecular and Translational Science, Monash University, Clayton, VIC, Australia*
Xiao-Bing Zhang • *Division of Regenerative Medicine MC1528B, Department of Medicine, Loma Linda University, Loma Linda, CA, USA*

Methods in Molecular Biology (2016) 1357: 1–22
DOI 10.1007/7651_2015_274

Published online: 01 July 2015

The *piggyBac* Transposon as a Platform Technology for Somatic Cell Reprogramming Studies in Mouse

Knut Woltjen, Shin-Il Kim, and Andras Nagy

Abstract

Somatic cell reprogramming to induced pluripotent stem cells (iPSCs) is a revolutionary technology, with repercussions affecting modern functional genomics and regenerative medicine. Still, relatively little is known about the processes underlying this dramatic cellular and molecular metamorphosis. Reprogramming technology based on the implementation of *piggyBac* (PB) transposons has enabled studies of iPSC reprogramming mechanisms, shedding an increasing light on these processes. Unique characteristics of PB transposons such as efficient genomic integration, unlimited cargo capacity, robust gene expression, and even seamless excision highlight the importance of this transgenic tool in advancing stem cell biology. In this chapter, we provide a detailed overview of versatile primary iPSC generation from mouse somatic cells using PB transposons, and the subsequent establishment of robust secondary reprogramming systems. These protocols are highlighted with examples from recent studies as to how *PB* has been, and continues to be, conducive to the dissection of reprogramming processes at the cellular and molecular levels.

Keywords: Induced pluripotent stem cells (iPSCs), Reprogramming, Secondary reprogramming, Transposon, *piggyBac*, Doxycycline regulation

1 Introduction

The first demonstration of induced pluripotent stem cell (iPSC) technology employed a battery of retroviral expression vectors to deliver the four Yamanaka factors, Oct3/4, Sox2, Klf4, and c-Myc [1]. Shortly thereafter, iPSC technology developments took two main routes: consolidation of the reprogramming factors into a single vector to ease both delivery and the number of required genomic integrations, and exploration of non-viral and non-integrating delivery methods [2]. Proceeding to address both these objectives, the *piggyBac* (PB) transposon rose as a unique reprogramming method [3, 4].

The PB transposon is a mobile DNA element, which can be integrated into the genome for stability during reprogramming. PB has no apparent species barrier, and is effective in the derivation of iPSCs from various sources [5]. Practically, the PB vector has a transgene cargo capacity far exceeding that of viruses. PB easily

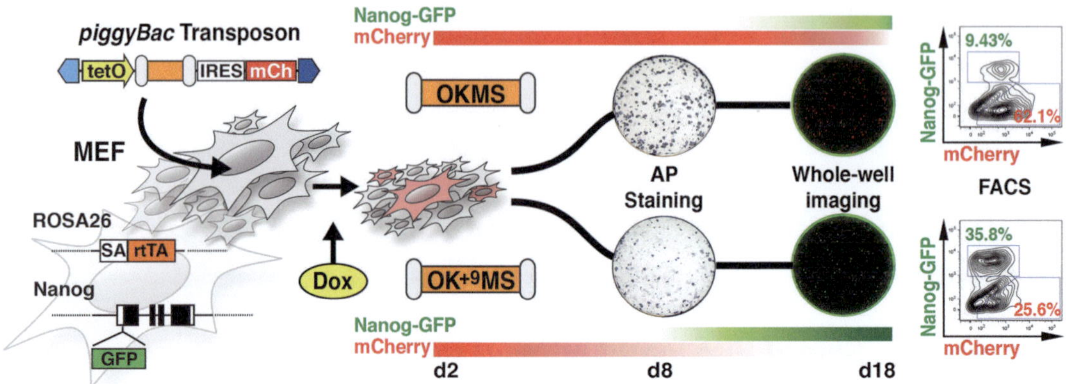

Fig. 1 Studying mouse iPSC reprogramming mechanisms using 1° *piggyBac* transposition. PB-TAC delivers reprogramming cassettes OKMS and $OK^{+9}MS$ into ROSA-rtTA; Nanog-GFP MEFs in a dox-inducible manner. PB transgenic cells are detected and monitored via mCherry, while pluripotency is indicated by activation of Nanog-GFP and eventual PB silencing (loss of mCherry). Reprogramming effects are assessed by colony formation (AP staining) and fluorescence microscopy of entire wells (composite 10 × 10 fields) for Nanog-GFP^+ and $mCherry^+$ on d8 and d18, respectively. Finally, FACS analysis of Nanog-GFP and mCherry expression in d18 SSEA-1^+ populations reveals that $OK^{+9}MS$ more effectively induces complete reprogramming

accepts Yamanaka factor cassettes linked with 2A peptides (polycistronic cassettes) in conjunction with reporters and selection markers. Moreover, delivery of PB transposons as plasmids using standard cell transfection methods provides a level of simplicity and safety unmatched by viruses. Uniquely, PB elements may be subsequently removed using transposase re-expression to generate footprint-free mouse and human iPSCs [6, 7].

These properties make PB a versatile tool. Primary (1°) transfection of PB elements is a straightforward and adaptable process, which, coupled with reporter cell lines and pluripotency assays, enables studies of reprogramming mechanisms (Fig. 1) [8], or the derivation of 1° *piggyBac* iPSC (PB-iPSC) clones (Fig. 2). As the PB system employs doxycycline (dox)-regulated expression of the Yamanaka factors, 1° PB-iPSCs can be withdrawn from dox and differentiated in vitro or in vivo to give rise to secondary (2°) reprogramming systems (Fig. 3). These 2° somatic cells derived from 1° PB-iPSCs are "primed" with transposons, meaning that they may be re-induced to reprogram simply by dox addition [9]. Experimental systems based on 2° reprogramming have enabled intensive studies of reprogramming processes towards pluripotency [10]. Moreover, 2° somatic cells from various tissue sources may ultimately lead to regenerative medicine approaches employing in situ reprogramming.

In this chapter, we describe the nature and application of current PB reprogramming vectors for mechanistic analyses during

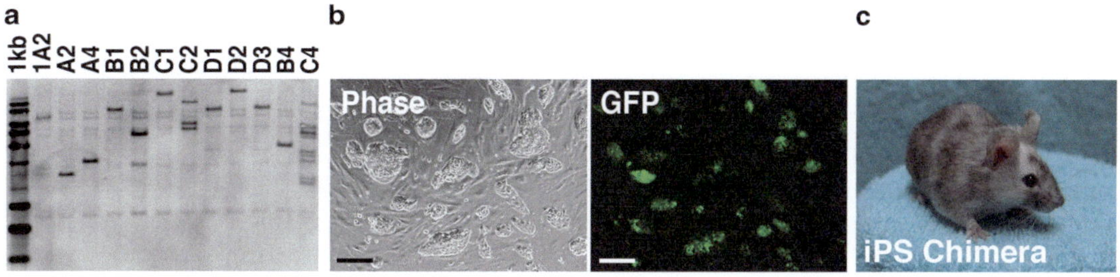

Fig. 2 Screening and validation steps in the selection of 1° PB-iPSC lines. (**a**) Southern blot analysis reveals PB copy numbers, and identifies unique clones (A2-C4). PB-iPSC genomic DNA was digested with *Hin*dIII and probed with a DIG-labeled mCherry probe. Note that mCherry also detects repetitive elements in the mouse genome, resulting in a common background pattern (*faint bands*) in 1A2 control sample. (**b**) Microscopy reveals Nanog-GFP expression in PB-iPSC clones maintained on feeders in media containing FBS and LIF. (**c**) Chimera contribution is a functional validation of pluripotency that can also act as source of cells for 2° reprogramming. Scale bars, 200 μM

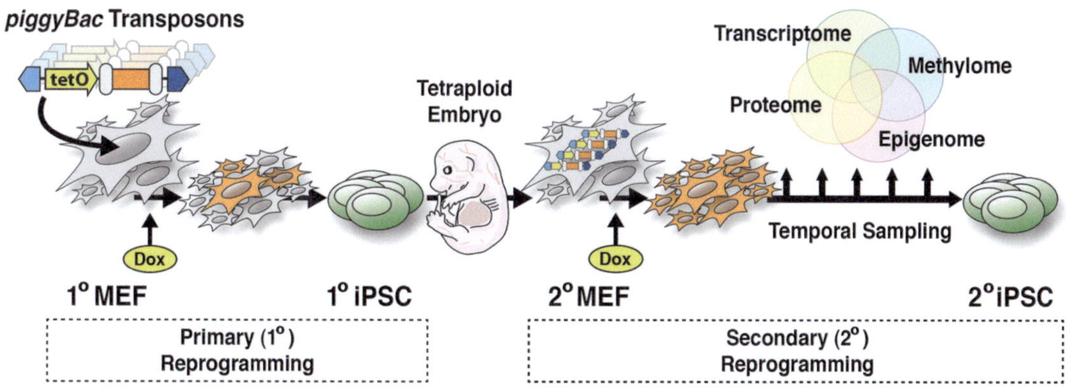

Fig. 3 Generation and application of 2° reprogramming systems to understand reprogramming processes. Select PB-iPSCs produced through 1° reprogramming are introduced into diploid or tetraploid host blastocysts to produce chimeric embryos. Isolation of 2° MEFs provides a population of transgenic cells that are primed for a homogenous response to dox treatment. Bulk reprogramming of 2° MEFs provides sufficient material for complex molecular assays, which can be performed temporally and in parallel

1° reprogramming, and considerations for clonal 1° PB-iPSC isolation for the purpose of establishing robust 2° reprogramming systems. We highlight strategies for the continued monitoring of intermediate reprogramming populations and stabilization of pluripotency by tracking the activity of reporters such as mCherry and Nanog-GFP. Analysis of reprogramming processes by whole-well imaging and flow cytometry using conventional pluripotency markers is described at intermediate and concluding steps in the protocol. Finally, we address strategies to derive and apply 2° reprogramming systems to the study of reprogramming processes either in vitro or in vivo.

2 Materials

Plasmids and reporter mouse strains described herein are available through the RIKEN Bio Resource Center DNA Bank (http://dna.brc.riken.jp/) or Experimental Animal Division (http://mus.brc.riken.jp/en/), respectively, and may be accessed using the database ID numbers in the sections below.

2.1 Mouse Embryonic Fibroblast Preparation

2.1.1 Mice

1. *Nanog*$^{GFP\text{-}IRES\text{-}Puro}$ mice [11] or appropriate transgenic or wild-type strain.

   Strain name: STOCK Tg (Nanog-GFP, Puro)1 Yam. RBRC No.: RBRC02290.
2. m2-rtTA mice [12] or equivalent with constitutive expression of the Tet-ON transactivator.
3. Pregnant dams at 13.5 days post-coitum (d.p.c) from natural mating or in vitro fertilization (IVF) using the two strains in **steps 1** and **2**.

2.1.2 MEF Medium

1. DMEM (high glucose) (Nacalai Tesque, #08488-55).
2. Penicillin–streptomycin solution (Gibco, #15140-122).
3. L-Glutamine 200 mM (100×) (Gibco, #25030-081).
4. Fetal bovine serum (FBS) (Gibco, #26140) (*see* **Note 1**).
5. MEF medium:
   (a) Add 3 mL penicillin–streptomycin, 6 mL L-glutamine, and 50 mL FBS into 500 mL DMEM medium.

2.1.3 MEF Isolation

1. 13.5 d.p.c. embryos produced by mating.
2. MEF medium (Section 2.1.2).
3. Dulbecco's phosphate-buffered saline (DPBS), Mg^{2+} and Ca^{2+} free (Nacalai Tesque, #14249-24).
4. 0.25 % Trypsin-EDTA (1×) (Gibco, #25200-072).
5. Disposable Scalpel (Kai Medical, #511-A).
6. 60 mm Tissue Culture Dish (FALCON, #353004).
7. EASY FLASK 175 F (Nunc, #159910).
8. 15 mL High-Clarity Polypropylene Conical Tube (BD Falcon, #352096).

2.1.4 MEF Passage

1. MEF medium (Section 2.1.2).
2. DPBS, Mg^{2+} and Ca^{2+} free (Nacalai Tesque, #14249-24).
3. 0.25 % Trypsin-EDTA (1×) (Gibco, #25200-072).
4. EASY FLASK 175 F (Nunc, #159910).
5. 50 mL Polypropylene Conical Tube (BD Falcon, #352070).

2.1.5 Cell Counting

1. Single-cell suspension (prepare as described for passage in Section 3.2.2).
2. Trypan Blue Stain 0.4 % (Gibco, #15250).
3. Microcentrifuge.
4. TC10™ Automated Cell Counter (BIO-RAD).
5. Counting Slides, Dual Chamber for Cell Counter (BIO-RAD, #145-0011).

2.1.6 Preparation of Frozen MEF Stocks

1. Single-cell suspension (2.5×10^6 cells/vial, Section 3.2.2).
2. Dimethyl sulfoxide (DMSO) HYBRI-MAX® (Sigma, #D2650).
3. Cell freezing solution (1×):
   (a) 10 % DMSO and 40 % FBS in 50 % MEF medium.
4. Nalgene® System 100™ Cryogenic Tubes (NALGENE, #5000-1020).
5. Nalgene™ Cryo 1 °C Freezing Container (NALGENE, #5100-0001).
6. 2-Propanol (isopropyl alcohol) (JUNSEI, #64605-0330).
7. MycoAlert® Mycoplasma Detection Kit (LONZA, #LT07).

2.2 Primary (1°) Reprogramming

2.2.1 Gelatin Coating

1. 100 mm Tissue Culture Dish (FALCON, #353003).
2. 6-Well Tissue Culture Plate (FALCON, #353046).
3. 60 mm Tissue Culture Dish (FALCON, #353004).
4. Gelatin, from porcine skin, Type A (Sigma, #G1890).
5. UltraPure™ DNase/RNase-Free Distilled Water (Invitrogen, #10977-015).
6. Gelatin stock solution (1 %):
   (a) Dissolve 5 g of gelatin in 500 mL of distilled water, autoclave, and store at room temperature.
7. Gelatin working solution (0.1 %):
   (a) Warm the 1 % gelatin stock solution to 37 °C and add 50 mL of the stock to 450 mL of distilled water. Filter the working solution with a bottle-top filter (0.22 μm) and store at 4 °C for up to 4 weeks.

2.2.2 MEF Defrost

1. Frozen stock (Section 2.1.6).
2. Gelatin-coated tissue culture plates, 100 mm (Section 2.2.1).
3. MEF medium (Section 2.1.2).

2.2.3 MEF Seeding for PB Transfection

1. Gelatin-coated tissue culture plates, 6-well.
2. MEF medium (Section 2.1.2).
3. DPBS, Mg^{2+} and Ca^{2+} free (Nacalai Tesque, #14249-24).
4. 0.25 % Trypsin-EDTA (1×) (Gibco, #25200-072).

2.2.4 PB Transfection

1. DNA (*see* **Note 2**):
   - (a) Transposase: pCAG-PBase (RDB13241) or pCyL43 (Wellcome Trust Sanger Center, http://www.sanger.ac.uk/form/Sanger_CloneRequests).
   - (b) Transposon: PB-TAC-OKMS (RDB13133), -OK^{+9}MS (RDB13135), or similar [8].
2. MEF medium (Section 2.1.2).
3. Opti-MEM I reduced-serum medium (Invitrogen, #31985-070).
4. FuGENE® HD transfection reagent (Promega, #E2311).

2.2.5 Mouse iPSC Culture Medium

1. DMEM (high glucose) (Nacalai Tesque, #08488-55).
2. Penicillin–streptomycin solution (Gibco, #15140-122).
3. Glutamax (100×) (Gibco, #35050-61).
4. Sodium-pyruvate 100 mM (100×) (Gibco, #11360-070).
5. MEM NEAA nonessential amino acids (100×) (Gibco, #11140-050).
6. 2-Mercaptoethanol, 1000× (Gibco, #21985-023).
7. LIF, human recombinant, culture supernatant (Wako, #125-05603).
8. FBS (Gibco, #26140).
9. Sartolab RF 500 Filter System (Sartorius, #180C2——————E).
10. Mouse iPSC culture medium:
    - (a) Thaw penicillin–streptomycin, Glutamax (100×), sodium pyruvate 100 mM (100×), and MEM NEAA nonessential amino acids (100×) at 4 °C overnight or room temperature.
    - (b) To make complete medium, add 1.2 mL 2-mercaptoethanol, 3 mL penicillin–streptomycin, 6 mL Glutamax (100×), 6 mL sodium pyruvate 100 mM (100×), 6 mL MEM NEAA nonessential amino acids (100×), 90 mL FBS, and 600 μL LIF to 500 mL DMEM medium and mix well. Filter sterilize. Complete mouse iPSC culture medium is stable for 3 weeks at 4 °C.

2.2.6 Doxycycline Solution

1. Doxycycline Hyclate (LKT Laboratories, Inc, #D5897).
2. UltraPure™ DNase/RNase-Free Distilled Water (Invitrogen, #10977-015).
3. Millex-GP Syringe Filter Unit, 0.22 μm (MERCK MILLIPORE, #SLGP033RS).
4. Dox solution (1 mg/mL):
   - (a) Dissolve doxycycline hyclate in distilled water to a final concentration of 10 mg/mL. Aliquot and store at −20 °C.
   - (b) Dilute concentrated stock 1:10 with distilled water and filter-sterilize through a 0.22 μm syringe filter. Aliquot and store at −20 °C. (*see* **Note 3**).

2.2.7 Alkaline Phosphatase Staining

1. Reprogramming culture (day 8).
2. BCIP/NBT Alkaline Phosphatase Substrate Kit IV (SK-5400, Vector Laboratories).
3. AP buffer: 100 mM Tris–HCl, pH 9.5 buffer.
4. BCIP/NBT substrate working solution (light sensitive).
   (a) Add two drops of Reagent 1 to 5 mL of AP Buffer. Mix well.
   (b) Add two drops of Reagent 2. Mix well.
   (c) Add two drops of Reagent 3. Mix well.
5. Color camera (Canon IXY Digital 900IS).

2.2.8 Surface Marker Analysis: Flow Cytometry

1. FACS buffer:
   (a) Add 1 mL FBS to 50 mL DPBS for a final concentration of 2 %.
2. 5 mL Polystyrene Round-Bottom Tube (BD Falcon, #352008).
3. 5 mL Polystyrene Round-Bottom Tube with Cell-Strainer Cap (BD Falcon, #352235).
4. Alexa Fluor® 647 SSEA-1 (480) (Santa Cruz Biotechnology, #sc-21702).
5. Alexa Fluor® 647 Mouse IgM, κ Isotype Control (BD Pharmingen, #560806).
6. BD LSR Fortessa™ Cell Analyzer (BD Biosciences).
7. FlowJo software (Tree Star Inc.).

2.2.9 Whole-Well Fluorescence Microscopy Imaging

1. Nikon BioStation CT (Nikon Corporation, Japan).

2.3 PB-iPSC Clone Screening/Validation for Secondary (2^0) Reprogramming Application

2.3.1 Cells

1. Mouse PB-iPSCs from *Nanog*$^{GFP\text{-}IRES\text{-}Puro}$/*m2-rtTA* MEF reprogramming.
2. Mouse feeder cells (DR4) [13]:
   (a) A day before seeding, defrost a vial of mitomycin C-treated DR4 feeder stock and seed the cells at a density of 4.4×10^4 cells per 10 cm^2 in gelatin-coated tissue culture plate with MEF medium (*see* **Note 4**).

2.3.2 PB-iPSC Colony Picking

1. 96-Well, Flat-bottom Tissue Culture Plate (FALCON, #353075) with DR4 feeders.
2. Stereomicroscope (Olympus SZ61).
3. Mouse iPSC culture medium (Section 2.2.5).
4. 96-Well, V-bottom Tissue Culture Plate (Nunc, #249935).
5. Single-channel pipette (1–10 μL volume range).

2.3.3 PB-iPSC Expansion

1. Two 96-well tissue culture plates with DR4 feeders.
2. Two 96-well tissue culture plates coated with gelatin.
3. DPBS, Mg^{2+} and Ca^{2+} free (Nacalai Tesque, #14249-24).
4. 0.25 % Trypsin-EDTA (1×) (Gibco, #25200-072).
5. Mouse iPSC culture medium (Section 2.2.5).

2.3.4 Nanog-GFP Reporter Analysis by Flow Cytometry

1. DPBS, Mg^{2+} and Ca^{2+} free (Nacalai Tesque, #14249-24).
2. 0.25 % Trypsin-EDTA (1×) (Gibco, #25200-072).
3. FACS buffer (Section 2.2.8).

2.3.5 PB-iPSC Freezing

1. DPBS, Mg^{2+} and Ca^{2+} free (Nacalai Tesque, #14249-24).
2. 0.25 % Trypsin-EDTA (1×) (Gibco, #25200-072).
3. Mouse iPSC culture medium (Section 2.2.5).
4. Thick paper towel.
5. Aluminum foil.
6. Cell freezing solution (2×):
   (a) 20 % DMSO in 80 % FBS.

2.3.6 Splinkerette PCR Screening

1. For a detailed protocol, please refer to reference [7].

2.3.7 Southern Blotting

1. For a detailed protocol, please refer to the manufacturer's online resource (https://lifescience.roche.com/dig) and technical manual (https://lifescience.roche.com/wcsstore/RASCatalogAssetStore/Articles/05353149001_08.08.pdf) for DIG-labeled nucleic acid detection.

2.3.8 Preparation of Candidate PB-iPSC Lines

1. Frozen 96-well plate (Section 3.4.4).
2. 24-Well tissue culture plate (FALCON, #353047) with DR4 feeders.
3. Mouse iPSC culture medium (Section 2.2.5).
4. 6-Well tissue culture plate with DR4 feeders.
5. 60 mm tissue culture dish with DR4 feeders.

3 Methods

3.1 The PB Vector 1° Reprogramming System

The first PB reprogramming experiments employed the PB-TAB transposon [4], which is a dox-inducible, reporter-linked (βgeo; a fusion of β-galactosidase and the neomycin-resistant gene) reprogramming system. The PB-TAB vector is the basis for the original

PB 2° reprogramming systems [4, 9, 14], where the four Yamanaka factors were delivered by four individual PB-TAB transposons.

In order to achieve constant live tracking of dox-responsive cells by microscopic imaging and flow cytometry, we produced the PB-TAC transposon, which delivers the Yamanaka factors linked to mCherry in a dox-inducible manner [8]. Note that for dox-inducible regulation of reprogramming, the target cells must be transgenic for the rtTA transactivator (see below). In order to overcome the need for such specialized transgenic backgrounds, we also developed PB transposons containing the entire dox-inducible system as a single, All-in-One vector as outlined in Kim, Chapter 9. The entire PB-TAC series and associated All-in-One vectors are available from the RIKEN BRC DNA Bank (http://dna.brc.riken.jp) using accession numbers RDB13131–RDB13138 and RDB13242–RDB13247, respectively. In this chapter, we describe the application of PB-TAC reprogramming vectors carrying OKMS and $OK^{+9}MS$ cassettes. For details on Gateway cloning of novel reprogramming cassettes, please refer to Section 3.1 of Chapter 9 in this volume.

3.2 Mouse Embryonic Fibroblast Preparation

Conventional reprogramming frequently makes use of mouse embryonic fibroblasts (MEFs) as a starting cell source. These cells are readily accessible, easily maintained in culture, and produced in abundance from wild-type, transgenic, or chimeric mice. For optimal reprogramming efficiencies, MEF passage numbers should be kept to a minimum. For the purposes of this chapter we focus on MEF isolation, expansion, and cryopreservation from transgenic mouse embryos which combine the constitutively expressed m2-rtTA Tet-ON transactivator [12] with a Nanog-GFP pluripotency reporter [11]. Procedures for mouse breeding and aseptic dissection of mouse embryos are as previously described [15].

3.2.1 Isolating MEFs

1. Set up timed matings and check vaginal plugs or simply separate females after 24 h of co-housing (*see* **Note 5**).
2. Observe until 13.5 d.p.c., and sacrifice pregnant dams.
3. Decapitate and eviscerate embryos (*see* **Note 6**).
4. Rinse the embryo tissue with DPBS and transfer to a 60 mm tissue culture dish (*see* **Note 7**).
5. Mince each embryo with a sterile scalpel to create small pieces (~1–2 mm).
6. Add 1 mL 0.25 % trypsin-EDTA and incubate for 15 min at 37 °C (*see* **Note 8**).
7. After 15 min, add 5 mL MEF medium and break any remaining clumps by pipetting up and down several times throughout the entire dish (*see* **Note 9**).

8. Transfer the cells and tissue into a 15 mL conical tube.
9. Wash the dishes with 5 mL MEF medium and pool into the same 15 mL conical tube.
10. Spin down the cells at 120 × *g* for 5 min.
11. Discard the medium and resuspend the cell pellet in 5 mL MEF medium. Transfer to a 175 cm^2 flask containing 15 mL MEF medium. Rinse the tube to collect residual cells before discarding.
12. Incubate the flask overnight in a humidified 37 °C, 5 % CO_2, incubator to allow the cells to adhere and grow.

3.2.2 Passaging MEFs

1. After 72-h culture, the cells should reach confluency (*see* **Note 10**). Discard the medium and wash the flask once with 6 mL DPBS.
2. Detach cells from the flask by incubating with 2 mL 0.25 % trypsin-EDTA for 5 min at 37 °C.
3. Add 10 mL MEF medium into the flask.
4. Generate a single-cell suspension by pipetting up and down several times throughout the entire flask. Observe the single-cell suspension under phase-contrast microscopy.
5. Pool the single-cell suspension into a 50 mL conical tube.
6. Wash the flask with 10 mL MEF medium and pool into the same 50 mL conical tube.
7. Spin down the cells at 120 × *g* for 5 min.
8. Discard the medium and resuspend the pellet in 5 mL MEF medium. Increase the media volume to 25 mL.
9. Add 5 mL each into four 175 cm^2 flasks containing 15 mL MEF medium (1:4 passage). Incubate the flasks overnight in a humidified 37 °C, 5 % CO_2 (*see* **Note 11**).

3.2.3 Freezing MEFs

Before proceeding to freezing steps, test a sample of the primary culture for contamination using the MycoAlert® Mycoplasma Detection Kit.

1. After 72-h culture, harvest cells as described in Section 3.2.2, **steps 2–7**, resuspending in 2 mL MEF medium per flask processed (*see* **Note 12**).
2. Count viable cells by trypan blue exclusion.
   (a) Add 11 μL of cell suspension into a 1.5 mL microcentrifuge tube containing 11 μL of 0.4 % trypan blue and mix by gently pipetting up and down several times.
   (b) Place 10 μL mixture into each side of the counting slide.

(c) Count viable cell numbers with the TC10 Automated Cell Counter.

3. Spin down the cells at 120 × *g* for 5 min.
4. Discard the medium and resuspend the pellet in an appropriate amount of cell freezing solution to obtain a concentration of 2.5 × 10^6 cells/mL (*see* **Note 13**).
5. Aliquot 1 mL of cell suspension in cell freezing solution into cryogenic tubes. Freeze at −80 °C overnight and transfer to liquid nitrogen for long-term storage.

3.3 Primary (1°) Reprogramming of MEFs with PB Transposons

Primary (1°) reprogramming refers to the de novo introduction of the Yamanaka factors into somatic cells for the purpose of 1° iPSC clone derivation (Figs. 1 and 2). These 1° iPSC clones may be used, in turn, to generate 2° MEFs for reprogramming analyses as described in Section 3.4. However, when applied directly to study the process of iPSC derivation, 1° reprogramming provides significant flexibility over 2° systems for adjusting reprogramming factor combinations [1] and concentrations [16, 17], as well as MEF genetic backgrounds, without the need for lengthy clonal characterization, in vitro or in vivo differentiation, or mouse breeding. Particularly, the flexibility of 1° reprogramming helped to study the effects of factor stoichiometry—the level of any given factor relative to the others—on reprogramming outcomes [8]. Employing a standardized PB-TAC reprogramming platform, direct comparison of publically available polycistronic constructs highlighted a difference in Klf4 cDNAs as an underlying cause of diverse reprogramming outcomes (Fig. 1) [8]. Through this approach, the expression level of Klf4 was correlated with colony formation efficiencies (favored by OKMS) and the ability of cells to achieve complete reprogramming (favored by $OK^{+9}MS$).

Standardized assays are used to quantify reprogramming initiation and stabilization at intermediate stages. Expression of alkaline phosphatase (AP) is an early but not definitive marker of reprogramming, and can be used in combination with whole-well imaging as a quantitative measure of initiation (Fig. 1). Activation of Nanog-GFP and silencing of mCherry may be monitored by imaging and flow cytometry. Silencing of the mCherry transgene in PB-TAC is an indicator of factor independence and full reprogramming [8, 18], with activation of pluripotency reporters often correlating with loss of mCherry (Fig. 1). Incorporating additional cell surface markers in flow cytometry can help resolve reprogramming subpopulations [19, 20], and custom analyses may be performed at key intermediate steps or more frequently (i.e., daily, or every second day). Imaging systems such as the BioStation CT or IM-Q (Nikon) that incorporate culture environments or even media perfusion can be employed to generate time-lapse images of reprogramming processes, even in multiple fluorescence channels. The following protocol exemplifies the basics of such reprogramming analyses.

3.3.1 Defrosting MEFs (−d4)

1. Into two 100 mm culture dishes, add 6 mL gelatin and leave for 30 min at room temperature. Discard the gelatin solution by aspiration and add 7 mL MEF medium.
2. Remove a cell stock vial of 2.5×10^6 cells/mL *Nanog*$^{GFP\text{-}IRES\text{-}Puro}$*/m2-rtTA* MEFs (Section 3.2.3) from the liquid N_2 tank and warm in a water bath at 37 °C until only a small suspended clump of ice remains (~1 min).
3. Wipe the vial with 70 % ethanol and transfer to the tissue culture hood.
4. Transfer cell suspension to a 15 mL conical tube with 5 mL MEF medium.
5. Centrifuge at $120 \times g$ to collect the cell pellet. Discard the medium and resuspend the cells in 6 mL MEF medium.
6. Add 3 mL cell suspension to each of the two 100 mm dishes prepared in **step 1**.

3.3.2 Seeding Nanog$^{GFP\text{-}IRES\text{-}Puro}$*/m2-rtTA MEFs (−d2)*

1. After 48-h culture, use 1 mL 0.25 % trypsin-EDTA per dish and a pooled 3 mL final resuspension volume to harvest and count two dishes as described in Section 3.2.3, **step 2**.
2. Seed MEFs at a density of 1.0×10^5 cells per 10 cm^2 surface area in MEF medium on a gelatinized 6-well tissue culture plate (*see* **Note 14**). Prepare enough dishes for long-term culture and intermediate analyses.
3. Incubate the plate overnight in a humidified 37 °C, 5 % CO_2. Change to fresh MEF medium before transfection.

3.3.3 piggyBac Transfection (−d1)

1. PB transposase and PB transposon plasmids are transfected using FuGENE® HD according to the manufacturer's instructions. Briefly:
   (a) Mix 0.5 μg of PB-TAC-OKMS or -OK^{+9}MS with 1.0 μg of pCAG-PBase in a 1.5 mL microcentrifuge tube.
   (b) Dilute the DNA to 100 μL with Opti-MEM I reduced-serum medium. Mix well with gentle vortexing.
   (c) Add 6 μL of FuGENE® HD (FuGENE® HD:DNA ratio of 6 μL:1.5 μg) into the 1.5 mL microcentrifuge tube containing DNA (*see* **Note 15**). Mix well with gentle vortexing. Incubate at room temperature for 15 min.
   (d) Add FuGENE® HD:DNA transfection complexes to the plate dropwise, and distribute by gentle swirling.

3.3.4 Dox Induction of Reprogramming Factors (d0)

After 24-h incubation, add mouse iPSC culture medium containing 1 μg/mL dox. Feed cells everyday with fresh dox-containing medium. Observe the dishes within 24 h for mCherry-positive cells and daily for colony formation (typically beginning between d3 and d5).

3.3.5 Colony Quantification by Alkaline Phosphatase Staining (d8)

1. Wash the reprogramming culture plate (6-well format) with DPBS once (*see* **Note 16**).
2. Add 1.5 mL BCIP/NBT substrate working solution (*see* **Note 17**).
3. Incubate the plate with the solution at RT in the dark until suitable staining develops, usually 20–30 min (*see* **Note 18**).
4. Acquire images with a color digital camera (Fig. 1).

3.3.6 Harvest for Analysis (d8) and Extended Culture (d8–d18)

1. Discard the medium and wash wells once with 2 mL DPBS.
2. Detach cells from the well by incubating cells with 500 μL 0.25 % trypsin-EDTA for 5 min at room temperature.
3. Add 3 mL mouse iPSC culture medium into the well.
4. Generate a single-cell suspension with a 5 mL pipette by pipetting up and down several times throughout the entire well.
5. Observe the single-cell suspension under phase-contrast microscopy.
6. Pool the single-cell suspension into a 15 mL conical tube.
7. Wash the well with 3 mL mouse iPSC culture medium and pool into the same 15 mL conical tube.
8. Spin down the cells at 120 × *g* for 5 min.
9. Discard the medium and resuspend the pellet in 1 mL mouse iPSC culture medium containing 1 μg/mL dox, as required.
10. Count cells as described in Section 3.2.3, **step 2**.
11. If FACS analysis or culture sub-fractionation using intermediate cell surface markers is planned, follow **steps 1–9** of Section 3.3.8 to analyze the harvested cells.
12. Seed the cells on gelatinized 6-well tissue culture plate at 3.0×10^5 cells in mouse iPSC culture medium containing 1.0 μg/mL dox, as required.
13. Feed cells everyday with fresh dox-containing medium until d18.

3.3.7 Whole-Well Fluorescence Microscopy Imaging (d18)

1. On d17 of reprogramming, change media and transfer the reprogramming cultures to the BioStation CT to allow for environmental equilibration before imaging.

2. Early on d18, before proceeding to Section 3.3.8, acquire whole-well images from GFP and mCherry fluorescence filters and phase contrast using 2× lenses.
3. The single-field images of each channel can be stitched using the automated image analysis software CL-Quant 3.0 to generate a whole-well image (Fig. 1). Further analysis of colony numbers and reporter gene expression requires the development of custom software macros.

3.3.8 Harvest for Analysis (d18) and Assessing Transgene Independence (d18–d25)

1. Harvest cells as described in Section 3.3.6, **steps 1–8**.
2. Discard the medium and resuspend the pellet in 1 mL mouse iPSC culture medium.
3. Count cells described in Section 3.2.3, **step 2**.

Conducting Flow Cytometry for Reprogramming Markers (d18)

1. Add 3×10^5 cells from Section 3.3.8 to a microcentrifuge tube.
2. Spin down the cells at 800 × *g* for 5 min at room temperature.
3. Discard the medium and resuspend the pellet in 50 μL FACS buffer.
4. Transfer 50 μL cell suspension to 5 mL polystyrene round-bottom tube.
5. Incubate the cells with 1 μL Alexa Fluor® 647 SSEA-1 antibody on ice in the dark for 30 min (*see* **Note 19**).
6. Add 2 mL FACS buffer and spin down the cells at 800 × *g* for 5 min at 4 °C.
7. Discard the FACS buffer and vortex the tube. Repeat **steps 6** and 7 twice.
8. Add 250 μL FACS buffer to the cell suspension and filter the cells using 5 mL polystyrene round-bottom tube with cell-strainer cap.
9. Acquire samples using a BD LSR Fortessa™ Cell Analyzer with BD FACSDiva Software. Flow cytometry data can be further analyzed using FlowJo software (Fig. 1).

Seeding Cells for Transgene-Independent iPSC Culture (d18–d25)

One day prior to seeding (d17), prepare two gelatin-coated 60 mm tissue culture dishes with DR4 feeders.

1. Seed the reprogrammed cells harvested in Section 3.3.8 on the DR4 feeder layer at a density of 2.5×10^4 cells in the absence of dox. Plate the same cells in the presence of 1.0 μg/mL dox on the second dish as a control (*see* **Note 20**).
2. Evaluate colony growth and reporter expression in the absence of dox using whole-well imaging (Section 3.3.7) and flow cytometry (Section 3.3.8), or pick well-isolated colonies to establish 1° PB-iPSC lines (Section 3.4.1).

3.4 Secondary (2°) Reprogramming Systems from Characterized 1° PB-iPSC Clones

The derivation of 1° PB-iPSC clones using dox-regulated transgenes opens the door to the development of robust 2° reprogramming systems, which provide homogeneity at the genetic level and thus have the potential for more uniform responses to dox treatment (Fig. 3). Additionally, the generation of adult mouse chimeras from 1° PB-iPSCs or gene-targeted ES cells provides a unique platform to study reprogramming of various cell types in vivo [21, 22]. Unlike gene-targeted systems that depend on *a priori* knowledge of commonly used transgene integration sites such as *Col1A1* [22–24] or *ROSA26* [25], reprogramming of 1° somatic cells by random integration of PB vectors acts as a phenotype-driven selective screen for loci that permit both a high fidelity and quality of reprogramming *a posteriori* [4].

2° reprogramming systems based on dox-inducible PB vectors have been employed to reveal fundamental aspects of iPSC derivation [26]. 2° PB systems revealed a critical role for BMP signaling in the mesenchymal-to-epithelial transition (MET) [27] and defined the initiation, maturation, and stabilization phases of reprogramming [18]. Alternative splicing patterns observed in ES and differentiated cells were also noted during 2° reprogramming [28]. More recently, a grandiose effort fueled by the PB-TAB 2° MEFs system [4] was used to establish time-course reprogramming data covering the transcriptome (including miRNA, lncRNA, and mRNA) [29], proteome (global and cell surface) [30], epigenome (H3K4me3, H3K27me3, and H3K36me3 chromatin marks) [14], and methylome (genome-wide CpG methylation) [31] of intermediate and fully reprogrammed cells. Moreover, the selection of unconventional cell morphologies during the course of dox induction suggested the existence of alternate pluripotent states [9]. This data collection has been made publically available as an interactive web resource (www.stemformatics.org), and should be considered as a reference and guide for future reprogramming studies.

Here, we highlight steps towards the isolation of 2° MEF populations, including screening 1° PB-iPSCs for dox independence, factor silencing and pluripotency reporter activation, PB copy number and positional mapping, and production of iPSC chimeras. For the purpose of this chapter, the application of 2° MEFs for reprogramming analysis succinctly follows the principles established for 1° PB-iPSC derivation: 2° MEFs are isolated (Section 3.2) from 1° PB-iPSC derived chimeras and induced to reprogram through the simple addition of dox (Section 3.3) without the need for PB transfection (Section 3.3.3).

3.4.1 PB-iPSC Colony Picking (d25)

One day before colony picking, prepare 96-well tissue culture plate with DR4 feeders.

1. Seven days after stably culturing cells in the absence of dox (Section 3.3.8), change to fresh mouse iPSC culture medium.
2. In a horizontal laminar flow hood, pick single colonies in 5 μL volume using a pipetman (P20). Transfer them sequentially into a 96-well V-bottom plate.
3. Add 25 μL 0.25 % trypsin-EDTA and incubate for 5 min at 37 °C.
4. Add 70 μL of mouse iPSC culture medium to the well.
5. Mix well by pipetting up and down several times and add 100 μL into 96-well-flat-bottom plate with DR4 in 50 μL mouse iPSC culture medium.
6. Incubate the plate overnight in a humidified 37 °C, 5 % CO_2 incubator. Change media early the next day. Continue culture for 2–3 days.

3.4.2 Expansion of PB-iPSC Single Clones for Screening (d28)

One day before passage, prepare two 96-well tissue culture plates with DR4 feeders for cryopreservation. On the day of passage, prepare two 96-well tissue culture plates coated with 50 μL gelatin for flow cytometry analysis (Section 3.4.3) and genomic DNA preparation (Section 3.4.5).

1. Approximately 3 days after colony picking, passage the cells by discarding the medium and washing once with 50 μL PBS.
2. Detach cells from the well by incubating with 30 μL 0.25 % trypsin-EDTA for 5–7 min at 37 °C.
3. Add 100 μL of mouse iPSC culture medium and generate a single-cell suspension by pipetting up and down several times.
4. Divide 30 μL cell suspension into each of the four 96-well tissue culture plates containing 70 μL mouse iPSC culture medium (1:4 expansion).

3.4.3 Flow Cytometry Analysis of PB-iPSC Clones (d30)

1. Two days after expansion, harvest the cells by washing once with 50 μL PBS and incubating with 30 μL 0.25 % trypsin-EDTA for 5–7 min at 37 °C.
2. Carefully aspirate the trypsin-EDTA, or simply add 200 μL FACS buffer directly to the wells.
3. Generate a single-cell suspension by pipetting up and down several times.
4. Conduct flow cytometry to measure Nanog-GFP expression from each PB-iPSC clone (*see* **Note 21**). Cell lines with >50 % Nanog-GFP positive cells with no detectable mCherry expression should be noted and prioritized.

3.4.4 Freezing PB-iPSC Clones in a 96-Well Plate (d30)

1. Two days later, harvest the cells by washing once with 50 μL PBS and incubating with 30 μL 0.25 % trypsin-EDTA for 5–7 min at 37 °C.

2. Add 70 μL of mouse iPSC culture medium and dissociate the cells well by pipetting.
3. Add 100 μL 2× cell freezing solution into the well.
4. Wrap the plate in a thick paper towel, and then one layer of aluminum foil. Label the foil layer.
5. Store the plate at −80 °C for no more than 2 months to avoid loss of viability. Perform the necessary genotyping steps during this storage window.

3.4.5 Verification of PB Insertions in 1° PB-iPSC Clones

In selecting appropriate 1° PB-iPSC clones to establish 2° systems, one should consider the copy number and position of PB insertions. Southern blotting may be used to identify single-copy clones (Fig. 2); however, the resolution of Southern blotting is limited by gel electrophoresis and therefore false positives (coincidentally similar band sizes) should be excluded by digestion with a second restriction enzyme. Note that under some conditions, optional passage steps during reprogramming can introduce clonal bias, which will be revealed in Southern blotting as common banding patterns shared amongst multiple clones. By virtue of their nearly invariable integration into "TTAA" tetranucleotide sequences [32, 33], the genomic sequence flanking PB insertions can act as a barcode to identify unique 1° PB-iPSC clones using Splinkerette.

For genotyping PB-iPSC clones, DNA is prepared from gelatin-coated 96-well plates as previously described for mouse ESCs [34].

1. For Splinkerette PCR screening:
   (a) Resuspend in 50 μL Tris-EDTA buffer pH 8.0 and use 2–5 μL of DNA.
2. For Southern blotting (1–2 rounds):
   (a) Resuspend in a minimal volume of Tris-EDTA buffer or directly in 30 μL restriction digest master mix, and digest in the 96-well plate before gel electrophoresis.

3.4.6 Defrosting, Expansion, and Cryopreservation of Candidate Lines

One day before defrosting, prepare a 24-well tissue culture plate with DR4 feeders as described in Section 3.3.6, **step 5**. On the day of defrost, change medium to 750 μL mouse iPSC culture medium per well.

1. Remove the frozen 96-well tissue culture plate from −80 °C and discard the foil and paper towel. Thaw the plate on a thermo-conductive or warmed surface (*see* **Note 22**).
2. Wipe any condensation from around the top and edges of the plate with 70 % ethanol and place in the tissue culture hood. Mark the desired wells on the underside of the plate.

3. Pipetting gently up and down to resuspend cells, transfer all 200 μL of the selected cell suspension into single wells of the 24-well plate.
4. The next day, replace with 500 μL mouse iPSC culture medium.
5. Observe the wells for growth, and expand confluent wells every 48–72 h stepwise from 24-well, to 6-well, and then 60 mm dishes with DR4 feeders. Each time, harvest cells by first washing in PBS (1.0, 1.5, or 3.0 mL) and incubating with 0.25 % trypsin-EDTA (0.25, 0.5, or 1.0 mL) for 5 min at 37 °C.
6. Reanalyze clones for Nanog-GFP by microscopy (Fig. 2) or flow cytometry (Section 3.4.3) and re-collect DNA from gelatin-coated dishes to verify recovery of the correct clones.
7. For cryopreservation, harvest and pellet the cells from a 60 mm tissue culture dish. Resuspend the iPSCs in 3 mL of 1× cell freezing solution, and aliquot to three separate vials. Store in liquid nitrogen.

3.4.7 Production of 2° MEFs Through Chimera or Tetraploid Complementation

The capacity for chimera contribution by mouse iPSC lines is a measure of reprogramming quality (Fig. 2). Moreover, it is useful in generating large quantities of material for 2° reprogramming systems (Section 3.2). Details on how to prepare iPSCs for blastocyst injection or morula aggregation are described in [15].

3.4.8 2° MEF Isolation and Reprogramming

1. Follow the protocol in Section 3.2 for MEF preparation.
2. Seed MEFs at a density of 1.0×10^5 cells per 10 cm^2 surface area in MEF medium on a gelatinized 6-well tissue culture plate. Prepare enough dishes for long-term culture and intermediate analyses.
3. Incubate the plate overnight in a humidified 37 °C, 5 % CO_2 incubator.
4. Replace with mouse iPSC culture medium containing 1 μg/mL dox (*see* **Note 23**).
5. Isolate and pellet cells over time for appropriate analyses (Fig. 3).

4 Notes

1. FBS should be thoroughly lot checked for reprogramming competence, ESC/iPSC chimera competence, and minimal tetracycline contamination, the latter of which can lead to high background expression from the tetO promoter.
2. Although Mini-prep DNA can be used for screening test vectors, we recommend Maxi-preparation or an endotoxin-free mini kit. Plasmid DNA stock concentrations should be maintained at 500–1000 ng/μL.

3. Dox is light sensitive. Dox-containing media should be protected from light during storage and use.
4. Frozen stocks of DR4 feeders are stored at 8×10^6 cells per vial, enough to coat three 6-well, 24-well, or 96-well dishes.
5. All animal care and experiments must be performed following the correct ethical guidelines as defined by the animal care committee of your research facility.
6. Use excess tissue for genotyping if required.
7. Use dedicated tissue culture hood for primary cells. Process one embryo at a time.
8. Trypsin-EDTA should fully cover minced embryo during incubation. Begin maceration of the next embryo.
9. A completely homogenous single-cell suspension is not necessary. Small clumps are tolerable.
10. Do not allow the cells to become overcrowded.
11. Minimize the passage numbers of MEFs to avoid senescence. Following one round of expansion, the cells should be frozen.
12. If you process many flasks, divide into two or three groups.
13. The number of MEFs per vial should be calculated based on future experiments. 2.5×10^6 cells/vial should provide enough cells for four 6-well tissue culture plates.
14. The 6-well scale is sufficient for most applications; however it can be adjusted as required. Reprogramming under the conditions described herein has also been tested in 24-well and 100 mm dishes.
15. Optimize the FuGENE® HD:DNA ratio for each cell line to determine the highest rate of transfection and the lowest toxicity.
16. Cells cannot be cultured following AP staining. Prepare a replica plate on −d2 to conduct AP staining if long-term reprogramming analyses are required.
17. Prepare the working solution immediately before use.
18. BCIP/NBT produces a dark blue/violet reaction product which is suitable for light photography. Incubation times should be determined empirically. Typically, 20–30 min results in low background and good staining intensity.
19. Control staining with appropriate isotype-matched control mAb (Alexa Fluor® 647 Mouse IgM, κ) should be included to establish thresholds for positive staining.
20. Dox removal will select for transgene-dependent PB-iPSC clones.
21. When screening a large number of clones, automated sample acquisition from the 96-well plate can be achieved using a BD™

High Throughput Sampler (HTS) Option (BD Bioscience) or MACSQuant® VYB (Miltenyi Biotec).

22. Avoid using a water bath for defrosting plates, as to avoid contamination inside the dish.
23. The response will be more uniform and robust.

Acknowledgments

We like to recognize all members of the Woltjen and Nagy laboratories for their contributions to the pioneering development of *piggyBac* materials and protocols for iPSC derivation. This work was supported by grants awarded to K.W. from the Cabinet Office, Government of Japan and the Japan Society for the Promotion of Science (JSPS) through the Funding Program for World-Leading Innovative R&D on Science and Technology (FIRST Program), the Core Center for iPS Cell Research, Research Center Network for Realization of Regenerative Medicine, and to K.W. and A.N. from the Strategic International Collaborative Research Program of the Japan Science and Technology Agency (JST) and Canadian Institutes of Health Research (CIHR). K.W. is a Hakubi Center Special Project Researcher. S-I.K. is a JSPS Fellowship recipient (2011–2013) and JST Researcher.

References

1. Takahashi K, Yamanaka S (2006) Induction of pluripotent stem cells from mouse embryonic and adult fibroblast cultures by defined factors. Cell 126(4):663–676. doi:10.1016/j.cell.2006.07.024
2. O'Malley J, Woltjen K, Kaji K (2009) New strategies to generate induced pluripotent stem cells. Curr Opin Biotechnol 20(5):516–521. doi:10.1016/j.copbio.2009.09.005
3. Kaji K, Norrby K, Paca A, Mileikovsky M, Mohseni P, Woltjen K (2009) Virus-free induction of pluripotency and subsequent excision of reprogramming factors. Nature 458(7239):771–775. doi:10.1038/nature07864
4. Woltjen K, Michael IP, Mohseni P, Desai R, Mileikovsky M, Hamalainen R, Cowling R, Wang W, Liu P, Gertsenstein M, Kaji K, Sung HK, Nagy A (2009) piggyBac transposition reprograms fibroblasts to induced pluripotent stem cells. Nature 458(7239):766–770. doi:10.1038/nature07863
5. Nagy K, Sung HK, Zhang P, Laflamme S, Vincent P, Agha-Mohammadi S, Woltjen K, Monetti C, Michael IP, Smith LC, Nagy A (2011) Induced pluripotent stem cell lines derived from equine fibroblasts. Stem Cell Rev 7(3):693–702. doi:10.1007/s12015-011-9239 5
6. Yusa K, Rad R, Takeda J, Bradley A (2009) Generation of transgene-free induced pluripotent mouse stem cells by the piggyBac transposon. Nat Methods 6(5):363–369. doi:10.1038/nmeth.1323
7. Woltjen K, Hamalainen R, Kibschull M, Mileikovsky M, Nagy A (2011) Transgene-free production of pluripotent stem cells using piggyBac transposons. Methods Mol Biol 767:87–103. doi:10.1007/978-1-61779-201-4_7
8. Kim SI, Oceguera-Yanez F, Hirohata R, Linker S, Okita K, Yamada Y, Yamamoto T, Yamanaka S, Woltjen K (2015) KLF4 N-terminal variance modulates induced reprogramming to pluripotency. Stem Cell Reports. doi:10.1016/j.stemcr.2015.02.004
9. Tonge PD, Corso AJ, Monetti C, Hussein SM, Puri MC, Michael IP, Li M, Lee DS, Mar JC, Cloonan N, Wood DL, Gauthier ME, Korn O, Clancy JL, Preiss T, Grimmond SM, Shin JY,

Seo JS, Wells CA, Rogers IM, Nagy A (2014) Divergent reprogramming routes lead to alternative stem-cell states. Nature 516 (7530):192–197. doi:10.1038/nature14047

10. David L, Polo JM (2014) Phases of reprogramming. Stem Cell Res 12(3):754–761. doi:10.1016/j.scr.2014.03.007
11. Okita K, Ichisaka T, Yamanaka S (2007) Generation of germline-competent induced pluripotent stem cells. Nature 448(7151):313–317. doi:10.1038/nature05934
12. Hochedlinger K, Yamada Y, Beard C, Jaenisch R (2005) Ectopic expression of Oct-4 blocks progenitor-cell differentiation and causes dysplasia in epithelial tissues. Cell 121 (3):465–477. doi:10.1016/j.cell.2005.02.018
13. Tucker KL, Wang Y, Dausman J, Jaenisch R (1997) A transgenic mouse strain expressing four drug-selectable marker genes. Nucleic Acids Res 25(18):3745–3746
14. Hussein SM, Puri MC, Tonge PD, Benevento M, Corso AJ, Clancy JL, Mosbergen R, Li M, Lee DS, Cloonan N, Wood DL, Munoz J, Middleton R, Korn O, Patel HR, White CA, Shin JY, Gauthier ME, Le Cao KA, Kim JI, Mar JC, Shakiba N, Ritchie W, Rasko JE, Grimmond SM, Zandstra PW, Wells CA, Preiss T, Seo JS, Heck AJ, Rogers IM, Nagy A (2014) Genome-wide characterization of the routes to pluripotency. Nature 516(7530):198–206. doi:10.1038/nature14046
15. Behringer R, Gertsenstein M, Nagy KV, Nagy A (2014) Manipulating the mouse embryo: a laboratory manual, 4th edn. Cold Spring Harbor Laboratory Press, Cold Spring Harbor, NY
16. Yamaguchi S, Hirano K, Nagata S, Tada T (2011) Sox2 expression effects on direct reprogramming efficiency as determined by alternative somatic cell fate. Stem Cell Res 6 (2):177–186. doi:10.1016/j.scr.2010.09.004
17. Papapetrou EP, Tomishima MJ, Chambers SM, Mica Y, Reed E, Menon J, Tabar V, Mo Q, Studer L, Sadelain M (2009) Stoichiometric and temporal requirements of Oct4, Sox2, Klf4, and c-Myc expression for efficient human iPSC induction and differentiation. Proc Natl Acad Sci U S A 106 (31):12759–12764. doi:10.1073/pnas.0904825106
18. Golipour A, David L, Liu Y, Jayakumaran G, Hirsch CL, Trcka D, Wrana JL (2012) A late transition in somatic cell reprogramming requires regulators distinct from the pluripotency network. Cell Stem Cell 11(6):769–782. doi:10.1016/j.stem.2012.11.008
19. Stadtfeld M, Maherali N, Breault DT, Hochedlinger K (2008) Defining molecular cornerstones during fibroblast to iPS cell reprogramming in mouse. Cell Stem Cell 2 (3):230–240. doi:10.1016/j.stem.2008.02.001
20. O'Malley J, Skylaki S, Iwabuchi KA, Chantzoura E, Ruetz T, Johnsson A, Tomlinson SR, Linnarsson S, Kaji K (2013) High-resolution analysis with novel cell-surface markers identifies routes to iPS cells. Nature 499 (7456):88–91. doi:10.1038/nature12243
21. Abad M, Mosteiro L, Pantoja C, Canamero M, Rayon T, Ors I, Grana O, Megias D, Dominguez O, Martinez D, Manzanares M, Ortega S, Serrano M (2013) Reprogramming in vivo produces teratomas and iPS cells with totipotency features. Nature 502(7471):340–345. doi:10.1038/nature12586
22. Ohnishi K, Semi K, Yamamoto T, Shimizu M, Tanaka A, Mitsunaga K, Okita K, Osafune K, Arioka Y, Maeda T, Soejima H, Moriwaki H, Yamanaka S, Woltjen K, Yamada Y (2014) Premature termination of reprogramming in vivo leads to cancer development through altered epigenetic regulation. Cell 156(4):663–677. doi:10.1016/j.cell.2014.01.005
23. Carey BW, Markoulaki S, Beard C, Hanna J, Jaenisch R (2010) Single-gene transgenic mouse strains for reprogramming adult somatic cells. Nat Methods 7(1):56–59. doi:10.1038/nmeth.1410
24. Stadtfeld M, Maherali N, Borkent M, Hochedlinger K (2010) A reprogrammable mouse strain from gene-targeted embryonic stem cells. Nat Methods 7(1):53–55. doi:10.1038/nmeth.1409
25. Haenebalcke L, Goossens S, Dierickx P, Bartunkova S, D'Hont J, Haigh K, Hochepied T, Wirth D, Nagy A, Haigh JJ (2013) The ROSA26-iPSC mouse: a conditional, inducible, and exchangeable resource for studying cellular (De)differentiation. Cell Rep 3 (2):335–341. doi:10.1016/j.celrep.2013.01.016
26. Nagy A (2013) Secondary cell reprogramming systems: as years go by. Curr Opin Genet Dev 23(5):534–539. doi:10.1016/j.gde.2013.07.004
27. Samavarchi-Tehrani P, Golipour A, David L, Sung HK, Beyer TA, Datti A, Woltjen K, Nagy A, Wrana JL (2010) Functional genomics reveals a BMP-driven mesenchymal-to-epithelial transition in the initiation of somatic cell reprogramming. Cell Stem Cell 7(1):64–77. doi:10.1016/j.stem.2010.04.015
28. Gabut M, Samavarchi-Tehrani P, Wang X, Slobodeniuc V, O'Hanlon D, Sung HK, Alvarez M, Talukder S, Pan Q, Mazzoni EO, Nedelec S, Wichterle H, Woltjen K, Hughes TR,

Zandstra PW, Nagy A, Wrana JL, Blencowe BJ (2011) An alternative splicing switch regulates embryonic stem cell pluripotency and reprogramming. Cell 147(1):132–146. doi:10.1016/j.cell.2011.08.023

29. Clancy JL, Patel HR, Hussein SM, Tonge PD, Cloonan N, Corso AJ, Li M, Lee DS, Shin JY, Wong JJ, Bailey CG, Benevento M, Munoz J, Chuah A, Wood D, Rasko JE, Heck AJ, Grimmond SM, Rogers IM, Seo JS, Wells CA, Puri MC, Nagy A, Preiss T (2014) Small RNA changes en route to distinct cellular states of induced pluripotency. Nat Commun 5:5522. doi:10.1038/ncomms6522
30. Benevento M, Tonge PD, Puri MC, Hussein SM, Cloonan N, Wood DL, Grimmond SM, Nagy A, Munoz J, Heck AJ (2014) Proteome adaptation in cell reprogramming proceeds via distinct transcriptional networks. Nat Commun 5:5613. doi:10.1038/ncomms6613
31. Lee DS, Shin JY, Tonge PD, Puri MC, Lee S, Park H, Lee WC, Hussein SM, Bleazard T, Yun JY, Kim J, Li M, Cloonan N, Wood D, Clancy JL, Mosbergen R, Yi JH, Yang KS, Kim H, Rhee H, Wells CA, Preiss T, Grimmond SM, Rogers IM, Nagy A, Seo JS (2014) An epigenomic roadmap to induced pluripotency reveals DNA methylation as a reprogramming modulator. Nat Commun 5:5619. doi:10.1038/ncomms6619
32. Ding S, Wu X, Li G, Han M, Zhuang Y, Xu T (2005) Efficient transposition of the piggyBac (PB) transposon in mammalian cells and mice. Cell 122(3):473–483. doi:10.1016/j.cell.2005.07.013
33. Liang Q, Kong J, Stalker J, Bradley A (2009) Chromosomal mobilization and reintegration of Sleeping Beauty and PiggyBac transposons. Genesis 47(6):404–408. doi:10.1002/dvg.20508
34. Ramirez-Solis R, Rivera-Perez J, Wallace JD, Wims M, Zheng H, Bradley A (1992) Genomic DNA microextraction: a method to screen numerous samples. Anal Biochem 201(2):331–335

Methods in Molecular Biology (2016) 1357: 23–31
DOI 10.1007/7651_2015_202

Published online: 17 February 2015

Generation of Human Induced Pluripotent Stem Cells from Peripheral Blood Mononuclear Cells Using Sendai Virus

Filipa A.C. Soares, Roger A. Pedersen, and Ludovic Vallier

Abstract

This protocol describes the efficient isolation of peripheral blood mononuclear cells from circulating blood via density gradient centrifugation and subsequent generation of integration-free human induced pluripotent stem cells. Peripheral blood mononuclear cells are cultured for 9 days to allow expansion of the erythroblast population. The erythroblasts are then used to derive human induced pluripotent stem cells using Sendai viral vectors, each expressing one of the four reprogramming factors Oct4, Sox2, Klf4, and c-Myc.

Keywords: Reprogramming, Human induced pluripotent stem cells, Sendai virus, Peripheral blood mononuclear cells, Erythroblast

1 Introduction

The discovery of human induced pluripotent stem cells (hiPSC) raised the exciting possibility of producing custom-tailored cells for disease modelling through the recapitulation of disease phenotype in differentiated hiPSCs, thereby enabling analysis of disease etiology and assessment of candidate therapeutic drugs. Furthermore, it opens the exciting possibility of personalized cell-based therapies. Since its discovery several somatic cell types and methods of reprogramming have been used to derive hiPSCs. So far fibroblasts remain the most popular donor cell type and have been used in more than 80 % of all published reprogramming experiments. However, peripheral blood is the most available adult tissue and would allow access to numerous frozen samples already stored at blood banks. Here, we describe an efficient and robust method of deriving hiPSCs from peripheral blood mononuclear cells using Sendai virus, a non-integrative method, expressing the four reprogramming factors Oct4, Sox2, Klf4, and c-Myc. This protocol is an increment of previous published methods (1–4).

2 Materials

2.1 Equipment

1. Inverted phase-contrast microscope.
2. Micropipette.
3. Laminar flow cabinet.
4. Cell counter.
5. Benchtop centrifuge.
6. Incubator at 37 °C, 5 % CO_2.

2.2 Disposables

1. 15 and 50 mL conical tubes.
2. Plastic disposable pipettes.
3. Sterile plastic transfer pipettes.
4. 12-well plate.
5. 100 mm plate.
6. Cryovials.

2.3 Chemicals

Chemical	Supplier	Catalogue number
1. Sodium citrate tubes (3.2 %)	Greiner Bio-One	455322
2. Ficoll paque premium (density 1.077 g/mL)	GE Healthcare	17-5442-02
3. D-PBS (without Ca^{2+} and Mg^{2+})	Sigma	D8537
4. Trypan blue	Lonza	17-942E
5. Knockout™ serum replacement	Invitrogen	10828028
6. DMSO	Sigma	D2650
7. StemSpan H3000	StemCell Technologies	9850
8. Pen/strep	Invitrogen	15140122
9. Recombinant human IL-3	Invitrogen	PHC0035
10. Recombinant human IGF-1	Miltenyi	130-093-885
11. Recombinant human SCF	Miltenyi	130-096-692
12. Recombinant human EPO	R&D Systems	287-TC-500
13. Dexamethasone	Sigma	D8893-1MG
14. L-ascorbic acid	Sigma	A4544-25G
15. CytoTune™-iPS Sendai Reprogramming Kit	Invitrogen	A1378001
16. Polybrene hexadimethrine bromide [10 mg/mL]	Sigma-Aldrich	107689

(continued)

(continued)

Chemical	Supplier	Catalogue number
17. Advanced DMEM F12	Life Technologies	12634010
18. L-glutamine	Life Technologies	25030024
19. β-mercaptoethanol	Sigma	M6250
20. Recombinant Human FGF basic	R&D Systems	233-FB-01 M
21. MEF CF-1 4 M IRR	GlobalStem	GSC-6001G
22. Fetal bovine serum (FBS)	Life Technologies	10500-064
23. Gelatin	Sigma	G1890
24. Water for embryo transfer	Sigma	W1503
25. Human serum albumin (HSA)	Sigma	A1653

2.4 Reagent Setup

1. StemSpan Medium (SM).

StemSpan medium (SM)	Stock concentration	Final concentration	Volume
1. StemSpan H3000			45 mL
2. Pen/Strep[a]	100×	1 %	450 μL

[a]Optional

Defrost StemSpan H3000 overnight at 4 °C and aliquot 45 mL in 50 mL tubes (the medium will expand once frozen). Store aliquots at −20 °C and defrost once needed. Do not freeze-thaw StemSpan H3000. Medium can be kept at 4 °C for up to 2 weeks.

2. Expansion Medium (EM).

Expansion medium (EM)	Stock concentration	Final concentration	Volume
1. StemSpan H3000			10 mL
2. Ascorbic acid	10 mg/mL	50 μg/mL	50 μL
3. SCF	50 μg/mL	50 ng/mL	10 μL
4. IL-3	10 μg/mL	10 ng/mL	10 μL
5. EPO	2 U/μL	2 U/mL	10 μL
6. IGF-1	100 μg/mL	40 ng/mL	4 μL
7. Dexamethasone[a]	1 mM	1 μM	10 μL

[a]Keep dexamethasone protected from light

Resuspend SCF, IL-3, EPO, IGF-1 in 0.1 % HSA (or BSA) in D-PBS to prepare stock solutions (*see* table for stock concentrations). Resuspend ascorbic acid in water for embryo transfer and filter to prepare 10 mg/mL stock solution. Dexamethasone is resuspended using DMSO to prepare 1 mM stock solution. Aliquot stock solutions in small volumes and store at −20 °C. When defrosted keep reagents at 4 °C and use within 5 days for the cytokines and 2 weeks for the ascorbic acid and dexamethasone. Also, keep dexamethasone away from light.

3. IPSC Medium.

IPSC medium	Stock concentration	Final concentration	Volume
1. Advanced DMEM F12			400 mL
2. Knockout serum replacement			100 mL
3. L-Glutamine			5 mL
4. β-Mercaptoethanol			3.5 μL
5. Pen/strep (optional)			5 mL
6. Human bFGF	4 μg/mL	4 ng/mL	500 μL

Mix reagents 1–5 and sterilize through 0.22 μm vacuum filter. Add human bFGF and use IPSC medium within a week. Store at 4 °C. Incomplete IPSC medium (reagents 1–5) can be prepared and stored up to 2 weeks at 4 °C without human bFGF.

4. IPSC Medium plus Cytokines (IPSC-Cy).

IPSC medium + cytokines (IPSC-Cy)	Stock concentration	Final concentration	Volume
1. IPSC medium			10 mL
2. Ascorbic acid (AA)	10 mg/mL	50 μg/mL	50 μL
3. SCF	50 μg/mL	50 ng/mL	10 μL
4. IL-3	10 μg/mL	10 ng/mL	10 μL
5. EPO	2 U/μL	2 U/mL	10 μL
6. IGF-1	100 μg/mL	40 ng/mL	4 μL
7. Dexamethasone[a]	1 mM	1 μM	10 μL

[a]Keep dexamethasone protected from light

Keep at 4 °C and use within 5 days.

5. MEF Medium.

MEF medium	Volume
1. Advanced DMEM F12	450 mL
2. FBS	50 mL
3. L-Glutamine	5 mL
4. β-Mercaptoethanol	3.5 μL
5. Pen/strep (optional)	5 mL

Sterilize through 0.22 μm vacuum filter and keep at 4 °C. Use within 2 weeks.

6. Mouse embryonic fibroblasts gelatin-coated 100 mm plates.

Gelatin solution	Volume
1. Gelatin	0.5 g
2. Water for embryo transfer	500 mL

Dissolve gelatin in water for embryo transfer by heating the solution in a 57 °C water bath for 30 min. Sterilize through 0.22 μm vacuum filter and keep at room temperature for up to 1 month. Coat the 100 mm plate with 6 mL of gelatin and leave at room temperature for 20 min. Aspirate excess of gelatin solution and add 6 mL of MEF medium. Defrost MEF CF-1 4 M IRR by adding the vial contents to 10 mL of MEF medium. Count and resuspend 1.2× 10^6 mouse embryonic fibroblasts per mL. Add 1 mL to each 100 mm plate and distribute for a homogeneous distribution. Place the plate in the incubator overnight before use.

3 Methods

3.1 Isolation of Peripheral Blood Mononuclear Cells from Peripheral Blood

1. Collect 50 mL of peripheral blood into sodium citrate tubes. Invert tube 8–10× and keep upright at room temperature (10 mL is sufficient when material is limited).
2. Invert Ficoll bottle several times to mix before use. In sterile flow cabinet add 15 mL of Ficoll to empty 50 mL falcon tubes (one tube for every 12.5 mL of blood, four tubes in total).
3. Dilute blood 1:1 with D-PBS and slowly layer 25 mL on top of the Ficoll.
4. Centrifuge 30 min at 500 × *g* at room temperature with the accelerator and BRAKE OFF (ideally within 2 h of collection).
5. Following density gradient spin, carefully collect PBMC layer using a transfer pipette (harvest about 15 mL from each tube).

Use a new falcon tube for every blood tube. Top up the volume to 50 mL with D-PBS. Invert several times and spin down at room temperature for 20 min at 300 × *g*.

6. Aspirate supernatant and resuspend pellet in 10 mL of D-PBS.
7. Count cells.
8. Expect ~60–90 × 10^6 PBMCs from 50 mL of peripheral blood (*see* Section 4.1).
9. Spin cells at 300 × *g* for 5 min and freeze down 2.5 × 10^6 cells/vial (Use 90 % KOSR, 10 % DMSO) or proceed with reprogramming (*see* Section 3.2).

3.2 Generation of Human Induced Pluripotent Stem Cells Using Sendai Virus

1. *Day −9* (Seed PBMC).
   (a) Thaw one vial of frozen PBMCs (*see* Section 3.1, Fig. 1) into 10 mL of *SM* and centrifuge at 300 × *g* for 5 min. Resuspend pellet in 2 mL of *EM*. For fresh isolated PBMCs add 2 × 10^6 cells to 2 mL of *EM*.
   (b) Transfer cells to 1 well of a 12-well plate and incubate at 37 °C, 5 % CO_2.
2. *Day −6 and Day −3* (Change medium).
   (a) Carefully transfer cells in suspension to 15 mL falcon tube and add 2 mL of *SM*. Add 1 mL of *EM* to well to avoid cells left in the plate drying (minimize disturbing the cells when replacing medium).

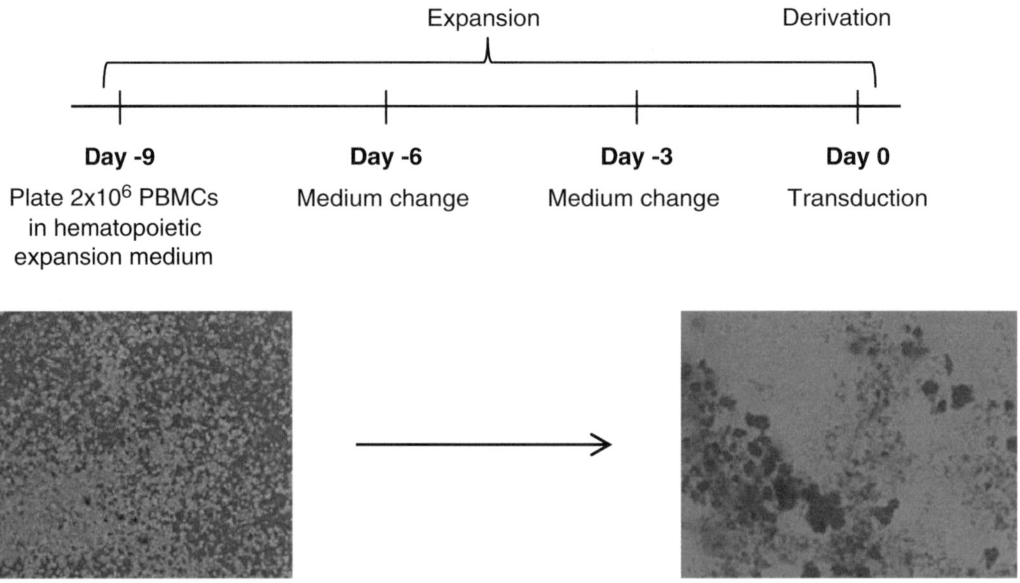

Fig. 1 Diagram describing the expansion of the erythroblasts population from peripheral blood mononuclear cells

(b) Spin cells at 300 × *g* for 5 min.

(c) Resuspend cells in 1 mL of fresh *EM* and carefully add cells back to same well of 12-well plate (total volume of well is 2 mL).

(d) Continue to culture at 37 °C, 5 % CO_2.

3. *Day 0* (Transduction).

(a) Transfer cells to sterile 15 mL conical tube and wash well 2× with 1 mL of *SM* to collect all cells.

(b) Count cells (*see* Section 4.2)

(c) Spin down 5 × 10^5 cells in 15 mL conical tube and add 0.3 mL of fresh *EM*.

(d) Remove one set of CytoTune™ Sendai tubes from the −80 °C storage. Thaw each tube by immersing the bottom of the tube in a 37 °C water bath for 5–10 s. Remove the tube from the water bath and allow thawing at room temperature. Once thawed, briefly centrifuge the tube and place it immediately on ice.

(e) Prepare 0.3 mL of fresh *EM* plus viruses at MOI 6 (*see* the CoA for the appropriate volume of virus) with 4 μg/mL of polybrene and transfer to one well of a 12-well plate (final volume/well 0.6 mL).

(f) Incubate overnight at 37 °C and 5 % CO_2.

4. *Day 1* (Wash Sendai virus).

(a) Collect transduced cells into 15 mL conical tube with 2 mL *SM* and spin at 300 × *g* for 5 min. Aspirate the supernatant and resupend cells in 2 mL of fresh *EM*. Return transduced cells to well and incubate at 37 °C and 5 % CO_2.

5. *Day 2* (Plate Mouse Embryonic Fibroblast feeders).

(a) Plate MEF feeders onto gelatin-coated 100 mm plate (*see* Section 2.4).

6. *Day 3* (Plate transduced cells).

(a) Collect transduced cells into 15 mL conical tube and spin at 300 × *g* for 5 min.

(b) Aspirate MEF medium from 100 mm plate and wash once with IPSC medium.

(c) Resuspend transduced cells in 7 mL of iPSC medium plus cytokines (*IPSC-Cy*) as above and transfer to 100 mm plate.

7. *Day 5 and Day 7*

(a) Feed cells every other day with 7 mL of IPSC medium.

8. *Day 9–Day 12* (Small colonies emerge).
   (a) Feed cells daily with 8 mL of IPSC medium (*see* Section 4.3).
   (b) Add additional MEF feeders as needed.
9. *~Day 14–Day 21* (Pick colonies).

4 Notes

4.1 Isolation of Pheripheral Blood Mononuclear Cells Not Working

1. Check the expiry date of all reagents and store the components appropriately as instructed on the datasheets.
2. Limit the time between bleeding and PBMC isolation to a minimum, ideally <2 h to maximize the chance of successful hiPSC generation.
3. Layer blood carefully on top of the Ficoll to avoid mixing prior to Ficoll gradient centrifugation.
4. If a clear separation of the PBMC layer is not achieved after Ficoll gradient centrifugation, check the centrifuge is balanced correctly and make sure the BRAKE is OFF.

4.2 Erythroblast Expansion Step Not Working

1. Check the expiry date of all cytokines and storage conditions (*see* Section 2.4).
2. Aliquot reagents in small aliquots to avoid repeated freeze–thaw cycles.
3. Minimize disturbing the cells at the expansion phase, by being gentle when moving the plate around and replacing the medium. When replacing medium centrifuge only the cells in suspension leaving a few microliters of medium behind in the well to avoid any cells that have adhered to the bottom from drying out.
4. Perform flow cytometry after 9 days of expansion (day 0) to validate nature of the cells. A good indicator of a successful expansion leading to subsequent reprogramming is induction of >70 % of double positive populations of CD235a/CD45 and CD36/CD71 (Fig. 2).

4.3 Sendai Virus Reprogramming Not Working

1. Check the expiry date of the Sendai virus and store the components appropriately as instructed on the datasheet.
2. Aliquot Svirus in small aliquots to avoid repeated freeze–thaw cycles. Discard any leftover when thawed more than twice.
3. Perform reprogramming of a control line (BJ fibroblast, Stemgent catalog number 08-0027) at MOI 3 to evaluate if the Sendai virus and other reprogramming reagents are working.

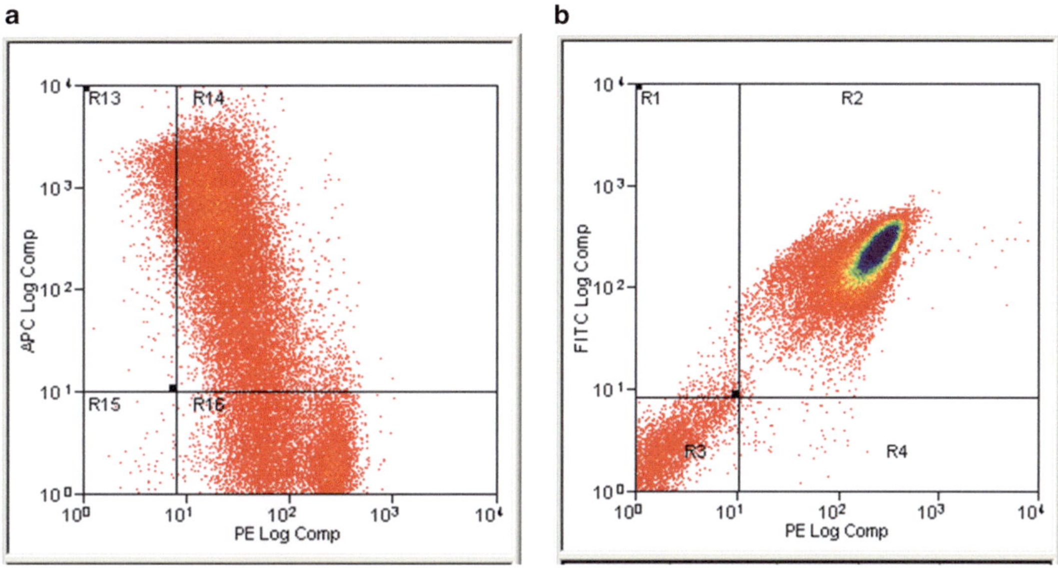

Fig. 2 Flow cytometry results after 9 days of expansion protocol (day 0). (**a**) CD235a APC/CD45 PE. (**b**) CD36 FITC/CD71 PE

Acknowledgements

The authors thank Dr. Thomas Moreau for expert advice and support. The authors also acknowledge the contribution of "Fundação para a Ciência e a Tecnologia" (SFRH/BD/69033/2010) and Cambridge Hospitals National Institute for Health Research Biomedical Research Center and ERC starting grant Relieve-IMDs.

References

1. Yang W, Mills JA, Sullivan S, Liu Y, French DL, Gadue P (2012) iPSC Reprogramming from human peripheral blood using sendai virus mediated gene transfer. In: StemBook (ed) The Stem Cell Research Community, StemBook
2. Chou B-K, Mali P, Huang X, Ye Z, Dowey SN, Resar LMS, Zou C, Zhang YA, Tong J, Cheng L (2011) Efficient human iPS cell derivation by a non-integrating plasmid from blood cells with unique epigenetic and gene expression signatures. Cell Res 21(3):518–529
3. van den Akker E, Satchwell TJ, Pellegrin S, Daniels G, Toye AM (2010) The majority of the in vitro erythroid expansion potential resides in CD34(-)cells, outweighing the contribution of CD34(+) cells and significantly increasing the erythroblast yield from peripheral blood samples. Haematologica 95(9):1594–1598
4. Leberbauer C, Boulmé F, Unfried G, Huber J, Beug H, Müllner EW (2005) Different steroids co-regulate long-term expansion versus terminal differentiation in primary human erythroid progenitors. Blood 105(1):85–94

Methods in Molecular Biology (2016) 1357: 33–44
DOI 10.1007/7651_2014_163

Published online: 06 August 2015

Sendai Virus-Based Reprogramming of Mesenchymal Stromal/Stem Cells from Umbilical Cord Wharton's Jelly into Induced Pluripotent Stem Cells

Cristian Miere, Liani Devito, and Dusko Ilic

Abstract

In an attempt to bring pluripotent stem cell biology closer to reaching its full potential, many groups have focused on improving reprogramming protocols over the past several years. The episomal modified Sendai virus-based vector has emerged as one of the most practical ones. Here we describe reprogramming of mesenchymal stromal/stem cells (MSC) derived from umbilical cord Wharton's Jelly into induced pluripotent stem cells (iPSC) using genome non-integrating Sendai virus-based vectors. The detailed protocols of iPSC colony cryopreservation (vitrification) and adaption to feeder-free culture conditions are also included.

Keywords: Umbilical cord, MSC, Sendai virus, Reprogramming, iPSC, Vitrification, Adaption to feeder-free culture conditions

1 Introduction

Since its emergence in 2006, reprogramming of somatic cells to pluripotency has revolutionized cell biology by opening new avenues and bypassing many hurdles researchers were struggling with. Significant progress has been made from basic research (1), drug development (2, 3), toxicity screening (4), and disease modeling (5, 6) to promise for cell therapy. The first patient with macular degeneration of retina has been treated with induced pluripotent stem cell (iPSC)-derived retinal pigment epithelium cells in September 2014, in the world's first iPSC clinical trial in Kobe, Japan.

The original proof-of-concept studies used retroviral transduction of four transcription factors, Oct4, Sox2, Klf-4, and c-Myc, later known as the Yamanaka factors, into human dermal fibroblasts (7). After several weeks in culture, colonies with embryonic stem cell (ESC)-like morphology and properties emerged in the culture dish. Known today as iPSC, these cells are the man-made equivalent to ESC. Much research has compared the two cell types (8), in an attempt to further pluripotent stem cell biology and to get closer to the promise of regenerative medicine. Several alternative methods have been attempted to increase efficiency or bypass problems with the original protocol based on genome-integrating viral vectors and presence of c-Myc oncogene (reviewed in (9)).

The modified Sendai virus method (10) is based on a non-integrating vector that is also replication deficient, so that upon several passages, the newly derived iPSC lines become transgene-free. The reasons for choosing this method include a straightforward, not labor-intensive protocol, the non-integrating nature of the vector, low cytotoxicity elicited in culture, broad range of target cells accessible, and highly reproducible protocol.

The CytoTune®-iPS Sendai Reprogramming Kit (Life Technologies) contains four SeV-based reprogramming vectors, each encoding one of the four Yamanaka factors. The kit has been shown to reprogram several cell types, including human CD34+ cord blood cells (11), nasal epithelial cells (12), dermal fibroblasts (13, 14), peripheral blood mononuclear cells (15), and T-lymphocytes (16).

Mesenchymal stromal/stem cells (MSC) are multipotent stem cells, capable of differentiating into osteoblasts, chondrocytes, and adipocytes. In the human body exist multiple sources of MSC and the most frequently used for clinical applications are bone marrow MSC. They are also present in extraembryonic perinatal tissue, having been isolated from placenta, fetal membrane, Wharton's Jelly (WJ) umbilical cord, cord blood, and amniotic fluid (17). Routinely discarded after birth, these tissues do not pose any ethical problems and are readily available for processing (18).

The characteristics of MSCs include plastic adherence and expression of a set of surface markers including CD90, CD73, and CD105. Although MSCs have been harvested from multiple different tissues, recent evidence points toward different properties based on the tissue of origin. Due to their developmental stage, extraembryonic derived MSCs are considered to have a broader differentiation ability and faster proliferation rate than adult tissue-derived MSCs (19).

Due to their low immunogenicity, one of the main uses of WJ-MSCs could be allogeneic transplantation. The multipotency and immunomodulatory properties make them ideal candidates for this purpose. WJ-MSCs have been reported to be useful in cancer therapy, liver disease, cardiovascular disease, cartilage regeneration, and peripheral nerve repair (reviewed in (17)).

We have recently reported a protocol for derivation of WJ-MSCs under chemically defined (20) and xeno-free conditions (21). Here we describe in detail the reprogramming protocol of WJ-MSC into iPSC using the CytoTune Kit from Life Technologies.

2 Materials

2.1 Feeder Cells

Reprogramming-Qualified Newborn Foreskin Fibroblasts (NuFF; Global Stem, GSC-3006G) (*see* **Note 1**).

2.2 Equipment

1. 100-μl pipettman.
2. 1,000-μl pipettman.
3. Biosafety cabinet.
4. Hemocytometer (Hausser).
5. Phase contrast microscope.
6. Pipette gun.
7. Tissue culture incubator.
8. Tube rack.

2.3 Plasticware and Other Disposables (See Note 2)

1. 75-cm^2 Flask (Corning; Cat. No.430641).
2. Six-well culture plate (Costar-VWR; Cat. No. 734-1596).
3. P60 (60-mm) TC-Treated Culture Dish (Corning; Cat. No. 430166).
4. Open pulled straws for vitrification (MTR; Cat. No. 19050/0050).
5. 0.22-μM syringe filter (VWR; Cat. No. 6896-2502).
6. Four-well dish (Nunc; Cat. No. 1256572).
7. 5-ml cryogenic vial (Nalgene; Cat. No. 5000-0050).
8. 100-μl Filter tips (SLS; Cat. No. 171403).
9. 1,000-μl Filter tips (SLS; Cat. No. 171703).
10. γ-Irradiated individually wrapped polystyrene 5-ml pipettes (Falcon; Cat. No. 357543).
11. γ-Irradiated individually wrapped polystyrene 10-ml pipettes (Falcon; Cat. No. 357551).
12. γ-Irradiated individually wrapped polystyrene 25-ml pipettes (Falcon; Cat. No. 357525).
13. γ-Irradiated 15-ml conical tube (BD Falcon; Cat. No. 352096).
14. Cryotube (Thermo Scientific; Cat No. 377224).
15. Waste container.

2.4 Reagents and Media

1. EmbryoMax® 0.1 % Gelatin Solution (Millipore; Cat. No. ES-006-B).

2. Pluriton™ Reprogramming Medium (Stemgent; Cat. No. 00-700).
3. Recombinant Human Fibroblast Growth Factor (FGF) basic (146 aa) (R&D Systems; Cat. No. 233-FB-025).
4. DMEM, high glucose, Glutamax™, pyruvate (Life Technologies; Cat. No. 31966-021).
5. MEM Non-Essential Amino Acid Solution (100×) (Life Technologies; Cat. No. 11140-035).
6. Fetal Bovine Serum, ES Cell Qualified (Life Technologies; Cat. No. 16141-079).
7. CytoTune®-iPS Sendai Reprogramming Kit (Life Technologies; Cat. No. A1378001).
8. DPBS, Ca^{2+}/Mg^{2+}-free (Life Technologies; Cat. No. 14190).
9. CTS™ TrypLE™ Select (Life Technologies; Cat. No. 12859-01).
10. KnockOut DMEM/F12 (Life Technologies; Cat. No. 12660-012).
11. KnockOut™ Serum replacement (KOSR) (Life Technologies; Cat. No. 10828010).
12. HEPES 1 M Buffer Solution (Life Technologies; Cat. No. 15630-056).
13. Sucrose (Sigma; Cat. No. S0389).
14. Ethylene glycol (Sigma; Cat. No. 10246-6).
15. Dimethyl sulfoxide (DMSO) (VWR; Cat. No. 23500.26).
16. Ethanol, absolute (Sigma; Cat. No. E7023).
17. Rock inhibitor Y-27632 (Sigma; Cat. No. Y0503).
18. Growth Factor Reduced (GFR) Matrigel (BD Biosciences; Cat. No. 354239).
19. PBS-EDTA (Lonza; Cat. No. BE02-017F).
20. TeSR™-E8™ basal medium (STEMCELL Technologies; Cat. No. 05941).
21. TeSR™-E8™ supplement (STEMCELL Technologies; Cat. No. 05942).
22. Alkaline Phosphatase Staining Kit II (Stemgent; Cat. No. 00-0055).
23. ReadyMix™ Taq PCR Reaction Mix (Sigma; Cat. No. P4600).
24. HyperLadder™ 100 bp (BioLine; Cat. No. BIO-33056).
25. Agarose (Life Technologies; Cat. No. 16500).
26. CryoStor CS10 (STEMCELL Technologies; Cat. No. A10142-01).

3 Methods

3.1 Conditioning Medium

1. Coat one T75 flask with 5 ml of 0.1 % Gelatin Solution for 1 h at 37 °C.
2. Plate 3–4 × 10^6 NuFF in 30 ml DMEM supplemented with 5 % FBS for 24 h. Wash the cells thoroughly 3–5x with DMEM without FBS before adding Pluriton medium supplemented with 4 ng/ml bFGF.
3. Change Pluriton medium supplemented with 4 ng/ml bFGF daily for 7–10 days. Keep spent medium at +4 °C.
4. The last day of conditioning, Pool all spent medium and pass through a 0.22-μm filter; aliquot the Pluriton-conditioned medium.
5. Aliquots are kept at −20 °C and should be used within Pluriton expiry date.

3.2 Reprogramming

Day: −2

6. Coat desired number of wells of a six-well plate dish with 0.1 % Gelatin Solution, 1 ml per well, for 1 h at 37 °C.
7. Plate MSC at several different densities (i.e., 1.5, 2.0, 2.5 × 10^5 cells/well) in 2 ml of DMEM high glucose, with Glutamax and pyruvate, supplemented with 10 % FCS and 1× Non-Essential Amino Acids.

Day: −1

8. Change medium.

Day 0

9. Choose 70 % confluent well for reprogramming. The cells from other wells can be collected and snap frozen as cell pellets serving as a negative control in Sendai virus detection RT-PCR later (Section 3.6).
10. Transduce the cells with Sendai virus encoding reprogramming factors at an multiplicity of infection (M.O.I.) of 3 per cell according to the certificate of analysis (*see* **Note 3**). Follow the manufacturer's protocol step by step exactly.

Day: 1–7

11. Refresh medium on a daily basis. High cell death is normal (*see* **Note 4**).

Day: 5

12. Coat with 2 ml of 0.1 % Gelatin Solution, for 1 h at 37 °C, two P60 culture dishes for each well of the six-well dish undergoing reprogramming.

13. Plate 3 × 10^5 NuFF feeders per P60 in 5 ml of DMEM high glucose, with Glutamax and pyruvate, supplemented with 10 % FCS and 1× Non-Essential Amino Acids.

Day: 7

14. Wash cells undergoing reprogramming with 3 ml of DPBS (Ca^{2+}/Mg^{2+}-free) and incubate in 1 ml TrypLE for 3–5 min at 37 °C. TrypLE is then diluted with 11 ml of pre-warmed DMEM. Mix well and separate cells into three tubes, 4 ml each. Centrifuge the cells at 1,200 rpm for 5 min.
15. Resuspend the cell pellet in 5 ml of conditioned Pluriton medium in each of two tubes and plate cell suspension from one tube into one P60 with NuFF feeders (*see* **Note 5**).
16. Remove supernatant from the third tube and snap freeze cell pellet. This will be your positive control in Sendai virus detection RT-PCR later (Section 3.6)

Day: 8–28

17. Exchange medium daily and allow iPSC colonies to grow and expand in culture.

Day: 18 (*see* **Note 6**)

18. Coat the vessel with 5 ml of 0.1 % Gelatin Solution for 1 h at 37 °C.
19. Plate 2 × 10^4 NuFF feeders per each well in 0.5 ml of DMEM high glucose, with Glutamax and pyruvate, supplemented with 10 % FCS and 1× Non-Essential Amino Acids.
20. Refresh culture medium every 2–3 days.

Day: 21–28

21. Primary iPSC colonies from one of two P60 dishes with the most uniform and compact cells are manually picked (*see* **Note 7**) and passaged onto NuFF feeder cells in 0.5 ml Pluriton (*see* **Note 8**).
22. The second P60 dish is used for assessing reprogramming efficiency.
23. Collect the cells from the third P60 dish and snap-freeze the cell pellets; use them as a positive control.

3.3 Cryopreservation (Vitrification) and Thawing of Vitrified Colonies

iPSCs cultured on feeder cells were cryopreserved in pieces of about 100–200 cells using the open pulled straw (OPS) vitrification method (22–24).

For vitrification you have to have prepared the following solutions:

ES-HEPES solution

Component	Volume (ml)
KnockOut™ DMEM/F12	15.6
KnockOut™ SR (KOSR)	4.0
HEPES 1 M buffer solution	0.4
	20.0

Mix the components, filter through a 0.22-μM syringe filter and store at 4 °C for up to 7 days.

1 M sucrose solution

1. Mix 3.42 g of sucrose with 6.0 ml ES-HEPES in a 15-ml centrifuge tube and place in the incubator at 37 °C.
2. When the sucrose had fully dissolved, add 2.0 ml of ES-HEPES and 2.0 ml KOSR. Mix well by flipping the tube.
3. Filter the solution through a 0.22-μM syringe filter and store at 4 °C for up to 7 days.

10 % vitrification solution (VS1)

Component	Volume (ml)
ES-HEPES	2.00
Ethylene glycol	0.25
Dimethyl sulfoxide	0.25
	2.50

Mix the components and store at 4 °C for up to 7 days.

20 % vitrification solution (VS2)

Component	Volume (ml)
ES-HEPES	0.75
1 M sucrose	0.75
Ethylene glycol	0.50
Dimethyl sulfoxide	0.50
	2.50

Mix the components and store at 4 °C for up to 7 days.

The procedure goes as follows:

1. Submerge straws in 70 % ethanol for at least 20 min and then allow them to dry in biosafety cabinet.

2. Label a 5-ml cryogenic vial with the cell line identification code, passage number, and the date of freezing, remove the top, and place in a cane. Place the cane into a container with liquid N_2, ensuring that the 5-ml cryogenic vial and most of the cane are submerged.
3. Place 500 µl of ES-HEPES in well #1 of four-well dish, CS1 in well #2, and VS2 in well #3. The well #4 remains empty.
4. Equilibrate the dish in the incubator at 37 °C for 10–15 min.
5. Place the dish in the biosafety cabinet and invert the lid.
6. Place several 30–40 µl drops of each VS1 and VS2 on the inner surface of the inverted lid of the four-well dish in two separate areas.
7. Cut the colonies to be vitrified with a 21 G syringe needle, generating a grid-like pattern, which ensures evenly sized pieces.
8. Detach the pieces from the dish by gentle scraping with the tip of a 200-µl micropipette
9. Transfer the detached pieces into well 1 containing ES-HEPES using a pipette tip primed with ES-HEPES solution.
10. Set timer for 1 min 25 s (*see* **Note 9**). Immediately transfer 4–8 pieces to one of VS1 drops on the lid.
11. When timer reaches 25 s, transfer them to VS2 using a pipette tip primed with VS2 solution.
12. When timer reaches 0, aspirate the pieces into a narrow end of the straw.
13. Hold the straw horizontally with a forceps and dip quickly into liquid N_2 for several seconds and then place into a 5-ml cryogenic vial (*see* **Note 10**).
14. Repeat the procedure until no pieces are left in the well #1.
15. Put lid on the 5-ml cryogenic vial and transfer into dewer with liquid N_2 for extended storage.

Thawing of the vitrified cells:

1. To thaw vitrified cells, you have to have ready four-well dishes with feeder cells (*see* Section 3.2, steps 17–21). A day before thawing replace feeder medium with Pluriton (*see* **Note 8**).
2. Just before thawing, add into the 24-h conditioned medium on feeders Rock inhibitor Y-27632 in a final concentration of 10 µM. Mix well.
3. Place 500 µl of DMEM into each of two wells of four-well dish.
4. Take straw to be thawed form the liquid N2 and place the tip directly into well 1 to release frozen pieces.
5. After 30 s, transfer the pieces into well 2.
6. After 5 min, transfer the cells into well with feeders.

7. Two days later, refresh the medium. There is no need for adding Y-27632.

3.4 Adapting iPSC to Feeder-Free Conditions

1. Thaw GFR-Matrigel stock overnight at 4 °C (*see* **Note 11**).
2. Check for the presence of any gel formation. If there were no visible clumps, GFR-Matrigel stock dilute to a final concentration of 0.34 mg/ml with cold DMEM.
3. Coat one well of a six-well plate with 750 μl of GFR-Matrigel working solution per well for 1 h at 37 °C.
4. Remove the GFR-Matrigel and replace with DMEM until use (*see* **Note 12**).
5. Rinse three to four wells of the iPSC cultured on feeders with 500 μl PBS-EDTA.
6. Incubate the cells with 500 μl PBS-EDTA per well for 3–5 min at 37 °C.
7. When the edges of the colonies visibly detach from the substrate (keep checking under the microscope), remove all PBS-EDTA.
8. Add 500 μl of complete TeSR-E8 supplemented with 10 μM Y-27632 and detach cells by pipetting up and down.
9. Transfer the suspension from all wells of the same cell line into one well of a six-well plate pre-coated with GFR-Matrigel.
10. Rinse the four-well dish wells with another 500 μl of medium and transfer into the same six-well plate well.
11. The following day change the culture medium with complete TeSR-E8 without Y-27632.
12. Passage cells using PBS-EDTA, by adjusting the volumes according to the volume of the plate being used (*see* **Note 13**).

3.5 Assessing Reprogramming Efficiency

We assess reprogramming approximately 20 days after passaging on NuFF feeder cells. Leaving the emerging colonies too long will lead to merger of some of them and will have a negative impact on the efficiency rate observed. We use Alkaline Phosphatase Staining Kit II and follow the manufacturer's recommendation. We count colonies using OpenCFU software.

3.6 Sendai Virus Detection PCR

1. In our experience Sendai virus becomes undetectable in 100 days of continuous cell culture since viral transduction on day 0 of reprogramming. Collect cells at day 100. Extract RNA with any commercially available column-based kit or alternative methods; perform reverse-transcription using any of kits suitable for nanogram-range of RNA. Negative control: non-transduced cells (Section 3.2, step 9); positive control: transduced cells 7 day post transduction (Section 3.2, step 16).

2. Perform end-point PCR with 5 μl of cDNA in 50 μl ReadyMix™ Taq PCR Reaction Mix.
3. Use the following primers to detect for the presence of Sendai virus vector (10):

β-Actin forward	CAA CCG CGA GAA GAT GAC
β-Actin reverse	AGG AAG GCT GGA AGA GTG
Sendai virus forward	GGA TCA CTA GGT GAT ATC GAG C
Sendai virus reverse	ACC AGA CAA GAG TTT AAG AGA TAT GTA TC

4. Cycling conditions: One cycle of 95 °C for 5 min, [95 °C 30 s, 55 °C 30 s, 72 °C 30 s] for 35 cycles, and one cycle of 72 °C for 5 min.
5. Analyze PCR products on 2 % agarose gel. To determine size of the product, we used 100 bp molecular weight ladder.
6. The expected size of the PCR product is 181 bp for SeV and 455 bp for β-actin.

4 Notes

1. Reprogramming efficiency depends also on quality of the feeders. It is important to choose reprogramming-qualified feeder cells.
2. Disposable culture and plasticware of the equivalent quality can be purchased from different manufacturers without altering the outcome of the procedure.
3. Viral titer varies from vector to vector and from batch to batch of the CytoTune kit. You will have to calculate for each vector independently, the specific volume to add per well.
4. Highest cytotoxicity is observed 48–72 h post transduction.
5. Before plating the cells, wash feeders 2× with DMEM/F12 to remove traces of the FBS.
6. This step does not have to be done exactly at day 18. It could be done a day or two earlier or later, depending on other workload. The main point is that it has to pass 72 h from the plating before the feeders could be used. The dishes with feeders should be used within the maximum of 7–10 days after plating.
7. Dissecting and picking colonies could be done with different tools (needle, Pasteur pipette, loading tip for Western blotting), it depends on personal preferences. Important: change tools between colonies.

8. For this step, we also validated NutriStem™ (Stemgent, Cat. No. 01-005), TeSR™-E8™ with its supplement, and KnockOut DMEM/F12 (Life Technologies, Cat No 10829-018) with its supplement (Life Technologies, Cat No 10828-028).
9. It is essential to do this process quickly, following indicated timing.
10. Do not take the straw out of liquid N_2 while doing that.
11. Work on ice. Precool tubes, pipettes, and tips.
12. Do not let GFR-Matrigel dry or results may vary. Just before use, remove DMEM and plate cells.
13. iPSCs adapted to feeder-free conditions do not require vitrification. They can be treated as any other adherent cells. We cryopreserve them in CryoStor CS10 following the manufacturer's protocol.

Acknowledgments

The study was supported by the studentship to C.M. from the Medical Research Council, UK.

References

1. Loh YH, Yang L, Yang JC et al (2011) Genomic approaches to deconstruct pluripotency. Annu Rev Genomics Hum Genet 12:165–185
2. Inoue H, Yamanaka S (2011) The use of induced pluripotent stem cells in drug development. Clin Pharmacol Ther 89:655–661
3. Egawa N, Kitaoka S, Tsukita K et al (2012) Drug screening for ALS using patient-specific induced pluripotent stem cells. Sci Transl Med 4:145ra104
4. Hou Z, Zhang J, Schwartz MP et al (2013) A human pluripotent stem cell platform for assessing developmental neural toxicity screening. Stem Cell Res Ther 4(Suppl 1):S12
5. Paşca SP, Portmann T, Voineagu I et al (2011) Using iPSC-derived neurons to uncover cellular phenotypes associated with Timothy syndrome. Nat Med 17:1657–1662
6. Miller JD, Ganat YM, Kishinevsky S et al (2013) Human iPSC-based modeling of late-onset disease via progerin-induced aging. Cell Stem Cell 13:691–705
7. Takahashi K, Tanabe K, Ohnuki M et al (2007) Induction of pluripotent stem cells from adult human fibroblasts by defined factors. Cell 131:861–872
8. Narsinh KH, Plews J, Wu JC (2011) Comparison of human induced pluripotent and embryonic stem cells: fraternal or identical twins? Mol Ther 19:635–638
9. Malik N, Rao M (2013) A review of the methods for human iPSC derivation. Methods Mol Biol 997:23–33
10. Fusaki N, Ban H, Nishiyama A (2009) Efficient induction of transgene-free human pluripotent stem cells using a vector based on Sendai virus, an RNA virus that does not integrate into the host genome. Proc Jpn Acad Ser B Phys Biol Sci 85:348–362
11. Nishishita N, Shikamura M, Takenaka C et al (2012) Generation of virus-free induced pluripotent stem cell clones on a synthetic matrix via a single cell subcloning in the naïve state. PLoS One 7(6):e38389
12. Ono M, Hamada Y, Horiuchi Y et al (2012) Generation of induced pluripotent stem cells from human nasal epithelial cells using a Sendai virus vector. PLoS One 7:e42855
13. Kudva YC, Ohmine S, Greder LV et al (2012) Transgene-free disease-specific induced pluripotent stem cells from patients with type 1 and type 2 diabetes. Stem Cells Transl Med 1:451–461

14. Jin ZB, Okamoto S, Xiang P et al (2012) Integration-free induced pluripotent stem cells derived from retinitis pigmentosa patient for disease modeling
15. Merling RK, Sweeney CL, Choi U et al (2013) Transgene-free iPSCs generated from small volume peripheral blood nonmobilized CD34 + cells. Blood 121:e98–e107
16. Wakao H, Yoshikiyo K, Koshimizu U et al (2013) Expansion of functional human mucosal-associated invariant T cells via reprogramming to pluripotency and redifferentiation. Cell Stem Cell 12:546–558
17. Kim DW, Staples M, Shinozuka K et al (2013) Wharton's Jelly-derived mesenchymal stem cells: phenotypic characterization and optimizing their therapeutic potential for clinical applications. Int J Mol Sci 14:11692–11712
18. Marcus AJ, Woodbury D (2008) Fetal stem cells from extra-embryonic tissues: do not discard. J Cell Mol Med 12(3):730–742
19. Pappa KI, Anagnou NP (2009) Novel sources of fetal stem cells: where do they fit on the developmental continuum? Regen Med 4:423–433
20. Badraiq H, Devito L, Ilic D (2014) Isolation and expansion of mesenchymal stromal/stem cells from umbilical cord under chemically defined conditions. Methods Mol Biol [Epub ahead of print]
21. Devito L, Badraiq H, Galleu A et al (2014) Wharton's Jelly MSC derived under chemically defined animal product-free low oxygen conditions are rich in MSCA-1+ subpopulation. Regen Med (in press)
22. Reubinoff BE, Pera MF, Vajta G et al (2001) Effective cryopreservation of human embryonic stem cells by the open pulled straw vitrification method. Hum Reprod 16:2187–2194
23. Ilic D, Stephenson E, Wood V et al (2012) Derivation and feeder-free propagation of human embryonic stem cells under xeno-free conditions. Cytotherapy 14:122–128
24. Stephenson E, Jacquet L, Miere C et al (2012) Derivation and propagation of human embryonic stem cell lines from frozen embryos in an animal product-free environment. Nat Protoc 7:1366–1381

Methods in Molecular Biology (2016) 1357: 45–56
DOI 10.1007/7651_2014_133

Published online: 20 November 2014

Very Rapid and Efficient Generation of Induced Pluripotent Stem Cells from Mouse Pre-B Cells

Bruno Di Stefano and Thomas Graf

Abstract

One of the major obstacles in generating induced pluripotent stem (iPS) cells suitable for therapeutic application is the low efficiency of the process and the long time required, with many iPS lines acquiring genomic aberrations. In this chapter we describe a highly efficient iPS reprogramming system based on the transient expression in pre-B cells of the transcription factor C/EBPα, followed by the induction of the four Yamanaka factors (OSKM). In addition, the process is very rapid, yielding Oct4 positive cells within 2 days and Nanog-positive iPS cell colonies within a week.

Keywords: Induced pluripotent stem cells, Cell reprogramming, B lymphocytes, Transcription factor C/EBPα, Deterministic process, Embryonic stem cells, Oct4, Sox2, Klf4, c-Myc

1 Introduction

Resetting the somatic epigenome to a pluripotent state has been achieved by several approaches, including somatic cell nuclear transfer, cell fusion, and overexpression of defined transcription factors such as Oct4, Sox2, Klf4, and c-Myc (OSKM, also known as the Yamanaka factors) (1). The induced pluripotent stem cells (iPS cells) derived by overexpression of the OSKM factors have opened up new opportunities for disease modeling and cell therapy applications (2, 3). However, the low efficiency of iPS cell derivation (0.1–1 %) (3, 4) has hindered these opportunities (5). Therefore, creating an efficient and rapid system to convert somatic cells into pluripotent cells would be beneficial for the generation of high-quality iPS cells suitable for cell therapy and also for the full understanding of the molecular mechanism underlying the reprogramming process.

Based on the observation that at any given time only a small proportion of somatic cells become reprogrammed by the Yamanaka factors, it has been proposed that reprogramming entails a stochastic phase followed by a more hierarchical phase that consists in the sequential activation of pluripotency genes (5). Thus, work by Jaenisch and colleagues documented the highly asynchronous reprogramming of individual murine B cells spanning a period from 3 to

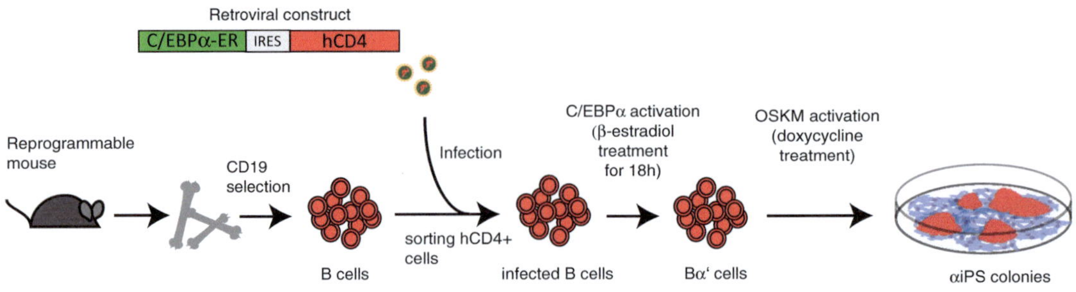

Fig. 1 Experimental design. Mouse primary B cells were purified from the bone marrow of reprogrammable mice and infected with a retrovirus carrying an estradiol-inducible form of C/EBPα. Infected cells were sorted based on the expression of the cell surface marker hCD4 and plated in B cell medium on OP9 stromal cells. B cells were then exposed for 18 h to β-estradiol, followed by washout of the inducer and addition of doxycycline to activate OSKM factors. iPS cell colonies are observed after 4–6 days

18 weeks (6). These findings raised the question whether it is possible to identify a factor or culture conditions able to overcome the stochastic phase, creating a rapid and synchronous cell reprogramming system optimally suited for mechanistic studies. In the past few years, multiple genetic factors have been tested for their ability to improve the reprogramming efficiency of the Yamanaka factors; however, none of them were found to have a substantial effect.

Here we report a method with which primary pre-B cell precursors (henceforth called B cells) derived from the bone marrow of the reprogrammable mouse (7) can be converted rapidly and efficiently into iPS cells (Fig. 1) (8). The method consists in the transient expression of the transcription factor C/EBPα followed by OSKM induction. This system allows the derivation of iPS cells within 1 week and at high efficiency. Pluripotency genes are activated as early as 2 days after OSKM induction and transgene-independent colonies are generated within 4 days. The method described here should help to fully understand the reprogramming mechanism of somatic cells into pluripotent cells and if applicable to humans could be beneficial for the generation of high-quality iPS cells suitable for disease modeling and cell therapy.

2 Materials

2.1 Mice

Reprogrammable mice (9).

2.1.1 Cells and C/EBPα Transgene

1. Mouse primary $CD19^+$ B cells.
2. OP9 stromal cells (10).
3. S17 stromal cells (11).
4. Mouse embryonic fibroblasts (MEFs).
5. Retroviral vector carrying C/EBPα-ER-hCD4 (12).

2.2 Media Components and Reagents

All reagents are from Life Technologies unless otherwise specified.

2.2.1 For Culture and Reprogramming of Mouse Primary CD19$^+$ B Cells

1. RPMI with L-glutamine (Lonza).
2. Fetal bovine serum (FBS), E.U. Approved (South American).
3. Interleukin 7 (Peprotech).
4. Interleukin 4 (Peprotech).
5. Interleukin 15 (Peprotech).
6. D-PBS without $Ca2^+$ and $Mg2^+$.
7. TrypLE™ select cell dissociation reagent.
8. KnockOut™ D-MEM.
9. Fetal bovine serum, ES cell-qualified.
10. Leukemia inhibitor factor (LIF) (ORF genetics).
11. β-Mercaptoethanol, penicillin-streptomycin, sodium pyruvate, NE amino acids.
12. Gelatin 0.1 % (Millipore).
13. 17β-estradiol (Calbiochem).
14. Doxycycline (Sigma-Aldrich).
15. Deionized water (B. Braun Medical SA).

2.2.2 For Culture of OP9 and S17 Stromal Cells

1. Alpha-MEM with GlutaMAX™.
2. Fetal bovine serum E.U. Approved (South American origin).
3. Penicillin-streptomycin.
4. TrypLE™ select cell dissociation reagent.

2.2.3 For Culture of Mouse Embryonic Fibroblasts (MEFs) and PlatE Cells

1. D-MEM (high glucose).
2. FBS, E.U. Approved (South American).
3. Penicillin-streptomycin.
4. L-Glutamine.
5. TrypLE™ select cell dissociation reagent.

2.2.4 For Culture of iPS Cells

1. KnockOut™ D-MEM.
2. Fetal bovine serum, ES cell-qualified.
3. Penicillin-streptomycin.
4. L-Glutamine.
5. β-Mercaptoethanol, penicillin-streptomycin, sodium pyruvate, NE amino acids.
6. Leukemia inhibitor factor (LIF) (Isokine).
7. TrypLE™ select cell dissociation reagent.

2.3 Equipment

1. Cell culture hood (i.e., biosafety cabinet).
2. Inverted microscope with 4× and 10× objectives.
3. Incubator set at 37 °C, 5 % CO_2.
4. Water bath set at 37 °C.
5. Centrifuge.
6. Micropipettes (1–10, 2–20, 20–200, 200–1,000 μl).
7. Pipettor.
8. Liquid nitrogen tank.
9. Cryovial storage rack.
10. Freezers: −20 and −80 °C.

2.4 Disposables

All materials are from BD Biosciences unless otherwise specified.

1. Sterile plastic pipettes (5 ml, 10 ml).
2. 15- and 50-ml conical tubes.
3. 100-mm tissue culture-treated dishes.
4. 96-, 24-, 12-, and 6-well tissue culture-treated plates.
5. Filter pipette tips (0.5–10, 2–20, 20–200, 200–1,000 μl).
6. 0.22-μm vacuum filtration (500 ml).
7. 0.45-μm pore size filter.
8. Glass Pasteur pipettes, 9 in., sterilized by autoclave.
9. Sterile syringes (1, 5, 20, 50 ml).
10. Cryovials.
11. 40-μm cell strainers.

2.5 Solutions

2.5.1 B Cell Medium

To prepare 100 ml of B cell medium, mix the following:

RPMI with L-glutamine 1×	88.9 ml
Fetal bovine serum 10 %	10 ml
Penicillin-streptomycin 100×	1 ml
β-Mercaptoethanol 1,000×	100 μl
Interleukin 7 1,000×	100 μl

The medium can be stored at 4 °C for up to 2 weeks.

2.5.2 Cell Reprogramming Medium

To prepare 100 ml of reprogramming medium, mix the following:

KnockOut™ D-MEM	80.86 ml
Fetal bovine serum, ES cell-qualified	15 ml
Penicillin-streptomycin 100×	1 ml

(continued)

β-Mercaptoethanol 1,000×	100 μl
Penicillin-streptomycin 100×	1 ml
Sodium pyruvate 100×	1 ml
NE amino acids 100×	1 ml
Leukemia inhibitor factor 10,000×	10 μl
Interleukin 7 1,000×	100 μl
Interleukin 4 500×	200 μl
Interleukin 15 5,000×	20 μl
β-Estradiol 1,000×[a]	100 μl
Doxycycline 500×[a]	200 μl

[a]Prepare the reprogramming medium without inducers and supplement 17β-estradiol and doxycycline when needed. Reprogramming medium can be stored at 4 °C for up to 4 days.

2.5.3 iPS Cell Medium

To prepare 100 ml of iPS cell medium, mix the following:

KnockOut™ D-MEM	80.89 ml
Fetal bovine serum, ES cell-qualified	15 ml
Penicillin-streptomycin 100×	1 ml
β-Mercaptoethanol 1,000×	100 μl
Penicillin-streptomycin 100×	1 ml
Sodium pyruvate 100×	1 ml
NE amino acids 100×	1 ml
Leukemia inhibitor factor 10,000×	10 μl

2.5.4 OP9/S17 Medium

To prepare 100 ml of OP9/S17 cell medium, mix the following:

Alpha-MEM with GlutaMAX™ 1×	89 ml
Fetal bovine serum 20 %	20 ml
Penicillin-streptomycin 100×	1 ml
β-Mercaptoethanol 1,000×	100 μl

OP9/S17 cell medium can be stored at 4 °C for up to 2 weeks.

2.5.5 Interleukin 7, 4, and 15 Stock Solutions

To prepare 1 ml of a 10 μg/ml interleukin solution, mix the following:

Interleukin	10 μg
Deionized water	1 ml

2.5.6 Leukemia Inhibitor Factor (LIF) Stock Solution

To prepare 1 ml of a 100 μg/ml LIF solution, mix the following:

Leukemia inhibitor factor	100 μg
Deionized water	1 ml

2.5.7 β-Estradiol Stock Solution

To prepare 10 mM 17β-estradiol solution, mix the following:

β-Estradiol	2.7 mg
EtOH	1 ml

2.5.8 Doxycycline Stock Solution

To prepare 1 ml of a 50 mg/ml doxycycline solution, mix the following:

Doxycycline	50 mg
Deionized water	1 ml

2.5.9 Labeling Buffer

To prepare 250 ml of labeling buffer, mix the following:

PBS 1×	239 ml
FBS	10 ml
EDTA 0.5 M	1 ml

2.6 Immunomagnetic Reagents for B Cell Separation

CD19 biotin antibody (BD Biosciences).

Fc Block (BD Biosciences).

Streptavidin beads (Miltenyi Biotec).

LS columns (Miltenyi Biotec).

hCD4 biotin antibody (eBioscience).

MACS manual separator (Miltenyi Biotec).

2.7 Preparation of Gelatin-Coated Dishes

1. Using sterile techniques in a tissue culture hood, prepare gelatinized plates by covering the entire dish surface with a 0.1 % gelatin solution and incubate for 20 min at 37 °C.
2. Completely remove the gelatin solution from the dishes. Gelatinized plates should be used immediately.

3 Methods

3.1 B Cell Isolation

1. One week before performing the reprogramming experiment, purify "B cells" from the bone marrow of two reprogrammable mice by sorting CD19 surface antigen-positive cells, which consist of a mixture of pre-B and pro-B cells.
2. Sacrifice two 8–16 week-old reprogrammable mice by ventilating them with CO_2.

3. Aseptically remove femurs and tibiae (*see* **Note 1**).
4. Use scissors to remove all tissues from the bones (*see* **Note 2**).
5. Crush the bones in a mortar in 10 ml of labeling buffer.
6. Thoroughly resuspend the cells by pipetting them up and down.
7. Filter the cells with a 40-μm cell strainer in a 50-ml conical tube.
8. Centrifuge the cells for 5 minutes at 300 × *g*, resuspend them in fresh labeling buffer, and count them in a hemocytometer (*see* **Note 3**).
9. Resuspend the cells in labeling buffer to a final concentration of 10×10^6 cells/ml.
10. Add Fc Block to the cell suspension at a concentration of $0.1\ \mu g/10^6$ cells and incubate for 10 min on ice (*see* **Note 4**).
11. Add CD19-biotin conjugated antibody and incubate for 20 min on ice.
12. Wash the cells with 10 ml of labeling buffer, and centrifuge for 5 min at 300 × *g*. Resuspend them in 90 μl of labeling buffer and 10 μl of streptavidin beads per 10×10^6 cells and incubate for 20 min on ice.
13. Wash the cells with 10 ml of labeling buffer and proceed with purification using the MACS LS column.
14. Place the LS column in the MACS separator.
15. Wash the column with labeling buffer.
16. Pipette 500 μl of cells in the LS column allowing the suspension to run through and collect the effluent as negative fraction (this fraction is enriched for myeloid progenitors).
17. Wash the column with 3 ml of labeling buffer.
18. Repeat STEP 17 three times.
19. Remove the column from the MACS separator and place it in a new 15-ml collection tube.
20. Apply 5 ml of labeling buffer and firmly flush out the cells using the plunger provided. This yields the CD19 positive B cell fraction.

3.2 Reprogramming of B Cells

3.2.1 Retrovirus Production and B Cell Infection

1. Two days before viral transduction, seed 9×10^6 PlatE cells (Cell Biolabs) into two 10-mm dishes. To detach PlatE cells, wash the cells once with PBS and then add 1–3 ml of TrypLE™ select cell dissociation reagent. After 3–5 min at 37 °C, add PlatE medium and count the cells in the hemocytometer.
2. The following morning transfect PlatE cells with 20 μg of C/EBPα-ER-hCD4 plasmid using the calcium phosphate protocol (*see* **Note 5**).

3. Replace the transfection medium after 8 h with 6 ml of fresh B cell medium per 10-mm plate. Culture the cells at 37 °C, 5 % CO_2 incubator.
4. After 18–20 h of culture at 37 °C, collect the supernatant, filter it using a 0.45-μm sterile filter, and use the filtered medium to infect the B cells (*see* **Note 6**). Use 2 ml of supernatant to infect 2×10^6 B cells per well of a 6-well plate. Add 6 ml of fresh B cell medium to the PlatE cells. Place the cells in a 37 °C, 5 % CO_2 incubator (*see* **Note 7**).
5. 24 h later collect the supernatant of the PlatE cells and use it to infect the B cells a second time.

3.2.2 Plating of B Cells on OP9 or S17 Stromal Cells

We use S17 stromal cells for the first days of B cell expansion before the reprogramming is initiated. OP9 cells are then used for the reprogramming experiments.

1. On the day of the viral transduction experiment, prepare gelatin-coated dishes. Then plate mitomycin C-inactivated S17 cells at a density of 3×10^4 cells/cm^2 onto 100-mm plates (*see* **Note 8**). To detach S17 cells, wash the cells with PBS and add the adequate amount of TrypLE™ select cell dissociation reagent to the cells. After 3–5 min, add S17 medium and count the cells in the hemocytometer.
2. Collect the infected B cells from the 6-well plate and centrifuge them at $300 \times g$ for 5 min. Then aspirate the supernatant, and resuspend the cells in 10 ml of B cell medium. Plate the B cells on the inactivated S17 cells.
3. Grow the cells 4 days or until they reach confluence (note that besides B cells that remain nonadherent, many of them move under the feeder, forming clusters of flattened cells with smooth edges that can easily be distinguished from the feeder cells).
4. Collect the B cells, stain them with human CD4 antibody, and sort hCD4$^+$ transgene containing cells using a MACS column or a fluorescence-activated cell sorter (Fig. 2) (*see* Section 3.1 for detail on the sorting procedure) (*see* **Note 9**).
5. On the day before the sorting, prepare gelatin-coated 6-well dishes. Plate mitomycin C-inactivated OP9 cells at a density of 3×10^4 cells/cm^2. To detach OP9 cells add the adequate amount of TrypLE™ select cell dissociation reagent. After 3–5 min, add OP9/S17 medium and count the cells in the hemocytometer.
6. After the sorting of the hCD4$^+$ B cells, count the cells using the desired method and seed the cells at a density of 1–10 $\times 10^4$ cells/cm^2. Let the cells grow for 2 days.
7. To activate C/EBPα add β-estradiol at a concentration of 10 μM.

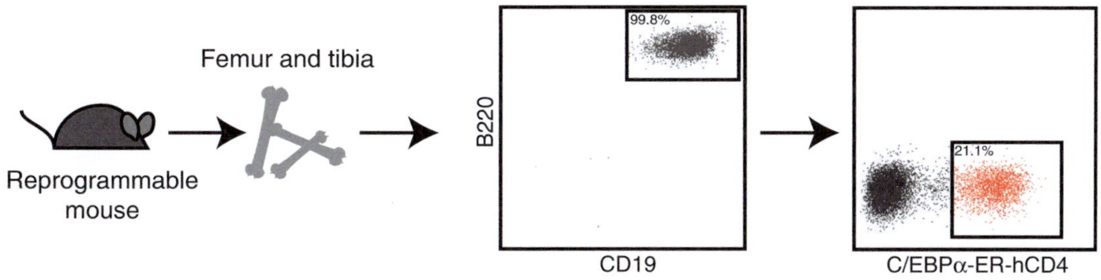

Fig. 2 Sorting strategy. B Cells are sorted based on the expression of CD19 surface antigen and infected cells are selected based on the expression of hCD4

8. After 18 h collect the supernatant and carefully wash the cultures twice with 500 μl of PBS. Then centrifuge the supernatant/PBS at 300 × *g* for 5 min (*see* **Note 10**).
9. Resuspend the pellet in fresh reprogramming medium containing doxycycline and reseed the cells into the original wells.
10. Gently replace the medium on a daily basis.
11. Starting at day 4 microscopically observe the cells daily for the emergence of small colonies indicative of reprogramming into iPS cells.
12. At day 8 change the medium to reprogramming medium without doxycycline to select transgene-independent iPS cell colonies (*see* **Note 11**).
13. At day 10–12 colonies should have grown to an appropriate size for isolation and reseeding. The day before isolating the colonies, prepare cultures of inactivated MEFs using gelatin-coated 12- or 24-well plates (*see* **Note 12**).

3.2.3 Isolating iPS Cell Colonies

1. Aliquot 20 μl of TrypLE™ select cell dissociation reagent into each well of a 96-well plate.
2. Replace the iPS cell medium with D-PBS.
3. Isolate colonies under the microscope by picking them with a 1,000-μl micropipette. Slowly suck the colony into the pipette tip.
4. Transfer the colony into the TrypLE™ select cell dissociation reagent in the 96-well plate and incubate at 37 °C for 20 min.
5. Disaggregate the colonies by adding iPS medium into the 96-well plates and pipette the cells up and down.
6. Transfer the dissociated cells into the 12-well plate containing mitomycin C-inactivated MEFs and incubate at 37 °C with 5 % CO_2.
7. After 3–4 days of culture, transfer the isolated outgrowing clones individually onto a 12-well dish containing fresh mitomycin C-inactivated MEFs (13). Cells are passaged in iPS

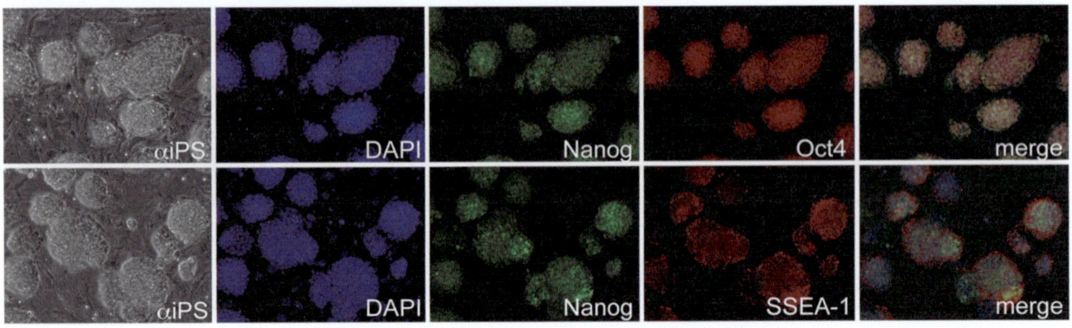

Fig. 3 iPS cell clones. iPS cells derived from B cells pulsed with C/EBPα (called αiPS cells) stained for Nanog, Oct4, and SSEA-1

medium every 2 days using TrypLE™ select cell dissociation reagent. Once expanded to 10-mm plates, iPS cells can be frozen/thawed as previously described (13).

3.3 Basic Characterization of iPS Cells

The overall quality of iPS cell formation can be easily assessed by immune fluorescence analysis to confirm the expression of pluripotency markers. iPS cells derived by using this protocol should express Nanog, Oct4, and SSEA-1 (Fig. 3). More detailed analyses are needed to confirm pluripotency of the iPS cell lines such as teratoma formation and generation of chimeric mice upon injection into the host blastocyst of the reprogrammed cells. Here we present a simple immunofluorescence protocol:

1. Wash the plate three times with D-PBS and fix them by adding 4 % paraformaldehyde solution for 20 min at room temperature.
2. Wash the plate three times with D-PBS.
3. Add blocking buffer (0.1 % Triton X-100/D-PBS/10 % goat serum) and incubate for 1 h at room temperature.
4. Incubate the cells with the primary antibody diluted in blocking buffer O/N at 4 °C.
5. Wash the plate three times with D-PBS.
6. Incubate the cells with secondary antibody diluted in blocking buffer and DAPI at RT for 1 h.
7. Wash the plate three times with D-PBS.
8. Observe the cells with a 20× objective under a fluorescence microscope.

4 Notes

1. Sterilize the abdomen and hind legs with 70 % ethanol. Make an incision in the midline of the abdomen. Clip outward to expose the hind legs.

2. Wash the bones carefully with PBS to avoid possible contamination with cells outside the bones.
3. We suggest to use Türk's solution to count the cells at this step as the erythrocytes are hemolyzed by the acetic acid and the leukocytes are stained by the dye that solution contains.
4. The use of Fc Block is suggested to avoid unspecific staining with the CD19 primary antibody.
5. We add 30 μM chloroquine to the medium used to transfect the PlatE cells to increase transfection efficiency.
6. We suggest infecting the B cells immediately after the CD19 purification.
7. We observed increased viability if 1 ml of fresh medium is added to the cells after infection. We do not recommend refreezing and thawing the virus samples since the titers decrease upon thawing.
8. We initially culture the B cells on S17 stromal cells to increase the number of cells as the proliferation is greatly enhanced in this condition compared to the OP9 feeder cells.
9. The hCD4 cell sorting can be performed as for the CD19 biotin (Section 3.1).
10. Maintaining the precise β-estradiol exposure time of 18 h is crucial to obtain an optimal reprogramming efficiency.
11. Longer time exposure to doxycycline may lead to an increase in the number of iPS cell colonies.
12. At this stage we suggest to calculate the efficiency of reprogramming by performing immunohistochemistry for Nanog and scoring as iPS colonies those that are >10 % positive for this marker.

References

1. Jaenisch R, Young R (2008) Stem cells, the molecular circuitry of pluripotency and nuclear reprogramming. Cell 132(4):567–582. doi:10.1016/j.cell.2008.01.015
2. Robinton DA, Daley GQ (2012) The promise of induced pluripotent stem cells in research and therapy. Nature 481(7381):295–305. doi:10.1038/nature10761
3. Takahashi K, Yamanaka S (2013) Induced pluripotent stem cells in medicine and biology. Development 140(12):2457–2461. doi:10.1242/dev.092551
4. Takahashi K, Yamanaka S (2006) Induction of pluripotent stem cells from mouse embryonic and adult fibroblast cultures by defined factors. Cell 126(4):663–676. doi:10.1016/j.cell.2006.07.024
5. Buganim Y, Faddah DA, Jaenisch R (2013) Mechanisms and models of somatic cell reprogramming. Nat Rev Genet 14(6):427–439. doi:10.1038/nrg3473
6. Hanna J, Saha K, Pando B, van Zon J, Lengner CJ, Creyghton MP, van Oudenaarden A, Jaenisch R (2009) Direct cell reprogramming is a stochastic process amenable to acceleration. Nature 462(7273):595–601. doi:10.1038/nature08592
7. Carey BW, Markoulaki S, Beard C, Hanna J, Jaenisch R (2010) Single-gene transgenic mouse strains for reprogramming adult

somatic cells. Nat Methods 7(1):56–59. doi:10.1038/nmeth.1410

8. Di Stefano B, Sardina JL, van Oevelen C, Collombet S, Kallin EM, Vicent GP, Lu J, Thieffry D, Beato M, Graf T (2014) C/EBPalpha poises B cells for rapid reprogramming into induced pluripotent stem cells. Nature 506 (7487):235–239. doi:10.1038/nature12885
9. Carey BW, Markoulaki S, Beard C, Hanna J, Jaenisch R (2009) Single-gene transgenic mouse strains for reprogramming adult somatic cells. Nat Methods 7(1):56–59, doi: nmeth.1410[pii]10.1038/nmeth.1410
10. Nakano T, Kodama H, Honjo T (1994) Generation of lymphohematopoietic cells from embryonic stem cells in culture. Science 265 (5175):1098–1101
11. Collins LS, Dorshkind K (1987) A stromal cell line from myeloid long-term bone marrow cultures can support myelopoiesis and B lymphopoiesis. J Immunol 138(4):1082–1087
12. Xie H, Ye M, Feng R, Graf T (2004) Stepwise reprogramming of B cells into macrophages. Cell 117(5):663–676
13. Takahashi K, Okita K, Nakagawa M, Yamanaka S (2007) Induction of pluripotent stem cells from fibroblast cultures. Nature protocols 2(12):3081–3089. doi:10.1038/nprot.2007.418

Methods in Molecular Biology (2016) 1357: 57–69
DOI 10.1007/7651_2014_139

Published online: 18 November 2014

Generation of iPS Cells from Human Peripheral Blood Mononuclear Cells Using Episomal Vectors

Ruijun Jeanna Su, Amanda Neises, and Xiao-Bing Zhang

Abstract

Peripheral blood is the easy-to-access, minimally invasive, and the most abundant cell source to use for cell reprogramming. The episomal vector is among the best approaches for generating integration-free induced pluripotent stem (iPS) cells due to its simplicity and affordability. Here we describe the detailed protocol for the efficient generation of integration-free iPS cells from peripheral blood mononuclear cells. With this optimized protocol, one can readily generate hundreds of iPS cell colonies from 1 ml of peripheral blood.

Keywords: Induced pluripotent stem cells, Reprogramming, Human peripheral blood, Mononuclear cells, Hematopoietic cells, Integration-free, Episomal vectors

1 Introduction

The generation of induced pluripotent stem cells (iPS cells or iPSCs) from somatic cells is one of the most exciting breakthroughs in the life sciences (1). This advance opens up many new opportunities for personalized cell therapy, disease modeling, and drug screening (2). Fibroblasts are widely used for reprogramming cells into pluripotency, however harvesting fibroblasts from a human skin biopsy is a painful procedure and expanding them to a sufficient amount takes several weeks. In addition, safety concerns arise from the skin's lifetime exposure to DNA-damaging insults in the environment. Peripheral blood has been widely accepted as the best cell source for cell reprogramming due to its easy accessibility and better quality and quantity (3). Mature white blood cells and a few progenitor cells in circulation are egressed from the stem cell niche located in the bone marrow, where they are best protected from environmental irradiation. An average adult has 4–11 $\times$ 10^6 white blood cells, the nucleated cells in blood, in 1 ml of blood. After depletion of mature granulocytes by gradient centrifugation using a reagent called Ficoll, 1–3 $\times$ 10^6 mononuclear cells (MNCs) can be harvested. MNCs are enriched in monocytes, T lymphocytes, B lymphocytes, and a few progenitor cells. Myeloid or erythroid progenitor cells are the preferable cell type for reprogramming due to their entirety in the genome and high plasticity for cell fate conversion. A recent study shows that culturing MSCs for 1 week in medium that favors expansion of erythroid progenitors is the best approach for generating iPSCs in large quantities (4).

Retroviral vectors or lentiviral vectors that express the Yamanaka factors OCT4, SOX2, MYC, and KLF4 are used for reprogramming fibroblasts into iPS cells. However, these vectors are less efficient in reprogramming blood cells, largely because the vector design does not lead to high-level transgene expression in hematopoietic cells. By comparing several commonly used promoters, we identified that SFFV, the promoter from spleen focus forming virus, is substantially stronger than other promoters like PGK, CAG, and EF1α (3, 5). The use of the SFFV promoter in lentiviral or episomal vectors leads to approximately a 100-fold increase in reprogramming (4–7). This seemingly simple but vitally critical innovation underlies the success of blood reprogramming.

For clinical use or unbiased disease modeling and drug screening, integration-free iPSCs are required. Many technology advances have been reported for generating integration-free iPSCs, such as using proteins, mRNAs, Sendai viral vectors, and other RNA vectors to deliver reprogramming factors (3). The combination of several small molecule compounds has been shown to be able to convert mouse fibroblasts into iPS cells; whether this approach works for human cells remains unknown. However, these approaches suffer from one or more of the following drawbacks, (1) requiring multiple administrations of agents and thus labor intensive, (2) packaging of vectors is technically challenging and thus most investigators have to purchase from vendors at a hefty price, (3) low reprogramming efficiency; and (4) some do not work for blood reprogramming, yet. In contrast, the EBNA1/oriP-based episomal vector, a plasmid developed and optimized by multiple labs, requires no packaging and one administration is sufficient for generating iPSCs. Building upon reports from James Thomson's lab (8), Linzhao Cheng's lab (9), Shinya Yamanaka's lab (10), and other colleagues, we have made several changes that have vastly increased the efficiency of blood reprogramming: (1) the use of the SFFV promoter, (2) inclusion of WPRE in the vector, (3) inclusion of BCL-XL as a reprogramming booster (6). With all these improvements, we can routinely generate hundreds of colonies using the episomal vector system, representing a 100- to 1,000-fold increase from ~1 colony per 1 ml of PB just 3 years ago. As follows we will detail the experimental procedures and tips.

2 Materials

2.1 Cells

1. Human peripheral blood. Draw from adult donors after obtaining informed consent, or obtain buffy coat white blood cells from a blood bank.
2. Rat embryonic fibroblast (REF) feeder cells (Applied Biological Materials Inc. or ABM; Richmond, BC, Canada) or made in-house from E14-16 rat embryos and inactivated by irradiation at 60 Gy.

2.2 MNC Isolation

1. Ficoll-Hypaque (1.077 g/ml) (G&E Healthcare; Cat. No. 17-1440-03).
2. PBS (Invitrogen; Cat. No. 10010-023).
3. (Optional) $CD34^+$ cells MACS purification kit (Miltenyi; Cat. No. VPA-1003).
4. (Optional) Red Blood Cell Lysis Buffer (RBC lysis buffer): Weigh out 4.14 g of NH_4Cl and 0.5 g of $KHCO_3$. Dissolve the NH_4Cl and $KHCO_3$ in 450 ml of water and add 0.9 ml of 5 % EDTA. Adjust the pH to 7.2–7.4. Add water to 500 ml and filter-sterilize through a 0.2 μm filter. Store at 4 °C, but bring to room temperature before use.

2.3 Cell Culture

1. Erythroid culture medium: Stemline® II Hematopoietic Stem Cell Expansion Medium (Sigma; Cat. No. S0192) supplemented with 100 ng/ml SCF, 10 ng/ml IL-3, 2 U/ml EPO, 20 ng/ml IGF-1, 1 μM dexamethasone (Sigma; Cat. No. D4902), 0.2 mM 1-thioglycerol (MTG) (Sigma; Cat. No. M6145). Cytokines were purchased from ProSpec (East Brunswick, NJ). Store the medium at 4 °C for up to 3 weeks.
2. iPSC generation medium: The iPSC medium is composed of Knockout DMEM/F12 medium (Invitrogen; Cat. No. 112660-012) supplemented with 1 % L-Glutamine (ABM, Cat. No. G275), 2 mM nonessential amino acids (Invitrogen; Cat. No. 11140-050), 1 % penicillin/streptomycin (ABM; Cat. No. G255), 50 ng/ml FGF2 (ABM; Cat. No. Z101455), 1 % ITS (BD; Cat No. 354350), and 50 μg/ml ascorbic acid 2-phosphate (Sigma; Cat. No. 49752). Store at 4 °C for up to 3 weeks.
3. Feeder cell culture medium: Knockout DMEM/F12 supplemented with 15 % fetal bovine serum (ABM; Cat. No. TM999-500), 1 % L-Glutamine and 0.1 mM β-mercaptoethanol (Sigma; Cat. No. M3148). Store at 4 °C for up to 4 weeks.
4. Feeder-free iPSC culture medium mTeSR™ 1 (Cat. No. 05850) or TeSR E8 (Cat. No. 05940) were purchased from Stemcell Technologies (*See* **Note 1**).

5. Accutase (Innovative Cell Technologies; Cat. No. AT-104) (*See* **Note 2**).
6. Matrigel (BD; Cat. No. 356235). Add 1 ml of Matrigel to 49 ml of Knockout DMEM/F12 medium supplemented with 1 % penicillin/streptomycin. Store at 4 °C for up to 2 months. Before iPS cell passage without feeder support, add 1 ml of diluted Matrigel into each of the TC-treated 6-well plates. Incubate at 37 °C for 2 h. Aspirate out the Matrigel and add medium and cells for culture.
7. Freezing medium. Dissolve 5 g of trehalose (Sigma Cat. No. T9531) in 30 ml of water in a 37 °C water bath, bring the temperature to 4 °C and then add 10 ml of FBS and 10 ml of DMSO (Fisher Scientific; Cat. No. BP231-1) (11). Filter-sterilize with a 0.22 μm syringe filter. Store at 4 °C for up to 3 months. Add equal volume of freezing medium to the cell suspension before use. Alternatively, commercial freezing medium can be purchased from Stemcell Technologies (Cat. No. 05855) or ABM (Cat. No. TM023).
8. Sodium butyrate (Sigma; Cat. No. B5887): Weigh out 1.1 g of sodium butyrate and dissolve in 40 ml of water to make 0.25 M sodium butyrate (*See* **Note 3**).
9. ROCK inhibitor Y-27632 (ABM; Cat. No. G604). Add a final concentration of 10 μM to the freezing medium and iPSC culture medium after cell thawing (*See* **Note 4**).

2.4 Episomal Vectors and Nucleofection Kits

1. Episomal vectors: pEV SFFV-OCT4-E2A-SOX2-wpre (pEV-OS); pEV SFFV-MYC-E2A-KLF4-wpre (pEV-MK); pEV SFFV-BCL-XL-wpre (pEV-B) (*See* **Notes 5** and **6**).
2. Nucleofector™ Kits for Human $CD34^+$ Cells (Lonza, Walkersville, MD; Cat. No. VPA-1003).

2.5 Plastics and Equipments

1. Non-TC treated 6-well plates (BD Falcon; Cat. No. 351146).
2. TC treated 6-well plates (BD Falcon; Cat. No. 353046).
3. TC treated 24-well plates (BD Falcon; Cat. No. 353047).
4. 5- or 15- or 50-ml polystyrene tubes (BD Falcon).
5. 10 ml syringes with long needle.
6. Pipettes and pipettors.
7. Nucleofector (Lonza; Amaxa II).
8. CO_2 incubator.
9. Bucket centrifuge.
10. Hypoxia chamber (Stemcell Technologies; Cat. No. 27310).

11. Mixed gas cylinder. Order a mixed gas composed of 3 % O_2, 5 % CO_2, and 92 % N_2. Flush the hypoxia chamber at 30 L/min for 1 min (*See* **Note 7**).
12. Flow cytometer.
13. Inverted microscope.
14. Confocal microscope.
15. Chamber slides (Nunc; Cat. No. 177429).

2.6 Immunohistochemistry and Flow Cytometry

1. Flow cytometry (FACS) buffer: PBS supplemented with 2 % FBS. Store at 4 °C. For long-term storage, add 0.05 % sodium azide (Sigma; Cat. No. S8032).
2. Fixation buffer and permeabilization buffer (eBiosciences, San Diego, CA; Cat. No. 00-8222-49 and 00-8333-56).
3. Antibodies: OCT4-PE (eBioscience; Cat. No. 12-5841-82), SOX2-FITC (BD; Cat. No. 560301), NANOG-PE (BD; Cat. No. 560483) and TRA-1-60-PE, SSEA3-PE, SSEA4-PE (eBioscience; Cat. Nos. 12-8863-82, 12-8833-73, and 12-8843-42).

3 Methods

Conduct all experiments at room temperature unless otherwise specified.

3.1 Ficoll

1. Add 10 ml of peripheral blood (PB) and 10 ml PBS in a 50 ml tube. Mix well with a 10 ml pipette. Slowly add 10 ml of Ficoll-Hypaque to the bottom of the tube using an 18 G needle attached to a 10 ml syringe (*See* **Note 8**).
2. Centrifuge at 400 × *g* for 30 min at a low acceleration and deceleration rate.
3. After centrifugation, there will be several layers of different cell types, the white layer in the middle is composed of platelets and MNCs. Place a 1,000 μl pipette tip at ~1 mm above the layer and carefully harvest cells in the white layer into a separate 50-ml tube. Once the PB MNCs are harvested, the total volume will be about 3–6 ml.
4. Add 4–6× volume of RBC lysis buffer, and the total volume should be less than 30 ml (*See* **Note 9**). If there is more than 30 ml, split into two tubes. If there is no red blood cell contamination at the buffy coat layer, add PBS instead of RBC lysis buffer. Mix by pipetting to obtain a single cell suspension. Centrifuge the cells at 400 × *g* for 10 min and remove the supernatant (*See* **Note 10**). Resuspend the cell pellet in 20 ml of IMDM or other culture medium, and spin down at 400 × *g* for 5 min to remove the majority of the platelets.

5. Resuspend the cells in 5 ml of erythroid culture medium and count the number of cells. Extra MNCs can be frozen down (*See* **Note 11**).

3.2 MNC Culture

1. Culture PB MNCs at 2–5 × 10^6/ml. Add 2 ml to each well of a non-TC treated 6-well plate (*See* **Note 12**). Add 1 ml of erythroid culture medium into each well every 2–3 days.
2. Culture the cells at 37 °C for 7–8 days in a CO_2 incubator with a water tray to maintain the humidity.
3. On the day of nucleofection, harvest cells to 5-ml tubes and count the number of cells. Spin down the cells at 400 × *g* for 5 min. Resuspend the cells in 2 ml of prewarmed PBS/0.1 % BSA. Spin down the cells at 200 × *g* for 8 min. Carefully aspirate out the medium before adding the nucleofection buffer.

3.3 REF Feeder Preparation

1. One day before nucleofection, thaw inactivated 2 × 10^6 REF feeder cells by adding 10–20 ml REF medium in a 50 ml tube (*See* **Note 13**).
2. Spin down the cells at 400 × *g* for 5 min. Resuspend the cells in 12 ml of REF medium and add 2 ml to each well of a TC-treated 6-well plate that has been precoated with 0.1 % gelatin.
3. Culture cells at 37 °C in CO_2 incubator.
4. Before nucleofection, aspirate out the REF medium and add 2 ml erythroid culture medium.

3.4 Nucleofection

1. Add 4 μg pEV-OS, 4 μg pEV-MK, and 2 μg pEV-B in a sterile Eppendorf tube.
2. Heat the tube at 50 °C for 5 min. Cool down the tube to room temperature and add 74 μl of nucleofection buffer and 16 μl of the supplement provided by the Nucleofector™ Kits.
3. Mix the DNA and the buffer with a 100 μl tip and add to the cell pellet. Flick the bottom of the tube, using your finger, 3–5 times to resuspend the cells in the buffer.
4. Use the kit-provided plastic pipette to transfer the mixture to the provided cuvette without producing any bubble. Use the program U-008 for nucleofection.
5. After nucleofection, use the same pipette to transfer 500 μl medium from the culture well to the cuvette. Place the pipette to the bottom of the cuvette and transfer the cells to the culture well. Alternatively, transfer 10–50 % of the transfected cells into each well.
6. Place the plate in a hypoxia chamber, flush with a mixed gas composed of 3 % O_2, 5 % CO_2, and 92 % N_2, and transfer into the CO_2 incubator (*See* **Note 14**).

3.5 iPSC Generation

1. On day 2, add 2 ml of iPSC generation medium into each well.
2. Starting on day 4, change the medium every 2 days for ~2 weeks by leaving 500 μl spent medium and adding 2 ml of fresh iPSC generation medium to the well (*See* **Note 15**).
3. Add sodium butyrate to a final concentration of 0.25 mM at days 2, 4, 6, 8, 10 and 12.
4. Every 5–7 days, when observing that the REF feeder cells are detached and deteriorating, add 3–5 × 10^5 REF feeder cells resuspended in 500 μl of REF medium directly into each well (*See* **Note 16**).
5. Small iPSC-like colonies will appear at ~1 week after nucleofection (*See* **Note 17**).

3.6 Picking iPSC Colonies

1. When the colonies become visible to the naked eye, start to pick the colonies. Mark the colonies with typical iPSC morphology at the bottom of the wells using a fine marker pen. Gently scratch the colonies using a 100 μl tip and transfer one colony to each well of the 24-well plates pre-seeded with 0.5–1 × 10^5 REF feeder cells. Pick 10–20 colonies for each donor.
2. Change the iPSC medium every 1–2 days. After 1 week, passage the cells to 6-well plates by adding 150 μl of Accutase into each well and incubate the cells at 37 °C for 5 min. Break down the colonies into clumps of 5–50 cells by scratching and/or gentle pipetting. Transfer cells directly to 6-well plates and add 2 ml of iPSC medium into each well (*See* **Note 18**).

3.7 Long-Term Feeder-Free Culture

1. iPS cells can be cultured with feeder support for long-term. Alternatively, iPS cells can be cultured in TeSR1 without feeder support.
2. After picking the colonies, iPS cells can be adapted to feeder-free culture by culturing cells in TeSR1 medium for several days before passaging.
3. For long-term passage of iPS cells with TeSR1, precoat the well plates with Matrigel. Change the medium every day. When the cells reach 30–60 % confluency, passage the cells by aspirating out the spent medium and adding 600 μl of Accutase to the well. Incubate at 37 °C for 5 min and break down the colonies into clumps of 5–50 cells by gentle pipetting. Transfer 10–20 % cells to a new well for continued culture (*See* **Note 19**).

3.8 Freezing Down iPSCs

1. After several passages, the iPSCs may be frozen down for long-term storage. At 4–5 days after the cell split, when iPSCs reach a confluency of 30–60 %, aspirate out the spent medium and add 500 μl of Accutase. Incubate the cells at 37 °C for ~5 min and then gently harvest cells to a 1 ml cryovial by pipetting 1–2 times.

2. Leave the cells in clumps because breaking down the clumps into single cells may substantially decrease the cell survival. Add 0.5 ml of freezing medium supplemented with ROCK inhibitor into the cryovial.
3. Mix well and transfer the vial to a −80 °C freezer for short-term storage (days to weeks). Several hours later when the cells are frozen, the vials may be transferred to a liquid nitrogen tank for long-term storage (years).

3.9 Thawing iPSCs

1. Thaw iPS cells by immersing the vial in a 37 °C water bath for 1–2 min. Add 5–10 ml of Knockout DMEM/F12 medium to a 15 or 50-ml tube. Gently pipette iPS cells 2–3 times and transfer all of the cells to the tube.
2. Spin down the cells at 200 × *g* for 3–5 min. Aspirate out the washing medium and resuspend the cells in 2 ml of TeSR1 medium supplemented with ROCK inhibitor. Culture the cells in Matrigel-precoated 6-well TC-plate.
3. Starting the next day, refresh the cells with 2 ml of TeSR1 medium. Long-term culture and passage the iPS cells as described above (*See* **Note 20**).

3.10 Selection of iPSC Clones Without Residual Episomal Plasmids

1. After culturing the iPSCs for 4–5 passages, some of the cells can be harvested for DNA extraction and the analysis of residual plasmid DNA in the cells by real-time PCR can be determined.
2. To make a standard of one copy per genome, we add 1.6 pg pEV-OS plasmid into 1 μg gDNA extracted from PB cells.
3. The primers used for PCR are: EBNA-F, TTTAATACGATTGAGGGCGTCT; EBNA-R, GGTTTTGAAGGATGCGATTAAG; WPRE-F, GGTTTAAACGCGTCGACAAT; WPRE-R, GTTGCGTCAGCAAACACAGT; ACTB-F, TCGTGCGTGACATTAAGGAG; ACTB-R, GGCAGCTCGTAGCTCTTCTC.
4. The amount of DNA was normalized by Beta-actin (gene name ACTB).
5. The PCR program is: 50 °C 2 min (holding stage); 95 °C 10 min (holding stage); 95 °C 15 s and 60 °C 1 min for 40 cycles (cycling stage); 95 °C 15 s, 60 °C 1 min, 95 °C 30 s, 60 °C 15 s (melt curve stage).
6. In most of the clones, the residual episomal plasmids are 0.001 copies or undetectable. Choose these clones for continued culture and you may discontinue the clones with more than 0.01 copies per cell.

3.11 ALP Staining

1. To determine the reprogramming efficiency, the iPSC colonies were stained with an ALP-staining kit (Stemgent; Cat No. 00-0055).
2. ALP-positive iPSC colonies were enumerated at 2–3 weeks after nucleofection.
3. To stain the cells, aspirate out the culture medium and add 1 ml of fixation buffer to each well. Incubate at room temperature for 3–5 min and then wash with 1 ml of PBS. Add the ALP staining solution to the well and incubate the plate for 15–20 min at room temperature.
4. Aspirate out all of the staining solution and wash the wells twice with PBS. You may add 1 ml of 20 % glycerol into each well for long-term storage at 4 °C.
5. Scan the plate or take a digital picture. Count the colonies using a picture processing software.

3.12 Flow Cytometry

1. For intracellular staining of OCT4, SOX2 and NANOG. Add 600 μl of Accutase to each well and incubate at 37 °C for 5 min. Pipette the cells up and down to make a single cell suspension. Use 100 μl for cell staining and FACS analysis. Add 900 μl of fixation buffer and 100 μl of 10× permeabilization buffer and incubate the cells at room temperature for 10 min. After spinning down the cells at 400 × *g* for 5 min, the cells were resuspended in 100 μl of FACS buffer with 10 % permeabilization buffer and stained with the antibody at room temperature for 2 h. Wash cells twice with 2 ml of FACS buffer supplemented with 10 % permeabilization buffer and resuspend the cells in 100 μl of FACS buffer before conducting FACS analysis.
2. For staining of cell surface markers TRA-1-60, SSEA3 and SSEA4. Cells resuspended in Accutase were incubated with antibodies for 30 min at room temperature. After washing with 2 ml of FACS buffer and resuspending in 100 μl of FACS buffer, flow cytometric analysis was performed using FACS Aria II (BD Biosciences, San Jose, CA). Thirty thousand events were collected for each sample.

3.13 Confocal Imaging

1. iPSCs were cultured in chamber slides for 4–5 days.
2. Cells were then treated with fixation buffer and permeabilization buffer for 30 min before being stained overnight at 4 °C with PE or FITC conjugated antibodies anti-OCT4, anti-SOX2, anti-NANOG, and anti-SSEA4.
3. Confocal imaging was performed using the Zeiss LSM 710 NLO laser scanning confocal microscope with a 20× objective. High resolution monochrome images were captured using a Zeiss HRm CCD camera.

3.14 Teratoma Assay

1. A teratoma assay was conducted to examine the pluripotency of iPS cells.
2. NOD/SCID/IL2RG^{null}/2 (NSG) immunodeficient mice were purchased from the Jackson Laboratory (Sacramento, CA) and maintained at Pathogen-free animal facility.
3. Approximately 1×10^6 iPSCs were suspended in 200 μl of DMEM/F12 diluted (1:1) Matrigel solution and injected into the subcutaneous tissue above the rear haunch of NSG mice.
4. At 2 months after the cell injection, teratomas were dissected and fixed in 10 % formalin. After microsectioning, samples were stained with hematoxylin and eosin (H and E) and analyzed by a pathologist.
5. Characterizations such as Karyotyping and G-banding and multilineage differentiation are covered by other chapters in detail.

4 Notes

1. Both TeSR1 medium and E8 medium support feeder-free culture of established iPS cells. However, anecdotal evidence shows that iPS cells maintained in TeSR1 medium have a better potential for differentiation than those cultured in E8 medium.
2. Accutase is used for cell splitting. Alternatively, dispase or 0.5 mM EDTA/PBS can also be used to passage iPS cells. The use of Accutase simplifies the cell passage procedure, because one does not need to remove the enzyme. Be cautious not to break down iPSC clumps into single cells.
3. Sodium butyrate can help convert intermediate cells into complete reprogrammed iPSCs. The use of sodium butyrate not only increases quantity but also improves the quality of the generated iPSCs due to its ability to open up chromatin allowing for more reprogramming factors to access it.
4. The use of ROCK inhibitor before cell freezing and right after cell thawing is critical for increased survival of iPS cells. The use of commercial freezing medium from Stemcell Technologies or ABM may increase the cell survival.
5. BCL-XL is critical for efficient reprogramming. A tenfold increase in reprogramming is seen with all the combinations we have tested (6). One may include other reprogramming factors such as NANOG, LIN28, ZSCAN, and miR-302 family to increase the quality of generated iPS cells. Including more factors do not necessarily increase the reprogramming efficiency, but a recent report demonstrates that it increases the quality of iPSCs by conferring the germ line transmission capacity (12).

6. Use endo-free Maxiprep Kits to obtain high-quality plasmid DNA (Qiagen; Cat. No. 12362).
7. We found that reprogramming efficiency is higher when using a hypoxia chamber flushed with 3 % O_2 and 5 % CO_2 than simply culturing cells in a 5 % O_2 hypoxia incubator, which is controlled by flushing with N_2 and CO_2.
8. Keep Ficoll-Hypaque at room temperature. If storing at 4 °C, bring it to room temperature before use. A common practice is adding Ficoll first and then carefully adding diluted blood on top of Ficoll. We prefer to add Ficoll later directly to the bottom of diluted blood using a syringe, which is less challenging and much more efficient. A common mistake in Ficoll gradient centrifugation is trying to harvest more buffy coat cells, which leads to contamination with too much Ficoll, leading to the density change of the washing solution. This will result in the loss of large quantities of MNCs, in particular progenitor/stem cells, because these cells are smaller and have less density.
9. Keeping the RBC lysis buffer at 4 °C is encouraged for long-term storage. Bring it to room temperature before use. Long-term storage at room temperature may lead to deterioration or degradation of the buffer.
10. Platelet contamination. Longer time and/or higher centrifugation force will lead to higher yield of MNCs, but this will increase the number of platelets in MNCs. Inexperienced researches may mistake platelets as bacteria. The platelets will die after several days in culture and it appears that the platelets do not affect the expansion of progenitors.
11. PB MNCs can be frozen down for later use. Freezing down whole peripheral blood is not recommended. Dying granulocytes will lead to the formation of clumps during Ficoll process. In case that only the whole blood is available for precious samples, one may use RBC lysis buffer to treat the sample and then culture cells for 1 day before conducting Ficoll gradient centrifugation.
12. Hematopoietic cells appear to proliferate better when they are cultured at higher densities. The cell number will decrease during the culture due to the death of mature cells.
13. Mouse embryonic fibroblasts (MEFs) are more commonly used. However, one study shows that rat embryonic fibroblasts support human iPSC generation better than MEFs (13). Due to the high reprogramming efficiency of our protocol, the use of fibronectin or Matrigel instead of feeder support system can also lead to the successful generation of iPSCs from blood cells, albeit at lower efficiencies. However, whether iPS cells

generated without feeder support have inferior quality is still an unanswered question.

14. Our experience suggests that the optimal cell density is 2–4 × 10^6 cells for nucleofection. Too low or too high a density may decrease the reprogramming efficiency. $CD34^+$ cells enriched from cultured PB MNCs may be used for reprogramming. Depletion of T and B cells is unnecessary because this culture system does not support reprogramming of T and B cells.
15. When changing the culture medium, leave ~500 μl per well to prevent the drying out of these cells in the wells. This can also increase the efficiency, in particular, for large scale cell culture.
16. Conditioned medium may be used starting 1 week after MNC and REF coculture. However, direct addition of REFs in the culture is simple and less labor-intensive. The use of TeSR1 or E8 medium during the reprogramming will substantially decrease the reprogramming efficiency due to the presence of TGF-β1 in the medium.
17. Colony numbers will continue to increase during the 3 weeks of reprogramming culture. Hundreds of iPSC-like colonies will be obtained from 1 ml peripheral blood. The reprogramming efficiency of PB from different donors is highly variable.
18. To pick small colonies, one may use an inverted microscope placed in the biosafety hood. However, this is more challenging and less efficient than picking large colonies without the use of microscope.
19. A common phenomenon of cells cultured in 6-well plates is that more cells are located in the center of wells. To prevent uneven spreading, gently rock front and back twice to mix the cells after adding medium and cells to the wells, and rock the plates again right before placing back the well plates into the incubator.
20. After thawing, iPS cells can be directly cultured in a feeder-free system. However, in case that the cell survival is poor, the use of feeders and conditioned medium will substantially increase the cell survival.

References

1. Takahashi K, Yamanaka S (2006) Induction of pluripotent stem cells from mouse embryonic and adult fibroblast cultures by defined factors. Cell 126(4):663–676, doi: S0092-8674(06)00976-7 [pii] 10.1016/j.cell.2006.07.024
2. Cherry AB, Daley GQ (2013) Reprogrammed cells for disease modeling and regenerative medicine. Annu Rev Med 64:277–290. doi:10.1146/annurev-med-050311-163324
3. Zhang XB (2013) Cellular reprogramming of human peripheral blood cells. Genomics Proteomics Bioinformatics 11(5):264–274, doi: S1672-0229(13)00092-2 [pii] 10.1016/j.gpb.2013.09.001

4. Liu SP, Li YX, Xu J, Gu HH, Zhang HY, Liang HY, Liu HZ, Zhang XB, Cheng T, Yuan WP (2014) An improved method for generating integration-free human induced pluripotent stem cells. Zhongguo Shi Yan Xue Ye Xue Za Zhi 22(3):580–587, doi: 1009-2137(2014) 03-0580-08 [pii]
5. Meng X, Neises A, Su R-J, Payne KJ, Ritter L, Gridley DS, Wang J, Sheng M, William Lau KH, Baylink DJ, Zhang X-B (2012) Efficient reprogramming of human cord blood CD34+ cells into induced pluripotent stem cells with OCT4 and SOX2 alone. Mol Ther 20 (2):408–416
6. Su R-J, Baylink DJ, Neises A, Kiroyan JB, Meng X, Payne KJ, Tschudy-Seney B, Duan Y, Appleby N, Kearns-Jonker M, Gridley DS, Wang J, Lau KHW, Zhang X-B (2013) Efficient generation of integration-free iPS cells from human adult peripheral blood using BCL-XL together with Yamanaka factors. PLoS One 8(5):e64496
7. Su RJ, Yang Y, Neises A, Payne KJ, Wang J, Viswanathan K, Wakeland EK, Fang X, Zhang XB (2013) Few single nucleotide variations in exomes of human cord blood induced pluripotent stem cells. PLoS One 8(4):e59908, doi: 10.1371/journal.pone.0059908 PONE-D-13-03145 [pii]
8. Yu J, Hu K, Smuga-Otto K, Tian S, Stewart R, Slukvin II, Thomson JA (2009) Human induced pluripotent stem cells free of vector and transgene sequences. Science 324 (5928):797–801, doi: 1172482 [pii] 10.1126/science.1172482
9. Chou BK, Mali P, Huang X, Ye Z, Dowey SN, Resar LM, Zou C, Zhang YA, Tong J, Cheng L (2011) Efficient human iPS cell derivation by a non-integrating plasmid from blood cells with unique epigenetic and gene expression signatures. Cell Res 21(3):518–529, doi: cr201112 [pii] 10.1038/cr.2011.12
10. Okita K, Matsumura Y, Sato Y, Okada A, Morizane A, Okamoto S, Hong H, Nakagawa M, Tanabe K, Tezuka K, Shibata T, Kunisada T, Takahashi M, Takahashi J, Saji H, Yamanaka S (2011) A more efficient method to generate integration-free human iPS cells. Nat Methods 8(5):409–412, doi: nmeth.1591 [pii] 10.1038/nmeth.1591
11. Zhang XB, Li K, Yau KH, Tsang KS, Fok TF, Li CK, Lee SM, Yuen PM (2003) Trehalose ameliorates the cryopreservation of cord blood in a preclinical system and increases the recovery of CFUs, long-term culture-initiating cells, and nonobese diabetic-SCID repopulating cells. Transfusion 43(2):265–272, doi: trf301 [pii]
12. Wu S, Wu Y, Zhang X, Capecchi MR (2014) Efficient germ-line transmission obtained with transgene-free induced pluripotent stem cells. Proc Natl Acad Sci U S A 111 (29):10678–10683, doi: 1409933111 [pii] 10.1073/pnas.1409933111
13. Linta L, Stockmann M, Kleinhans KN, Bockers A, Storch A, Zaehres H, Lin Q, Barbi G, Bockers TM, Kleger A, Liebau S (2012) Rat embryonic fibroblasts improve reprogramming of human keratinocytes into induced pluripotent stem cells. Stem Cells Dev 21(6):965–976. doi:10.1089/scd.2011.0026

Methods in Molecular Biology (2016) 1357: 71–84
DOI 10.1007/7651_2015_201

Published online: 17 February 2015

Enhancing Induced Pluripotent Stem Cell Generation by MicroRNA

Jason Dang and Tariq M. Rana

Abstract

Somatic reprogramming to generate induced pluripotent stem cells, or iPSC, is a powerful tool in developmental biology, disease modeling, and regenerative medicine. microRNAs have been shown to regulate many key pathways in iPSC induction. Here we describe a microRNA mimic enhanced somatic reprogramming process starting from mouse embryonic fibroblast isolation to iPSC induction to colony derivation and characterization.

Keywords: Induced pluripotent stem cell, MicroRNA, Somatic reprogramming, Transfection, Mouse embryonic fibroblasts, Differentiation, Embryoid body, Teratoma

1 Introduction

In 2006, Yamanaka et al. showed that somatic cells could be reprogrammed into an embryonic stem cell like state; these cells were termed induced pluripotent stem cells, or iPSCs (1). With the expansion of regenerative medicine and need for advanced disease modeling, the field of iPSCs has exploded with reports showing the reprogramming of various cell types of both mouse and human origin (2–10). Because of their embryonic stem cell-like pluripotency, iPSCs are capable of differentiating into virtually any lineage and cell type including neuronal cells (11, 12), hematopoietic cells (13), and cardiac muscle tissue (14). Thus, iPSCs possess the potential to revolutionize the way we study disease pathogenesis and develop novel therapeutics.

MicroRNAs are a class of endogenous small noncoding RNAs involved in posttranscriptional regulation (15). MicroRNAs are associated with the RNA induce silencing complex, or RISC, and bind imperfectly to their mRNA targets to facilitate degradation or destabilization of the mRNAs (16, 17). MicroRNAs are known to enhance reprogramming (18, 19) through regulation of many pathways associated with iPSC induction such as the mesenchymal to epithelial transition (20, 21), the p53 pathway (22) and extracellular matrix formation (23). Understanding the ways microRNAs

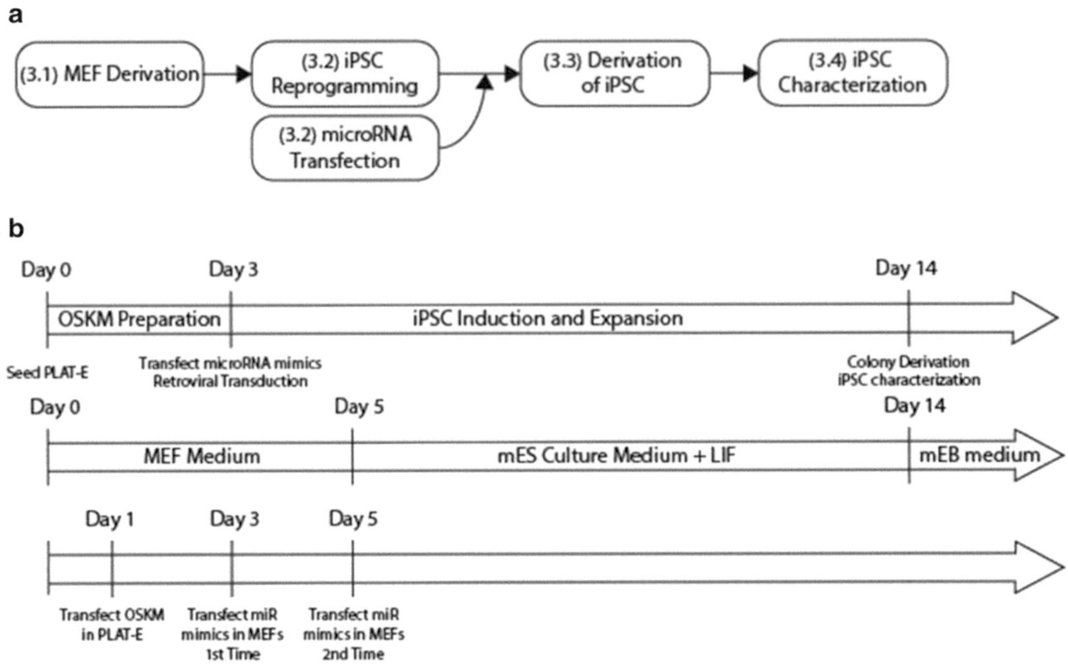

Fig. 1 (**a**) Schematic for iPSC reprogramming starting from MEF isolation to OSKM retroviral reprogramming to iPSC derivation and characterization. (**b**) Timeline for experiments from Day 0 of PLAT-E seeding to Day 14 of iPSC derivation and characterization. PLAT-E cells are seeded on Day 0. At Day 1, Oct4, Sox2, Klf4, and cMyc retroviral vectors are transfected into PLAT-E to generate reprogramming viral supernatant. On Day 3, microRNA mimics are transfected into MEFs 3 h before MEFs are transduced with OSKM. microRNA mimics are again transfected in reprogrammed MEFs at Day 5

and their putative targets regulate iPSC induction will provide insights into developmental biology, stem cell biology, disease pathogenesis, and targets for therapeutics (24).

Here we describe a retroviral reprogramming protocol for iPSC generation enhanced with the use of microRNAs. We begin with the harvesting of primary mouse embryonic fibroblasts, or MEFs, which will be used as the starting material for iPSCs (Fig. 1a). Briefly, MEFs will be collected from E13.5 time mated female mice and cultured or frozen for future use. Next, we detail the reprogramming process, including Oct4, Sox2, Klf4, and cMyc (OSKM) retrovirus production and microRNA mimic transfection. We use a commercially available retrovirus packaging cell line, PLAT-E, to produce the reprogramming retroviruses. Prior to transduction with OSKM virus, microRNA mimics or inhibitors can be transfected into MEFs to enhance or knock down specific microRNA pathways. Lastly, in vitro embryoid body formation assays and in vivo teratoma generation assays are outlined to characterize the derived iPSC.

2 Materials

2.1 Preparation of Mouse Embryonic Fibroblasts (MEFs)

1. MEF derivation and culture medium: Filter 450 ml DMEM (Mediatech, cat. no. 10017CV), 50 ml fetal bovine serum (Hyclone), 5 ml 100× L-glutamine solution (Invitrogen), 5 ml 100× nonessential amino acids (Invitrogen), and 5 ml 100× penicillin–streptomycin (Lonza) through a 0.22 μm filter unit (Millipore). Store at 4 °C for up to 1 month.
2. Cryopreservation medium, 2×: Filter 60 ml DMEM (Mediatech, cat. no. 10017CV), 20 ml DMSO (Sigma), 20 ml fetal bovine serum (Hyclone) through a 0.22 μm filter syringe. Store on ice until use.
3. Phosphate buffered saline (PBS, Invitrogen, cat. no. 10010023)
4. 0.25 % trypsin (Invitrogen, cat. no. 25200)
5. Female mice, B6;129S4-$Pou5f1^{tm2Jae}$/J (time mated, 13–14 days gestation)
6. Liquid nitrogen tank
7. Ethanol
8. Betadine
9. Sterile dissecting scissors
10. Sterile disposable petri dishes (Sarstedt, cat. no. 82.1473.001)
11. Watchmaker's forceps
12. Sterile disposable 10 ml pipets (Costar, cat. no. 4488)
13. Castro-Viejo scissors
14. 37 °C, 5 % CO_2 incubator
15. 50 ml conical tubes (Costar, cat. no. 430828)
16. 75 cm^2 flasks (BD Falcon, cat. no. 353136)
17. 1.7 ml cryopreservation tubes (Corning, cat. no. 430488)
18. Isopropanol freezing container

2.2 MicroRNA Enhanced iPSC Reprogramming

1. PLAT-E cells (Cell Biolabs, cat. no. RV-101)
2. PLAT-E culture medium: Filter 450 ml DMEM (Invitrogen, cat. no. 10995-065), 50 ml fetal bovine serum (Hyclone), 500 μl blasticidin S hydrochloride (10 mg/ml), 50 μl puromycin (10 mg/ml) through a 0.22 μm filter unit (Millipore). Store at 4 °C for up to 1 month.
3. MEF derivation and culture medium: Filter 450 ml DMEM (Mediatech, cat. no. 10017CV), 50 ml fetal bovine serum (Hyclone), 5 ml 100× L-glutamine solution (Invitrogen), 5 ml 100× nonessential amino acids (Invitrogen), and 5 ml 100× penicillin–streptomycin (Lonza) through a 0.22 μm filter unit (Millipore). Store at 4 °C for up to 1 month.

4. mES culture medium: Filter 450 ml DMEM (Invitrogen, cat. no. 10995-065), 75 ml ES-screened fetal bovine serum (Hyclone), 5 ml 100× L-glutamine (Invitrogen), 5 ml 100× nonessential amino acids (Invitrogen), 4.3 μl 1-thioglycerol (Sigma-Aldrich), 50 μl LIF (Invitrogen) through a 0.22 μm filter unit (Millipore). Store at 4 °C for up to 1 month.
5. pMXs vectors (Oct4, Sox2, Klf4, cMyc; Addgene)
6. Lipofectamine (Invitrogen, cat. no. 18324012)
7. PLUS reagent (Invitrogen, cat. no. 11514015)
8. Opti-MEM medium
9. 0.1 % (w/v) gelatin solution
10. MEFs (CF-1, Oct4-GFP, etc. from Preparation of Mouse Embryonic Fibroblasts)
11. microRNA mimics (Dharmacon)
12. Opti-MEM (Invitrogen)
13. Polybrene (Millipore, cat. no. TR-1003-G)
14. Lipofectamine 2000 (Invitrogen, cat. no. 11668019)
15. 15 ml conical tubes (Costar, cat. no. 430790)
16. 37 °C water bath
17. Benchtop centrifuge
18. 150 cm^2 tissue culture flasks (BD Biosciences, cat. no. 355001)
19. 37 °C, 5 % CO_2 incubator
20. 10 cm tissue culture plates (Corning, cat. no. 430167)
21. Syringe filters, 0.22 and 0.45 μm (Millipore)
22. 12-well tissue culture plates (BD Biosciences, cat. no. 353043)

2.3 Derivation and Characterization of iPSC

1. MEF derivation and culture medium: Filter 450 ml DMEM (Mediatech, cat. no. 10017CV), 50 ml fetal bovine serum (Hyclone), 5 ml 100× L-glutamine solution (Invitrogen), 5 ml 100× nonessential amino acids (Invitrogen), and 5 ml 100× penicillin–streptomycin (Lonza) through a 0.22 μm filter unit (Millipore). Store at 4 °C for up to 1 month.
2. mES culture medium: Filter 450 ml DMEM (Invitrogen, cat. no. 10995-065), 75 ml ES-screened fetal bovine serum (Hyclone), 5 ml 100× L-glutamine (Invitrogen), 5 ml 100× nonessential amino acids (Invitrogen), 4.3 μl 1-thioglycerol (Sigma-Aldrich), 50 μl LIF (Invitrogen) through a 0.22 μm filter unit (Millipore). Store at 4 °C for up to 1 month.
3. 0.25 % trypsin–EDTA
4. MEFs (CF-1, Oct4-GFP, etc. from Preparation of Mouse Embryonic Fibroblasts)

5. Phosphate buffered saline (PBS, Invitrogen, cat. no. 10010023)
6. 0.1 % (w/v) gelatin solution
7. 15 % ES-screened FBS, 10 % DMSO in DMEM

2.4 Characterization of iPSC

1. mEB formation medium: Filter 415 ml DMEM (Invitrogen, cat. no. 10995-065), 75 ml ES-screened fetal bovine serum (Hyclone), 5 ml 100× L-glutamine (Invitrogen), 5 ml 100× nonessential amino acids (Invitrogen), 4.3 μl 1-thioglycerol (Sigma-Aldrich) through a 0.22 μm filter unit (Millipore). Store at 4 °C for up to 1 month.
2. 4 % paraformaldehyde
3. 4–6-week-old female athymus nude mice
4. Ketamine
5. Xylazine
6. Alcohol pads
7. Zinc formalin solution (Fisher, cat. no. 23313096)
8. 150 cm^2 flasks
9. 50 ml conical tubes
10. 12-well tissue culture plates (BD Biosciences, cat. no. 353043)
11. Microscope
12. Pasteur pipet
13. 37 °C incubator
14. 96-well tissue culture plates
15. 20 μl Gilson pipet
16. 10 cm petri dishes (SARSTED, cat. no. 821473001)
17. 6-well tissue culture plates (BD Biosciences, cat. no. 353046)

2.5 Irradiating MEFs for Feeder Layer

1. Irradiator (RS2000, Rad Source)

3 Methods

3.1 Preparation of Mouse Embryonic Fibroblasts (MEFs)

1. Sacrifice E13.5 time mated mice by CO_2 followed by cervical dislocation in a sterile biosafety hood. B6;129S4-*Pou5f1*^{tm2Jae}/J mice have GFP expressed under the control of the Pou5f1 promoter, which serves as a marker for fully reprogrammed cells.
2. Face the mouse belly up, and clean the abdomen of the mouse with Betadine and ethanol for further sterilization.
3. Using sterile scissors cut open the peritoneal wall to expose the uterine horns. Place scissors in ethanol and do not reuse until

sterilized. Scissors can be reused if autoclaved or sterilized in a glass bead sterilizer after use.

4. Cut and remove uterine horns with new sterile scissors and place into a sterile petri dish containing 20 ml PBS.
5. Wash uterine horns with PBS three times to remove blood cells and place in a new sterile petri dish.
6. Release embryos from sacs using sterile Watchmaker's forceps and transfer them to a new dish containing PBS.
7. Wash embryos three times with PBS. Record the number of embryos to roughly estimate the number of MEF cells harvested for freezing.
8. Remove visceral tissue, identified as the dark red tissue in the abdomen, and transfer embryos to a fresh petri dish.
9. Again, wash embryos three times with PBS.
10. To aid the digestion of the tissue into a single cell suspension, remove excess PBS and mince tissue with Castro-Viejo scissors. This should continue until tissue is approximately 1-mm in diameter.
11. Add 2 ml of 0.05 % trypsin to the embryos and continue mincing for an additional 5 min. Use diluted 0.05 % trypsin in PBS for a more gentle digestion.
12. Add an additional 5 ml of 0.05 % trypsin and digest tissue at 37 °C, 5 % CO_2 for 30 min.
13. Collect digested tissue in a 50 ml conical tube and pipet vigorously to further dissociate tissue into a single cell suspension.
14. Neutralize trypsin by adding 20 ml MEF derivation medium and allow to sit for 10 min. Remaining tissue debris will settle to the bottom of the tube.
15. Aliquot supernatant from three embryos into a 75 cm^2 flask to culture overnight at 37 °C, 5 % CO_2.
16. Refresh MEF derivation medium the next day.
17. After 2 days of culture or until flasks are at least 90 % confluent, harvest cells to establish a frozen stock.
    (a) Aspirate the medium and wash cells with warmed PBS.
    (b) Add 5 ml 0.25 % trypsin and incubate for 5 min at 37 °C, 5 % CO_2 to detach cells.
    (c) Neutralize trypsin with 5 ml of MEF culture medium and transfer to a 50 ml conical tube.
    (d) Pellet cells by centrifugation at 150 × *g* for 5 min at room temperature.

18. For each flask, containing approximately 3 embryos, resuspend cells in 1.5 ml MEF culture medium and 1.5 ml 2× cryopreservation medium.
19. Aliquot 1 ml of resuspended MEFs into 1.7 ml cryovials and place in isopropanol freezing containers at −80 °C. Transfer vials to liquid nitrogen the following day for long-term storage.

3.2 MicroRNA Enhanced iPSC Reprogramming

1. Thaw a frozen vial of PLAT-E cells in a 37 °C water bath (Fig. 1b).
2. Resuspend the contents of the thawed vial in 5 ml of fresh 10 % FBS medium in a 15 ml conical tube and centrifuge for 5 min at 150 × *g*.
3. Aspirate supernatant and resuspend the PLAT-E cell pellet in 15 ml of fresh 10 % FBS medium. Culture cells in a 150 cm^2 flask at 37 °C, 5 % CO_2 incubator.
4. When cells reach 90 % confluency, wash cells with PBS, and harvest cells with 0.25 % trypsin. Neutralize trypsin solution with 10 % FBS medium.
5. Seed four plates of 5.5×10^6 PLAT-E cells per 10 cm plate. Shake plate gently for even distribution of cells. Incubate at 37 °C, 5 % CO_2 incubator overnight (Day 0).
6. To prepare Oct4, Sox2, Klf4, and cMyc reprogramming retroviruses, transfect PLAT-E plates with 9 μg of pMXs-Oct4, Sox2, Klf4, or cMyc vectors the following day (Fig. 2a). *See* **Note 1**.
   (a) Prepare four individual 1.5 ml tubes for each plasmid and label accordingly (Day 1).
   (b) Dilute 9 μg of DNA in Opti-MEM and add 20 μl PLUS reagent for final volume of 450 μl. Dilute the DNA before adding PLUS reagent to prevent precipitation. Mix and incubate for 15 min at room temperature. *See* **Note 2**.
   (c) Prepare four additional 1.5 ml tubes with 25 μl Lipofectamine in a final volume of 450 μl in Opti-MEM. Mix and incubate for ~10 min at room temperature.
   (d) Add the Lipofectamine solution from (c) to the contents of (b) and mix. Incubate for an additional 15 min at room temperature.
   (e) Add the 900 μl solution to the 10 cm dishes of PLAT-E in a dropwise manner. Return cells to the 37 °C, 5 % CO_2 incubator overnight.
   (f) Refresh the PLAT-E medium the following day with MEF culture medium (Day 2).
7. Coat 12-well tissue culture plates with a 0.1 % gelatin solution for 30 min at room temperature to improve cell adhesion (Day 2).

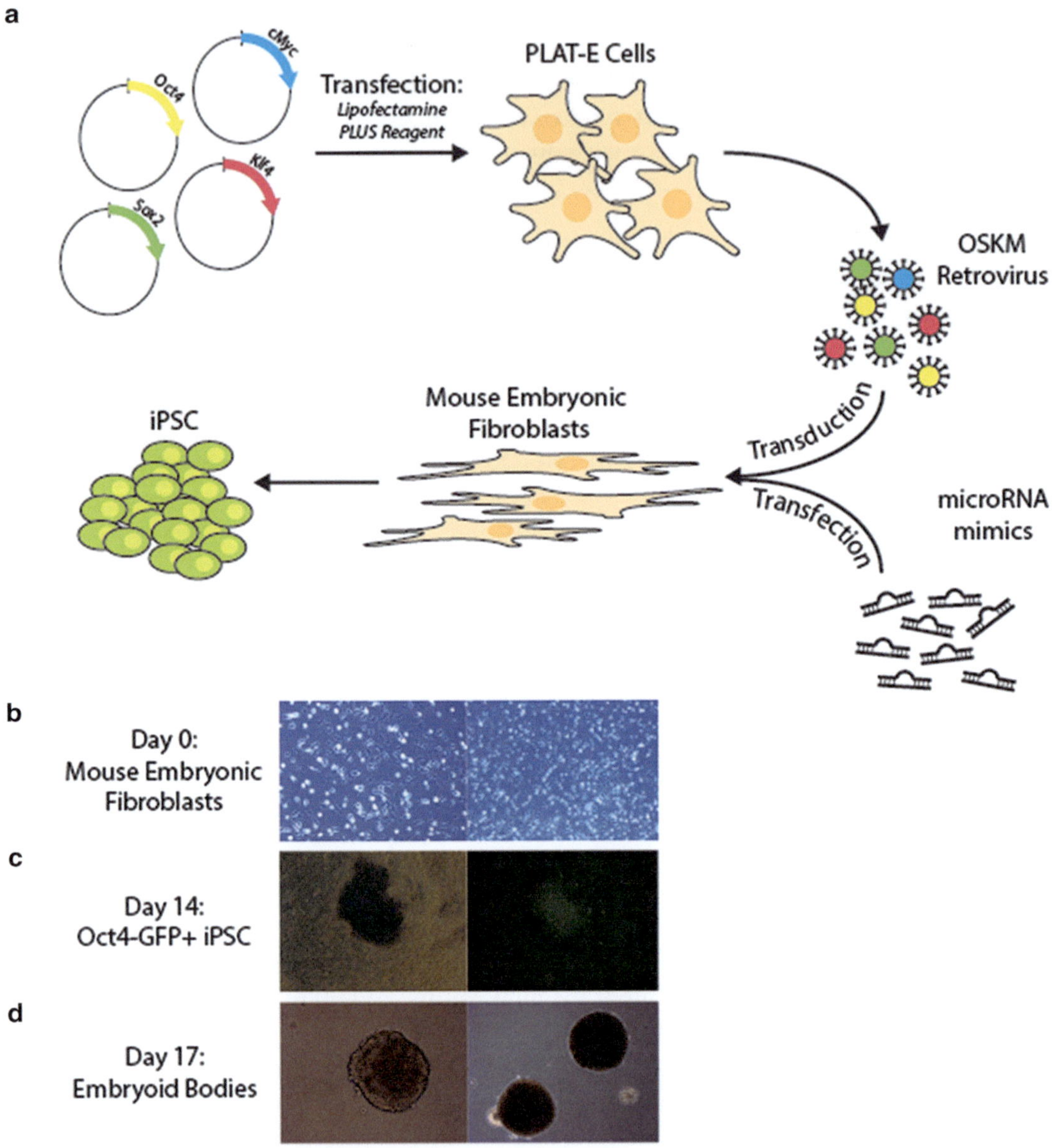

Fig. 2 (**a**) Schematic for iPSC reprogramming. Oct4, Sox2, Klf4, and cMyc expressing retroviral vectors are transfected into PLAT-E packaging cell line. MEFs are transfected with microRNA mimics and transduced with OSKM retrovirus supernatant to generate iPSC. (**b**) Bright-field images of MEF seeded for reprogramming. (**c**) Fully reprogrammed Oct4-GFP+ iPSC colonies. (**d**) Embryoid bodies created from iPSC for in vitro differentiation and characterization

8. Harvest MEF, count and seed 4×10^4 cells per well in 12-well plate. Shake to evenly distribute cells in the well and incubate overnight at 37 °C, 5 % CO_2 (Fig. 2b). *See* **Note 3**.
9. Transfect MEFs with 50 nM microRNA mimics/inhibitors prior to transduction with reprogramming factors (Day 3).

(a) Prepare individual 1.5 ml tubes for each microRNA mimic/inhibitor and label accordingly.

(b) Dilute microRNA mimics in 40 μl Opti-MEM. Mix and incubate for 15 min at room temperature. *See* **Note 4**.

(c) Prepare additional 1.5 ml tubes with 2 μl Lipofectamine 2000 in a final volume of 40 μl in Opti-MEM. Mix and incubate for ~10 min at room temperature.

(d) Add the Lipofectamine solution from (c) to the contents of (b) and mix. Incubate for an additional 15 min at room temperature.

(e) Wash MEFs with Opti-MEM and then add the 80 μl microRNA mimic/inhibitor solution to the well of MEFs. Return plate to the incubator.

10. Three hours after transfection with microRNA mimics/inhibitors, transduce MEFs with OSKM reprogramming viruses.

(a) To transduce MEFs with Oct4, Sox2, Klf4, and cMYC, collect viral supernatant from PLAT-E dishes and combine four reprogramming viruses together. Centrifuge supernatant at 2,000 × *g* for 5 min at room temperature then filter using a 0.22 μm syringe filter to remove cell debris. Use viral supernatant for transduction. *See* **Note 5**.

(b) Add polybrene to viral supernatant for a final concentration of 4–8 μg/ml.

(c) Aspirate medium from MEF and replace with 0.8 ml viral supernatant to transduce cells overnight. Incubate overnight at 37 °C, 5 % CO_2. *See* **Note 6**.

11. Change medium the day following transduction to fresh MEF culture medium (Day 4).

12. Change medium to mES culture medium with LIF 2 days post-transduction (Day 5).

13. Repeat microRNA mimic/inhibitor transfection on Day 5. Change medium back to mES culture medium with LIF 3 h after transfection. *See* **Note 7**.

14. Culture mouse iPSC until colonies form around Day 14. Refresh mES medium every other day or every day if needed (Fig. 2c). *See* **Notes 8** and **9**.

15. To assess microRNA mimic/inhibitor efficiency, RNA can be collected by direct addition of TRIzol reagent to a well of transfected cells 2 days post-transfection. Standard phenol–chloroform RNA extraction can be done to isolate RNA. microRNAs mimic/inhibitor efficiency can be determined by northern blot or microRNA qPCR.

3.3 Derivation of iPSC

1. At around Day 14, iPSC colonies should be large enough to pick. Identify and mark iPSC colonies to be picked under the microscope. If using Oct4-GFP reporter mice (B6;129S4-*Pou5f1*^{tm2Jae}/J), fully reprogrammed colonies can be identified by GFP signal. If not using Oct4-GFP reporter MEFs, alkaline phosphatase live staining (Life Technologies) can be used to determine reprogrammed colonies of interest.
2. Prepare colony picking tools by pulling Pasteur pipets under a flame to form a sharp, curved tip. Alternatively, sterile p10 tips may be used instead.
3. Prepare a 96-well plate with 50 μl PBS in each well.
4. Seed 1.5×10^5 cells/ml of irradiated MEFs (*See* Section 3.5) onto a 0.1 % gelatin coated 12-well and 24-well plate. Incubate overnight in a 37 °C, 5 % CO_2 incubator.
5. Using the sterile pulled Pasteur pipet or p10 tip, dissect out the marked iPSC clones and transfer colonies into individual wells in the 96-well plate. Repeat this procedure in 20 min intervals to ensure that colonies at not kept in PBS for prolonged periods.
6. Add 20 μl of 0.25 % trypsin to each well to dissociate the iPSC colonies at 37 °C for 10 min. During this step, change the medium from the 24-well plate of irradiated MEF to mES culture medium.
7. Pipet the trypsinized iPSC colonies to create single cell suspensions and transfer the contents to the 24-well plate to expand the colonies from each picked iPSC clone.
8. Incubate overnight and refresh the medium the following day. Change medium daily to improve colony growth.
9. Expand colonies showing ES morphology, and GFP signal if using Oct4-GFP reporter MEFs, to the 12-well plate containing the irradiate MEF feeder layer.
10. Again, change medium daily until colonies are formed. These colonies can be picked for further characterization or frozen using equal parts 2× cryopreservation medium and mES medium as previously outlined.

3.4 Characterization of iPSC

1. iPSC colonies can be characterized by in vitro and in vivo differentiation assays. Embryoid bodies, or EB, can be formed to assess the pluripotency and differentiation capabilities of iPSC in vitro.
   (a) Collect iPSC and trypsinize with 0.25 % trypsin solution. Neutralize with mES medium and transfer to a conical tube.
   (b) Centrifuge cells at $150 \times g$ for 5 min to pellet cells.

(c) Resuspend and count cells.

(d) Centrifuge cells once again and resuspend in mEB formation medium to reach a concentration of $1–1.5 \times 10^5$ cells/ml.

(e) Pipet 20 μl droplets on the inside lid of a 10 cm low adhesion petri dish and culture for 2–3 days in a 37 °C, 5 % CO_2 incubator. 15 cm petri dishes or additional dishes can be used depending on the number of embryoid bodies needed.

(f) Collect EB with a 1 ml pipet and transfer to a 50 ml conical tube and add additional mEB medium (~20 ml).

(g) Allow EB to settle to the bottom and carefully aspirate the medium.

(h) Add 5 ml of mEB medium and transfer to a 0.1 % gelatin coated 6-well plate. Incubate for 2 days in a 37 °C, 5 % CO_2 incubator as EB attach to the plates (Fig. 2d).

(i) Refresh mEB medium after the EB have attached, approximately 2 days later.

(j) Change medium every other day until beating colonies appear, approximately 14 days.

(k) When ready to character differentiated cells, aspirate medium, wash with PBS, and fix with 4 % paraformaldehyde for 20 min at room temperature to prepare cells for immunostaining. Alternatively, protein can be collected by adding M-PER buffer directly to wells and shaking for 15 min at room temperature and analyzed by western blot. RNA can be collected by directly adding TRIzol reagent.

2. Teratoma can be generated in athymic mice to characterize the differentiation potential of iPSC in vivo. Tissues from all three germ layers should be present and can be analyzed by histology.

(a) Anesthetize 4–6 week old female athymic nude mice by intraperitoneal injection of a mixture of ketamine (80–100 mg/kg) with xylazine (5–10 mg/kg). Check reflexes by toe pinching.

(b) Collect iPSC and trypsinize with 0.25 % trypsin solution.

(c) Collect and resuspend in mES culture medium to a final concentration of 1×10^7 cells/ml.

(d) Inject 1×10^6 iPSCs into the dorsal neck region of the recipient mice. *See* **Note 10**.

(e) Monitor tumor growth once a week at the site of injection. Once teratomas are visible or palpable under the skin, mice should be monitored twice a week.

(f) Teratomas can be harvested by dissection when they have reached a diameter of ~1 cm, approximately 3–4 weeks after injection.

(g) Fix teratoma tissue in zinc formalin solution overnight. Tissue can now be embedded in paraffin, sectioned, and stained for the presence of all three germ layers.

3.5 Irradiating MEFs for Feeder Layers

1. Remove medium from 90 % confluent plates of MEFs and wash with 10 ml PBS. Trypsinize cells and collect in a 50 ml conical tube.
2. Centrifuge cells at 150 × *g* for 5 min at room temperature.
3. Aspirate supernatant, resuspend the cell pellet and count the cells.
4. Place the conical tube with MEF cell suspension in an irradiator and expose with 7,000 rad. This step will prevent the proliferation of cells used for feeder layers.
5. Centrifuge irradiated cells at 150 × *g* for 5 min at room temperature and resuspend in equal parts MEF culture medium and 2× cryopreservation medium.
6. Aliquot 1 ml of resuspended MEFs into 1.7 ml cryovials and place in isopropanol freezing containers at −80 °C. Transfer vials to liquid nitrogen the following day for long-term storage.
7. Alternatively, MEF feeder layers can be created by mitotically inactivating MEFs with 10 μg/ml of mitomycin-C for 2 h. Treated MEFs can then be washed with PBS and frozen in cryopreservation medium.

4 Notes

1. It is best to prepare fresh retroviruses for each set of reprogramming. Reprogramming is a low yield process, and freezing and long-term storage will further reduce the titer of the virus.
2. During the transfection step, be sure to dilute DNA before adding Lipofectamine or PLUS reagent. This will decrease the possibility of the DNA precipitating from the solution. If DNA precipitates, the transfection process should be restarted.
3. The starting condition of MEFs is crucial to the success of the iPSC induction. MEFs should have a spindle like morphology and they should multiply rapidly. MEFs are typically viable for only 5 passages.
4. siRNA can be substituted for microRNA mimics for knockdown of specific genes during iPSC induction.

5. To assess transduction efficiency, a retrovirus expressing the red fluorescent protein dsRed (pMXs-dsRed available through Addgene) can be used as a positive control for transfection and transduction efficiency.
6. Decontaminate all tips and pipets used for handling viral supernatant in 10 % bleach before discarding.
7. Because of the length of the reprogramming process, iPSC induction can be assessed early on by close observation of changes in MEF morphology. Reprogrammed MEFs should undergo a change in morphology to a more rounded, clustered, less elongated shape at around Day 5–6. Early stage colony formation should become visible at this point in the reprogramming process. This change in morphology can be used to roughly assess the reprogramming efficiency.
8. Change medium daily if necessary to ensure that reprogrammed cells are healthy. The mES medium should be changed before it turns yellow as this may decrease the iPSC yield.
9. Other systems such as episomal reprogramming, mRNA reprogramming and Sendai virus can be used for nonintegrating somatic reprogramming as an alternative to the traditional retroviral based reprogramming process described.
10. Inject teratomas into the dorsal neck region because mice are not able to scratch or bite the area. A single mouse can be injected with two teratomas for replication.

References

1. Takahashi K, Yamanaka S (2006) Induction of pluripotent stem cells from mouse embryonic and adult fibroblast cultures by defined factors. Cell 126(4):663–676. doi:10.1016/j.cell.2006.07.024
2. Giorgetti A, Montserrat N, Aasen T, Gonzalez F, Rodriguez-Piza I, Vassena R, Raya A, Boue S, Barrero MJ, Corbella BA, Torrabadella M, Veiga A, Izpisua Belmonte JC (2009) Generation of induced pluripotent stem cells from human cord blood using OCT4 and SOX2. Cell Stem Cell 5(4):353–357. doi:10.1016/j.stem.2009.09.008
3. Takahashi K, Tanabe K, Ohnuki M, Narita M, Ichisaka T, Tomoda K, Yamanaka S (2007) Induction of pluripotent stem cells from adult human fibroblasts by defined factors. Cell 131(5):861–872. doi:10.1016/j.cell.2007.11.019
4. Park IH, Zhao R, West JA, Yabuuchi A, Huo H, Ince TA, Lerou PH, Lensch MW, Daley GQ (2008) Reprogramming of human somatic cells to pluripotency with defined factors. Nature 451(7175):141–146. doi:10.1038/nature06534
5. Meissner A, Wernig M, Jaenisch R (2007) Direct reprogramming of genetically unmodified fibroblasts into pluripotent stem cells. Nat Biotechnol 25(10):1177–1181. doi:10.1038/nbt1335
6. Yu J, Vodyanik MA, Smuga-Otto K, Antosiewicz-Bourget J, Frane JL, Tian S, Nie J, Jonsdottir GA, Ruotti V, Stewart R, Slukvin II, Thomson JA (2007) Induced pluripotent stem cell lines derived from human somatic cells. Science 318(5858):1917–1920. doi:10.1126/science.1151526
7. Aoi T, Yae K, Nakagawa M, Ichisaka T, Okita K, Takahashi K, Chiba T, Yamanaka S (2008) Generation of pluripotent stem cells from adult mouse liver and stomach cells. Science 321(5889):699–702. doi:10.1126/science.1154884

8. Yamanaka S (2012) Induced pluripotent stem cells: past, present, and future. Cell Stem Cell 10(6):678–684. doi:10.1016/j.stem.2012.05.005
9. Loh YH, Agarwal S, Park IH, Urbach A, Huo H, Heffner GC, Kim K, Miller JD, Ng K, Daley GQ (2009) Generation of induced pluripotent stem cells from human blood. Blood 113 (22):5476–5479. doi:10.1182/blood-2009-02-204800
10. Park IH, Arora N, Huo H, Maherali N, Ahfeldt T, Shimamura A, Lensch MW, Cowan C, Hochedlinger K, Daley GQ (2008) Disease-specific induced pluripotent stem cells. Cell 134(5):877–886. doi:10.1016/j.cell.2008.07.041
11. Chen H, Qian K, Du Z, Cao J, Petersen A, Liu H, Blackbourn LW 4th, Huang CL, Errigo A, Yin Y, Lu J, Ayala M, Zhang SC (2014) Modeling ALS with iPSCs reveals that mutant SOD1 misregulates neurofilament balance in motor neurons. Cell Stem Cell 14 (6):796–809. doi:10.1016/j.stem.2014.02.004
12. Marchetto MC, Carromeu C, Acab A, Yu D, Yeo GW, Mu Y, Chen G, Gage FH, Muotri AR (2010) A model for neural development and treatment of Rett syndrome using human induced pluripotent stem cells. Cell 143 (4):527–539. doi:10.1016/j.cell.2010.10.016
13. Hanna J, Wernig M, Markoulaki S, Sun CW, Meissner A, Cassady JP, Beard C, Brambrink T, Wu LC, Townes TM, Jaenisch R (2007) Treatment of sickle cell anemia mouse model with iPS cells generated from autologous skin. Science 318(5858):1920–1923. doi:10.1126/science.1152092
14. Itzhaki I, Maizels L, Huber I, Zwi-Dantsis L, Caspi O, Winterstern A, Feldman O, Gepstein A, Arbel G, Hammerman H, Boulos M, Gepstein L (2011) Modelling the long QT syndrome with induced pluripotent stem cells. Nature 471(7337):225–229. doi:10.1038/nature09747
15. Rana TM (2007) Illuminating the silence: understanding the structure and function of small RNAs. Nat Rev Mol Cell Biol 8 (1):23–36. doi:10.1038/nrm2085
16. Ambros V (2004) The functions of animal microRNAs. Nature 431:6
17. Li Z, Rana T (2012) Molecular mechanisms of RNA-triggered gene silencing machineries. Acc Chem Res 45(7):10
18. Yang CS, Li Z, Rana TM (2011) microRNAs modulate iPS cell generation. RNA 17 (8):1451–1460. doi:10.1261/rna.2664111
19. Henzler C, Li Z, Dang J, Arcila ML, Zhou H, Liu J, Chang K-Y, Bassett D, Rana T, Kosik K (2013) Staged miRNA re-regulation patterns during reprogramming. Genome Biol 14: R149
20. Li Z, Yang CS, Nakashima K, Rana TM (2011) Small RNA-mediated regulation of iPS cell generation. EMBO J 30(5):823–834. doi:10.1038/emboj.2011.2
21. Liao B, Bao X, Liu L, Feng S, Zovoilis A, Liu W, Xue Y, Cai J, Guo X, Qin B, Zhang R, Wu J, Lai L, Teng M, Niu L, Zhang B, Esteban MA, Pei D (2011) MicroRNA cluster 302-367 enhances somatic cell reprogramming by accelerating a mesenchymal-to-epithelial transition. J Biol Chem 286(19):17359–17364. doi:10.1074/jbc.C111.235960
22. Choi YJ, Lin CP, Ho JJ, He X, Okada N, Bu P, Zhong Y, Kim SY, Bennett MJ, Chen C, Ozturk A, Hicks GG, Hannon GJ, He L (2011) miR-34 miRNAs provide a barrier for somatic cell reprogramming. Nat Cell Biol 13 (11):1353–1360. doi:10.1038/ncb2366
23. Li Z, Dang J, Rana T (2014) MicroRNA-mediated regulation of extracellular matrix formation modulates somatic cell reprogramming. RNA 20:1900
24. Li Z, Rana TM (2014) Therapeutic targeting of microRNAs: current status and future challenges. Nat Rev Drug Discov 13(8):622–638. doi:10.1038/nrd4359

Methods in Molecular Biology (2016) 1357: 85–95
DOI 10.1007/7651_2014_161

Published online: 05 December 2015

Generation of Partially Reprogrammed Cells and Fully Reprogrammed iPS Cells by Plasmid Transfection

Jong Soo Kim, Hyun Woo Choi, Yean Ju Hong, and Jeong Tae Do

Abstract

Induced pluripotent stem (iPS) cells can be directly generated from somatic cells by overexpression of defined transcription factors. iPS cells can perpetually self-renew and differentiate into all cell types of an organism. iPS cells were first generated through infection with retroviruses that contain reprogramming factors. However, development of an exogene-free iPS cell generation method is crucial for future therapeutic applications, because integrated exogenes result in the formation of tumors in chimeras and regain pluripotency after differentiation in vitro. Here, we describe a method to generate iPS cells by transfection of plasmid vectors and to convert partially reprogrammed cells into fully reprogrammed iPS cells by switching from mouse ESC culture conditions to KOSR-based media with bFGF. We also describe basic methods used to characterize fully reprogrammed iPS cells.

Keywords: Induced pluripotent stem (iPS) cells, Pluripotency, Partially reprogrammed cells, Fully reprogrammed iPS cells, Conversion

1 Introduction

Yamanaka and colleagues showed that induced pluripotent stem (iPS) cells could be generated from mouse and human differentiated somatic cells by transduction of four defined transcription factors (Oct4, Sox2, Klf4, and c-Myc) (1, 2). iPS cells are very similar to embryonic stem (ES) cells in morphology, gene expression profile, epigenetic state, differentiation potential into all three germ layers, germline chimera formation, and the ability to develop to term after tetraploid complementation (1, 3, 4). Of note, iPS cells can perpetually self-renew and differentiate into all cell types of the body. This potential of iPS cells could provide a valuable tool for studying mechanisms of development, drug discovery, and personalized regenerative medicine to treat incurable diseases (5–8). iPS cells were first generated through infection of the retrovirus-containing reprogramming factors Oct4, Sox2, Klf4, and c-Myc. However, iPS cells induced by viral infection frequently form tumors in chimeras (3) and integrated transgenes that can reactivate when the iPS cells differentiate in vitro (9). Therefore, an exogene-free iPS cell generation method is crucial for future therapeutic applications.

During reprogramming, most cells stably accumulate in the intermediate stage (partially reprogrammed cells). Partially reprogrammed cells are not pluripotent but show several features of pluripotency such as increase of several pluripotent markers, decrease of fibroblast markers, formation of an ES cell-like morphology, and histone modification (10–12). Partially reprogrammed cells barely form embryoid bodies, fail to differentiate in vitro (12), and do not induce reactivation of the inactive X chromosome in females (13). Studying the stepwise process during which partially reprogrammed cells are converted into fully reprogrammed iPS cells would help to elucidate the basic mechanism underlying reprogramming and pluripotent stem cells. In this chapter, we describe a reprogramming method that uses a plasmid vector to generate partially reprogrammed cells and fully reprogrammed iPS cells. Partially reprogrammed cells were converted into fully reprogrammed iPS cells by changing the culture system.

2 Materials

2.1 Preparation of Mouse Embryonic Fibroblasts from Embryos

1. Naturally mated 13.5 days postcoitus (dpc) pregnant C57BL/6 or CD1 mice that were mated with OG2 homozygote male mice or wild-type male mice.
2. Phosphate-buffered saline (PBS) without calcium and magnesium. PBS is stored at room temperature.
3. Culture dish coating solution: 0.1 % gelatin type B from bovine skin in PBS.
4. Mouse embryonic fibroblast (MEF) culture medium: Dulbecco's Modified Eagle Medium (DMEM) with high glucose supplementation containing 15 % fetal bovine serum (FBS), 0.5 % penicillin/streptomycin/glutamine, 0.1 mM nonessential amino acids, and 0.1 mM β-mercaptoethanol. The medium is filtered with a 0.22-μm syringe filter.
5. A 1× freezing medium: 20 % FBS, 10 % dimethyl sulfoxide (DMSO), and 70 % MEF culture medium. The medium is filtered with a 0.22-μm syringe filter. Store at −20 °C (*see* **Note 1**).
6. Mitomycin C solution: 10 μg/ml mitomycin C (Sigma, cat. no. M4287) in MEF medium; this solution is filtered with a 0.22-μ m filter system.
7. Passaging medium: 0.25 % trypsin/1 mM EDTA (Gibco, cat. no. 25200-056). MEF passaging medium is stored at −20 °C.

2.2 Generation of iPS Cells

1. Expression vectors: pCX-OKS-2A (Oct4, Klf4, and Sox2; Addgene, plasmid no. 19771) and pCX-c-Myc (Addgene, plasmid no. 19772).
2. Transfection reagent: Xfect™ transfection reagent (Clontech, cat. no. 631317). Thaw Xfect Polymer at room temperature just prior to use. Once thawed, Xfect Polymer can be used for up to 12 months when stored at 4 °C. Thaw Xfect Reaction Buffer at room temperature just prior to use. Once thawed, Xfect Reaction Buffer can be used for up to 12 months when stored at 4 °C.
3. Mouse embryonic stem cell (mESC) culture medium: DMEM with high glucose supplementation containing 15 % FBS, 0.5 % penicillin/streptomycin/glutamine, 0.1 mM nonessential amino acids, 0.1 mM β-mercaptoethanol, and 1,000 U/ml LIF (ESGR; Chemicon). The medium is filtered with a 0.22-μm syringe filter.
4. Passaging medium: 0.25 % trypsin/1 mM EDTA. MEF passaging medium is stored at −20 °C.

2.3 Conversion of Partially Reprogrammed Cells into iPS Cells

1. Conversion medium: DMEM/nutrient mixture F-12 (DMEM/F-12) containing 20 % KnockOut serum replacement (KOSR), 2 mM glutamine, 0.5 % penicillin/streptomycin/glutamine, 0.1 mM nonessential amino acids, 0.1 mM β-mercaptoethanol, and 5 ng/ml bFGF.
2. Culture dish coating solution: 0.1 % gelatin type B from bovine skin in PBS.
3. Passaging medium: 0.25 % trypsin/1 mM EDTA. Passaging medium is stored at −20 °C.

2.4 Characterization of iPS Cells

2.4.1 Quantitative Real-Time PCR (Endo-Oct4, Endo-Sox2, and Nanog)

1. SuperScript III first-strand synthesis system (Invitrogen). This system includes SuperScript III reverse transcriptase (200 U/μl), 25 mM $MgCl_2$, 0.1 M dithiothreitol, and 10× RT buffer, which is comprised of 200 mM Tris–HCl (pH 8.4) and 500 mM KCl. Store the SuperScript III RT at −20 °C.
2. Oligo(dT)20 (50 μM) primers are stored at −20 °C.
3. SYBR Premix EX Taq II (Tli RNaseH Plus), Rox Plus system (Takara). This product contains EX Taq HS, Mg^{2+}, Tli RNaseH I, dNTP mixture (2.5 mM each), and ROX reference dye. This product is stable for up to 6 months when stored at 4 °C.
4. Sterilized distilled water.
5. Forward and reverse PCR primers for endo-*Oct4*, endo-*Sox2*, *Nanog*, and *β-actin*. Endo-*Oct4* forward (60 °C): cca atc agc ttg ggc tag ag; endo-*Oct4* reverse: ctg gga aag gtg tcc ctg ta; endo-*Sox2* forward (60 °C): cac aac tcg gag atc agc aa; endo-*Sox2* reverse: ctc cgg gaa gcg tgt act ta; *Nanog* forward (60 °C):

agg ctg att tgg ttg gtg tc; *Nanog* reverse: ccc agg aag acc cac act cat; *β-actin* forward (56 °C): cgc cat gga tga cga tat cg; and *β-actin* reverse: cga agc cgg ctt tgc aca tg.

6. LightCycler 5480 (Roche).

2.4.2 Chimera Formation Analysis

1. Naturally mated 1.5 dpc pregnant BDF1 female mice.
2. Vasectomized male mice. These mice retain their sexual potency and leave a copulatory plug after mating but fail to transmit any sperm.
3. Recipient pseudopregnant female mice.
4. Mineral oil (Sigma, cat. no. M8410). This product is tested as an overlay in microdrop embryo cultures. Store at room temperature.
5. Embryo culture medium: EmbryoMax KSOM medium (1×) with 1/2 amino acids and phenol red can be stored at −20 °C; once thawed, it should be used within 2 weeks.
6. Acid Tyrode's solution (Sigma, cat. no. T1788).
7. Passaging medium: 0.25 % trypsin/1 mM EDTA. Passaging medium is stored at −20 °C.
8. Aggregation needles for generating depression wells in aggregation plates.
9. Transfer pipettes.
10. Aggregation plates (*see* **Note 2**).

3 Methods

3.1 Preparation of Fibroblasts from OG2$^{+/-}$ and Wild-Type Embryos

1. Prewarm PBS and MEF medium in a water bath at 37 °C.
2. Transfer the uterus from each female mouse at 13.5 dpc to a 100-mm dish filled with PBS.
3. Transfer the uterus to a new 100-mm Petri dish filled with PBS and separate the embryos from the uterine muscle layer and extraembryonic membranes with scissors and forceps.
4. Transfer the embryos to a clean dish of PBS. Pinch off the heads and remove the spinal cords and all internal organ primordia.
5. Transfer bodies to a 10-ml syringe and add 5 ml PBS.
6. Gently insert the plunger into a syringe. Remove the plunger and add 5 ml PBS containing the dissociated bodies (*see* **Note 3**).
7. Repeat steps 5–6 about five to ten times.
8. Transfer PBS-containing dissociated bodies into a 15-ml tube and triturate with a 5-ml pipette.
9. The dissociated cells are passed through a 70-μm nylon mesh to remove the large cell clusters.

10. Add 10 ml MEF medium and then centrifuge at 200 × g for 5 min.
11. Remove supernatant medium and add new MEF medium, thoroughly pipetting the cell suspension up and down.
12. Evenly distribute the cell suspension into a 150-mm culture dish that is pre-coated with 0.1 % gelatin and add additional MEF medium to the dishes to a total volume of 20 ml.
13. Culture the cells in a 5 % CO_2 incubator.
14. After 3 days of culture, trypsinize each dish and replate onto a new 35-mm culture dish that is pre-coated with 0.1 % gelatin.

3.2 Mitotic Inactivation of MEFs

1. Remove medium from a nearly confluent 100-mm dish of MEFs, rinse dish with 5 ml of PBS, and add 5 ml of mitomycin C solution.
2. Incubate at 37 °C in a 5 % CO_2 incubator for 2 h (*see* **Note 4**).
3. Remove mitomycin C solution and rinse dish with 5 ml of PBS three times.
4. Add 5 ml MEF passaging medium and incubate at 37 °C for 1 min. Add 10 ml MEF culture medium to inactivate the trypsin, triturate to a single-cell suspension, and spin down at 200 × g for 5 min.
5. Remove supernatant medium and add fresh MEF culture medium, thoroughly pipetting the cell suspension up and down.
6. Evenly distribute the cell suspension into each 100-mm culture dish that is pre-coated with 0.1 % gelatin.

3.3 Reprogramming into the Pluripotent State with Plasmid Vectors

An overview of the schematic representation of iPS cell generation is provided in Fig. 1a.

3.3.1 Induction of Reprogramming by Transfection with Plasmid Vectors

1. One day prior to the transfection, plate MEFs at 1×10^5 cells/35-mm culture dish in 1 ml of MEF culture medium (Day 0). For each 35-mm culture dish, use 3 μg of pCX-OKS and 1 μg of pCX-c-Myc plasmid vectors (*see* **Note 5**).
2. (Day 1) In a 1.5-ml tube, mix 3 μg of pCX-OKS and 1 μg of pCX-c-Myc plasmid vectors with Xfect Reaction Buffer to a final volume of 98.5 μl and vortex for 5 s at high speed (*see* **Note 6**).
3. Add 1.5 μl of Xfect Polymer to the mixture with plasmid vectors and buffer. Vortex for 10 s at high speed. Incubate for 10 min at room temperature.
4. Add the entire 100 μl of mixture dropwise to the cell culture medium. Rock the plate gently back and forth to mix.

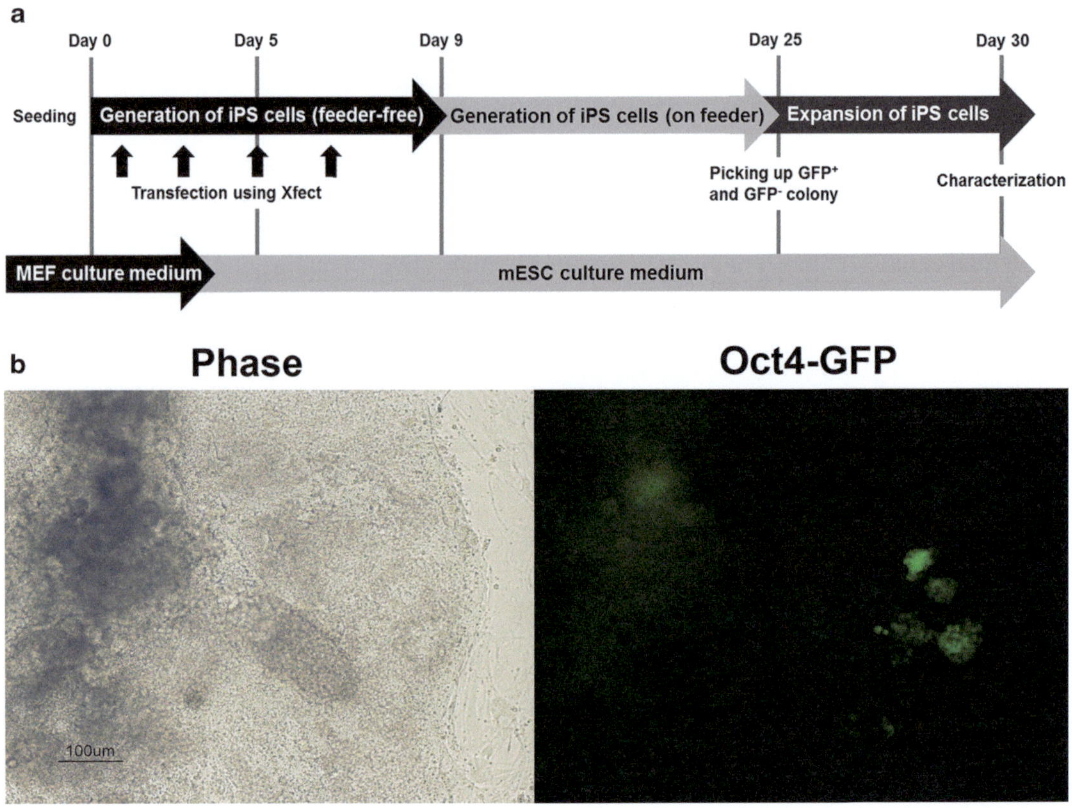

Fig. 1 (**a**) Schematic time schedule for iPS cell generation using pCX-OKS and pCX-c-Myc. (**b**) Morphology of iPS colonies before selection. The phase contrast (*left*) and fluorescence (Oct4–GFP reporter, *right*) images on day 25 after the first transfection. Scale bar = 100 μm

5. Incubate the plate at 37 °C for 4 h. Aspirate the medium from a 35-mm culture dish and replace with 2 ml of fresh MEF culture medium.
6. (Day 2) Remove the medium by aspiration and replace with 2 ml of fresh MEF culture medium.
7. (Day 3) Repeat steps 2–5.
8. (Day 4) Remove the medium by aspiration and replace with 2 ml of fresh mESC culture medium instead of MEF culture medium.
9. (Day 5) Repeat steps 2–5.
10. (Day 6) Remove the medium by aspiration and replace with 2 ml of fresh mESC culture medium.
11. (Day 7) Repeat steps 2–5.
12. (Day 8) Remove the medium by aspiration and replace with 2 ml of fresh mESC culture medium.

13. (Day 9) Remove the medium by aspiration and wash the cells with 2 ml of PBS. Discard the PBS, add 0.5 ml of 0.25 % trypsin/1 mM EDTA, and incubate for 3 min at 37 °C. Add 4.5 ml mESC culture medium to inactive the trypsin, triturate to a single-cell suspension, and spin down at 200 × g for 5 min. Remove supernatant medium and add 5 ml of fresh mESC culture medium, pipetting the cell suspension up and down thoroughly. Count the cells and seed 1×10^6 cells to a 100-mm culture dish covered with feeder cells (mitotic inactivated MEF).
14. (Days 10–28) Change the medium with fresh mESC culture medium once per day.

3.3.2 Selecting the iPS Colonies and Partially Reprogrammed Cells

1. On days 25–28, Oct4–GFP-positive and Oct4–GFP-negative colonies are observed with feeder cells (Fig. 1b).
2. (Day 0) Select Oct4–GFP-positive or Oct4–GFP-negative colonies using a transfer pipette and transfer to four-well dishes that are coated with feeder cells in iPS cell medium (*see* **Note 7**).
3. (Day 1) Remove iPS cell medium and rinse dish with 250 μl PBS. Discard PBS, add passaging medium, and incubate for 1 min at room temperature. Add MEF culture medium, and then dissociate cells by pipetting (*see* **Note 8**).
4. Transfer to a conical tube and centrifuge at 200 × g for 5 min.
5. Aspirate supernatant medium and add fresh mESC culture medium. Transfer to four-well dishes that are coated with feeder cells.

3.4 Conversion of Partially Reprogrammed Cells into Fully Reprogrammed iPS Cells

1. (Day 0) Partially reprogrammed cells that were cultured on feeder cells in mESC culture medium did not express Oct4–GFP after 20 passages (Fig. 2a).
2. (Day 0) Remove the mESC culture medium by aspiration and wash the cells with 2 ml of PBS. Discard the PBS and add 2 ml of conversion medium.
3. (Days 1–5) Change medium with fresh conversion medium once per day.
4. On days 3–5, Oct4–GFP-positive cells appear on feeder cells (Fig. 2b, c).
5. Select the Oct4–GFP-positive colonies and maintain in mESC culture medium.

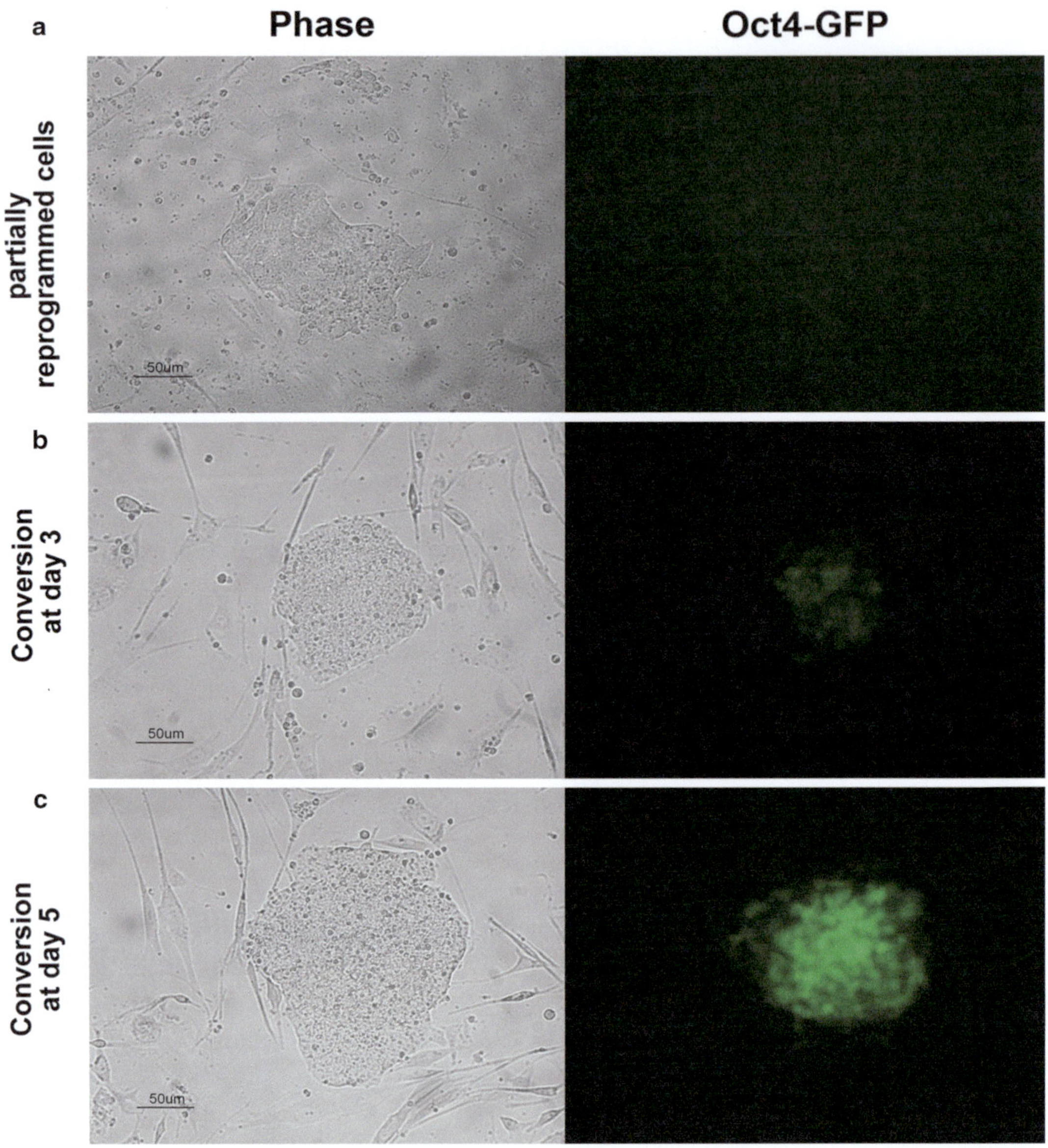

Fig. 2 (**a**) Partially reprogrammed cells (Oct4–GFP negative) observed during somatic cell reprogramming. (**b**, **c**) Partially reprogrammed cells were converted to Oct4–GFP-positive cells in conversion medium. The phase (*left*) and fluorescence (Oct4–GFP reporter) images on day 3 (**b**) and day 5 (**c**) in conversion medium. Scale bar = 50 μm

3.5 Characterization of iPS Cells

3.5.1 Quantitative Real-Time PCR

1. One microliter of cDNA is used for each PCR. Prepare a master mix for each primer set that is used for target detection. Each reaction consists of 10 μl of SYBR Premix EX Taq II (Tli RNaseH Plus), 7 μl of PCR-grade water, 1 μl of 10 pM forward primer, and 1 μl of 10 pM reverse primer. Add 1 μl of cDNA to the RT-PCR plate followed by 19 μl of the reaction master mix. Mix by pipetting and spin briefly in an RT-PCR plate by centrifugation.

2. Program the Roche LightCycler 5480 (Roche) with the following generic program:

   1 initial cycle of 95 °C for 5 min (denaturation).

   40 cycles of 95 °C for 10 s (denaturation), 60 °C for 10 s (annealing), 72 °C for 10 s (extension).

   1 melting cycle of 95 °C for 5 s, 65 °C for 10 s, 97 °C for 1 s.

3.5.2 Chimera Formation

1. (Day 0) Prepare microdrop embryo culture dishes by placing three drops of embryo culture medium in a 35-mm Petri dish. Cover all of the drops with mineral oil and incubate at 37 °C (*see* **Note 9**).
2. (Day 1) Flush the host embryos from oviducts of 1.5 dpc pregnant mice, transfer to microdrop embryo culture dishes, and incubate at 37 °C (*see* **Note 10**).
3. (Day 2) In a 6-cm Petri dish, place three separate drops of embryo culture medium and three separate drops of acid Tyrode's solution.
4. Transfer the embryo to the first drop of acid Tyrode's solution; then, transfer the second and third embryos to the second and third drops of acid Tyrode's solution.
5. Transfer zona-free embryos to depression wells of aggregation plates.
6. To prepare iPS cells, first remove the medium from the iPS cells. Rinse dish with PBS, add iPS cell passaging medium, and incubate for 1 min at room temperature. Add ESC culture medium. Using a transfer pipette, transfer a clump of iPS cells to the embryo culture dish.
7. Using a transfer pipette, select a small (4–10 cells) clump of iPS cells and transfer the clump to the depression wells of the aggregation plates.
8. (Day 3) Transfer aggregation embryos to a recipient pseudopregnant female mouse.

4 Notes

1. Prepare this freezing medium fresh each time.
2. Place eight drops of the KSOM medium on a 35-mm Petri dish, and the drops should be covered with a layer of mineral oil. Press the aggregation needle carefully into the plastic, making nine depressions per drop. Incubate in a 5 % CO_2 incubator 24 h prior to aggregation.
3. The bodies of the embryos should be in very small clumps.
4. We recommend using a process time of 2 h for mitotic inactivation of MEF.

5. We recommend that the prepared concentration of plasmid vectors be at least 1 μg/μl using the EndoFree Plasmid Maxi kit according to the manufacturer's protocol.
6. Thoroughly vortex Xfect Polymer before use. Always add plasmid vector to the buffer before adding Xfect Polymer. At least 50 μl of the solution must be Xfect Reaction Buffer.
7. Partially reprogrammed cells were morphologically indistinguishable from fully reprogrammed iPS cells. Select only Oct4–GFP-negative cells using a fluorescent microscope; select these colonies and then transfer them to new feeder cells.
8. Prepare fresh passaging medium to efficiently generate single cells.
9. Mineral oil should be carefully added to the drop dishes.
10. Wash embryos with KSOM medium three times before incubation.

Acknowledgment

This work was supported by the Biomedical Technology Development Program and Basic Science Research Program through the National Research Foundation of Korea funded by the Ministry of Science, ICT and Future Planning (grant nos. 20110019489 and 2013R1A1A2011394).

References

1. Takahashi K, Yamanaka S (2006) Induction of pluripotent stem cells from mouse embryonic and adult fibroblast cultures by defined factor. Cell 126(4):663–676
2. Takahashi K, Tanabe K, Ohnuki M, Narita M, Ichisaka T, Tomoda K, Yamanaka S (2007) Induction of pluripotent stem cells from adult human fibroblasts by defined factors. Cell 131 (5):861–872
3. Okita K, Ichisaka T, Yamanaka S (2007) Generation of germline-competent induced pluripotent stem cells. Nature 448(7151):313–317
4. Kim MJ, Choi HW, Jang HJ, Chung HM, Arauzo-Bravo MJ, Schöler HR, Do JT (2013) Conversion of genomic imprinting by reprogramming and redifferentiation. J Cell Sci 126(11):2516–2524
5. Park IH, Arora N, Huo H, Maherali N, Ahfeldt T, Shimamura A, Lensch MW, Cowan C, Hochedlinger K, Daley GQ (2008) Disease-specific induced pluripotent stem cells. Cell 134(5):877–886
6. Soldner F, Hockemeyer D, Beard C, Gao Q, Bell GW, Cook EG, Hargus G, Blak A, Cooper O, Mitalipova M, Isacson O, Jaenisch R (2009) Parkinson's disease patient-derived induced pluripotent stem cells free of viral reprogramming factors. Cell 136(5):964–977
7. Dimos JT, Rodolfa KT, Niakan KK, Weisenthal LM, Mitsumoto H, Chung W, Croft GF, Saphier G, Leibel R, Goland R, Wichterle H, Henderson CE, Eggan K (2008) Induced pluripotent stem cells generated from patients with ALS can be differentiated into motor neurons. Science 321(5893):1218–1221
8. Ebert AD, Yu J, Rose FF Jr, Mattis VB, Lorson CL, Thomson JA, Svendsen CN (2009) Induced pluripotent stem cells from a spinal muscular atrophy patient. Nature 457 (7227):277–280
9. Choi HW, Kim JS, Choi S, Hong YJ, Kim MJ, Seo HG, Do JT (2014) Neural stem cells differentiated from iPS cells spontaneously regain pluripotency. Stem Cells 32(10):2596–2604

10. Mikkelsen TS, Hanna J, Zhang X, Ku M, Wernig M, Schorderet P, Bernstein BE, Jaenisch R, Lander ES, Meissner A (2008) Dissecting direct reprogramming through integrative genomic analysis. Nature 454(7200):49–55
11. Chan EM, Ratanasirintrawoot S, Park IH, Manos PD, Loh YH, Huo H, Miller JD, Hartung O, Rho J, Ince TA, Daley GQ, Schlaeger TM (2009) Live cell imaging distinguishes bona fide human iPS cells from partially reprogrammed cells. Nat Biotechnol 27 (11):1033–1037
12. Kim JS, Choi HW, Choi S, Seo HG, Moon SH, Chung HM, Do JT (2014) Conversion of partially reprogrammed cells to fully pluripotent stem cells is associated with further activation of stem cell maintenance- and gamete generation-related genes. Stem Cells Dev. doi:10.1089/scd.2014.0020
13. Sridharan R, Tchieu J, Mason MJ, Yachechko R, Kuoy E, Horvath S, Zhou Q, Plath K (2009) Role of the murine reprogramming factors in the induction of pluripotency. Cell 136(2):364–377

Methods in Molecular Biology (2016) 1357: 97–110
DOI 10.1007/7651_2015_198

Published online: 17 February 2015

Using Oct4:MerCreMer Lineage Tracing to Monitor Endogenous *Oct4* Expression During the Reprogramming of Fibroblasts into Induced Pluripotent Stem Cells (iPSCs)

Lucas V. Greder, Jason Post, and James R. Dutton

Abstract

The reprogramming of somatic cells into induced pluripotent stem cells (iPSCs) using a combination of defined transcription factors has become one of the most widely used techniques in stem cell biology. A critical, early event in iPSC reprogramming is the induction of the endogenous transcription factor network that maintains pluripotency in iPSCs. Here we describe using a transgenic, conditional Oct4-Cre construct to investigate the spatial and temporal induction of endogenous *Oct4* expression during the reprogramming of mouse fibroblasts into iPS cells.

Keywords: Oct4, Oct4-Cre, Induced pluripotent stem cell, iPSC, Reprogramming, Lineage tracing

1 Introduction

The reprogramming of somatic cells into induced pluripotent stem cells (iPSCs) requires the initial overexpression of the transcription factor Oct4, usually with a cohort of other transcription factors, to induce and initially support a network of endogenous genes that functions to maintain the pluripotent stem cell phenotype in the absence of the exogenous reprogramming factors (1–4). A key early step in iPSC reprogramming is the activation of endogenous *Oct4* gene expression and without this induction taking place the reprogramming does not occur (5, 6). Investigating the molecular steps in iPSC reprogramming has been greatly enhanced by the introduction of new research tools, and the Oct4 lineage tracing system described in this chapter allows examination of one of the earliest key events in this transformation.

The transcription factor Oct4, (also known as POU5F1 POU domain, class 5, transcription factor 1) is highly expressed in pluripotent cells of the developing mammalian blastocyst and during development is rapidly downregulated until it is maintained only in primordial germ cells (7, 8). *Oct4* expression is also critical for the self-renewal of pluripotent embryonic stem (ES) cells and induced pluripotent stem (iPS) cells. However, the ubiquitous expression of

Oct4 in cells of the early embryo prohibits the use of unmodified Cre recombinase for lineage tracing *Oct4* gene expression in post-embryonic cells. We recently generated a transgenic mouse with the conditional MerCreMer cassette inserted into the 3′ UTR of the *Oct4* locus that, when combined with a Cre reporter mouse strain, enables the detection of cells expressing endogenous *Oct4* and the subsequent lineage tracing of their progeny (9, 10). In the Oct4-MerCreMer mTmG mouse the transgenic allele is present in the dual dTomato—eGFP reporter mouse (11). The recombinase is present in all Oct4-MerCreMer mTmG cells that are expressing endogenous *Oct4* but it is only active when tamoxifen is present. If tamoxifen is present, the *Oct4* expressing cell loses membrane bound dTomato protein expression and instead expresses membrane bound green fluorescent protein (eGFP), as do all subsequent cell progeny (*see* Fig. 1). If cells are characterized shortly after the addition of tamoxifen, eGFP expression identifies cells actively expressing *Oct4*. eGFP expression following a chase period after tamoxifen addition marks the progeny of cells that were actively expressing endogenous *Oct4* at the time of tamoxifen addition.

Cells from Oct4-MerCreMer mTmG mice provide an ideal platform to investigate the timing and consequence of the induction of endogenous *Oct4* expression during the reprogramming of somatic cells into pluripotent stem cells. In the method described in this chapter, fibroblasts are first cultured from Oct4-MerCreMer mTmG transgenic embryos and then reprogrammed into iPSCs using the Yamanaka factors (2). Tamoxifen is added during discrete time windows during the reprogramming. Any cell that responds to the reprogramming factors by inducing endogenous *Oct4* gene expression will express eGFP if tamoxifen is present at the time of

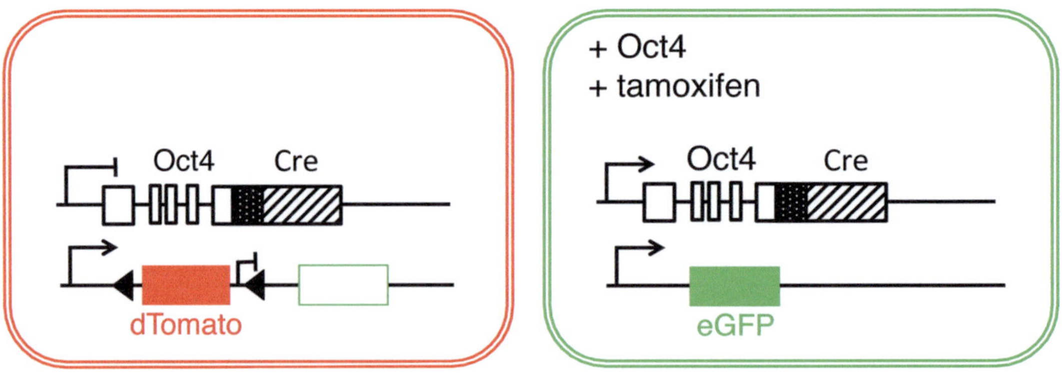

Fig. 1 Oct4 lineage tracing with Oct4-MerCreMer mTmG fibroblasts. The conditional MerCreMer cassette is inserted in the 3′UTR of the Oct4 locus and combined with the dual membrane-bound fluorescent protein mTmG reporter. Cre recombination only occurs in cells that are actively expressing Oct4 and when tamoxifen is present. In these cells there is an irreversible switch from expressing membrane bound dTomato to expressing membrane bound eGFP

Oct4 induction, and the progeny of these cells retain eGFP expression even if the *Oct4* expression is not maintained. With this pulse-chase method, the timing and progressive induction of endogenous Oct4 expression can be determined by the presence of eGFP expressing colonies in each tamoxifen window. At end of the experiment eGFP and Nanog protein co-expression is used to determine the proportion of cells that induced endogenous *Oct4* expression during each time-point that in turn reprogrammed to form an iPSC colony. This system can be used to investigate the influence of different factors on the efficiency and eventual outcome of iPSC reprogramming (9) and is also suitable for discovering new gene and chemical combinations able to initiate iPSC conversion. We have also used iPS cells derived from this mouse strain in the absence of tamoxifen to investigate the spatiotemporal loss of endogenous Oct4 expression during the differentiation of iPSC embryoid bodies (10).

2 Materials

2.1 Reagents and Supplies

1. DMEM containing 4 g/l glucose and with sodium pyruvate (Life Technologies 11995-073).
2. Fetal bovine serum (GE Healthcare HyClone SH30071.03).
3. Nonessential amino acids (Life Technologies 11140-076).
4. Antibiotic–antimycotic (Life Technologies 15240-112).
5. dPBS without calcium and magnesium (Millipore BSS-1006-B).
6. 0.25 % trypsin (Life Technologies 25200-114).
7. DMSO (Sigma-Aldrich D2650).
8. 0.1 % gelatin (Millipore ES-006-B).
9. Freezing medium (10 % DMSO in FBS, filter-sterilized).
10. Knockout DMEM (Life Technologies 10829-018).
11. Knockout serum replacer (KSR) (Life Technologies 10828-028).
12. 2-mercaptoethanol (Life Technologies 21985-023).
13. LIF (Millipore ESG1106).
14. 4-hydroxytamoxifen (Sigma H7904-5MG). Store as 10 mM stock in DMSO at −20 °C.

    Dilute to 100 μM working stock in DMSO and store at −20 °C.
15. FuGENE 6 transfection reagent (Promega E2691).
16. Polybrene (Hexadimethrine bromide) (Sigma H9268). Stock solution 8 mg/ml in water. Filter-sterilize and store at −20 °C.
17. 10 % buffered formalin (Fisher Scientific 23-305-510).

18. Disposable scalpels (Feather 2975-21).
19. 100 mm cell culture dishes (Thermo Fisher Scientific 150679).
20. T-25 flasks (Thermo Fisher Scientific 1012628).
21. T 75 flasks (Thermo Fisher Scientific 1012637).
22. 6-well plates (Thermo Fisher Scientific 0720080).
23. 0.45 μ syringe filters (Millipore SLHV033RS).
24. 10 ml disposable syringes with Luer-Lok (BD 309653).
25. 15 ml conical tubes (Thermo Fisher 0553859A).
26. 50 ml conical tubes (Thermo Fisher 0553860).

2.2 Antibodies

1. Rabbit anti-Nanog antibody (Abcam ab80892).
2. Donkey anti-rabbit antibody conjugated to Alexa Fluor 647 (Life Technologies A-31573).

2.3 Cell Lines and Plasmid Vectors

1. PlatE cells (12).
2. pMXs vectors (Available from Addgene).

2.4 Mouse Strains

1. Oct4-MerCreMer (The Jackson Laboratory, Stock No. 016829) (9).
2. Double-fluorescent Cre reporter strain mT/mG (The Jackson Laboratory, Stock No. 007576) (11).

2.5 Media

1. Fibroblast media: DMEM, 1× antibiotic–antimycotic, 1 % non-essential amino acids, 10 % FBS.
2. Mouse ESC media: Knockout-DMEM, 10 % FBS, 10 % KSR, 2 mM L-glutamine, 1 % nonessential amino acids, 0.1 mM 2-mercaptoethanol, and 1,000 units/ml ESGRO-mLIF.

3 Methods

3.1 Culturing Fibroblasts from Oct4-MerCreMer mTmG e13.5 Embryos

1. Oct4-MerCreMer mTmG transgenic embryos are obtained by mating heterozygous Oct4-MerCreMer mice with homozygous dual fluorescent mTmG reporter mice. The presence of a vaginal plug in the female mouse determines day e0.5 of gestation and embryos are collected at e13.5. Euthanize the female mouse according to institutional guidelines, spray the abdomen with 70 % ethanol, and place in a laminar flow hood. Excise the uterus in one piece and place it in a 100 mm tissue culture dish containing dPBS on ice.
2. Dissect each individual embryo from the uterus, remove the placenta but keep the yolk sac intact, and place each embryo into a separate 35 mm dish on top of ice in an ice bucket (*see* **Note 1**).

3. Move the dishes containing the embryos into a BSL2 culture hood culture hood and keep the dishes on ice. Using sterile watchmakers forceps carefully remove the yolk sac from the embryo and place the sac into a labeled microfuge tube for genotyping. These tubes can be kept on ice before proceeding to the genotyping when all the embryos have been processed (*see* **Note 5**. Genotyping embryos). Number the plate with the embryo and yolk sac to enable future matching of the embryo derived fibroblasts and genotype. If working alone, prepare all the embryos to this point before proceeding but if working as a team, process each embryo as soon as it is removed from the yolk sac.
4. Working with one embryo at a time transfer the embryo to an empty 100 mm culture dish. Using sterile dissecting tools remove the head and viscera from the embryo and transfer the body into a 60 mm dish containing a 200 μl 0.25 % trypsin.
5. Using two sterile scalpels chop the embryo into pieces until individual pieces are not obvious (*see* **Note 2**). Replace the lid on the dish and place in a 5 % CO_2, 37 °C incubator for 5 min.
6. Return the dish to the culture hood, add 1 ml of Fibroblast medium and pipette the embryo pieces gently with a 1 ml pipette. Place all the tissue suspension into a 15 ml conical tube. Add 10 ml of fibroblast medium and pipette to mix (*see* **Note 3**).
7. Centrifuge the tube at 100 × *g* for 4 min. Aspirate the medium from the tissue pellet and resuspend the pellet in 1 ml of fibroblast medium. Add an additional 11 ml of fibroblast medium and place the tissue suspension into a gelatin-coated T-75 flask. Place the flask in a 5 % CO_2, 37 °C incubator and shake the flask to disperse the suspension evenly.
8. Prepare cultures for each remaining embryo the same way. Incubate for 24 h without disturbing the flasks (*see* **Note 4**).
9. When available use the genotyping data to identify flasks containing cells from Oct4-MerCreMer mTmG transgenic embryos. Cells isolated from non-transgenic embryos can be discarded.
10. Examine the cell cultures using an inverted microscope. 24 h after plating the cell culture medium should be replaced. Passage the cells when they reach 80 % confluency.
11. At this point passage the cultures to expand and freeze stocks of early passage cells (*see* **Note 6**).
12. For each passage. Aspirate media from the cells and replace it with 10 ml of PBS without calcium or magnesium at room temperature. Allow the PBS to cover the cells and then aspirate the PBS.

13. Add 4 ml of 0.25 % trypsin at room temperature and return the flask to the incubator. View the cells every 2 min until the fibroblasts are just starting to round up. Tap the flask sharply on the bench to detach the cells, add 4 ml of fibroblast medium and wash the cells into the corner of the flask.
14. Collect the cells and media into a 15 ml conical tube. Rinse the flask with 4 ml of fibroblast media and combine with the cells in the tube. Centrifuge the cells at 100 × *g* for 4 min at room temperature.
15. Aspirate the medium and resuspend the cells in 1 ml of fibroblast medium. Add an additional 2 ml of media and transfer 1 ml into a new T-75 flask. Add 11 ml of fibroblast medium and return the flask to the 5 % CO_2, 37 °C incubator and shake the flask to disperse the cells evenly.
16. Centrifuge the tube containing the remaining cells at 100 × *g* for 4 min. Aspirate the media and resuspend the cells in 2 ml of freezing media and divide evenly into two cryovials. Freeze the cells in a controlled freezing device to −80 °C and then place in liquid nitrogen for long-term storage.
17. Continue to expand and freeze the early passage cells until you have banked cells at passage 4. Cells up to and including passage 4 can be used routinely to seed reprogramming experiments (*see* **Note 6**).

3.2 Preparing PlatE Cells to Produce Retrovirus

1. Prepare maxiprep plasmid DNA preparations of each retroviral plasmid pMXs-Oct3/4, pMXs-Sox2, pMXs-Klf4, pMXs-cMyc, pMXs-M_3O (13), pMXs-GFP (*see* **Note 7**). Determine the concentration of each plasmid DNA preparation and dilute each preparation to 1 μg/μl in TE buffer (10 mM Tris–HCl 1 mM EDTA pH 8.0).
2. You must coordinate the culture of the fibroblasts to be reprogrammed and the culture of PlatE cells that will be generating the retroviruses. Once transfected the PlatE cell culture supernatant containing the retroviruses is collected 48 h later and at this time the cells to be reprogrammed must also be plated and ready for viral infection (*see* section below).
3. Remove a vial of frozen PlatE cells from liquid nitrogen storage (*see* **Note 8**). Hold the vial in a 37 °C water bath until the contents are just thawed. Be sure to not allow the vial cap to be submersed. As the last ice in the vial is just melting spray the vial with 70 % ethanol and place in the BSL2 culture hood.
4. Transfer the contents of the vial into a 15 ml conical tube and add 1 ml of fibroblast culture medium dropwise to the 15 ml tube with gentle shaking. Add an additional 10 ml of fibroblast medium to the tube and centrifuge at 100 × *g* for 4 min at room temperature.

5. Aspirate the media from the tube and resuspend the cells gently in 1 ml of fibroblast medium. Add an additional 4 ml of fibroblast medium and count the cells using a hemocytometer or automated cell counter. Plate PlatE cells into a T-175 flask at 2.5×10^4 cells/cm^2 and culture in 15 ml fibroblast medium in a 5 % CO_2, 37 °C incubator.
6. The PlatE cells are ready to passage for retrovirus production when they are 80 % confluent. This should be considered Day −3 of the experiment.
7. Aspirate media from the PlatE cells and replace it with 10 ml of PBS at room temperature. Allow the PBS to cover the cells and then aspirate the PBS.
8. Add 4 ml of 0.25 % trypsin at room temperature and return the flask to the incubator. View the cells every minute until the cells are just starting to round up. Tap the flask sharply on the bench to detach the cells, add 4 ml of fibroblast medium and wash the cells into the corner of the flask.
9. Collect the cells and media into a 15 ml conical tube. Rinse the flask with 4 ml of fibroblast media and combine with the cells in the tube. Centrifuge the cells at $100 \times g$ for 4 min at room temperature.
10. Aspirate the media and resuspend the cells in 1 ml of high glucose DMEM with 1× NEAA and containing 10 % FBS. Add a further 4 ml of this media and count the cells.
11. Plate the PlatE cells into gelatin coated 100 mm culture dishes using 1.5×10^6 cells per dish with 10 ml of DMEM + 10 % FBS. Prepare one 100 mm dish per virus. Return the plates to a 5 % CO_2, 37 °C incubator and shake to disperse the cells evenly. Incubate cells overnight.

3.3 Culturing Oct4-MerCreMer Fibroblasts for Reprogramming

1. On Day −3 thaw a vial of frozen Oct4-MerCreMer mTmG fibroblasts from liquid nitrogen storage.
2. Transfer the contents of the vial into a 15 ml conical tube and add 1 ml of fibroblast culture medium dropwise to the 15 ml tube with gentle shaking. Add an additional 10 ml of fibroblast medium to this tube and centrifuge at $100 \times g$ for 4 min at room temperature.
3. Aspirate the media from the tube and gently resuspend the cells in 1 ml of fibroblast medium. Add an additional 4 ml of fibroblast medium and count the cells. Plate the Oct4 fibroblasts into a T25 flask at 1.5×10^4 cells/cm^2 and culture in 5 ml fibroblast medium in a 5 % CO_2, 37 °C incubator.
4. The Oct4 fibroblasts are ready to passage when they are ~70–80 % confluent. The cells must be ready to passage for retroviral infection on Day −1.

5. On Day −1 aspirate media from the Oct4 fibroblasts and wash once with PBS at room temperature.
6. Add 2 ml of 0.25 % trypsin at room temperature and return the flask to the incubator. Observe the cells closely to ensure cells are harvested once the cells begin to round up. Use a 1000 μl Gilson pipette or similar to gently disperse the cells into single cells.
7. Place the cells into a 15 ml conical tube with an additional 6 ml of fibroblast media and centrifuge the cells at 100 × *g* for 4 min at room temperature.
8. Resuspend the cells in 1 ml fibroblast growth media and count the cells.
9. Plate the fibroblasts at 5×10^4 cells per well in a gelatin coated 12-well dish in 1 ml fibroblast media/well. Return the plates to a 5 % CO_2, 37 °C incubator and shake to disperse the cells evenly. Incubate the cells overnight before viral transduction (*see* **Note 9**).

3.4 Transfecting PlatE Cells

1. On Day −2 before transfecting the PlatE cells with the retroviral vector plasmid DNA ensure that the cells are ~50 % confluent. If the cells reach 100 % confluency before transfection is complete they are likely to produce low titer virus.
2. For each transfection place 300 μl of DMEM without serum into a sterile 1.5 ml microfuge tube (*see* **Note 10**). Add 27 μl of FuGENE 6 transfection reagent directly into the medium dropwise and gently shake to mix. Incubate at room temperature for 5 min.
3. Add 9 μg of plasmid DNA directly into the media in the microfuge tube and again shake gently to mix. Incubate at room temperature for 15 min.
4. Replace the media on each plate of PlatE cells with 10 ml DMEM + 10 % FBS. For each separate transfection add the media/FuGENE 6/plasmid directly from the microfuge tube into the culture media dropwise around the plate. Gently swirl the plate to mix and return the plate to the incubator for 24 h.
5. On Day −1. Aspirate the media from the transfected PlatE cell cultures. Wash each transfected plate once with 10 ml PBS and add 8 ml DMEM + 2 % FBS medium to each plate. Return the cells to a 5 % CO_2, 37 °C incubator and incubate cells for a further 24 h (*see* **Note 11**).

3.5 Retrovirus Infection

1. On Day 0. For each virus plate in turn collect the medium from the PlatE cell culture and, using a 10 ml syringe and a Luer lock 0.45-micron filter, filter the viral supernatant into 15 ml conical tubes. Remove the plunger from the syringe and firmly attach a

45 μm filter. Stand the syringe and filter upright on a 15 ml conical tube. Pipette the media from the virus plate into the syringe and gently reinsert the plunger and filter the virus supernatant into the tube. To avoid potential exposure to viral supernatant from residual pressure in the syringe, do not remove the filter before appropriate waste disposal.

2. Prepare the viral supernatant cocktail: For each well of cells to be reprogrammed add 250 μl of each pMXs-Oct4 (or pMXs-M_3O) (13), pMXs-Sox2, pMXs-Klf4, and pMXs-c-Myc viral supernatant into a 15 ml conical tube. Prepare a separate tube from pMXs-GFP if you are using it as an infection control.
3. Supplement each viral supernatant mix with FBS to 10 % FBS and add 1 μl/ml of 8 μg/ml polybrene.
4. Aspirate the medium from each well of fibroblasts to be reprogrammed and add 1 ml of viral supernatant mix to each well.
5. Return the 12 well plate to the 5 % CO_2, 37 °C incubator and incubate Oct4-MerCreMer mTmG fibroblasts with the viral supernatant mix for 18 h.

3.6 Tamoxifen Addition Time Points During Reprogramming

1. On Day +1 replace the viral supernatant containing medium with 1 ml mESC medium (*see* **Note 12**) and incubate in 5 % CO_2, 37 °C for an additional 48 h.
2. On Day +3: Prepare mESC growth medium containing 100 ng/ml 4 hydroxytamoxifen (4-OHT). Add 1 ml of mESC + 4OHT to day 3/4 wells tamoxifen time point wells. For all the other wells replace the mESC medium. Return the plate to the incubator and incubate for 24 h.
3. On Day +4: Remove media from the day 3/4 well. Wash 2× with PBS and replace with fresh mESC growth medium without 4-OHT.
4. Day +4: Remove medium from day 4/5 well and replace with mESC + 4-OHT medium for 24 h.
5. Day +5 to Day +11: Repeat mESC + 4-OHT addition in turn to each of the tamoxifen time points (*see* **Notes 13** and **14**).
6. Replace medium with fresh mESC growth medium every other day for wells that are not associated with that day's tamoxifen window.
7. Day +12: Continue to Section 3.7. This is the end point of this experiment (*see* **Note 15**).

3.7 Fixing and Nanog Protein Detection

1. Aspirate the medium from each well and wash each well twice with PBS.
2. Aspirate the PBS and replace with 10 % buffered formalin for 10 min at room temperature.

3. Wash cells twice with PBS and replace with 1 ml PBS containing 0.1 % Triton X100 (PBS-T).
4. Permeabilize the cells by adding 0.5 ml of PBS + 1 % Triton-X100 to each well for 5 min at room temperature (*see* **Note 16**). Wash the cells twice with 1 ml PBS.
5. Incubate the cells with 0.5 ml of blocking solution (PBS-T + 1 % bovine serum albumin) and incubate for at least 30 min at room temperature.
6. Dilute the anti-Nanog antibody in blocking solution 1:200. Aspirate the blocking solution from each well and add 300 μl of diluted antibody to each well. Incubate overnight at 4 °C.
7. Remove the blocking solution and antibody and wash each well 3× with 1 ml PBS-T.
8. Before removing the third wash dilute the donkey anti-rabbit Alexa Fluor^{647} 1:500 in blocking solution. Add 300 μl of diluted secondary antibody to each well. Incubate for 1 h at room temperature in the dark.
9. Remove the secondary antibody and wash the cells 3× with PBS-T.
10. Add 1 ml of PBS-T to each time point well.
11. Visualize the cells using a fluorescent inverted microscope fitted with a camera and with filters to excite eGFP, dTomato, and Alexa Fluor 647.

3.8 Expected Results

1. Oct4-MerCreMer mTmG cells expressing endogenous *Oct4* will, in the presence of tamoxifen, lose dTomato expression and express transmembrane bound eGFP, as will all their subsequent cell progeny.
2. If a cell is expressing endogenous *Oct4* during the time tamoxifen is present in the well that cell and its progeny will be eGFP expressing at the experiment end point.
3. In the early time point wells there may be no cells that have induced endogenous *Oct4* so no eGFP expressing cells or colonies will be present at day 12 in these wells. All cells will express dTomato.
4. As the experiment progresses the number of cells that induce endogenous *Oct4* expression should increase and the number of eGFP expressing colonies in later tamoxifen time point wells will be greater at day 12.
5. Not all cells that induce endogenous *Oct4* reprogram fully to an iPSC phenotype. If these cells were expressing endogenous *Oct4* in a well where tamoxifen was present, a colony generated from them will express eGFP at day 12 but the cells will not co-express Nanog protein.

6. During the experiment cells also reprogram to an iPSC phenotype at times where tamoxifen is not present in their well. If this reprogramming occurs after the tamoxifen window for that well the colonies from these cells will co-express dTomato and Nanog at day 12. If however tamoxifen is added after the cell reprograms then the *Oct4* expressing iPS cells in that colony and their progeny will express eGFP and Nanog when examined at day 12.
7. Over the course of this type of experiment the number of eGFP expressing colonies increases in each well during the early time windows (*see* Fig. 3 in Greder et al. (9)).
8. The proportion of cells that induce endogenous *Oct4* during each time point and proceed to reprogram fully to an iPSC phenotype can be determined by calculating the percentage of eGFP expressing colonies that co-express Nanog. In Fig. 2 we show typical results from three different transgenic embryos reprogrammed with four factors using either unmodified Oct4 or $M_3$0 where the Oct4 protein is fused to the myoD activation domain (13). Although reprogrammed simultaneously with the same viral preparations, each embryo has a different proportion of *Oct4* inducing cells that reprogram fully to express Nanog.

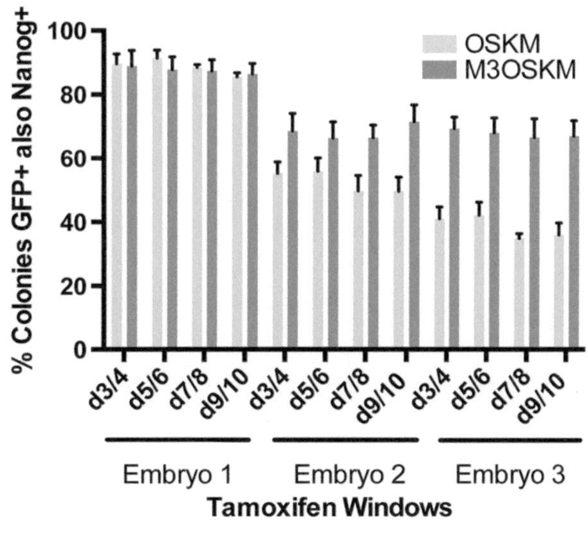

Fig. 2 Results from reprogramming Oct4-MerCreMer mTmG fibroblasts with discrete time points of tamoxifen addition. Embryonic Oct4-MerCreMer mTmG fibroblasts from three different embryos from the same litter were reprogrammed simultaneously using OSKM or M_3OSKM retroviruses. Tamoxifen was present in the media during the time windows shown. 12 days after infection the number of colonies co-expressing GFP and Nanog protein was determined

4 Notes

1. Be careful to keep the uterus on ice in PBS while you dissect each individual embryo. With large litters it is preferable to include multiple people in the embryo dissection to reduce the time for processing all the embryos for fibroblast preparation. As the genotype of each embryo is not yet known all the embryos must be prepared separately.
2. Chopping the embryos can take 5 min before incubation. Chop the tissue in multiple directions to ensure cutting the pieces as small as possible.
3. The tube can be placed on ice at this point until all the embryo preparations are completed to this point.
4. Over this time the tissue pieces and dispersed cells will attach to the flask and cells will begin to multiply. Use this time to complete the embryo genotyping.
5. To genotype embryonic yolk sacs prepare DNA for PCR genotyping from the individual yolk sacs according to your preferred method. We routinely use the Promega Genomic DNA isolation kit (Promega A1120). Perform PCR amplification with DNA from each yolk sac with the following primers together. R1 (5′ GCT TTC TCC AAC CGC AGG CTC TC 3′), R2 (5′ GCC CTC ACA TTG CCA AAA GAC GG 3′), F1 (5′ CCA AGG CAA GGG AGG TAG ACA AG 3′) using an annealing temperature of 55 °C. DNA from wild-type embryos should amplify only a single band of approximately 200 bp. Transgenic samples will display an additional slightly smaller band.
6. Fibroblasts for reprogramming experiments should be used before Passage 6. When banking fibroblasts from new embryos we aim to passage P0 cells 1:3, P2 cells 1:4 and P3 cells 1:3. We bank 1/3 T-75 flask per vial and aim to bank 2 vials of P0 cells, 2 vials of P1 cells, 3 vials of P2 cell, 6 vials of P3 cells, and 9 vials of P4 cells. Each vial of P4 cells should seed a T-75 flask for use in reprogramming experiments. Each vial of preceding passage cells can be plated and expanded 1:3 to generate new Pn + 1 cells.
7. The quality of plasmid DNA has a major influence on the success of the transfection to generate the retroviral vector. We routinely use the PowerPrep® HP Maxiprep Kits with Prefilters from Origene (NP100024). Parallel retroviral production with pMXs-GFP can be used to monitor both the transfection efficiency and the subsequent success of retroviral transduction.
8. PlatE cells are best used at the low passage number available. Expand your lowest passage cells and freeze away a number of

vials each containing 5×10^6 cells. Using a new vial each time to start the culture to generate the retroviruses will ensure consistency of virus production.

9. When culturing early passage Oct4-MerCreMer mTmG fibroblasts prior to reprogramming it is important to thaw and plate the cells at the correct density to ensure they are actively dividing after the passage for retroviral transduction.
10. We routinely prepare pMXs-GFP retrovirus as a both a transfection and viral transduction control. The transfection efficiency can be determined live by viewing the GFP expression in the transfected PlatE cells after 36 h. This is usually indicative of the efficiency of transfection of the other factors. If the number of PlatE cells expressing GFP is less than 70 % do not continue with the experiment. Recheck the plasmid DNA quality and PlatE cell culture and perform test transfections until this level of transfection is obtained.
11. PlatE cells that are producing virus will start fusing together 24–36 h after transfection.
12. Non-reprogramming fibroblasts provide adequate "feeder" support for Oct4-MerCreMer mTmG fibroblasts that reprogram to form iPSCs so there is no need to transfer the infected cells onto irradiated feeder cells.
13. Although *Oct4* expressing cells will begin expressing eGFP within 4 h of 4-OHT treatment, single cells or small eGFP expressing colonies are unlikely to be immediately visible under a microscope. For example, the progeny of a cell that would have expressed *Oct4* during the day 3/4 window is likely to appear as a small eGFP expressing cell cluster by day 5.
14. In later time points, *Oct4*+ iPSC colonies will begin to express visible eGFP 4 h after 4-OHT treatment.
15. The termination day is determined by the experimental question and longer chase periods may be appropriate.
16. Do not allow the permeabilization step to exceed 5 min.

References

1. Okita K, Ichisaka T, Yamanaka S (2007) Generation of germline-competent induced pluripotent stem cells. Nature 448(7151):313–317
2. Takahashi K, Yamanaka S (2006) Induction of pluripotent stem cells from mouse embryonic and adult fibroblast cultures by defined factors. Cell 126(4):663–676
3. Park IH et al (2008) Reprogramming of human somatic cells to pluripotency with defined factors. Nature 451(7175):141–146
4. Takahashi K et al (2007) Induction of pluripotent stem cells from adult human fibroblasts by defined factors. Cell 131 (5):861–872

5. Brambrink T et al (2008) Sequential expression of pluripotency markers during direct reprogramming of mouse somatic cells. Cell Stem Cell 2(2):151–159
6. Jaenisch R, Young R (2008) Stem cells, the molecular circuitry of pluripotency and nuclear reprogramming. Cell 132(4):567–582
7. Ovitt CE, Scholer HR (1998) The molecular biology of Oct-4 in the early mouse embryo. Mol Hum Reprod 4(11):1021–1031
8. Scholer HR et al (1990) Oct-4: a germline-specific transcription factor mapping to the mouse t-complex. EMBO J 9(7):2185–2195
9. Greder LV et al (2012) Analysis of endogenous Oct4 activation during induced pluripotent stem cell reprogramming using an inducible Oct4 lineage label. Stem Cells 30 (11):2596–2601
10. Sajini AA et al (2012) Loss of Oct4 expression during the development of murine embryoid bodies. Dev Biol 371(2):170–179
11. Muzumdar MD et al (2007) A global double-fluorescent Cre reporter mouse. Genesis 45(9): 593–605
12. Morita S, Kojima T, Kitamura T (2000) Plat-E: an efficient and stable system for transient packaging of retroviruses. Gene Ther 7(12): 1063–1066
13. Hirai H et al (2011) Radical acceleration of nuclear reprogramming by chromatin remodeling with the transactivation domain of MyoD. Stem Cells 29(9):1349–1361

Methods in Molecular Biology (2016) 1357: 111–131
DOI 10.1007/7651_2015_251

Published online: 30 May 2015

Inducible Transgene Expression in Human iPS Cells Using Versatile All-in-One *piggyBac* Transposons

Shin-Il Kim, Fabian Oceguera-Yanez, Chiho Sakurai, Masato Nakagawa, Shinya Yamanaka, and Knut Woltjen

Abstract

Transgenics is a mainstay of functional genomics. Conditionally overexpressing genes of interest (GOIs) helps to reveal their roles in the control of complex biological processes. Complemented by findings in classic animal model systems, recent advances in human embryonic stem cell (hESC) and patient-specific induced pluripotent stem cell (hiPSC) differentiation have led to sophisticated in vitro models of human development and disease. Yet, as transgenic elements encoding inducible systems must be introduced de novo into each genetically unique human stem cell line, robust and straightforward solutions to gene delivery are required. Transposons are a family of mobile DNA elements that have been adapted as experimental tools for stable genomic integration of transgenes. The *piggyBac* (PB) transposon from *Trichoplusia ni* presents a number of benefits over classic viral or BAC transgenesis: ease of application, simple integration-site mapping, and the unique capacity for traceless excision. Moreover, their large capacity permits the consolidation of multiple transgene components in a single vector system. In this chapter, we outline the features of a panel of "All-in-One" PB transposons designed for drug-inducible gene expression and provide guidelines to establish and validate populations or clones of transgenic hiPSCs.

Keywords: Human induced pluripotent stem cell (hiPSC), Reprogramming, Transposon, *piggyBac*, Doxycycline regulated, Stable transgenesis, Gene expression

1 Introduction

The ability to reprogram human somatic cells into induced pluripotent stem cells (1) has revolutionized human functional genomics studies, and will certainly have a long-lasting impact on the emerging field of personalized regenerative medicine. As models of development and disease, hiPSC lines carry with them the genetic identity of the patient, including disease-causing mutations and a plethora of subtle genetic variations. These variations define individuality and contribute to patient-specific phenotypes, highlighting the importance of maintaining an isogenic background for comparative studies. Thus, unlike inbred model organisms, a "standard" human iPSC line bearing preintegrated transgenic elements is currently not feasible.

Given the diversity of iPSC lines available, it is pertinent to consider methods of transgenesis that permit the simple introduction of complex transgene systems in a consolidated manner. Stable transgenesis of human cells by *piggyBac* (PB) transposons (2) is efficient, versatile, and simple to implement, with many advantages over viral delivery. PB transposons are deployed in cells as plasmid DNA, and as such, the preparation of vectors is technically accessible to labs with a basic capacity for molecular biology. As transposon mobilization requires a helper plasmid expressing the PB transposase enzyme (PBase) (3), there is no requirement for specialized biohazard containment in preparation or application. Moreover, transposon vectors can be stored and distributed easily. Integration into the genome occurs in a large proportion of transfected cells, where transposon copy numbers may be scaled through control of the amount of PB transposon plasmid (4). PB elements integrate as intact monomers, and nearly always at TTAA sites (5, 6), making their mapping in the genome simple using PCR-based approaches (7). Importantly, the cargo capacity of PB is massive compared to virus; BAC-sized elements have been shown to undergo full-length integration (8), a major advantage over the unpredictable end resection or fragmentation seen with random BAC integration.

PB transposon systems have been applied to address many aspects of stem cell biology. Permanent cell labeling with GFP reporter PB transposons traced the in vivo behavior of mouse fetal neural stem cells through adulthood following maternal stimuli (9). Using a multitransposon approach, a synthetic Notch-Delta signaling network was reconstituted, providing an in vitro model of lateral inhibition (10). In order to produce human iPSCs through somatic cell reprogramming, large-capacity PB vectors delivering the four reprogramming factors (Oct3/4, Sox2, Klf4, and c-Myc) have been employed (11, 12) (Chapter 11). Moreover, methods describing re-exposure of stable iPSC lines to PBase in order to drive stochastic excision of PB elements, resulted in transgene-free iPSC lines (13, 11, 14). Using dox-inducible PB vectors in normal and patient-derived iPSCs, temporal control of master regulators such as MYOD1 induced differentiation along the myogenic lineage (15). Finally, drug-regulated expression of SYT-SSX from integrated PB vectors addressed the nature of oncogene-regulated gene expression in human cancer (16). Thus, PB vectors are applicable to studies of stem cell derivation, differentiation, and disease modeling.

Here we present a PB transposon system employing doxycycline (dox)-inducible transgene regulation from a single "All-in-One" vector (Fig. 1). Gateway-compatibility supports the rapid construction of PB vectors expressing a chosen GOI, or 2A-peptide linked GOI series (Chapter 11). PB variants incorporating different reporter genes and selection markers provide flexibility in application (Table 1). Transgene expression may be monitored indirectly

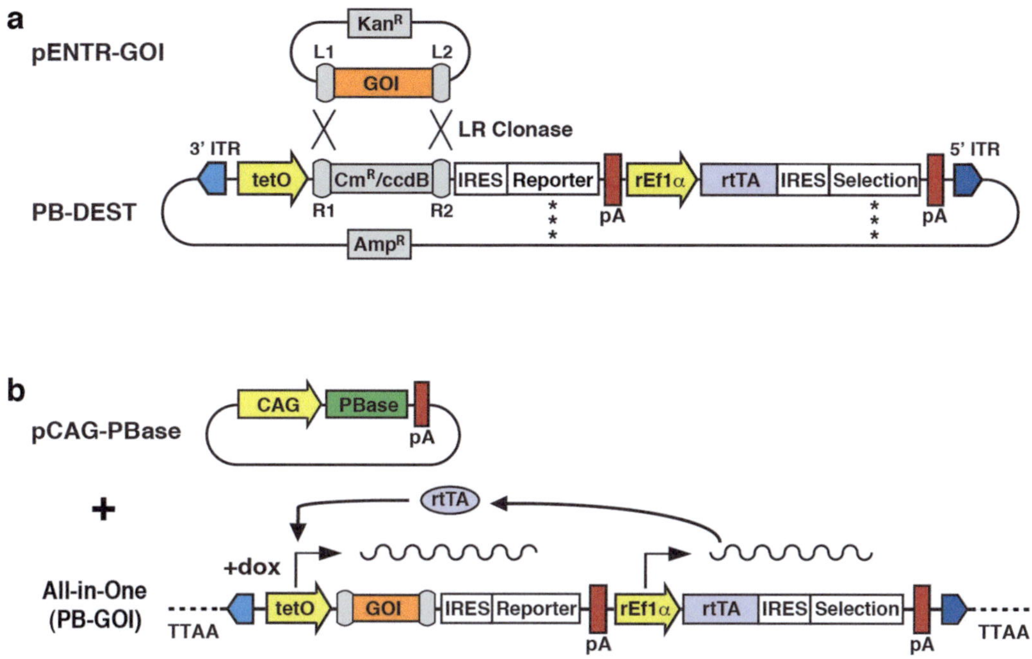

Fig. 1 Diagram of the *piggyBac* transposon system and function of the All-in-One inducible vector in response to dox. (**a**) An All-in-One PB vector (PB-DEST) (Table 1) allows drug inducible expression of a Gateway-cloned GOI. An IRES-linked fluorescent "Reporter" assists in imaging and FACS sorting. Co-transfection of an additional vector encoding rtTA is not required, as the PB-DEST vectors provide constitutive expression of the rtTA transactivator in *cis*. "Selection" markers permit enrichment of transgenic cells. (**b**) pCAG-PBase constitutively expresses *piggyBac* transposase (PBase) during transfection (*top*). Co-transfection of an All-in-One transposon with the PBase expression vector results in stable genomic PB integration in TTAA sites throughout the genome (transposition), after which, dox treatment induces expression of the GOI. *Kan^R*, Kanamycin resistance gene; *L1/L2*, Gateway attL1/L2 recombination sites; Blue polygons, PB 3' (*left*) and 5' (*right*) inverted terminal repeats (ITRs); *tetO*, dox-responsive promoter; *R1/R2*, Gateway attR1/R2 recombination sites; *Cm^R*, Chloramphenicol resistance gene; *ccdB*, negative selection marker; *IRES*, internal ribosome entry site; *pA*, polyadenylation signal; *rEf1α*, rat elongation factor 1α; *rtTA*, reverse tetracycline transactivator; *Amp^R*, Ampicillin resistance gene; *CAG*, chicken β-actin/rabbit β-globin promoter; *PBase*, *piggyBac* transposase gene; *dox*, doxycycline

Table 1
All-in-One PB-DEST vectors

Vector name	Reporter	Selection	RIKEN ID# PB-DEST vector
PB-TA-ERN	None	Neomycin	RDB13242
PB-TAC-ERN	mCherry		RDB13243
PB-TAG-ERN	GFP		RDB13244
PB-TA-ERP2	None	Puromycin	RDB13245
PB-TAC-ERP2	mCherry		RDB13246
PB-TAG-ERP2	GFP		RDB13247

All plasmids require cotransfection with the pCAG-PBase expression vector (KW158, RDB13241) for stable genomic integration

through the reporter, and fine-tuned by adjusting drug concentrations. Thus, a spectrum of gene expression levels, including levels that mimic endogenous expression, may be tested experimentally. Protocols describing both clonal- and population-based establishment of transgenic lines using a feeder-free culture method are detailed.

2 Materials

2.1 PB Vector Preparation by Gateway Cloning

The PB-DEST vectors described in the Table 1 are available through the RIKEN Bio Resource Center DNA Bank (dna.brc.riken.jp). All PB-DEST vectors are Destination vectors designed to be compatible with cloning of your GOI by Gateway (Invitrogen). The PB-GOI control vector (PB-TAC-luc2-ERN) is available upon request.

2.1.1 LR Recombination Reaction

1. Entry clone containing a cDNA for the sequence-verified GOI (pENTR-GOI), prepared as high-quality plasmid DNA (*see* **Note 1**).
2. PB Destination vector (Table 1; PB-DEST) prepared as high-quality plasmid DNA.
3. Tris-EDTA (TE) Buffer, pH 8.0
4. Gateway® LR Clonase™ II Enzyme Mix (Invitrogen, #11791).
5. Competent *E. coli* cells (DH5α or equivalent) and standard microbiology culture materials and reagents (*see* **Note 2**).
6. Antibiotics:
   (a) Kanamycin Sulfate (nacalai tesque, #19860-44), used for the growth of pENTR-GOI vectors.
   (b) Ampicillin Sodium Salt (nacalai tesque, #02739-74), used for the growth of PB-DEST and PB-GOI vectors.
   (c) Chloramphenicol (nacalai tesque, #08027-14), used in combination with Ampicillin for the growth of pre-Gateway PB-DEST vectors (Table 1).
7. Restriction enzymes and gel electrophoresis equipment (*see* **Note 3**).

2.1.2 DNA Preparation for Transfection

1. Wizard® Plus SV Miniprep DNA Purification System (Promega, #A1460).
2. HiSpeed® Plasmid Maxi Kit (QIAGEN, #12663) (*see* **Note 4**).

2.2 Routine Maintenance of Feeder-Free Human iPSCs

The derivation of transgenic hiPSC populations or clonal cell lines is described using culture under feeder-free (Ff) conditions (17). The culture conditions may be adapted to be appropriate for your target iPSC line. Recommended alternative materials are provided at each step. The 201B7 hiPSC line is available through the RIKEN BRC Cell Bank (http://www.brc.riken.jp/lab/cell/english/).

2.2.1 Cells

1. 201B7 hiPSCs (1), RIKEN BRC Cell No. HPS0063.
2. 201B7 Ff-hiPSCs, adapted from HPS0063.
3. *Optional*: Feeder cells (*see* **Note 5**).

2.2.2 Ff-Human iPSC Culture Medium

1. StemFit, AK03 (AJINOMOTO) (*see* **Note 6**).
2. Penicillin Streptomycin solution (Gibco, #15140-122).
3. Ff-Human iPSC Culture Medium:
   (a) Thaw supplements B (100 mL) and C (2 mL) at 4 °C overnight.
   (b) To make complete medium, add supplements B and C to 400 mL of StemFit AK03 basal medium and mix well, avoiding the introduction of bubbles. Supplement with 5 mL Penicillin Streptomycin (1:100) as required. Complete AK03 medium is stable for ~2–3 weeks at 4 °C.
   (c) For small-scale long-term use, aliquot 45 mL of media into 50-mL conical tubes and freeze at –80 °C (stable up to at least 6 months).

2.2.3 Laminin-Coated Tissue Culture Plates

1. 6-well Tissue Culture Plate (FALCON, #353046).
2. 60-mm Tissue Culture Plate (FALCON, #353004).
3. Dulbecco's phosphate buffered saline (DPBS), Mg^{2+} and Ca^{2+} free (Nacalai Tesque, #14249-24).
4. UltraPure™ DNase/RNase-Free Distilled Water (Invitrogen, #10977-015).
5. iMatrix-511, Recombinant Human Laminin-511 E8 Fragment (175 μg/ tube) (Nippi, #892004) (*see* **Note 7**).
6. Laminin solution (0.5 mg/mL):
   (a) Add 350 μL UltraPure™ distilled water into the tube of lyophilized iMatrix-511 (175 μg/tube). Store at 4 °C for up to 4 weeks.

2.2.4 hiPSC Passage

1. Ff-hiPSC culture, maximum 70–80 % confluency.
2. Ff-hiPSC Culture Medium (described in Subheading 2.2.2).
3. UltraPure™ DNase/RNase-Free Distilled Water (Invitrogen, #10977-015).

4. ROCK inhibitor solution 10 μM, 1,000× stock (Y-27632, Wako, #253-00513):
   (a) Dissolve 5 mg Y-27632 in 1,478 μL distilled water. Aliquot 100 μL into microfuge tubes and store at −20 °C (protected from light).
5. Passage Medium: Culture Medium containing 1× ROCK inhibitor.
6. Laminin-coated tissue culture plates, 6-well or 35-mm plates (prepare as described in Subheading 3.2.1).
7. Dulbecco's phosphate buffered saline (DPBS), Mg^{2+} and Ca^{2+} free (Nacalai Tesque, #14249-24).
8. Accumax®, Cell/Tissue Dissociation Solution (Innovative Cell Technologies, Inc., #AM-105) (*see* **Note 8**).
9. 15-mL Polypropylene Conical Tube (BD Falcon, #352096).
10. 50-mL Polypropylene Conical Tube (BD Falcon, #352070).
11. *Optional*: Cell Scraper (IWAKI, #9000-220).

2.2.5 Cell Counting

1. Single cell suspension (prepare as described in Subheadings 2.2.4 and 3.2.2).
2. Trypan Blue Stain 0.4 % (Gibco, #15250).
3. TC10™ Automated Cell Counter (Bio-Rad).
4. Counting Slides, Dual Chamber for Cell Counter (Bio-Rad, #145-0011).

2.2.6 Preparation of Frozen Ff-hiPSC Stocks

1. Single cell suspension, 2×10^5 cells/vial (prepare as described in Subheading 3.2.2).
2. STEM-CELLBANKER® (Chemically Defined Cryopreservation Solution) (TaKaRa, #CB043).
3. Nalgene® System 100™ Cryogenic Tubes (NALGENE, #5000-1020).
4. Nalgene™ Cryo 1 °C Freezing Container (NALGENE, #5100-0001).
5. −80 °C Freezer.
6. Liquid nitrogen storage tank.

2.3 Establishing Transgenic hiPSCs

2.3.1 Electroporation (EP)

1. DNA:
   (a) pCAG-PBase plasmid DNA, RDB13241.
   (b) PB-TAC-luc2-ERN (control GOI) or other appropriate post-Gateway PB-GOI plasmid DNA.
2. Single cell suspension (Subheadings 2.2.4 and 3.2.2).
3. Laminin-coated tissue culture plates, 60-mm plates (prepare as described in Subheading 3.2.1).

4. Passage Medium (Subheading 2.2.4, **Step 5**).
5. Opti-MEM I reduced-serum medium (Invitrogen, #31985-070).
6. Electroporation Cuvettes (NEPA GENE CO., LTD, #EC-002S).
7. Super Electroporator NEPA21 Type II, In Vitro, and In Vivo Electroporation (Nepa Gene Co., Ltd).

2.3.2 Drug Selection

1. Ff-hiPSC Culture Medium (Subheading 2.2.2).
2. UltraPure™ DNase/RNase-Free Distilled Water (Invitrogen, #10977-015).
3. Millex-GP Syringe Filter Unit, 0.22 μm (Merck Millipore, #SLGP033RS).
4. Puromycin dihydrochloride from *Streptomyces alboniger* (Sigma-Aldrich, #P7255).
5. Puromycin solution (1 mg/mL):
   (a) Dissolve puromycin in distilled water to a final concentration of 1 mg/mL.
   (b) Filter-sterilize through a 0.22-μm syringe filter. Aliquot and store at −20 °C.
6. G418 Sulfate, Cell Culture Tested (Calbiochem, #345810).
7. G418 solution (50 mg/mL, activity):
   (a) Dissolve G418 in distilled water to a final concentration of 50 mg/mL.
   (b) Filter-sterilize through a 0.22-μm syringe filter. Aliquot and store at −20 °C.
8. Selection Medium: Culture Medium (Subheading 2.2.2) containing appropriate antibiotics. Recommended: Puromycin, 0.5–1.0 μg/mL; G418, 50–200 μg/mL (*see* **Note 9**).

2.3.3 Population Expansion/Colony Picking and Expansion

1. Micropipette (1–10 μL volume range).
2. Stereomicroscope (Olympus SZ61).
3. Passage Medium (Subheading 2.2.4).
4. Puromycin or G418 solution.
5. Laminin-coated tissue culture plates, 6-well or 35-mm plates (prepare as described in Subheading 3.2.1).
6. 96-well, Flat-bottom Tissue Culture Plate (FALCON, #353075).
7. Dulbecco's phosphate buffered saline (DPBS), Mg^{2+} and Ca^{2+} free (Nacalai Tesque, #14249-24).

7. Accumax®, Cell/Tissue Dissociation Solution (Innovative Cell Technologies, Inc., #AM-105).
8. Multichannel micropipette and aspirator.
9. 25 mL Disposable Reagent Reservoir (Thermo SCIENTIFIC, #8094).

2.4 Validation of hiPSCs with Inducible Transgene Expression

2.4.1 Doxycycline Treatment

1. Culture Medium (described in Subheading 2.2.2).
2. Doxycycline Hyclate (LKT Laboratories, Inc, #D5897).
3. UltraPure™ DNase/RNase-Free Distilled Water (Invitrogen, #10977-015).
4. Millex-GP Syringe Filter Unit, 0.22 μm (Merck Millipore, #SLGP033RS).
5. Dox solution (1 mg/mL):
   (a) Dissolve doxycycline in distilled water to a final concentration of 10 mg/mL. Aliquot and store at −20 °C.
   (b) Dilute concentrated stock 1:10 with distilled water and filter-sterilize through a 0.22-μm syringe filter. Aliquot and store at −20 °C.

2.4.2 Reporter Gene Analysis: Imaging

1. Fluorescence microscopy (BZ-X710, KEYENCE).
2. *Optional*:
   (a) 96-well, Black/Clear, Tissue Culture Treated Plate, Flat Bottom with Lid (FALCON, #353219).

2.4.3 Reporter Gene Analysis: Luciferase Assay

1. Dual-Glo® Luciferase Assay System (Promega, #E2940).
   (a) Thaw Dual-Glo® Luciferase Reagent overnight at 4 °C and equilibrate to room temperature before use (1–2 h).
2. 96 W White Polystyrene, Cell Culture, with lid, Sterile (Thermo Scientific or Nunc, #136101).
3. 2104 EnVision Multilabel Plate Reader (Perkin Elmer).

2.4.4 Reporter Gene Analysis: Flow Cytometry

1. FACS buffer:
   (a) Add 1 mL Fetal Bovine Serum (FBS) to 50 mL PBS for a final concentration of 2 %.
2. 5-mL Polystyrene Round-Bottom Tube (BD Falcon, #352008).
3. 5-mL Polystyrene Round-Bottom Tube with Cell-Strainer Cap (BD Falcon, #352235).
4. BD LSRFortessa™ Cell Analyzer (BD Biosciences).
5. FlowJo software (Tree Star Inc.).

6. *Optional*:
   (a) 7-Aminoactinomycin D (7-AAD) (BD Pharmingen, #51-68981E).
   (b) Alexa Fluor® 647 Mouse anti-Human TRA-1-60 Antigen (BD Pharmingen, #560850).
   (c) Alexa Fluor® 647 Mouse IgM, κ Isotype Control (BD Pharmingen, #560806).

3 Methods

The methods presented herein outline the process of transgenic hiPSC production using All-in-One PB vectors (Fig. 1 and Table 1). The process involves three main steps: (1) Gateway cloning of your GOI, (2) transfection, selection, and isolation of transgenic hiPSC populations or clones, and (3) validation of inducible gene expression by reporter or GOI detection (Fig. 2).

3.1 Gateway Construction of PB Vectors

Gateway Cloning has a long legacy as a reliable method for recurrent cloning schemes, such as shuffling cDNAs from Entry vector (pENTR) intermediates into various Destination (pDEST) expression vectors. A series of PB-DEST vectors is outlined in Table 1. Extensive public cDNA libraries such as the Mammalian Gene Collection (http://mgc.nci.nih.gov/) incorporate Gateway cloned genes. In contrast to popular new cloning methods such as Golden Gate or Gibson Assembly, Gateway LR recombination steps are not PCR-based. Thus, following initial GOI validation in pENTR vectors, there is no requirement for DNA sequencing post-Gateway products, saving considerable time and effort. Moreover, pENTR-GOI vectors can be easily adapted into sophisticated Multisite Gateway reactions for nearly unlimited shuffling of promoter, GOI, and reporter/selection elements. Details on the production of custom pENTR-GOI vectors may be found in the Invitrogen product resources (http://www.lifetechnologies.com/jp/en/home/life-science/cloning/gateway-cloning/gateway-technology.html). For the purposes of this chapter, we will assume that pENTR-GOI construction and validation have been completed prior.

3.1.1 Combining pENTR-GOI and PB-DEST Vectors by LR Clonase Recombination

1. Prepare the following reaction (*see* **Note 10**).
   (a) Mix 100 ng pENTR-GOI and 100 ng of PB-DEST plasmid.
   (b) Increase the volume to 8 μL with TE Buffer pH 8.0.
   (c) Add 2 μL of LR Clonase™ II Enzyme Mix.
   (d) Incubate the reaction at room temperature for 1 h.
   (e) Stop the reaction with 1 μL Proteinase K solution (provided in the Gateway Cloning kit). Incubate at 37 °C for 10 min.

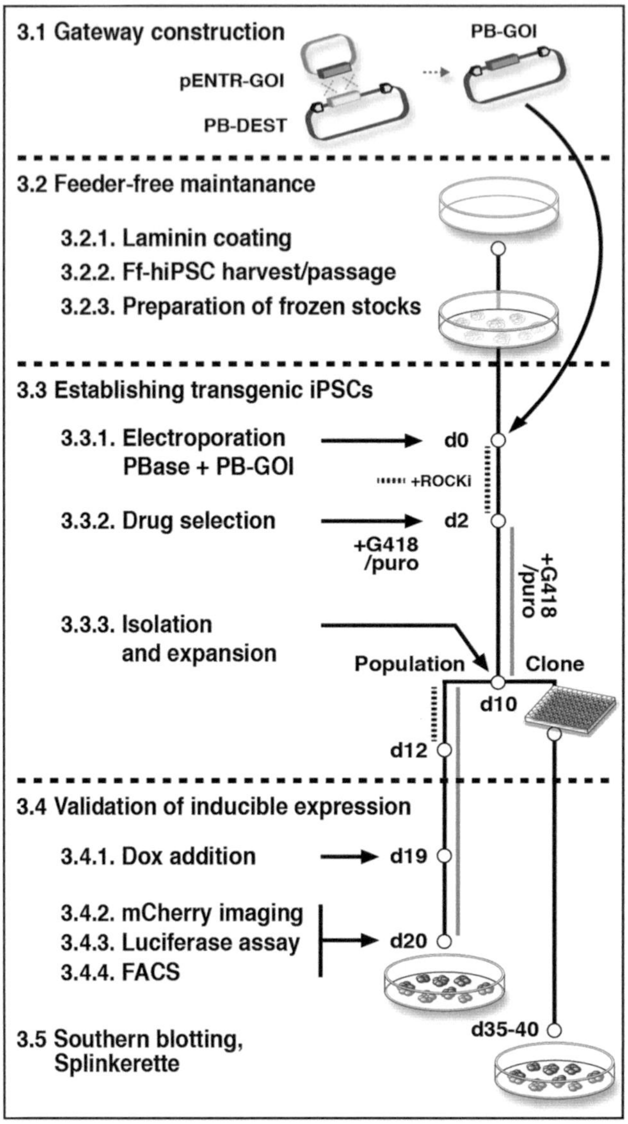

Fig. 2 Overview of generation of hiPSC lines expressing inducible transgene with *piggyBac* transposon system. Refer to the Subheading 3 for details

2. Transform 2–5 μL of the reaction mix into competent bacteria, such as XL1Blue or DH5α. Select for recombinant clones on Ampicillin (100 μg/mL) at 37 °C.
3. Pick and culture 3–4 bacterial clones per LR Gateway reaction. Verify final PB-GOI constructs by conventional plasmid mapping by restriction endonuclease digestion (*see* **Note 11**).

3.1.2 DNA Preparation for hiPSC Electroporation

1. Although Miniprep DNA can be used for screening test vectors, we recommend Maxi preparation or an endotoxin-free kit.
2. Plasmid DNA stock concentrations should be maintained at 500–1,000 ng/μL.

3.2 Feeder-Free Maintenance of Human iPSCs

Basic protocols for hiPSC maintenance, passage, and cryopreservation are outlined in this section. Mastery of these steps is critical, as they form the foundation of hiPSC transgenesis and validation.

3.2.1 Preparation of Laminin-Coated 6-Well Tissue Culture Plate

1. Select tissue culture plates of appropriate well size. Typically 35-mm or 6-well dishes are appropriate for maintenance ($1–2 \times 10^6$ cell yield); whereas, 60-mm dishes yield $3–5 \times 10^6$ cells, suitable for electroporation.
2. Add 9.6 μL of Laminin solution (0.5 mg/mL) into 1.5 mL DPBS for one well of a 6-well tissue culture plate (0.5 μg/cm^2). Immediately agitate the plate in order to cover the entire plastic surface (*see* **Note 12**).
3. Incubate the plate at 37 °C for 1 h (or overnight, use within 24 h).
4. Before use, aspirate the Laminin solution from the plate and replace immediately with Passage Medium.

3.2.2 Harvesting Ff-hiPSCs

1. Aspirate the culture medium and wash once with 2 mL DPBS. Aspirate.
2. Incubate the cells with 300 μL Accumax® for 5–10 min at 37 °C. Check the cells periodically for disruption of cell–cell adhesions (Fig. 3).
3. With a 1 mL micropipette, detach the cells from the plate by washing the Accumax® several times across the entire surface of the well (*see* **Note 13**).
4. Add 700 μL Passage Medium into the well and pipette up and down 6–12 times to generate a single cell suspension. Observe under phase-contrast microscopy.
5. Transfer the single cell suspension into a 1.5 mL microfuge tube.
6. Spin down the cells at 120 g for 5 min.

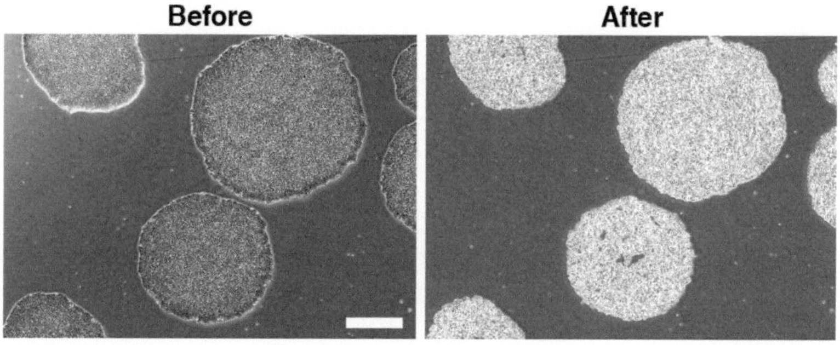

Fig. 3 Morphology of Ff-hiPSCs before (*left*) and after (*right*) Accutase treatment. Scale bar, 500 μm

7. Discard the medium and resuspend the pellet in 1 mL fresh Passage Medium.
8. Determine the viable cell number:
   (a) Add 11 μL cell suspension into a 1.5 mL microcentrifuge tube containing 11 μL 0.4 % Trypan blue and mix by gently pipetting up and down several times.
   (b) Apply 10 μL of cell suspension to each side of the counting slide.
   (c) Count cell numbers with the TC10 Automated Cell Counter (*see* **Note 14**).
9. Seed the cells on a Laminin-coated plate in Passage Medium at 1–4 × 10^3 cells per cm^2 (*see* **Note 15**).
10. The next day, observe cell growth. If colonies are composed of >4–8 cells, change Passage Media with fresh Culture Media. Otherwise, allow the cells to grow for an additional 24 h (*see* **Note 16**).
11. Feed with Culture Media at least every second day until the culture reaches ~50–80 % confluency. Do not allow the cells to overgrow.

3.2.3 Preparation of Frozen Ff-hiPSC Stocks

1. Harvest cells as described in Subheading 3.2.2, and count viable cells (**Step 8**).
2. Resuspend cells at a density of 1 × 10^6 viable cells per 1 mL STEM-CELLBANKER®.
3. Transfer 200–500 μL of cell suspension (2–5 × 10^5 hiPSC) to a Nalgene Cryogenic Tube.
4. Slow-freeze in a Nalgene™ Cryo 1 °C Freezing Container at −80 °C.
5. After 24 h, transfer the cells to liquid nitrogen for long-term storage.

3.3 Establishing of PB-Transgenic hiPSCs

Delivery of PB transposons to hiPSCs can be achieved by any high-efficiency DNA transfection method, amongst which we find electroporation (EP) preferable. All-in-One PB vectors provide constitutive expression of the rtTA transactivator via the rat EF1α promoter (Fig. 1). rtTA is linked to ires-Neo or ires-Puro in *cis* (Table 1), such that transgenic hiPSC clones with active PB integrations are selected. Thus, dox-induced gene expression may be achieved from a single vector integration in any genetic background (15, 16). It should be noted that, like any randomly inserted transgene, transposons might be subject to pleiotropic position effects. We encourage the use of multicopy PB transgenesis or cell populations, when permissible, to obviate the need for clonal screening. Protocols describing the isolation of either populations or clones are derived from transposition of a PB-GOI control vector expressing luciferase (PB-TAC-luc2-ERN) are provided.

3.3.1 Electroporation (EP)

1. Prepare pCAG-PBase (1 μg) and PB-TAC-luc2-ERN (1 μg) DNA (Subheading 3.1.2) (*see* **Note 17**). Bring the volume to a maximum of 10 μL with TE (pH 8.0).
2. Harvest and count cells as described in Subheading 3.2.2.
3. Transfer 1×10^6 cells (per EP) to a 15-mL conical tube and spin down cells at 120 g for 5 min. For multiple EPs, prepare the cells in bulk to avoid technical variation in cell numbers (*see* **Note 18**).
4. Discard the medium and resuspend the cell pellets at a concentration of 1×10^6 cells per 100 μL in Opti-MEM I reduced-serum medium.
5. Add 100 μL of cell suspension into a 1.5 mL microcentrifuge tube containing transposon and transposase DNA (**Step 1**) and mix by gently pipetting up and down three to five times. Minimize the incubation time before electroporation.
6. Transfer the cell/DNA mixture into an electroporation cuvette (*see* **Note 19**).
7. Load the cuvette into the NEPA21 cuvette stand. Check impedance (*see* **Note 20**).
8. Electroporate the cells using the following settings: Poring pulse; 125 v, 5 ms pulse, 50 ms gap, 2 pulse, 10 % decay (+ pulse orientation), Transfer pulse; 20 v, 50 ms pulse, 50 ms gap, 5 pulse, 40 % decay (+/− pulse orientation).
9. Gently resuspend the electroporated cells by adding 900 μL of Passage Medium directly into the cuvette (*see* **Note 21**). Transfer the cells to a 1.5 mL microcentrifuge tube.
10. Plate cells in Passage Medium at a density of ~$1–5 \times 10^4$ cells per cm^2. Plating densities should be adjusted according to the experiment, taking into account PB vector length, DNA amount, cell viability, and transfection efficiency. We suggest plating over a range of three dilutions, choosing more dense plates for population assays and picking clones from plates where colonies are well isolated.
11. Observe cell viability 24 h after electroporation. Replace the medium with Culture Medium if cell–cell adhesions have formed.

3.3.2 Drug Selection

1. 48 h after EP, replace the medium with 1.5 mL Selection Medium containing puromycin (0.5 μg/mL) or G418 (175 μg/mL) (*see* **Note 9**). Feed everyday with fresh drug-containing Selection Medium. Under feeder-free conditions, emerging drug-resistant colonies should be observed early in the process.

3.3.3 Isolation and Expansion of PB transgenic hiPSCs

1. After 8–10 days of drug selection, colonies should reach a diameter of >1,000 μm, and are ready to be harvested in bulk (population), or picked to generate clonal cell lines.
2. Population:
   (a) Harvest all drug-resistant colonies as described in Subheading 3.2.2.
   (b) Seed the cells on Laminin-coated plates at 5×10^3 cells per cm^2.
   (c) Feed everyday with fresh Selection Medium and passage as usual.
   (d) Maintain drug selective pressure for two passages to ensure non-transgenic cells are eliminated (*see* **Note 22**).
3. Clones:
   (a) Prepare a Laminin-coated 96-well tissue culture plate as described in Subheading 3.2.1 with the following changes:
   - For each well of a 96-well tissue culture plate, add 0.66 μL (2×) of Laminin solution (0.5 mg/mL) in 50 μL PBS.
   - After 1 h of incubation at 37 °C, replace the Laminin solution with Passage Medium.

   (b) Wash the 60-mm drug selection plates once with 3.5 mL DPBS and add 3.5 mL fresh Culture Medium (*see* **Note 23**).
   (c) Under a stereo microscope, choose colonies that are round and well isolated. Pick colonies from the dish with a small volume of media using a micropipette (set at 5–10 μL) and transfer them sequentially into the 96-well tissue culture plate containing Passage Medium.
   (d) Once all clones are picked, use a multichannel micropipette to gently disrupt the colonies into a few small clumps.
   (e) Incubate the plate in 37 °C, 5 % CO_2 incubator so that the colony fragments attach and expand. Feed with Culture Medium for 6–10 days (*see* **Note 24**).
   (f) Passage the plate 1:1 such that the wells reach 80–90 % confluency. Follow the procedure outlined in Subheading 3.2.2 with the following changes:
   - Use a 50 μL DPBS wash and 30 μL Accumax per well.
   - Incubate the plate until cell-cell adhesions are disrupted, but cells are not yet detached from the plate (Fig. 3).
   - Aspirate all the Accumax before mechanically detaching the cells (*see* **Note 25**).
   - Use a multichannel pipette to dissociate the cells, and split to new dishes.

(g) At this point the clones may be passaged into 12- or 24-well plates for further expansion, cryopreservation, and analysis (*see* **Note 26**).

4. Prepare frozen stocks of populations or clones at an early passage. If screening can be completed beforehand (e.g., simple dox induction and FACS analysis, Subheading 3.4.4), do so prior to extensive expansion and focus on promising clones to reduce subsequent screening efforts. If time consuming screening is required (Southern blotting or Splinkerette, Subheading 3.5), store 1–3 vials of each clone temporarily, and chose positive clones for expansion at a later time.

3.4 Validation of Human iPSCs with Inducible Transgene Expression

Fluorescent reporter versions of the All-in-One PB vectors (Table 1), provide an indirect readout of dox-inducible transgene expression, and are appropriate for applications such as live imaging or FACS sorting. Fluorescence-free versions are suitable for use with GOIs with an epitope tag or for which an appropriate antibody is available. We encourage researchers to perform a dox titration to determine the concentration that induces appropriate transgene expression levels. For the purposes of demonstration, luciferase (luc2) is used as an example GOI (Fig. 4), and three analysis methods—imaging, FACS, and luminescence—are employed.

3.4.1 Doxycycline Induction

At the concentrations used to induce expression from PB vectors (0.1–1.0 μg/mL), dox has no known cytotoxicity. PB transgenic cell cultures of any scale can be induced 24 h after seeding, by adding fresh Culture Medium containing appropriate concentrations of dox (*see* **Note 27**).

3.4.2 Fluorescence Microscopy for Reporter Gene Expression

1. Examine cells daily following dox induction. When colonies display a compact rounded morphology with defined edges, or when wells reach confluency, acquire images under fluorescent illumination (Fig. 4b).
2. After image analysis, cells can be harvested for subsequent analyses Subheadings 3.4.3 and 3.4.4).

3.4.3 Luciferase Assay as an Example of GOI Induction

1. Prepare PB-transgenic cells (96-well format).
   (a) Harvest cells as described in Subheading 3.3.
   (b) Seed 5×10^4 cells per well of a Laminin-coated 96-well plate in Passage Medium.
2. Replace Passage Medium with Culture Medium containing dox across a range of concentrations (0, 0.01, 0.05, 0.1, 0.25, 0.5, 0.75, 1.0, 1.5, and 2.0 μg/mL).
3. 24 h after dox addition, the cells should reach 80–90 % confluency.

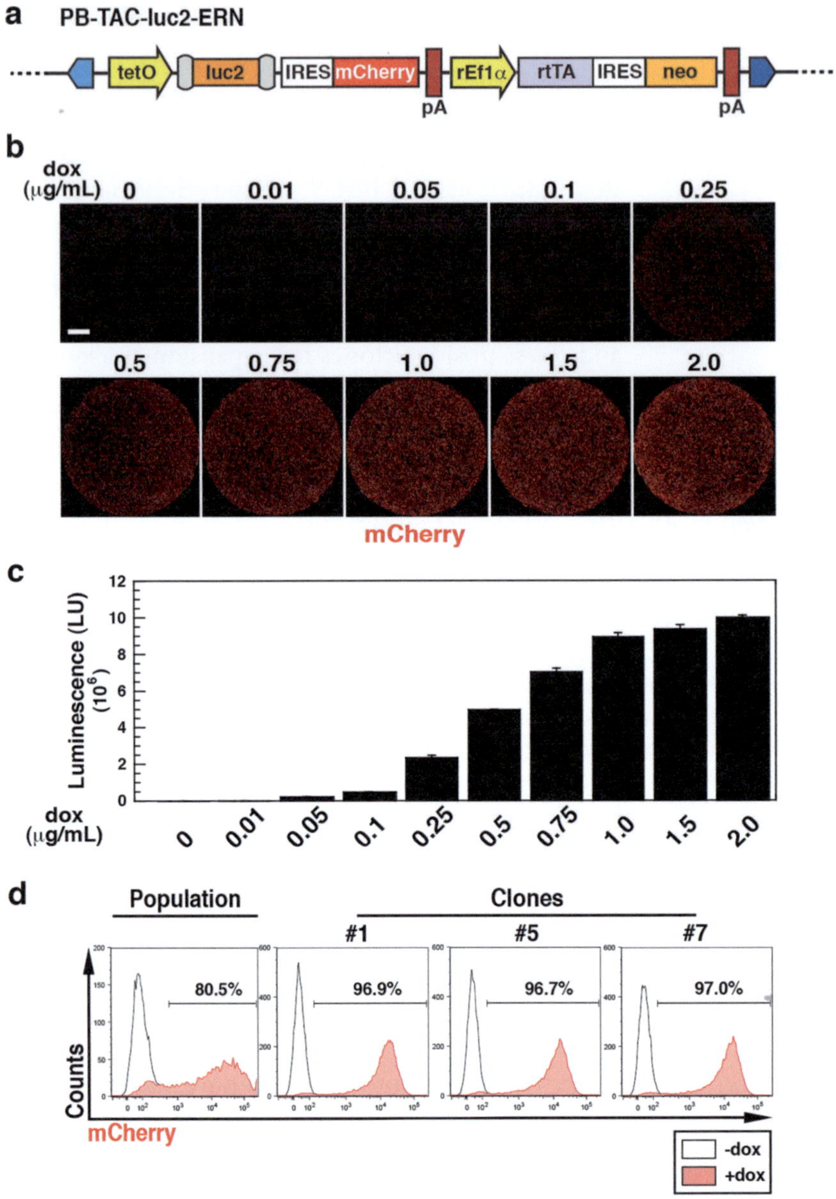

Fig. 4 Validation of inducible gene expression in Ff-hiPSCs transfected with an All-in-One PB vector. (**a**) Diagram of the PB-TAC-luc2-ERN vector used in validation. (**b**) Fluorescence microscopy images of mCherry expression and (**c**) measurement of luciferase activity in transgenic Ff-hiPSCs. Expression of the GOI and IRES-linked reporter are dependent on dox concentration. Scale bar, 1,000 μm. *luc2*, luciferase reporter gene. (**d**) FACS analysis of mCherry expression from a transgenic hiPSC population (*left*) and single clones (#1, #5, and #7) which were cultured with (*shaded*) or without (*open*) 1 μg/mL dox for 24 h

4. Examine the plate and take images using fluorescence microscopy (Subheading 3.4.2).
5. Rinse the cells with DPBS and aspirate thoroughly without allowing the cells to dry.

6. Add a volume of Dual-Glo® Luciferase Reagent equal to the culture medium volume to each well and mix. For 96-well plates, 75 μL of reagent can be added to cells grown in 75 μL of medium. Alternatively, you can replace the Culture Medium with 150 μL of a 1:1 mixture of reagent and medium.
7. Wait 10 min and measure luminescence on the microplate reader (Fig. 4c).

3.4.4 Flow Cytometry of hiPSC Populations and Clones

1. Harvest and count cells as described in Subheading 3.2.2.
2. Be certain to include a sample of parental hiPSCs as a negative control.
3. Maintain the cells on ice until the end of the FACS analysis, especially if antibody staining is to be performed.
4. Add 2×10^5 cells to a pre-chilled 1.5 mL microcentrifuge tube.
5. Spin down the cells at 120 g for 5 min at 4 °C.
6. Discard the medium and resuspend the pellet in 100 μL pre-chilled FACS buffer.
7. *Optional*: Staining for the viable pluripotent hiPSC fraction.
   (a) Incubate the cells with 5 μL Alexa Fluor® 647 Mouse anti-Human TRA-1-60 Antigen antibody on ice in the dark for 30 min (*see* **Note 28**).
   (b) Add 1 mL FACS buffer and spin down the cells at 120 g for 5 min at 4 °C. Aspirate FACS buffer and repeat washes twice to remove unbound antibody.
   (c) Resuspend the cell pellet in 95 μL FACS buffer and add 5 μL of 7-AAD staining solution.
   (d) Leave for 10 min on ice in the dark.
8. Transfer into a FACS tube and dilute the sample to 1×10^6 cells/mL (200 μL).
9. Analyze on a flow cytometer (Fig. 4d).

3.5 Optional Validation Protocols

For certain applications, it may be important to define the number and location of PB integration sites in transgenic hiPSCs.

For the validation of copy number, genomic TaqMan assays have been reported (18). However, direct analysis of PB elements by Southern blotting typically provides more clear results. For Southern blotting, we suggest using at least two separate restriction analyses to reduce the possibility of overlapping genomic fragments obscuring copy numbers. Suitable probes for the PB vectors described herein derive homology from GFP, mCherry, Neo, or rtTA. Avoid using Puro gene fragments, which have high GC-content and result in excessive background binding. Please refer to the manufacturer's online resource (https://lifescience.roche.com/dig)

and technical manual (https://lifescience.roche.com/wcsstore/RASCatalogAssetStore/Articles/05353149001_08.08.pdf) for details on DNA preparation and probe hybridization.

To determine the location of integrated PB elements, we suggest using PCR-based Splinkerette technique (7). A detailed protocol for hiPSCs is available in (13). Splinkerette combined with next-generation sequencing strategies may be helpful to deconvolve complex integration patterns resulting from the analysis of populations or clones containing multi-copy PB integrations. Finally, subsequent introduction of PB vectors in hiPSCs already containing PB transgenes should be performed with caution, as re-exposure to pCAG-PBase will stimulate transposon mobilization or loss from the genome (13).

4 Notes

1. Many cDNA clones from public repositories (such as I.M.A.G.E.) are available in Gateway-compatible expression vectors. We suggest that Entry vectors contain the GOI cDNA with a minimal 3'UTR, since PB expression vectors with IRES-linked fluorescence may function unpredictably if the GOI 3'UTR already contains a polyadenylation signal.
2. Expansion of PB-DEST vectors requires an *E. coli* host resistant to the *ccdB* gene, such as DB3.1.
3. Restriction mapping of the final Gateway product is recommended. cDNA sequence verification should be performed on the initial Entry vector, and is not required post-Gateway cloning.
4. Endotoxin-free plasmid preparation may be required for particularly sensitive applications.
5. Feeder cells must be resistant to the drugs used for selection of PB transgenes (Table 1 and Subheading 3.3.2). The SNL feeder cell line (19) or DR4 MEFs (20) are commonly used alternatives. If such feeders are not readily available, we suggest adapting your cell line to feeder-free conditions.
6. Appropriate alternative media: mTeSR™1 (STEMCELL™ TECHNOLOGIES, #05850), Essential 8™ Medium (Gibco, #A1517001).
7. Appropriate alternative surface matrices: Laminin-521 Stem Cell Matrix (Biolamina, #BL001), BD Matrigel™ hESC-qualified Matrix, 5 mL vial (BD Biosciences, #354277), Geltrex® LDEV-Free, hESC-Qualified, Reduced Growth Factor Basement Membrane Matrix (Invitrogen, #A1413301).
8. As an alternative to Accumax®, 0.5× TrypLE™ may be used. Dilute TrypLE™ Select (1x) (Gibco, #12563-011) 1:1 with 0.5 mM EDTA/PBS and store at room temperature.

9. Optimum drug concentrations should be determined empirically for each cell line. Set up a kill curve using ~1×10^4 cells per cm^2 and a drug titration as follows: Puromycin—0, 0.25, 0.5, 0.75, 1.0, 1.5 μg/mL. Observe cell death over a period of ~7 days. Cells should undergo massive death on d1-3 of selection; G418—0, 50, 100, 150, 200, 300 μg/mL. Observe cell death over a period of ~7 days. Cells should undergo massive death on d2-5 of selection.
10. It is critical to use high quality and accurately measured plasmid DNA for Gateway reactions.
11. Choose restriction enzymes that will cut both pENTR-GOI- and PB-DEST-derived elements. Sequence verification of Gateway products after Gateway Cloning is not required.
12. A Laminin-DPBS master-mix may be prepared in order to coat multiple wells. It should be used immediately, and not stored.
13. If cells do not readily detach, a cell scraper may be used. However cell viability will be decreased by harsh mechanical removal. In our experience, it is more effective to extend the Accumax® incubation time.
14. Record cell viability and review the optimal harvest procedure if survival is <80 %.
15. The plating cell density significantly affects growth and time required until the next passage. Plating densities for 5–7 days passage; 1×10^4 (7 days), 4×10^4 (5 days).
16. Single cells have low survival in the absence of ROCK inhibitor. Apoptosis can be detected by cell rounding and membrane ruffling.
17. The amount of input transposon DNA can be titrated to control integration copy numbers. Include a control sample without pCAG-PBase to determine PB-GOI random integration versus transposition.
18. Always prepare enough cells for one additional EP to account for pipetting error.
19. Avoid introducing air bubbles during **Steps 5** and **6**. Gaps between cuvette electrodes can adversely affect electrical discharge.
20. Impedance should be measured within ~0.036–0.042 Ω (0.04 Ω is optimum). Low readings can indicate the presence of excess salt in the sample. Resistance normally decreases with increased DNA concentrations.
21. Cell death from electroporation using the described conditions is expected to be ~30 %. It is normal for a whitish clump of dead cells to form after electroporation.

22. Continued selective pressure can enrich for cells that do not undergo transgene silencing. However, with continued selective pressure, even drug resistant cells may experience a slower growth rate.
23. If picking colonies from G418 Selection Medium, replace with Culture Medium 24 h before picking to improve clonal survival.
24. Drug selective pressure can be continued after growth of colony fragments is confirmed. If contamination with non-transgenic cells is suspected, maintain drug selection through two expansion passages.
25. Residual Accumax carry-over has a severe effect on subsequent attachment and growth of hiPSCs in 96-well format. Ensure that all Accumax is removed by aspiration, or gently wash the wells with DPBS or Passage Media before mechanical detachment.
26. If clones are numerous, all subsequent screening and storage steps may be performed in 96-well culture dishes.
27. Dox concentrations to induce GOI expression levels suitable for your experiments must be determined for each hiPSC clone or population empirically.
28. Control staining with appropriate isotype-matched control antibody, such as Alexa Fluor® 647 Mouse IgM, κ Isotype Control, should be included to establish thresholds for positive staining.

Acknowledgments

We like to recognize Ryoko Hirohata and Tomoko Matsumoto for their valuable technical support in developing these materials and protocols. Our appreciation to all our colleagues who continue to make use of these transposon resources, in various research applications. This work was supported by the Research Center Network for Realization of Regenerative Medicine, Program for Intractable Diseases Research Utilizing Disease-Specific iPS Cells, of the Japan Science and Technology Agency (JST), and the Strategic International Collaborative Research Program of the JST. K.W. is a Hakubi Center Special Project Researcher. S.Y. is a non-salaried scientific advisor of iPS Academia Japan. S-I.K. is a JSPS Fellowship recipient (2011–2013) and JST Researcher.

References

1. Takahashi K, Tanabe K, Ohnuki M, Narita M, Ichisaka T, Tomoda K, Yamanaka S (2007) Induction of pluripotent stem cells from adult human fibroblasts by defined factors. Cell 131(5):861–872. doi:10.1016/j.cell.2007.11.019
2. Cary LC, Goebel M, Corsaro BG, Wang HG, Rosen E, Fraser MJ (1989) Transposon

mutagenesis of baculoviruses: analysis of Trichoplusia ni transposon IFP2 insertions within the FP-locus of nuclear polyhedrosis viruses. Virology 172(1):156–169

3. Fraser MJ, Cary L, Boonvisudhi K, Wang HG (1995) Assay for movement of Lepidopteran transposon IFP2 in insect cells using a baculovirus genome as a target DNA. Virology 211 (2):397–407. doi:10.1006/viro.1995.1422
4. Wang W, Lin C, Lu D, Ning Z, Cox T, Melvin D, Wang X, Bradley A, Liu P (2008) Chromosomal transposition of piggyBac in mouse embryonic stem cells. Proc Natl Acad Sci U S A 105(27):9290–9295. doi:10.1073/pnas.0801017105
5. Ding S, Wu X, Li G, Han M, Zhuang Y, Xu T (2005) Efficient transposition of the piggyBac (PB) transposon in mammalian cells and mice. Cell 122(3):473–483. doi:10.1016/j.cell.2005.07.013
6. Liang Q, Kong J, Stalker J, Bradley A (2009) Chromosomal mobilization and reintegration of sleeping beauty and PiggyBac transposons. Genesis 47(6):404–408. doi:10.1002/dvg.20508
7. Uren AG, Mikkers H, Kool J, van der Weyden L, Lund AH, Wilson CH, Rance R, Jonkers J, van Lohuizen M, Berns A, Adams DJ (2009) A high-throughput splinkerette-PCR method for the isolation and sequencing of retroviral insertion sites. Nat Protoc 4(5):789–798. doi:10.1038/nprot.2009.64
8. Rostovskaya M, Fu J, Obst M, Baer I, Weidlich S, Wang H, Smith AJ, Anastassiadis K, Stewart AF (2012) Transposon-mediated BAC transgenesis in human ES cells. Nucleic Acids Res 40(19):e150. doi:10.1093/nar/gks643
9. Gallagher D, Norman AA, Woodard CL, Yang G, Gauthier-Fisher A, Fujitani M, Vessey JP, Cancino GI, Sachewsky N, Woltjen K, Fatt MP, Morshead CM, Kaplan DR, Miller FD (2013) Transient maternal IL-6 mediates long-lasting changes in neural stem cell pools by deregulating an endogenous self-renewal pathway. Cell Stem Cell 13(5):564–576. doi:10.1016/j.stem.2013.10.002
10. Matsuda M, Koga M, Woltjen K, Nishida E, Ebisuya M (2015) Synthetic lateral inhibition governs cell-type bifurcation with robust ratios. Nat Commun 6:6195. doi:10.1038/ncomms7195
11. Yusa K, Rad R, Takeda J, Bradley A (2009) Generation of transgene-free induced pluripotent mouse stem cells by the PiggyBac transposon. Nat Methods 6(5):363–369. doi:10.1038/nmeth.1323
12. Woltjen K, Michael IP, Mohseni P, Desai R, Mileikovsky M, Hamalainen R, Cowling R, Wang W, Liu P, Gertsenstein M, Kaji K, Sung HK, Nagy A (2009) PiggyBac transposition reprograms fibroblasts to induced pluripotent stem cells. Nature 458(7239):766–770. doi:10.1038/nature07863
13. Woltjen K, Hamalainen R, Kibschull M, Mileikovsky M, Nagy A (2011) Transgene-free production of pluripotent stem cells using PiggyBac transposons. Methods Mol Biol 767:87–103. doi:10.1007/978-1-61779-201-4_7
14. Igawa K, Kokubu C, Yusa K, Horie K, Yoshimura Y, Yamauchi K, Suemori H, Yokozeki H, Toyoda M, Kiyokawa N, Okita H, Miyagawa Y, Akutsu H, Umezawa A, Katayama I, Takeda J (2014) Removal of reprogramming transgenes improves the tissue reconstitution potential of keratinocytes generated from human induced pluripotent stem cells. Stem Cells Transl Med 3(9):992–1001. doi:10.5966/sctm.2013-0179
15. Tanaka A, Woltjen K, Miyake K, Hotta A, Ikeya M, Yamamoto T, Nishino T, Shoji E, Sehara-Fujisawa A, Manabe Y, Fujii N, Hanaoka K, Era T, Yamashita S, Isobe K, Kimura E, Sakurai H (2013) Efficient and reproducible myogenic differentiation from human iPS cells: prospects for modeling Miyoshi Myopathy in vitro. PLoS One 8(4):e61540. doi:10.1371/journal.pone.0061540
16. Hayakawa K, Ikeya M, Fukuta M, Woltjen K, Tamaki S, Takahara N, Kato T Jr, Sato S, Otsuka T, Toguchida J (2013) Identification of target genes of synovial sarcoma-associated fusion oncoprotein using human pluripotent stem cells. Biochem Biophys Res Commun 432(4):713–719. doi:10.1016/j.bbrc.2013.01.003
17. Nakagawa M, Taniguchi Y, Senda S, Takizawa N, Ichisaka T, Asano K, Morizane A, Doi D, Takahashi J, Nishizawa M, Yoshida Y, Toyoda T, Osafune K, Sekiguchi K, Yamanaka S (2014) A novel efficient feeder-free culture system for the derivation of human induced pluripotent stem cells. Sci Rep 4:3594. doi:10.1038/srep03594
18. Kolacsek O, Krizsik V, Schamberger A, Erdei Z, Apati A, Varady G, Mates L, Izsvak Z, Ivics Z, Sarkadi B, Orban TI (2011) Reliable transgene-independent method for determining Sleeping Beauty transposon copy numbers. Mobile DNA 2(1):5. doi:10.1186/1759-8753-2-5
19. Tsubooka N, Ichisaka T, Okita K, Takahashi K, Nakagawa M, Yamanaka S (2009) Roles of Sall4 in the generation of pluripotent stem cells from blastocysts and fibroblasts. Genes Cells 14(6):683–694. doi:10.1111/j.1365-2443.2009.01301.x
20. Tucker KL, Wang Y, Dausman J, Jaenisch R (1997) A transgenic mouse strain expressing four drug-selectable marker genes. Nucleic Acids Res 25(18):3745–3746

Methods in Molecular Biology (2016) 1357: 133–148
DOI 10.1007/7651_2015_200

Published online: 17 February 2015

Generation and Characterization of Rat iPSCs

Jing Liao* and Chun Cui*

Abstract

The laboratory rat (*Rattus norvegicus*) is now on the leading edge used as a laboratory model system to study pharmacology, toxicology, immunology, nutrition, behavior, and numerous other topics. Therefore, generation of rat induced pluripotent stem cells (iPSCs) through somatic cells reprogramming is a powerful tool for establishing in vitro disease model, development of new protocols for treatment of different diseases, and creating transgenic rat models. Here, we describe a simple adopted protocol for establishing rat iPSCs from different types of somatic cells including rat primary ear fibroblast (PEF) and primary bone marrow cells (BMC).

Keywords: Rat, iPSCs, Reprogramming, Somatic cells, Disease model

1 Introduction

The laboratory rat (*Rattus norvegicus*) was the first domesticated mammalian species for scientific research. It has been extensively used in studies including physiology, pharmacology, toxicology, nutrition, immunology, and behavior for over 150 years (1). Compared to mouse and other organisms, rat has many advantages including their reasonable bigger size, intelligence, similar physiology to human, hormone responsive manner, and so on (2).

Rat ESCs were established in 2008 (3–5) that quickly expanded their utility in generating transgenic rat models. However, there are still a lot of pathophysiological and metabolic characteristics that remain to be clarified in established rat disease models. Although, in 2006, Yamanaka et al. have already reported the generation of induced pluripotent stem cells (iPSCs) from mouse somatic cells by transduction of four transcription factors (Oct3/4, Sox2, Klf4, and Myc) (6), this technology hadn't been widely applied to other species other than mouse and human in the next few years. Therefore, iPSCs derivation from rat somatic cells became an urgent technique to develop. And it offers great opportunities to promote

*Author contributed equally with all other contributors.

disease investigations in vitro and search for new protocols for diseases treatment (7).

In 2009, Dr. Lei Xiao's group first reported the establishment of rat iPSCs from somatic cells (8). Almost at the same time, another group also reported the successful generation of rat iPSCs (9). Afterwards, a lot of efforts have been made on the efficiency improvement (10), developing non-viral-based method (11) and the generation of germline-competent rat iPSCs (12). It is also proved that rat iPSCs can be used to generate transgenic rat (13).

Here, we describe in detail the protocol to generate and characterize rat iPSCs from rat primary ear fibroblast (PEF) and primary bone marrow cells (BMC) (8). This protocol is cost-effective, easy to adopt, and very reproducible. It will be very helpful to quickly generate the rat iPSCs in order to set up disease models in vitro.

2 Materials

Rat primary bone marrow medium, specifically consists of: 90 % of α-MEM culture medium (Invitrogen); 10 % fetal bovine serum (HyClone).

Rat primary ear fibroblasts medium, specifically consists of: 69 % D-MEM culture medium (Invitrogen); 30 % fetal calf serum (HyClone); 1 mM L-glutamic acid (Invitrogen); 1 % nonessential amino acids (Invitrogen).

Rat primary passaged fibroblasts medium, specifically consists of: 89 % D-MEM culture medium (Invitrogen); 10 % fetal calf serum (HyClone); 1 mM L-glutamic acid (Invitrogen); 1 % nonessential amino acids (Invitrogen).

Rat iPS cell culture medium, specifically consists of: 79 % Knockout D-MEM/F12 medium (Invitrogen); 10 % fetal calf serum (HyClone); 10 % of the Knockout SR (Invitrogen); 1 mM L-glutamic acid (Invitrogen); 1 % nonessential amino acids (Invitrogen); 0.1 mM β-mercaptoethanol (Sigma).

Lentiviral vector LV-ef1α-IRES-EGFP used in the following description were purchased from Invitrogen, the structure shown in Fig. 1, with ampicillin resistance. The following common cell culture products were purchased from Invitrogen.

3 Methods

3.1 Construction of Lentiviral Vector

3.1.1 Reprogramming Factors

From the NCBI web site (http://www.ncbi.nlm.nih.gov/), query these specific genes (Oct4, Sox2, c-Myc, Klf4, Lin28, Nanog). According to the coding region sequences, primers were designed and the restriction sites were introduced. Primer sequences are shown in Table 1 (where F denotes the forward primer, R is the reverse primer).

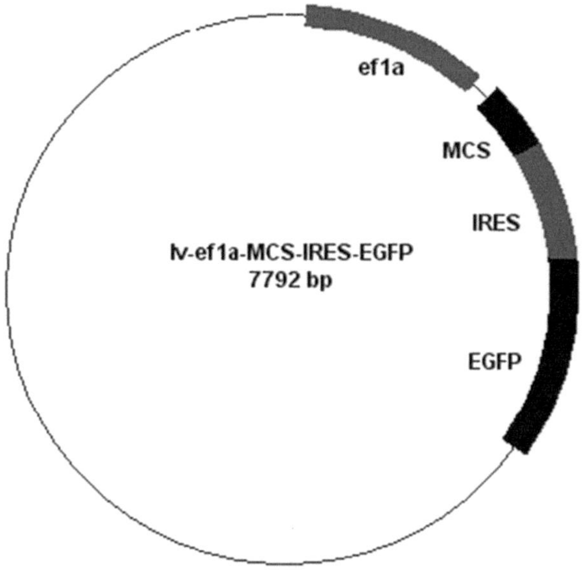

Fig. 1 Viral vector of LV-ef1α-IRES-EGFP

Table 1
PCR primers for amplification of the target genes

Gene name	GenBank ID	Primer name	Primer sequence	Restriction site
Lin28	NP_078950	1F 1R	ggggGGATCCgccaccatgggctccgtgtccaaccag gcgcGCTAGCtcaattctgtgcctccgggag	BamHI NheI
c-Myc	NP_002458	2F 2R	ggggGGATCCgccaccatgcccctcaacgttagcttca gcgcGCTAGCttacgcacaagagttccgtagc	BamHI NheI
K1f4	NP_004226	3F 3R	ggggGGATCCgccaccatgaggcagccacctggcgag gcgcGCTAGCttaaaaatgcctcttcatgtg	BamHI NheI
Nanog	NP_079141	4F 4R	ggggAGATCTgccaccatgagtgtggatccagcttgtc gcgcGCTAGCtcacacgtcttcaggttgcatg	BglII NheI
Pou5F1 (Oct4)	NM_002701	5F 5R	ggggGGATCCgccaccatggcgggacacctggcttc ggggGGATCCgccaccatggcgggacacctggcttc	BamHI NheI
Sox2	NP_003097	6F 6R	ggggGGATCCgccaccatgtacaacatgatggagacgg gcgcGCTAGCtcacatgtgtgagaggggcag	BamHI NheI

3.1.2 PCR Amplification

Total human cDNA was used as a template and the primers for each gene were listed in Table 1. PCR amplifications were carried out as follows:

Reaction system (25 μl): 10× pfx Mix 2.5 μl, AccuPrime pfx enzyme 0.2 μl, each primer (50 μM) 0.25 μl, template 0.25 μl, ddH_2O 21.55 μl.

Reaction conditions: 95 °C 2 min; 95 °C 20 s, 66 °C 20 s, 68 °C 30 s, 35 cycles; 68 °C 10 min.

3.1.3 Construction of Lentiviral Vector

PCR products were inserted into lentiviral vector LV-ef1α-IRES-EGFP. The ligation product was transformed in GBE180 competent bacteria. Overall, there are six lentiviral vectors including LV-ef1α-Oct4-IRES-EGFP, LV-ef1α-Sox2-IRES-EGFP, LV-ef1α-c-Myc-IRES-EGFP, LV-ef1α-Klf4-IRES-EGFP, LV-ef1α-Lin28-IRES-EGFP, and LV-ef1α-Nanog-IRES-EGFP.

3.2 Cell Culture

3.2.1 Primary Rat Bone Marrow Cells (BMC) Culture

1. Take the femur and tibia from 6-week-old SD rat strain. Immerse them into 75 % ethanol for 10 min. Wash with PBS containing penicillin and streptomycin three times.
2. Cut the end of the bone. Flush everything out from the bone into 15 ml sterile centrifuge tube with 10 % fetal calf serum in α-MEM medium. Centrifuge at 300 × *g* for 5 min.
3. Wash cells once with culture medium. Centrifuge at 300 × *g* for 5 min. Discard the supernatant.
4. Cells were resuspended and plated into T75 culture flask. Three days later, the medium was changed. Tiny clones can be observed.
5. Cells were passaged a week after plating by the ratio of 1:3. Cells were first washed twice with PBS and passaged using 0.25 % trypsin. 5 min after treatment of trypsin, primary rat bone marrow cells (BMS) can be trypsinized into single cells. The first four generations can be maintained in very good condition.

3.2.2 Primary Rat Fibroblasts (PEF) Culture

1. Take the rat ears. Shave carefully after cleaning with 75 % alcohol (*see* **Note 1**). Soak in PBS (containing penicillin and streptomycin) for 15 min.
2. Clean the ears several times with PBS or serum-free medium (D-MEM). Then the ears were soaked in a small amount of D-MEM containing 30 % FBS.
3. Cut the ears into small pieces with sterile scissors (*see* **Note 2**).
4. Move the pieces of ears to culture dish. Leave a small distance between each piece. Keep the dish inverted to make sure the tissue will attach well to the bottom. After 6–8 h, invert the dish back and add D-MEM containing 30 % FBS. Medium was changed daily.
5. 3–4 days afterwards, fibroblast cells can be observed.
6. Cells were passaged a week after plating. Wash the cells with PBS twice. Incubate at 37 °C with 0.25 % trypsin for 5 min. Terminate the reaction with 10 % FBS in D-MEM. Passage the cells by a ratio of 1:1–1:2 at the first passage.
7. In the later passages, cells can be passaged every 3–4 days by a ratio of 1:3–1:4. Rat primary fibroblasts (PEF) can be maintained in good condition for more than 10 passages.

3.3 Virus Infection

1. Infect 5 × 10^4 PEF or BMC with the MOI 50 for each virus (*see* **Note 3**).
2. 48 h after infection, trypsinize the infected cells with 0.25 % Trypsin. Replate the single cells into one plate of 6-well plate precoated with feeder cells.
3. Change the medium (Rat iPS cell culture medium) every other day until the formation of the clones.

3.4 Screening for Positive Cells After Reprogramming

3.4.1 Detection of the Infected Cell with Fluorescence and Alkaline Phosphatase (AP) Staining

1. 48 h after infection, green fluorescence can be observed under fluorescence microscope.
2. 2–3 days after replating, the shape of cells changed. There was obvious cell aggregation.
3. 4–6 days after replating, there are two major types of clone morphology (Fig. 2). One resembles the morphology of mouse embryonic stem cells. Clones are with condensed cells and have smooth surface and bright edges (Fig. 2a, b from infected PEFs; Fig. 2c, d from infected BMCs). Another type of clone morphology is that cells are relatively loose and clones have vague boundaries (Fig. 2e, f). These are non-ES-like clones.
4. 12 days after infection, alkaline phosphatase (AP) staining can be carried out to determine the reprogramming efficiency. The reprogramming efficiency from PEF is about double of the one from BMC (Fig. 3).

3.4.2 Picking Positive Cells After Reprogramming

1. 12 days after infection, clones were selected randomly (*see* **Note 4**).
2. Pick clones manually. Trypsinize the clones into single cells with 0.25 % trypsin at 37 °C for 5 min. Stop the reaction with rat iPS cell culture medium and replate the cells into two wells of 96-well plate (*see* **Note 5**). One well is used for maintaining while the other well is used for AP staining.
3. Medium was changed every day. Cells proliferated very fast and should be passaged every 3–4 days. The passage ratio is 1:5–1:50.
4. Rat iPS cells after picking and passaging (Fig. 4) are much more like the ES-like morphology (Fig. 2a, b).

3.5 Characterization of Rat iPS Cells

3.5.1 Alkaline Phosphatase (AP) Staining

1. Cells were washed twice with PBS and then fixed with 4 % PFA (paraformaldehyde) at room temperature for 1–2 min (*see* **Note 6**).
2. AP staining was carried out using the Chemicon Alkaline Phosphatase Detection Kit (Millipore Corporation).
3. Rat iPS cells clones have a strong alkaline phosphatase activity (Fig. 5).

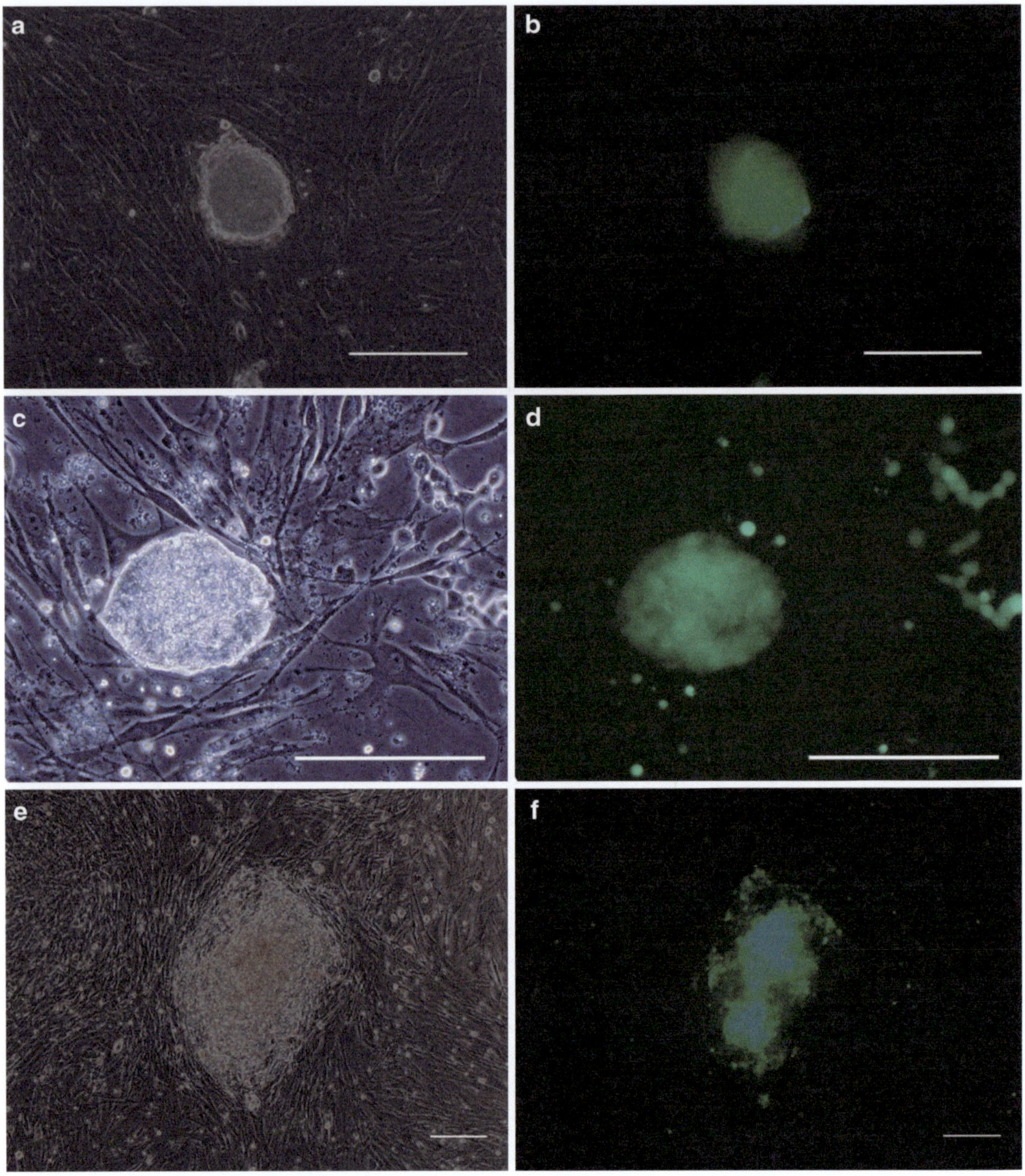

Fig. 2 Rat iPS cell morphology and fluorescence microscopy results. (**a**) Bright field of ES-like clone form from PEF. (**b**) Green fluorescence microscopy result of ES-like clones from PEF. (**c**) Clones of ES-like morphology from BMC. (**d**) Green fluorescence microscopy result of ES-like clones from BMC. (**e**) Non-ES-like clone. (**f**) Green fluorescence microscopy result of non-ES-like clones

3.5.2 Detection of Genes Highly Expressed in Stem Cells

1. The expression of Oct4, Sox2, Nanog, Nodal, Fgf4, Gal, Leftyb, Lin28, and Gabra was determined by qPCR. qPCR primers are listed in Table 2.
2. These genes are highly expressed in rat iPS cells, clone F65 and M13 (Fig. 6).

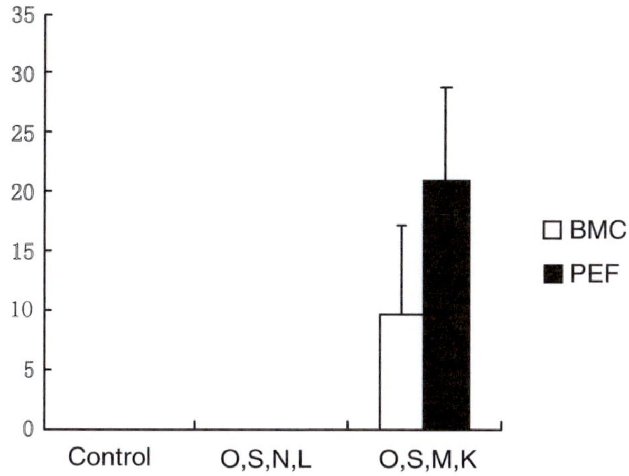

Fig. 3 iPS cell clones tested positive for alkaline phosphatase

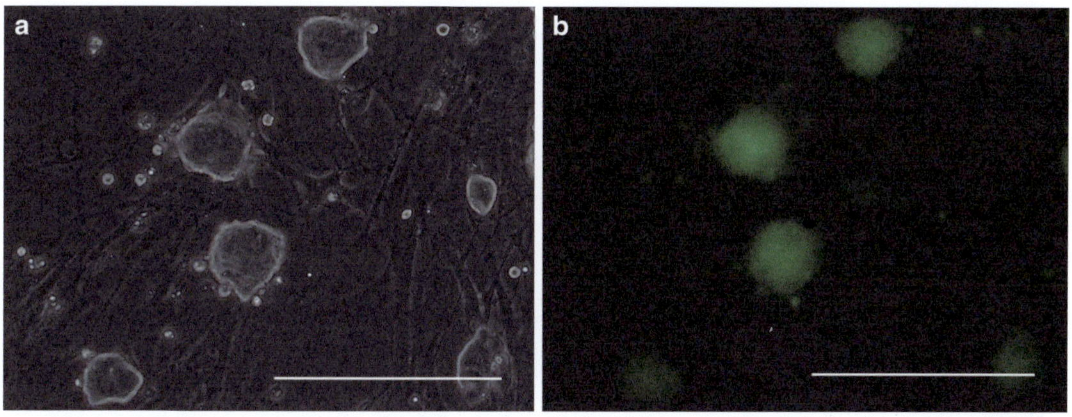

Fig. 4 Morphology of selected iPS cell clones. (**a**) Bright field. (**b**) Green fluorescence microscopy result

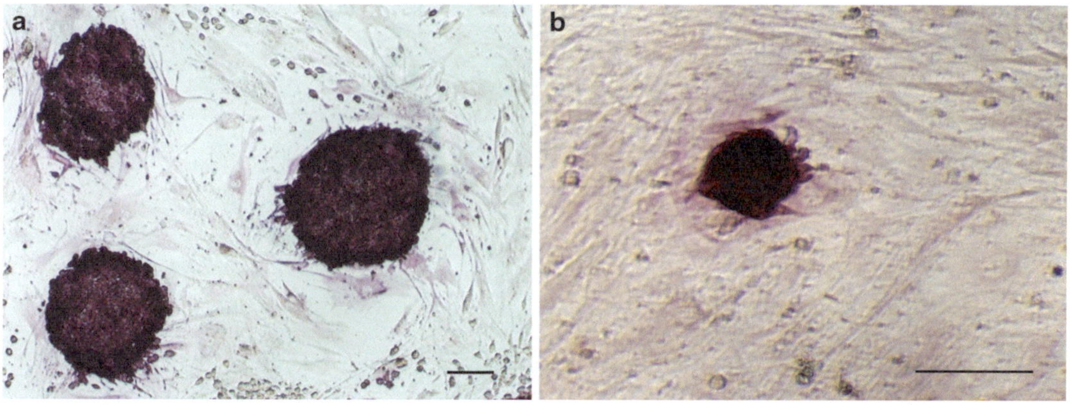

Fig. 5 Alkaline phosphatase activity of rat iPS cells. (**a**) iPS cells obtained from PEF. (**b**) iPS cells obtained from BMC

Table 2
qPCR primers for determination of the pluripotent genes

Gene name	Primer sequence
Oct4 3′UTR-f	CGAGGCCTTTCCCTCTGTTCCT
Oct4 3′UTR-r	TCTCTTTGTCTACCTCCCTTCCTTGC
Sox2 3′UTR-f	GGCCATTAACGGCACACTGCC
Sox2 3′UTR-r	TTACTCTCCTCTTTTGCACCCCTCC
Nanog-f	ACCTACCTCTTCAAGATAGCCCTG
Nanog-r	ACCTTTGCCTCTGAAACCTATCCT
Nodal-f	GAGCGTGTTTGGATGGAGAGG
Nodal-r	ATGCCAACACTTTCCTGCTTGAC
Fgf4-f	GTGTGCCTTTCTTTACCGACGAGTG
Fgf4-r	GGAAGTGGGTTACCTTCATGGTCG
Gal-f	TGGAAGTGGAGGAAGGGAGACTAGG
Gal-r	GGGATGCCAGGCAGGCTGTC
Leftyb-f	TGACCATCAGGTGGCCATTTCTG
Leftyb-r	TGTTGGGCAGGCTGACCACTTG
Lin28 3′UTR-f	CGGGAGGAGGAAGAAGAGATCCAC
Lin28 3′UTR-r	CCACTCTGCGGATTGATGCCTC
Gabrb-f	AATCAACCGGGTGGATGCTC
Gabrb-r	AATCAACCGGGTGGATGCTC

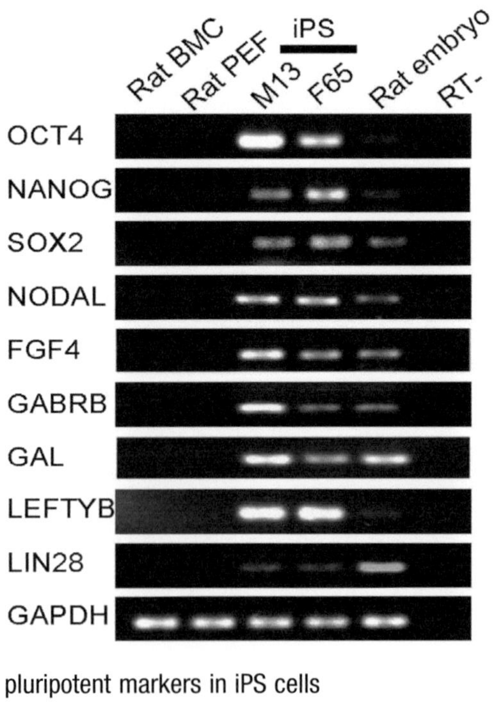

Fig. 6 qPCR of pluripotent markers in iPS cells

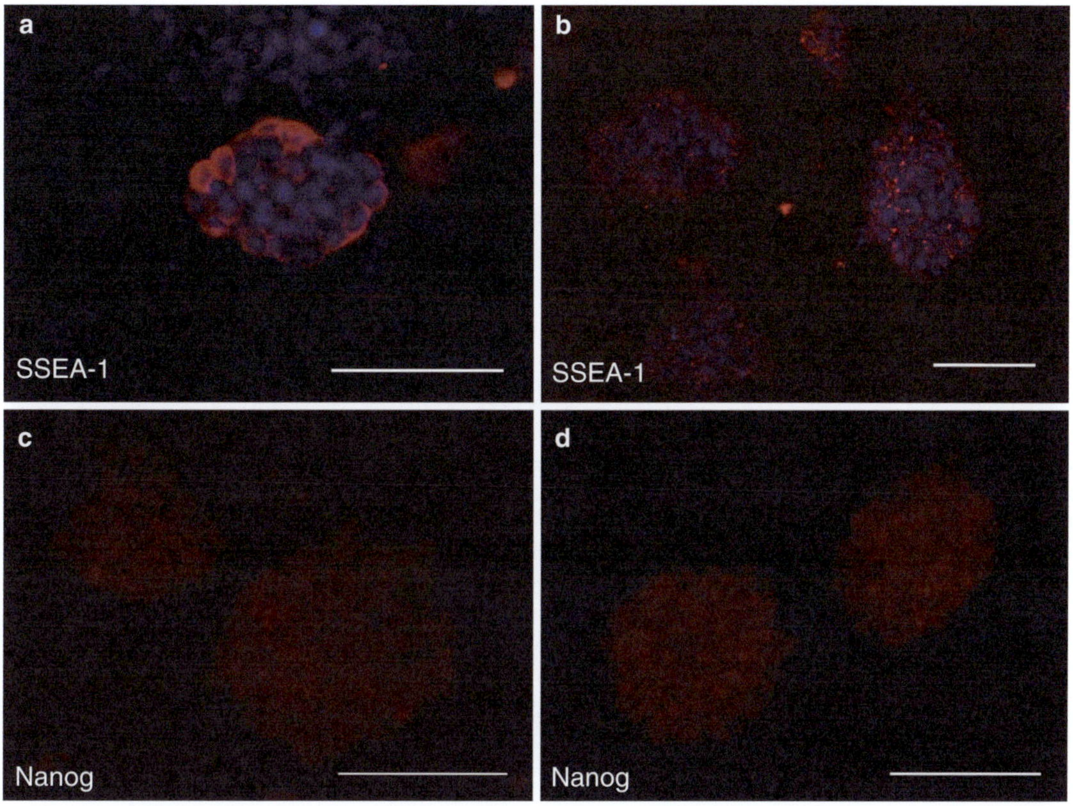

Fig. 7 Immunofluorescence staining of rat iPS cells. (**a**) SSEA-1 staining of rat iPS cells obtained from BMC. (**b**) SSEA-1 staining of rat iPS cells obtained from PEF. (**c**) Nanog staining of rat iPS cells obtained from BMC. (**d**) Nanog staining of rat iPS cells obtained from PEF. Scare bar, 100 μm

3.5.3 Immunostaining for Pluripotent Markers Nanog and SSEA-1

From Fig. 7, rat iPS cells are positive for Nanog and SSEA-1.

3.5.4 Determination of the Endogenous and Exogenous Gene Expression

1. qPCR primers are listed in Table 3.
2. From the results shown in Fig. 8, iPS clones M13 and F65 have higher exogenous Oct4 expression level than endogenous level. The exogenous Sox2 gene expression in F65 was silenced. And there is still only endogenous Sox2 expressed in F65. M13 still had exogenous Sox2 gene expression. But both F65 and M13 didn't have exogenous Nanog expression (*see* **Note** 7).

3.5.5 Stem Cell-Specific Promoter Demethylation Detection

1. Choose Oct4 promoter regions for detection.
2. Bisulfite treatment was performed using the CpGenome modification kit (Chemicon) according to the manufacturer's recommendations. PCR primers are listed in Table 4.

Table 3
qPCR primers for determination of the exogenous and endogenous gene expression

	Gene name	Primer sequence
Endogenous genes	Oct4 3′UTR-f	CGAGGCCTTTCCCTCTGTTCCT
	Oct4 3′UTR-r	TCTCTTTGTCTACCTCCCTTCCTTGC
	Sox2 3′UTR-f	GGCCATTAACGGCACACTGCC
	Sox2 3′UTR-r	TTACTCTCCTCTTTTGCACCCCTCC
	Nanog-f	ACCTACCTCTTCAAGATAGCCCTG
	Nanog-r	ACCTTTGCCTCTGAAACCTATCCT
Exogenous genes	Oct4 3′UTR-f	AGAAGGATGTGGTCCGAGTGTG
	Oct4 3′UTR-r	CAGAGTGGTGACAGAGACAGGG
	Sox2-f	TCTTGGCTCCATGGGTTCGG
	Sox2-r	AGTGCTGGGACATGTGAAGTCTG
	Nanog-f	GCTGAGATGCCTCACACGGA
	Nanog-r	GGTCTTCACCTGTTTGTAGCTGAG

3. Amplified products were cloned into T-vector, and at least ten randomly selected clones were sequenced. The results can be seen from Fig. 9. Oct4 promoter region of BMC and PEF cells was hypermethylated while got demethylated after reprogramming.

3.5.6 Determination of the Telomerase Activity

1. The telomerase activity of the iPS cells and ES cells was determined with the TRAPEZE telomerase detection kit (Chemicon) according to the manufacturer's recommendations.
2. The lysates were heated at 85 °C for 10 min and used as negative controls. Reactions were separated with nondenaturing TBE-based 12 % polyacrylamide gel electrophoresis and visualized with SYBR Gold staining.
3. Figure 10 shows that iPS cells from M13 and F65 have higher telomerase activity compared to PEF and BMC cells. Cells after heat inactivation had no detectable telomerase activity, indicating that the specificity of RNA-dependent telomerase activity.

3.5.7 Determination of the In Vitro Differentiation Potential by Embryonic Body (EB) Formation

1. Differentiate the cells in vitro by EB formation in Knockout DMEM supplemented with 10 % ES cell qualified fetal bovine serum.
2. Determine the lineage-specific gene expression of three germ layers by qPCR. qPCR primers are listed in Table 5.
3. From the result in Fig. 11, EBs from rat iPS cells expressed rat ectoderm genes Ncam and Pax6, mesoderm genes Sm22-a and Myod, and endoderm genes Sox17 and Afp. This proved that rat iPS cells can differentiate into all three germ layers in vitro.

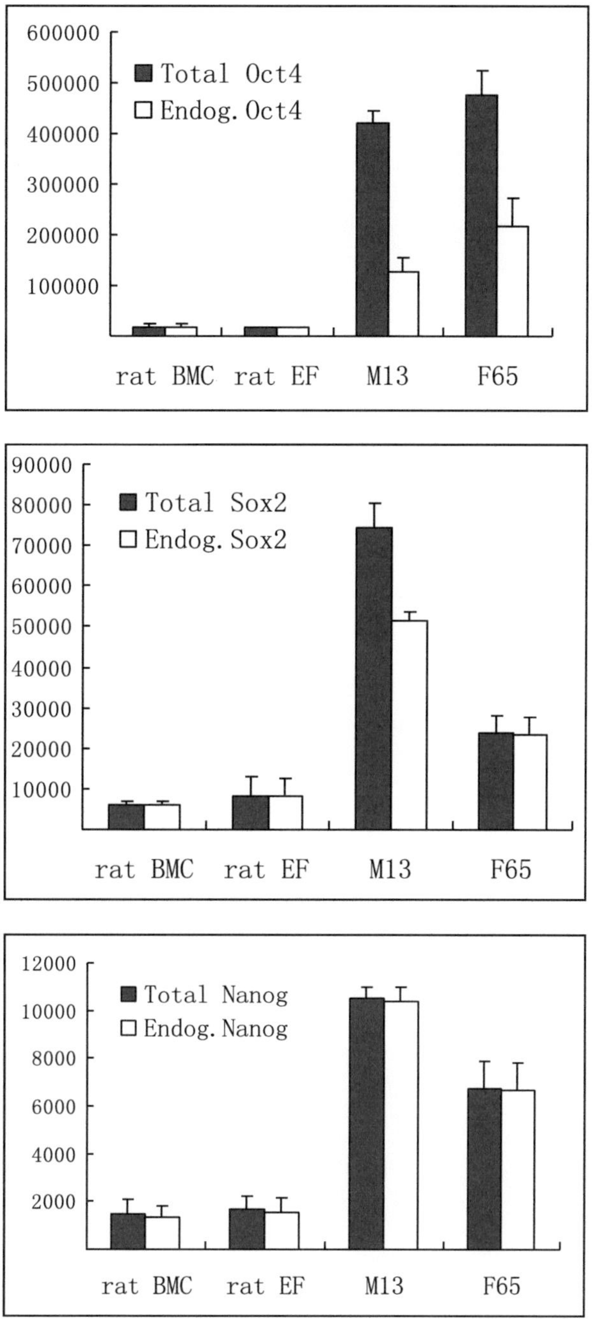

Fig. 8 qPCR detection of exogenous and endogenous gene expression in rat iPS cells. From *top* to *bottom* are: Oct4, Nanog, Sox2

Table 4
PCR primers for bisulfite genomic sequencing

Name	Sequence	Region (ATG + 1)
rOCT4-OF-1	TTATAGTGAATGTATAGATATTGGGAG	−1,794 bp−−1,202 bp
rOCT4-OR-1	CCTCCAAAATCCCATTACTAAC	−1,794 bp−−1,202 bp
rOCT4-OF-1	TTATAGTGAATGTATAGATATTGGGAG	−1,794 bp−−1,202 bp
rOCT4-IR-1	ACTAACCCAATACTTAATTTATCTTTAC	−1,794 bp−−1,202 bp
rOCT4-OF-2	TTTAAGATTTAGGAGGTAAGAAGTCG	−2,357 bp−−1,932 bp
rOCT4-OR-2	AAAACTAAAACCACCCAATTC	−2,357 bp−−1,932 bp
rOCT4-IF-2	TTCGTGGTTATCGGGTTAATATTAG	−2,357 bp−−1,932 bp
rOCT4-IR-2	CTACCCCCAAAACAAAACTATC-	−2,357 bp−−1,932 bp

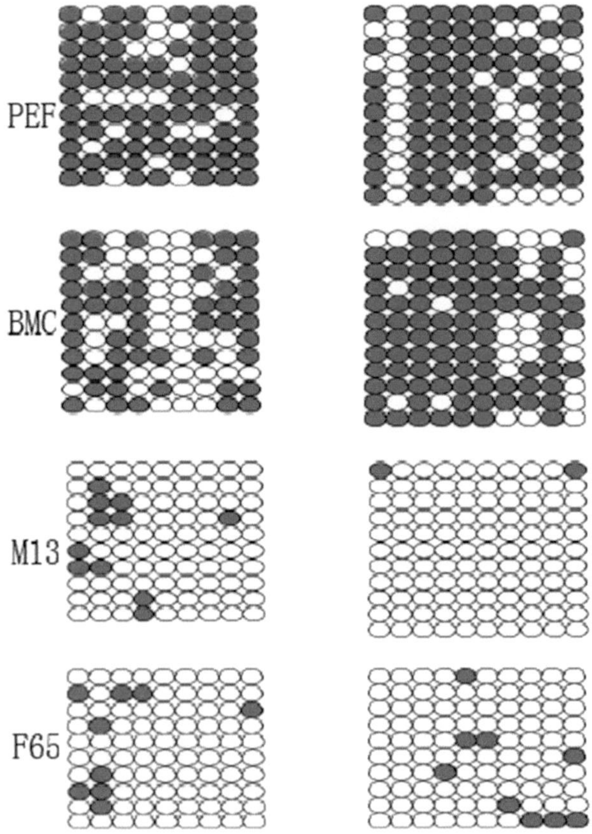

Fig. 9 Bisulfite genomic sequencing analysis of iPS cells. From *top* to *bottom* are: rat fibroblasts, rat bone marrow cells, clone M13 iPS cells, and clone F65 iPS cells

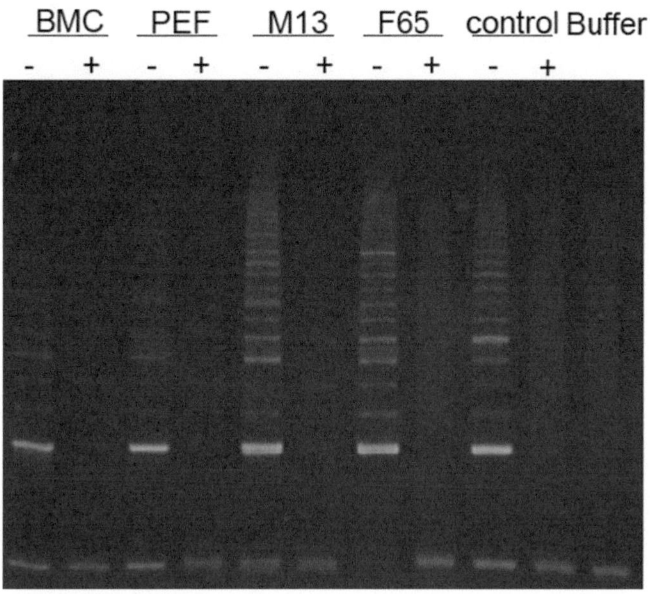

Fig. 10 Telomerase activity of rat iPS cells. From *left* to *right* lanes are: rat bone marrow cells, rat fibroblasts, clone M13 iPS cells, clone F65 iPS cells, the positive control cells, heat-inactivated control cell lysates. "−" indicates no heat inactivation, "+" indicates thermal inactivation

Table 5
qPCR primers for determination of the lineage-specific genes in EBs

Name	Primer sequence
Sox17-f	GGCACGGAACCCAACCAGC
Sox17-r	CAGTCGTGTCCCTGGTAGGGAAGAC
AFP-f	TCTGAAACGCCATCGAAATGCC
AFP-r	AATGTAAATGTCGGCCAGTCCCT
NCAM-f	TGCTCAAGTCCCTAGACTGGAACG
NCAM-r	CTTCTCGGGCTCTGTCAGTGGTGTGG
PAX6-f	TGCTCAAGTCCCTAGACTGGAACG
PAX6-r	GGACGGGAACTGACACTCCAGG
SM22-a-f	GCTGAAGAATGGCGTGATTCTGAG
SM22-a-r	CCTTCAAAGAGGTCAACAGTCTGG
Myod-f	GCTCAGGAAGATTGCTGTGTCCA
Myod-r	CCGCAACACTCCATGCATATCTCC

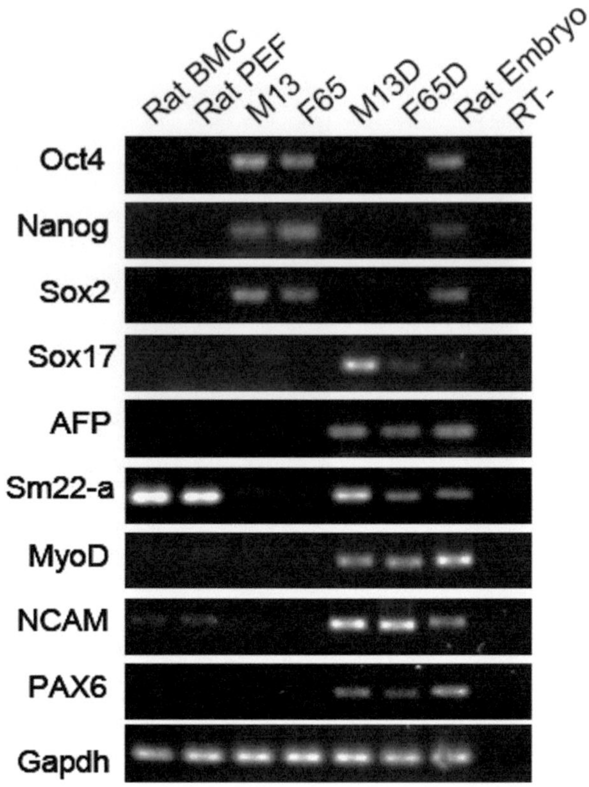

Fig. 11 qPCR analysis of EBs from rat iPS cells. From *left* to *right* lanes are: rat bone marrow cells, rat fibroblasts, clone M13 iPS cells, clone F65 iPS cells, EBs of M13 iPS cells, EBs of F65 iPS cells, rat embryo cells, negative control

3.5.8 Determination of the In Vivo Differentiation Potential by Teratoma Formation

1. Cells were injected intramuscularly into nonobese diabetic/severe combined immune deficient (NOD/SCID) mice ($\sim 5 \times 10^6$ cells per site) (*see* **Note 8**). After 4–6 weeks, tumors were processed for hematoxylin-eosin staining. All animal experiments were conducted in accordance with the Guide for the Care and Use of Animals for Research Purposes.
2. Result from Fig. 12 shows that iPS cells can form three layers of cells in vivo, including ectoderm rosette-like junction neuroepithelial cells, cartilage mesoderm, and intestinal-like endoderm epithelial cells.

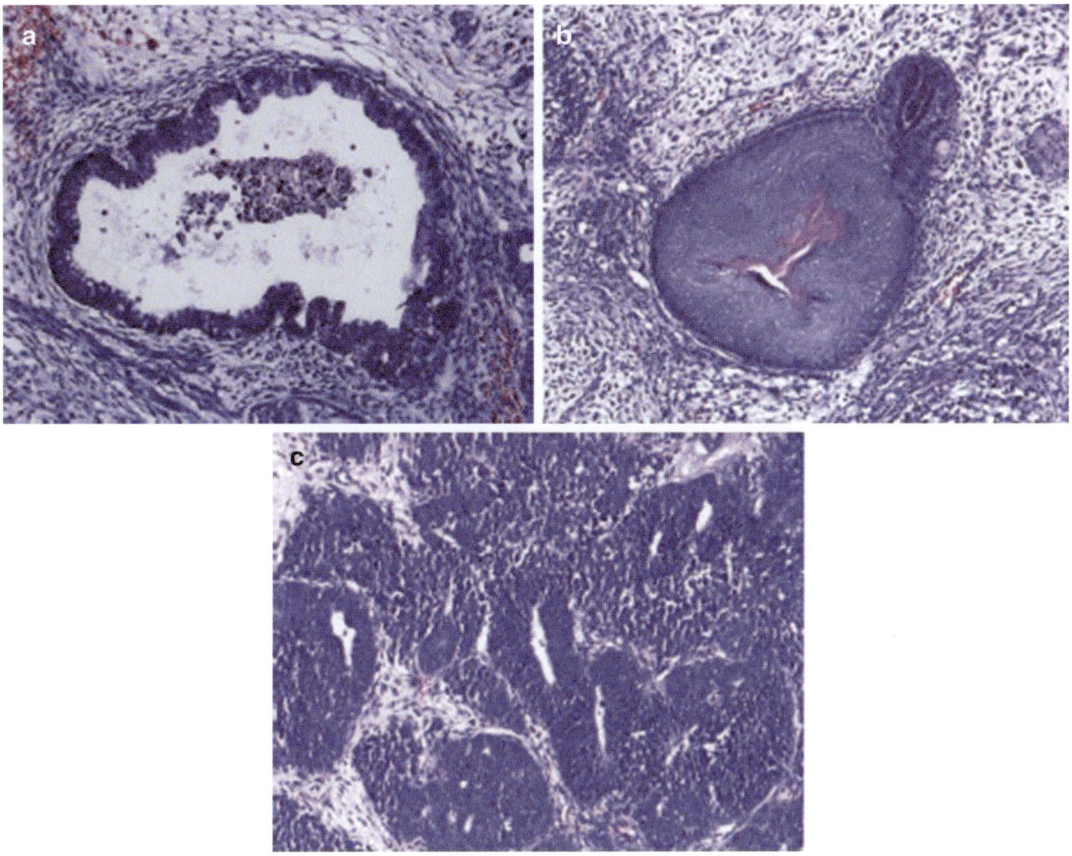

Fig. 12 Teratoma formation of iPS. (**a**) neural epithelium (ectoderm). (**b**) cartilage (mesoderm). (**c**) intestinal-like epithelium (endoderm)

4 Notes

1. Make sure that the ears are shaved carefully and very clean. If not, it is easy to get contaminated.
2. Cut the ears into small pieces. The smaller the pieces are, the easier for the fibroblast to grow out.
3. The reprogramming efficiency using the combination of O, S, M, K (Oct4, Sox2, Myc, Klf4) is much efficient than using O, S, N, L (Oct4, Sox2, Nanog, Lin28).
4. Pick the clones that resemble the ES-like morphology.
5. It is not necessary to centrifuge the cells. Cells can be replated into two wells directly.
6. Control the fixation within 1–2 min. The AP signal will decrease if cells were fixed for too long.

7. The expression level differs from clone to clone. M13 and F65 are the two examples of reprogrammed cells.
8. Cells were injected as single cells instead of clumps.

Acknowledgments

This research was finished in Dr. Lei Xiao's lab and was partially supported by the National Key Basic Research and Development Program of China Grants (2007CB947902, 2007CB948003, 2009CB940900, and 2009CB941100) and the National Natural Science Foundation of China (30600306).

References

1. Jacob HJ (1999) Functional genomics and rat models. Genome Res 9:1013–1016
2. Iannaccone PM, Jacob HJ (2009) Rats! Dis Model Mech 2:206–210
3. Buehr M, Meek S, Blair K, Yang J, Ure J, Silva J, McLay R, Hall J, Ying QL, Smith A (2008) Capture of authentic embryonic stem cells from rat blastocysts. Cell 135:1287–1298
4. Li P, Tong C, Mehrian-Shai R, Jia L, Wu N, Yan Y, Maxson RE, Schulze EN, Song H, Hsieh CL, Pera MF, Ying QL (2008) Germline competent embryonic stem cells derived from rat blastocysts. Cell 135:1299–1310
5. Ueda S, Kawamata M, Teratani T, Shimizu T, Tamai Y, Ogawa H, Hayashi K, Tsuda H, Ochiya T (2008) Establishment of rat embryonic stem cells and making of chimera rats. PLoS One 3:e2800
6. Takahashi K, Yamanaka S (2006) Induction of pluripotent stem cells from mouse embryonic and adult fibroblast cultures by defined factors. Cell 126:663–676
7. Takenaka-Ninagawa N, Kawabata Y, Watanabe S, Nagata K, Torihashi S (2014) Generation of rat-induced pluripotent stem cells from a new model of metabolic syndrome. PLoS One 9: e104462
8. Liao J, Cui C, Chen S, Ren J, Chen J, Gao Y, Li H, Jia N, Cheng L, Xiao H, Xiao L (2009) Generation of induced pluripotent stem cell lines from adult rat cells. Cell Stem Cell 4:11–15
9. Li W, Wei W, Zhu S, Zhu J, Shi Y, Lin T, Hao E, Hayek A, Deng H, Ding S (2009) Generation of rat and human induced pluripotent stem cells by combining genetic reprogramming and chemical inhibitors. Cell Stem Cell 4:16–19
10. Liskovykh MA, Chuikin IA, Ranjan A, Safina DA, Tolkunova EN, Minina Iu M, Zhdanova NS, Dyban PA, Mullins J, Kostyleva EI, Chikhirzhina EV, Bader M, Alenina N, Tomilin AN (2011) Generation of rat induced pluripotent stem cells: the analysis of reprogramming and culturing media. Tsitologiia 53:939–945
11. Merkl C, Saalfrank A, Riesen N, Kuhn R, Pertek A, Eser S, Hardt MS, Kind A, Saur D, Wurst W, Iglesias A, Schnieke A (2013) Efficient generation of rat induced pluripotent stem cells using a non-viral inducible vector. PLoS One 8:e55170
12. Hamanaka S, Yamaguchi T, Kobayashi T, Kato-Itoh M, Yamazaki S, Sato H, Umino A, Wakiyama Y, Arai M, Sanbo M, Hirabayashi M, Nakauchi H (2011) Generation of germline-competent rat induced pluripotent stem cells. PLoS One 6:e22008
13. Jiang MG, Li T, Feng C, Fu R, Yuan Y, Zhou Q, Li X, Wan H, Wang L, Li W, Xiao Y, Zhao XY, Zhou Q (2013) Generation of transgenic rats through induced pluripotent stem cells. J Biol Chem 288:27150–27158

Methods in Molecular Biology (2016) 1357: 149–172
DOI 10.1007/7651_2014_140

Published online: 21 November 2014

Generation of Induced Pluripotent Stem Cells in Rabbits

Marielle Afanassieff, Yann Tapponnier, and Pierre Savatier

Abstract

We describe a procedure for generating induced pluripotent stem cell lines in rabbits, using retroviral vectors expressing *Oct4, Sox2, Klf4,* and *c-Myc* of human origin to reprogram rabbit fibroblasts prepared from an ear skin biopsy. We also provide detailed procedures for characterizing the resulting iPSC lines, including the analysis of pluripotency marker expression by RT-qPCR, immunolabeling, and fluorescent-associated cell sorting, the evaluation of pluripotency by teratoma production and genetic stability by karyotyping.

Keywords: Induced pluripotent stem cells, Reprogramming, Rabbit, Skin fibroblasts, Retroviral vectors, Pluripotency genes, *Oct4, Sox2, Klf4, c-Myc*

1 Introduction

In 2006, Takahashi and Yamanaka demonstrated that fibroblasts could be reprogrammed into *bona fide* pluripotent stem cells by overexpressing four transcription factors, namely *Oct4, Sox2, Klf4,* and *c-Myc* (1). The resulting "induced pluripotent stem cells," or iPS cells, displayed the cardinal features of their embryonic counterpart, the embryonic stem (ES) cells. This major discovery was made in mice, and was soon applied to human with a similar outcome (2, 3). Since these pioneering studies, the iPS cell technology was implemented in several other species, including rhesus macaque (4), pig (5), and rabbits (6–8). We generated iPS cell lines from a New Zealand White breed of rabbits, making use of retroviral vectors expressing human *Oct4, Sox2, Klf4,* and *c-Myc* to reprogram rabbit ear fibroblasts. Interestingly, contrary to rabbit ES cells, the resulting iPS cell lines were capable of colonizing the rabbit pre-implantation embryo after aggregation with rabbit morulas, albeit with a low efficiency (8). This finding opens the path to the generation of germline chimeras in the rabbit. Here, we describe in detail the protocol set up for generating high quality, genetically stable iPS cells in rabbits.

2 Materials

2.1 Rabbit Fibroblast (rbF) Derivation

1. Sterilized small scissor, scalpels, and forceps (sterilize in dry-heat oven for 2 h at 180 °C).
2. Petri dishes, 15- and 50-mL Falcon tubes, 60- and 100-mm Corning cell culture plates.
3. Water bath maintained at 37 °C, low-speed centrifuge, incubator maintained at 38 °C and 5 % CO_2.
4. Sterilized PBS (1× phosphate-buffered saline without calcium and magnesium, Life Technologies, 14190-169).
5. PBS-PSG: for 100 mL, dilute 1 mL of 100× PSG (penicillin, streptomycin, glutamine, 50 mg/mL, Life Technologies, 10378-016) with 99 mL PBS.
6. Dispase II, 1×: For 10 mL, dilute 1 mL of 10× dispase II with 9 mL PBS. Prepare 10× dispase II by dissolving 1 g of dispase II powder (Life Technologies, 17105-041) in 8 mL PBS and filtering the solution through a 0.22-μM PES syringe filter (Merck-Millipore, SLGP033RS).
7. Trypsin, 2×: dilute 2 mL 10× trypsin (Trypsin-EDTA, 0.5 %, w/o phenol red, Life Technologies 15400-054) with 8 mL PBS.
8. FBS (fetal bovine serum, Hyclone, Fisher Scientific, SH30073003).
9. Gelatin, 0.1 %: for 100 mL, dilute 5 mL of 2 % gelatin (Gelatin solution Type B, Sigma-Aldrich, G1393) with 95 mL PBS.
10. Fibroblast culture medium: to 500 mL of DMEM (DMEM with high glucose, L-glutamine, phenol red, and sodium pyruvate, Life Technologies, 41966-052), add 55 mL of FBS (10 %), 5.5 mL of 100× PSG, 5.5 mL of 100× NEAA (non essential amino acids, 100×, Life Technologies, 11140-035), and 1.1 mL of 50 mM β-mercaptoethanol (Life Technologies, 31350-010).
11. DMSO: sterile-filtered dimethyl sulfoxide (Sigma-Aldrich, D2650).

2.2 Retrovirus Production

1. Eppendorf tubes, 15- and 50-mL Falcon tubes, 100-mm Corning cell culture plates, 0.8-μM Nalgene acrylic syringe filter unit (Thermo Scientific, 190-2580), 0.22-μM PES syringe filter unit (Merck-Millipore, SLGP033RS).
2. Vortex, low-speed centrifuge, incubator maintained at 38 °C and 5 % CO_2.
3. 293FT cell line (ATCC, CRL 11268).

4. pMX retroviral vectors harboring coding sequences for the human pluripotency factors OCT4, SOX2, KLF4, and cMYC: pMXs-hOCT3/4 (Addgene, 17217), pMXs-hSOX2 (Addgene, 17218), pMXs-hKLF4 (17219), and pMXs-hc-MYC (17220) (9).
5. Packaging vectors for producing amphotropic retroviral particles (obtained from Dr Cosset, INSERM U758, Lyon, France): pTG5349 (plasmid encoding gag and pol proteins) and phCMV-GP [plasmid encoding the vesicular stomatitis virus glycoprotein (VSV-G) envelope] (10, 11) (*see* **Note 1**).
6. Fibroblast culture medium (*see* Section 2.1, item 10).
7. TE Buffer: 10 mM Tris–HCl containing 1 mM disodium EDTA, pH 8.0 (Sigma-Aldrich, 93282).
8. $CaCl_2$, 2 M: dissolve 14.7 g of anhydrous $CaCl_2$ (Sigma-Aldrich, 746495) in 50 mL H_2O.
9. HBS, 10×: dissolve 5.975 g HEPES (Sigma-Aldrich, H3375) and 8.181 g NaCl (Sigma-Aldrich, S7653) in 50 mL H_2O.
10. Phosphate buffer: prepare two separate solutions. Solution 1 is 140 mM $NaH_2PO_4{\cdot}H_2O$ (Sigma-Aldrich, 71507: dissolve 0.965 g $NaH_2PO_4{\cdot}H_2O$ in 50 mL H_2O) and solution 2 is 140 mM $Na_2HPO_4{\cdot}2H_2O$ (Sigma-Aldrich, 71643: dissolve 1.246 g $Na_2HPO_4{\cdot}2H_2O$ in 50 mL H_2O). Mix solutions 1 and 2 to obtain phosphate buffer of pH 6.8.
11. NaOH, 1 N: 1 M sodium hydroxide solution (Sigma-Aldrich, 71463).
12. G418 disulfate salt solution, 50 mg/mL (Sigma-Aldrich, G8168).

2.3 Feeder Cell Preparation

1. 96-Well, 24-well, 12-well, and 6-well and 100-mm Corning cell culture plates, 15- and 50-mL Falcon tubes.
2. Low-speed centrifuge, incubator maintained at 38 °C and 5 % CO_2.
3. Fibroblast culture medium, 0.1 % gelatin, 1× trypsin, PBS (*see* Section 2.1).
4. Mitomycin-C, 5 μg/mL (Sigma-Aldrich, M4287): dissolve one vial containing 2 mg of mitomycin-C in DMEM to prepare a 10× stock solution of 50 μg/mL concentration. Store aliquots of 1 mL at −20 °C. Before use, dilute 1 aliquot containing 1 mL of 10× mitomycin-C with 9 mL of DMEM or fibroblast medium to a final concentration of 5 μg/mL (1×).
5. MEF: Mouse Embryonic Fibroblasts prepared from 12.5-day-old embryos collected from the OF1 strain (Charles River) (*see* **Note 2**).

2.4 Fibroblast Infection

1. 100-mm Corning cell culture plates, gelatin (0.1 %)-coated Corning cell culture dishes plated with 9×10^5 mitomycin-inactivated MEF.
2. Low-speed centrifuge, incubator maintained at 37 °C and 5 % CO_2, tri-gas incubator maintained at 38 °C, 5 % CO_2, and 5 % O_2.
3. Ear skin rbFs.
4. Amphotropic pMX retroviral particles expressing human *Oct4, Sox2, Klf4,* and *c-Myc.*
5. Polybrene: hexadimethrine bromide (Sigma-Aldrich, H9268), 1 mg/mL solution in 0.9 % NaCl. Store at 4 °C.
6. KOSR: knockout serum replacement (Life Technologies, 10828-028).
7. FGF2: human recombinant basic fibroblast growth factor (Merck-Millipore, 01-106). Prepare a 100 ng/μL stock solution by dissolving 1 vial (25 μg) of FGF2 in 250 μL PBS. Prepare 10-μL aliquots and store at −20 °C. Prepare a 10-ng/μL working solution by diluting the stock solution with PBS. The working solution can be stored at 4 °C for a maximum of 1 week. Immediately before use, add desired volume of the working solution of FGF2 to warm RbiPSC medium to a final concentration of 10 ng/mL.
8. Sodium pyruvate, 100 mM (Life Technologies, 11360-039). Working concentration is 1 mM.
9. PSG, NEAA, and β-mercaptoethanol (*see* Section 2.1, item 10).
10. RbiPSC culture medium: to 500 mL DMEM/F12 (DMEM/F12 with phenol red, without L-Glutamine, Life Technologies, 21331-020), add 130 mL of KOSR (20 %), 6.5 mL of 100× PSG, 6.5 mL of 100× NEAA, 6.5 mL of 100 mM sodium pyruvate, and 1.3 mL of 50 mM β-mercaptoethanol. Extemporaneously add FGF2 to a final concentration of 10 ng/mL.

2.5 RbiPSC Clone Isolation and Culture

1. Gelatin (0.1 %)-coated 96-well, 24-well, 12-well, and 6-well Corning culture plates seeded with mitomycin-inactivated MEFs, Eppendorf tubes, 15-mL Falcon tubes.
2. Low-speed centrifuge, tri-gas incubator maintained at 38 °C, 5 % CO_2, and 5 % O_2.
3. Glass capillary of diameter 1.5 mm/1.17 mm (Phymep, GC150T10).
4. Trysin (1×), PBS, RbiPSC culture medium (*see* Sections 2.1 and 2.4, items 10).
5. Methanol, 100 % (Sigma-Aldrich, 322415).
6. Alkaline phosphatase substrate: prepare a 2× solution of Fast Red TR Salt (Santa Cruz, sc-215024) by dissolving 20 mg of the substance in 10 mL of 0.1 M Tris–HCl (pH 9.2). Prepare a

2× solution of Naphthol AS phosphate disodium salt (Santa Cruz, sc-206043) by dissolving 4 g of the substance in 10 mL of 0.1 M Tris–HCl (pH 9.2). Aliquot these solutions and store at −20 °C. Extemporaneously mix equal volumes of the two solutions to obtain the alkaline phosphatase substrate. This substrate is sensitive to light and should not be stored.

2.6 Test of Transgene and Endogenous Gene Expression by RT-PCR

1. Reagents for mRNA extraction: RNeasy mini kit (Qiagen, 74106), RNase-free DNAse Set (Qiagen, 79254).
2. Reagents for reverse transcription of mRNA into cDNA: M-MLV Reverse Transcriptase (Promega, M1701), M-MLV Reverse Transcriptase buffer pack (Promega, M5313), and Random primer kit (Promega, C1181).
3. Reagents for semiquantitative PCR: Euroblue Taq DNA Polymerase (Eurobio, GAETAQ024D), 100 mM dNTPs (Eurobio, GAEPCR 115D), 50 mM $MgCl_2$ (Eurobio, BIO-37026), and rabbit and human gene-specific primers (Table 1). The primers specific for rabbit genes will amplify a fragment located along the boundary of two exons, distinguishing the amplification of cDNA from that of genomic DNA in case of DNA contamination in the RNA extracts. The primers specific of human transgenes will amplify a fragment positioned at the frontier between the 3′ end of the transgene and the transcribed sequence of the retroviral vector preceding the 3′-LTR.

Table 1
List of primers used for endogene and transgene analysis by RT-PCR

Specie	Gene	Primers	Annealing temperature (°C)	Size of fragment (pb)
Rabbit	*Gadph*	Forward: GAGCTGAACGGGAAACTCAC Reverse: CCCTGTTGCTGTAGCCAAAT	56	304
	Nanog	Forward: CCTCAGCCTTCAGCAGATGCAAGAACTC Reverse: GGCAGTGGTGTAGGCAGCCCC	58	363
	Oct4	Forward: GCAGATCAGCCACATCGCCCAGC Reverse: GCGTCTCCCCTGACCTCTGCCTC	60	619
Human	*OCT4*	Forward: CAGGGCCCCATTTTGGTACC Reverse[a]: TTATCGTCGACCACTGTGCTGGCG	60	213
	SOX2	Forward: CCTCCGGGACATGATCAGC Reverse[a]: TTATCGTCGACCACTGTGCTGGCG	60	247
	KLF4	Forward: TTCGCCCGCTCAGATGAACTG Reverse[a]: TTATCGTCGACCACTGTGCTGGCG	60	231
	C-MYC	Forward: CAGCATACATCCTGTCCGTCC Reverse[a]: TTATCGTCGACCACTGTGCTGGCG	60	221

[a]Reverse primer localized in the sequence of retroviral vector and common to all four human transgenes

The reverse primer is common to all four pMX vectors, but the forward primer is specific for each human pluripotency gene.

4. Reagents for agarose gel electrophoresis: high-melt/wide-range agarose (Dutscher, 4905009), TAE (Tris-Acetate-EDTA) buffer (Promega, V4281), ethidium bromide (10 mg/mL, Sigma, E2515), and Smart Ladder (Molecular Weight Marker, Eurogentec MW-1700-10).

2.7 Analysis of Pluripotency Markers by Immunofluorescence Microscopy and Flow Cytometry

1. Glass coverslips (Dutsher, 140540), Glass slides (Dutscher, 100258), 24-well culture plates, tubes for flow cytometry.
2. 0.1 % gelatin, FBS, PBS, 1× Trysin (*see* Section 2.1).
3. PFA: 37 % aqueous solution of paraformaldehyde (Delta Microscopy Sciences, 15714). For a 4 % working solution, dilute 5.4 mL of 37 % PFA with 44.6 mL PBS.
4. BSA: 30 % solution of bovine serum albumin in PBS (Sigma-Aldrich, A9576).
5. TBS (Tris Buffered Salt), 10×: 200 mM Tris–HCl (pH 7.4), 0.9 % NaCl (Sigma-Aldrich, T5912). Dilute tenfold with water to obtain a working solution.
6. TBS-Triton: 0.4 % Triton X-100 (Sigma-Aldrich, T8787) in TBS. For a 1× solution, dilute 0.4 mL Triton X-100 with 100 mL TBS.
7. TBS-Tween: For preparing 0.1 % Tween-20 (Sigma-Aldrich, P5927) in TBS, dilute 1 mL of Tween-20 with 1 L TBS.
8. Blocking solution: For preparing TBS containing 10 % FBS and 0.1 % BSA, dilute 10 mL of FBS with 90 mL of TBS and add 0.34 mL of 30 % BSA.
9. Hoechst 33342: dissolve 25 mg of bis-benzimide H 33342 trihydrochloride (Sigma-Aldrich, B2261) in 25 mL water to prepare aliquots of 1 mg/mL and store at −20 °C. Dilute 1 μL of the stock solution with 1 mL PBS to obtain a 1-μg/mL working solution.
10. Antibodies for OCT4, SSEA-1, SSEA-4, TRA-1-60, E-CADHERIN, N-CADHERIN, and CD90: *see* Table 2.
11. Mounting medium (Sigma-Aldrich, M1289).
12. DPX medium (CellPath, SEA-1304-00A).

2.8 Analysis of the Expression of Pluripotency Genes by RT-qPCR

1. StepOnePlus real-time PCR system (Applied Biosystems), low speed centrifuge.
2. 96-Well PCR microplates (Life Technologies, 4346906), Optical adhesive covers (Life Technologies, 4311971).
3. RNA to cDNA kit (Life Technologies, 4387406).
4. Fast SYBR Green Master MIX (Life Technologies, 4385618).

Table 2
List of antibodies

Molecule	Antibodies	Dilution	Manufacturer	Reference	Method
OCT-4	1: Anti-Oct-4 Rabbit IgG 2: AF555[a] Goat Anti-Rabbit IgG	1/300 1/1,000	Santa Cruz Invitrogen	SC-9081 A21429	IF[f]
SSEA1	PE[b] Anti-SSEA1 Mouse IgM AF647[c] Anti-SSEA1 Mouse IgM	1/50 1/100	R&D Systems Santa Cruz	FAB2155P SC-21702	IF FC[g]
SSEA4	1: Anti-SSEA4 Mouse IgG 2: AF555 Goat Anti-Mouse IgG 1: Anti-SSEA4 Mouse IgG 2: AF488[d] Goat anti-Mouse IgG	1/100 1/100 1/200 1/1,000	Millipore Invitrogen Santa Cruz Invitrogen	MAB4304 A21422 SC-21704 A11017	IF FC
Tra-1-60	1: Anti-Tra-1-60 Mouse IgM 2: AF555 Goat Anti-Mouse IgM	1/50 1/500	Millipore Invitrogen	MAB4360 A21426	IF
E-CADH	PE Anti-E-Cadherin Rat IgG2A	1/50	R&D Systems	FAB7481P	IF
N-CADH	1: Anti-*N*-Cadherin Rabbit IgG 2: Rhod[e] Goat anti-Rabbit IgG	1/200 1/1,000	Santa Cruz Molecular Probes	SC-7939 RG314	IF
CD90	1: Anti-CD90 Mouse IgG 2: AF555 Goat Anti-Mouse IgM	1/50 1/500	PharmingenTM: Invitrogen	550402 A21426	IF

Fluorochromes: [a]Alexa Fluor 555; [b]Phycoerythrin; [c]Alexa Fluor 647; [d]Alexa Fluor 488, [e]Rhodamine
Methods: [f]Immunostaining; [g]Flow cytometry

5. Primers (14 μM) specific for pluripotency genes (Table 3). Prepare a working solution of primer pairs (each 1.4 μM) by mixing 10 μL of each reverse and forward primers with 80 μL water. Primers are chosen based on the known mouse pluripotency genes (12, 13) and rabbit homologs annotated in sequence databases (*Oct4*, *Nanog*, *Klf4*, *Dazl*, *Tbx3*, *Blimp1*, *Lefty2*, *Dax1*, *Fbxo15*, *Pecam1*, *Piwil2*, *Gbx2*, *Rex1*, *Fgf4*, *Otx2*, *Clnd6*, *Pitx2*, *Cdx2*, *Cdh1*, *Cdh2*, *Esrrb*, *Tcfcp2l1*). To avoid the amplification of mouse genes originating from contaminating feeder cells, these primers must be designed to specifically amplify the rabbit genes.

2.9 Teratoma Formation

1. Eight to ten-week-old immunodeficient mouse strain, Fox Chase SCID (CB17/lcr-P*rdc*^{scid}/lcrlcoCrl, Charles River, strain 236): 2 mice per tested rbiPSC line.
2. Glass capillary, glass slides (*see* Sections 2.5, item 3 and 2.7, item 1).
3. Sterilized small scissor, scalpels, screed clips, and forceps (dry-heat oven, 2 h, 180 °C).
4. Anesthesia solution: mix ¼ volume of 2 % Rompun solution (Bayer Healthcare) with ¾ volume of Imalgene 1000 (Merial).
5. 4 % PFA, PBS (*see* Section 2.7, item 3).

Table 3
List of primers used for RT-qPCR analysis

Gene	Forward primer	Reverse primer
Tbp	CTTGGCTCCTGTGCACACCATT	ATCCCAAGCGGTTTGCTGCTGT
Gadph	TTCCACGGCACGGTCAAGGC	GGGCACCAGCATCACCCCAC
Nanog	CACTGATGCCCGTGGTGCCC	AGCGGAGAGGCGGTGTCTGT
Oct4	CCTGCTCTGGGCTCCCCCAT	TGACCTCTGCCTCCACCCCG
Klf4	TCCGGCAGGTGCCCCGAATA	CTCCGCCGCTCTCCAGGTCT
Dazl	CACAGTGGCCTACTGGGGAACA	TTCGGCGCCTGGGTCAACTT
Tbx3	TGGATTCGGGCTCGGAACTGA	AGCCGCTGGATGCTCTGAAGT
Blimp1	AGCGGCGAACGGCCTTTCAAAT	GACCTGGCATTCATGCGGCTTT
Lefty2	ACTGCCGCATTGCCCATGAT	AGCTGCACTGCTTCACCCTCAT
Dax1	GCCTGCAGTGCGTGAAGTACA	CGGTGCGTCATCCTGACGTG
Fbxo15	AGGCTCGGCCACTGTTCCTT	CACGTGGAGCTGGTAGCCATGT
Pecam1	AGAGGAGCTGGAGCAGGTGTTAAT	GCTGATGTGGAACTTCGGAACAGA
Piwil2	TGACCTTTCCGGATCCTTCAGTGT	TCCGAACTCCCTCTTCCAAGCATT
Gbx2	AACGCGTGAAGGCGGGCAAT	TGCTGGTGCTGGCTCCGAAT
Rex1	AGCCCAGCAGGCAGAAATGGAA	TGGTCAGTCTCACAGGGCACAT
Fgf4	ACGCAGACACGAGCGACAGC	CGGCTGGCCACGCCAAAGAT
Otx2	TCCGGCTCGGGAAGTGAGTT	GGAGCACTGCTGCTCGCAAT
Cldn6	GCAGCCTCGGGCCTTTTGTTG	TCGGGCCAGACGCTGAGTAG
Pitx2	ACTAGCGCGCAGCTCAAGGA	CAGCTCCTCGCGCGTGTAAA
Cdx2	CTCAACCTGGCGCCGCAGAA	GCGCGCTGTCCAAGTTCGC
Cdh1	TGCACAGGCCGGAAACCAGT	ACGGCCTTCAGCGTGACCTT
Cdh2	CCGTGGCAGCTGGACTGGAT	GATGACGGCCGTGGCTGTGT
Esrrb	CGTGGAGGCCGCCAGAAGTA	TCTGGCTCGGCCACCAAGAG
Tcfcp2l1	AGAAAAGGGCGTGCCGTTCC	TGGCAGCTGGCTGAGTGCAA

6. OCT: Tissue-TekTM CRYO-OCT Compound (Fisher Scientific, 14-373-65).
7. HPS: Hematoxylin–Phloxine–Saffron Stain (Poly Scientific R&D Corp., k023).

2.10 Karyotyping

1. 175-cm^2 Nunc culture flask (Dutscher, 055421).
2. KaryoMAX Colcemid solution: 10-µg/mL solution of *N*-desacetyl-*N*-methylocolchicine in PBS (Life Technologies, 15212-012). For preparing a 0.4-µg/mL working solution, dilute 2 mL of karyoMAX with 48 mL of rbiPSC culture medium.

3. Shock solution: For preparing 75 mM KCl (Sigma-Aldrich, P9333), dissolve 0.28 g KCl in 50 mL water.
4. Fixation solution: Mix 1 volume of glacial acetic acid (Sigma-Aldrich, 1005706) with 3 volumes of ethanol (Sigma-Aldrich, 02860). Store at −20 °C.
5. Digestion solution: 0.045 % trypsin in Ca^{++}- and Mg^{++}-free Tyrode Ringer's saline. Prepare Ringer's buffer by dissolving 8 g NaCl (Sigma-Aldrich, S7653), 0.3 g KCl (Sigma-Aldrich, P9333), 0.093 g $NaH_2PO_4{\cdot}5H_2O$ (Sigma-Aldrich, 71505), 0.025 g KH_2PO_4 (Sigma-Aldrich, P5655), 1 g $NaHCO_3$ (Sigma-Aldrich, S5761), and 2 g glucose (Sigma-Aldrich, G0350500) in 1 L water and adjust the pH to 7.6–7.7. Dilute 1.8 mL of 2.5 % trypsin solution (Sigma-Aldrich, 59427C) with 100 mL of ringer's buffer to obtain 0.045 % trypsin.
6. Staining solution: Mix 90 mL water, 3 mL of KaryoMAX Giemsa (Life Technologies, 10092-054), 3 mL methanol (Sigma-Aldrich, 322415), and 3 mL of 0.1 M citric acid (5.35 g in 250 mL water) (Sigma-Aldrich, 251275) in order. Verify that the pH of the solution is between 2 and 3, and adjust the pH to 6.8 with 0.5 M Na_2HPO_4 (17.9 g in 250 mL water) (Sigma-Aldrich, 255793).
7. Glass slides (*see* Section 2.7, item 1): degrease the glass slides with soap solution, rinse with water, and store immersed in 33 % ethanol (33 mL ethanol and 67 % water) at 4 °C.

2.11 Freezing and Thawing rbFs, MEFs or rbiPSCs

1. Freezing solution for MEFs or rbFs: a mixture of 90 % FBS and 10 % DMSO (Sigma-Aldrich, D2650).
2. Freezing solution for rbiPSCs: a mixture of 90 % KOSR and 10 % DMSO.
3. 1.8-mL Nunc Cryotubes (Dutscher, 055003).
4. Nalgene cell-freezing container (Sigma-Aldrich, C1562).
5. Isopropanol (Sigma-Aldrich, I9516).

3 Methods

Carry out all procedures under a Class II Biological Safety Cabinet at room temperature unless otherwise specified. Virus production, and infection must be performed in a level three bio-containment culture room.

3.1 Rabbit Fibroblast Derivation

1. Perform an ear skin biopsy of 1-cm^2 on a 5-month old rabbit.
2. Place the skin biopsy in a Petri dish and wash twice with 5 mL PBS-PSG.
3. To remove dirt, hair, and fatty tissue, gently scrape the upper and lower sides of the skin with a scalpel.

4. Place the skin biopsy in a new Petri dish and wash twice with 5 mL PBS-PSG.
5. With the help of a scalpel, cut the skin biopsy into fine pieces (about 1 mm^2) under PBS-PSG.
6. Transfer the tissue fragments into a 50-mL Falcon tube and wash twice with 5 mL PBS-PSG.
7. Remove the PBS-PSG, add 10 mL of 1× dispase II, and incubate overnight at 4 °C.
8. Transfer the pieces of the skin tissue into a Petri dish, remove dispase II, and wash the pieces of skin four times with 5 mL PBS-PSG.
9. Transfer the pieces of skin tissue into a 50-mL Falcon tube, add 5 mL of 10× trypsin, triturate, and incubate for 20–30 min in a water bath at 37 °C.
10. Stop the digestion by transferring the supernatant into a 50-mL Falcon tube containing 20 mL of warm FBS. Trypsinize again the remaining pieces of skin two times (repeat steps 9 and 10).
11. At the end of third trypsinization, add the remaining pieces of skin tissue to the tube containing FBS and centrifuge the skin cells and remaining fragments at 130 × *g* for 5 min.
12. Wash the pellet twice with 20 mL PBS, centrifuging the cells at 130 × *g* for 5 min after each wash.
13. For coating the plates with gelatin, incubate a 60-mm culture plate with 2 mL of 0.1 % gelatin for 10 min.
14. Resuspend the skin cells and pieces in 5 mL of fibroblast culture medium and transfer the suspension into a gelatin-coated 60-mm culture dish.
15. Incubate the dish overnight at 38 °C and 5 % CO_2.
16. Wash the adherent skin cells and pieces five times with PBS and add 5 mL of fresh fibroblast culture medium containing 20 % FBS.
17. Culture the cells at 38 °C and 5 % CO_2 until they reach confluence (approximately 7–10 days, Fig. 1), changing the medium once every 2–3 days. The medium containing 20 % FBS should be used only for the first few days (until the first passage).
18. Passage the cells using 2× trypsin (*see* **Note 3**). Seed a 100 mm culture plate with 1×10^6 cells in fresh culture medium (cell density = 1.8×10^4 cells/cm^2).
19. Amplify the rabbit fibroblasts for 3–4 passages and then freeze at a density of 2×10^6 cells/0.5 mL freezing medium (90 % FBS + 10 % DMSO) (*see* Section 3.11).

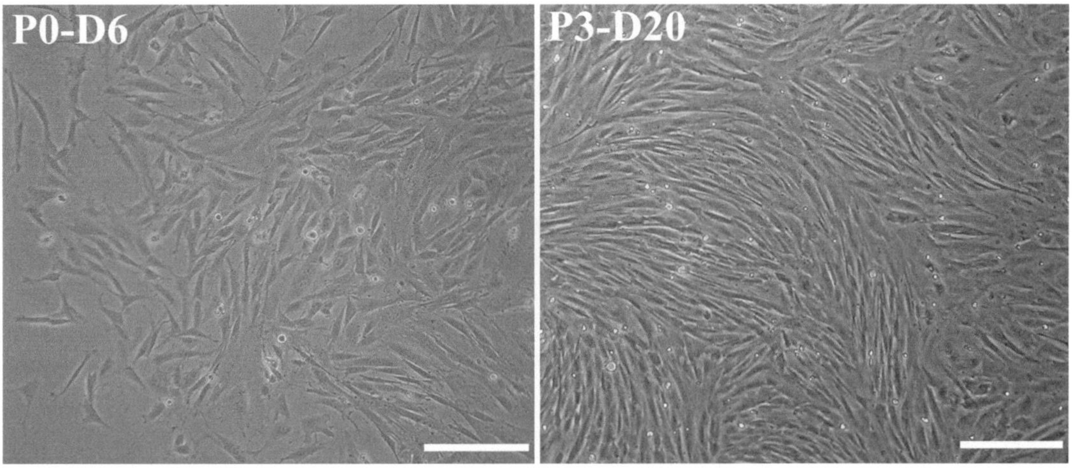

Fig. 1 Derivation of rbFs. Phase-contrast images of rbFs at P0 (6 days after derivation) and P3 (after 20 days of culture). Scale bar = 50 μm

3.2 Retrovirus Production and Titration

1. Prepare the transfection reagents ($CaCl_2$, 10× HBS, phosphate buffer), filter through a 0.22-μM PES syringe filter unit, and store at 4 °C.
2. To insure the stability of the cell line and high virus titer, culture the 293FT cells at 37 °C and 5 % CO_2 in fibroblast medium supplemented with 400 μg/mL G418.
3. Day 0: plate 2×10^6 293FT cells in 12 mL of fibroblast medium without G418 onto a 100-mm dish. Prepare four 100 mm-dish cultures. Incubate the cells overnight at 37 °C and 5 % CO_2.
4. Day 1: to produce the four pMX retroviruses, transfect the 293FT cells using the calcium phosphate precipitation method. Extemporaneously prepare 2 mL of 1× HBS by mixing 200 μL of 10× HBS, 40 μL of phosphate buffer, 30 μL of 1 N NaOH, and 1,730 μL filtered H_2O. In four separate Eppendorf tubes, mix 6.5 μg of pTG5349 DNA, 3.5 μg of phCMV-GP DNA, and 10 μg of the desired pMX vector in 393 μL TE buffer. To each tube, add 11 μL of 2 M $CaCl_2$, mix well, add 46 μL of 2 M $CaCl_2$, and mix once again. Prepare four distinct 15 mL Falcon tubes, each containing 450 μL of 1× HBS, and gently vortex the tubes. Add the DNA–$CaCl_2$ mixture dropwise into HBS to induce precipitation of the calcium phosphate–DNA complexes. The smaller the size of the precipitated particles is, the more efficient the transfection is. The solution will appear cloudy without any flocculation or precipitate visible to the naked eye. Incubate the four mixtures for 20 min and add each solution dropwise into the 293FT cultures. Incubate the cells overnight at 37 °C and 5 % CO_2.

5. Day 2: remove the supernatants from the transfected 293FT cell cultures and to each dish, slowly add 5 mL of fresh fibroblast medium. Incubate the cells for 24 h at 37 °C and 5 % CO_2.
6. Day 3: collect the supernatants containing four different viruses from the transfected 293FT cultures and pool them in a 50-mL Falcon tube. Incubate the virus solution on ice until transduction. Remove the debris by centrifuging the virus suspension at 1,800 × *g* for 10 min at 4 °C. Filter the supernatant using a 0.8-μM acrylic syringe filter unit. The resulting virus solution is ready for use. It can be concentrated, aliquoted, frozen, and stored at −80 °C.
7. The titer of the Oct4 retrovirus suspension can be calculated by counting the number of infected fibroblasts after immunolabeling for hOCT4 expression. Plate 2×10^4 infected rbFs on glass coverslips placed at the bottom of 24-well plates.
8. Perform immunolabeling as described in Section 3.7 using the anti-Oct4 Rabbit IgG as the primary antibody and Alexa Fluor 555-conjugated Goat anti-Rabbit IgG as the secondary antibody (Table 2).
9. Count the labeled cells under a fluorescence microscope and compare this number with the total number of cells to determine the infection rate [(number of labeled cells/2×10^4 infected cells) × 100]. Calculate the viral titer using the following expression: number of infectious particles (i.p.)/mL = number of infected cells × infection rate/volume of virus used (mL). Generally, the titer of pMX-hOCT4 retrovirus obtained using the protocol described below (*see* Section 3.4) is 2×10^5 i.p./mL.

3.3 Feeder Cell Preparation

1. Day 1: thaw a vial of OF1 MEFs (*see* Section 3.11) and plate the cells in three 100-mm culture dishes, each containing 10 mL of fresh culture medium. Incubate the cells at 37 °C and 5 % CO_2 for 72 h.
2. Day 4: replace the culture medium over MEFs with 5 mL of 1× mitomycin-C. Incubate the cells at 37 °C and 5 % CO_2 for 2–3 h. Remove the mitomycin-C, and rinse the cells five times with 5 mL of PBS. Then, to each dish, add 1 mL of 1× trypsin and incubate the dishes for 5 min at 37 °C. After incubation, add 1 mL of fibroblast medium to each dish to stop the enzymatic reaction. Dislodge and dissociate the cells by repeated pipetting. Transfer the cell suspension into a 15-mL Falcon tube containing 10 mL of fibroblast medium, centrifuge for 5 min at 300 × *g*, and resuspend the cell pellet in 10 mL of fresh medium. Count the cells using a Malassez counting chamber and plate onto gelatin-coated dishes at a density of 1.6×10^4 cells/cm^2 (9×10^5 cells in 10 mL medium for a 100-mm dish, 1.5×10^5 cells in 2 mL medium

for one well of a 6-well plate, 7.5×10^4 cells in 1 mL medium for one well of a 12-well plate, 3.8×10^4 cells in 0.5 mL medium for one well of a 24-well plate, or 5×10^3 cells in 150 μL medium for one well of a 96-well plate). Incubate the cells overnight at 37 °C and 5 % CO_2 before use. Inactivated MEFs must be used within 3 days.

3.4 Fibroblast Infection

1. To generate rbiPSCs, infect 5×10^5 rbFs in suspension, twice at 2-day intervals, with 3×10^6 viral particles (equal amounts of the four freshly produced hOct4, hSox2, hKlf4, and hMyc retroviruses) [multiplicity of infection (m.o.i.) = number of viral particles/cell = 2 for each retrovirus and 8 for the mixture of the 4 retroviruses], in the presence of 4 μg/mL polybrene. The production of retroviruses, infection of rbFs, and the feeder cell preparation should be carefully coordinated and must follow the schedule presented in Fig. 2.
2. Day 0: dissociate the rbFs using 1× trypsin, count the cells, and prepare a suspension containing 5×10^6 fibroblasts/mL.

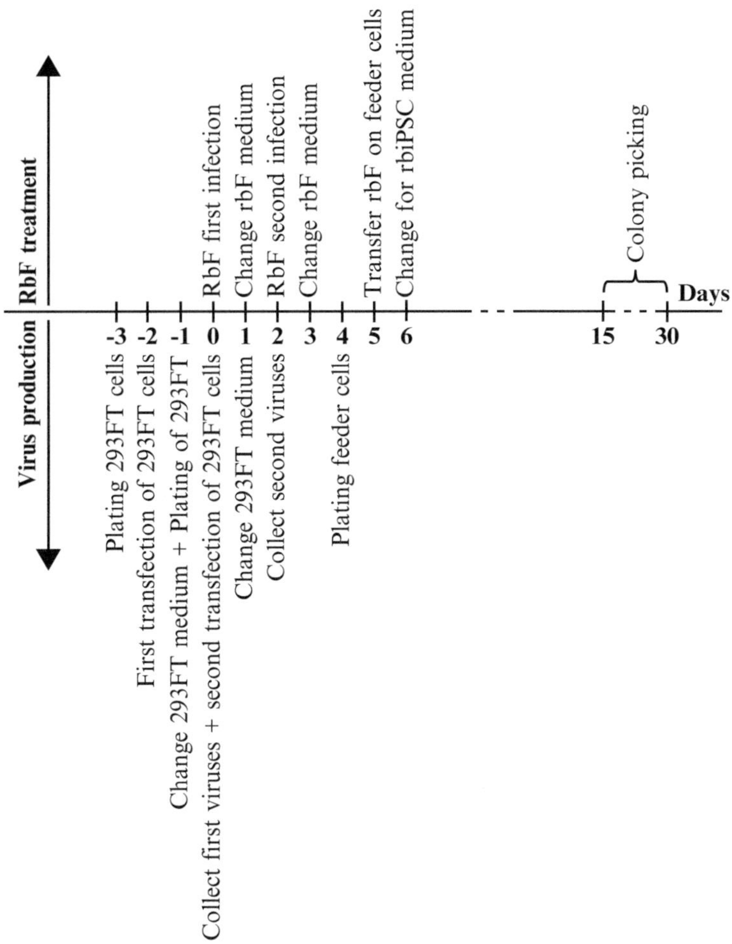

Fig. 2 Time schedule of rbF reprogramming procedure

Transfer 100 μL of the cell suspension (5×10^5 rbFs) into a 100-mm Corning dish containing 10 mL of freshly prepared mixture of the four pMX retroviruses and 80 μL polybrene (final concentration 4 μg/mL). Incubate the cells at 37 °C and 5 % CO_2 for 24 h.

3. Day 1: replace the medium of the infected and plated rbFs with 10 mL of fresh fibroblast medium and allow the cells recover at 37 °C and 5 % CO_2 for 24 h.
4. Day 2: dissociate the infected RbFs with 1× trypsin, centrifuge, and resuspend in 100 μL of fibroblast medium. Mix the cell suspension with 10 mL of a freshly prepared mixture of the four retroviruses and 80 μL polybrene, and plate onto a 100-mm Corning dish. Incubate the cells at 37 °C and 5 % CO_2 for 24 h.
5. Day 3: replace the medium of the infected rbFs with 10 mL of fresh fibroblast medium and allow the cells to recover by incubating at 37 °C and 5 % CO_2 for 48 h.
6. Day 5: dissociate the infected rbFs with 1× trypsin, centrifuge, and resuspend in 100 mL of fibroblast medium. Plate the cells (5×10^4 infected rbFs in 10 mL per dish) on feeder cells maintained in ten 100-mm Corning dishes (*see* **Note 4**). Incubate the dishes at 38 °C in a humidified atmosphere of 5 % CO_2 and 5 % O_2.
7. Day 6: replace the spent medium with 10 mL of rbiPSC culture medium supplemented with 10 ng/mL FGF2. Change the medium every day for 1 week. To avoid acidification, increase the volume of the medium gradually as the cell density increases.

3.5 RbiPSC Clone Isolation and Culture

1. Colonies with compact morphologies will begin to appear approximately 14 days after the infection. Carefully examine the cultures daily. This is because it is often hard to distinguish the putative rbiPSC colonies from the growing and compact rbFs. The iPSC colonies composed of tightly packed small cells will have an elongated, egg-like shape and will appear darker than the rbFs (Fig. 3a).
2. Between day 15 and day 30 post-infection, as soon as they become large enough (approximately 50 μm in width and 100 μm in length), pick the colonies with a glass capillary and place individually in separate Eppendorf tubes containing 10 μL of 1× trypsin. Incubate the tubes for 5 min at 37 °C, add 100 μL of rbiPSC medium into each tube, and gently dissociate the cells by repeated pipetting. Centrifuge the tubes for 5 min at 300 × *g*, discard the supernatant, resuspend the dissociated cells in 200 μL of rbiPSC medium, and plate the cells on feeder cells maintained in a 96-well culture plate (passage 1 = P1).

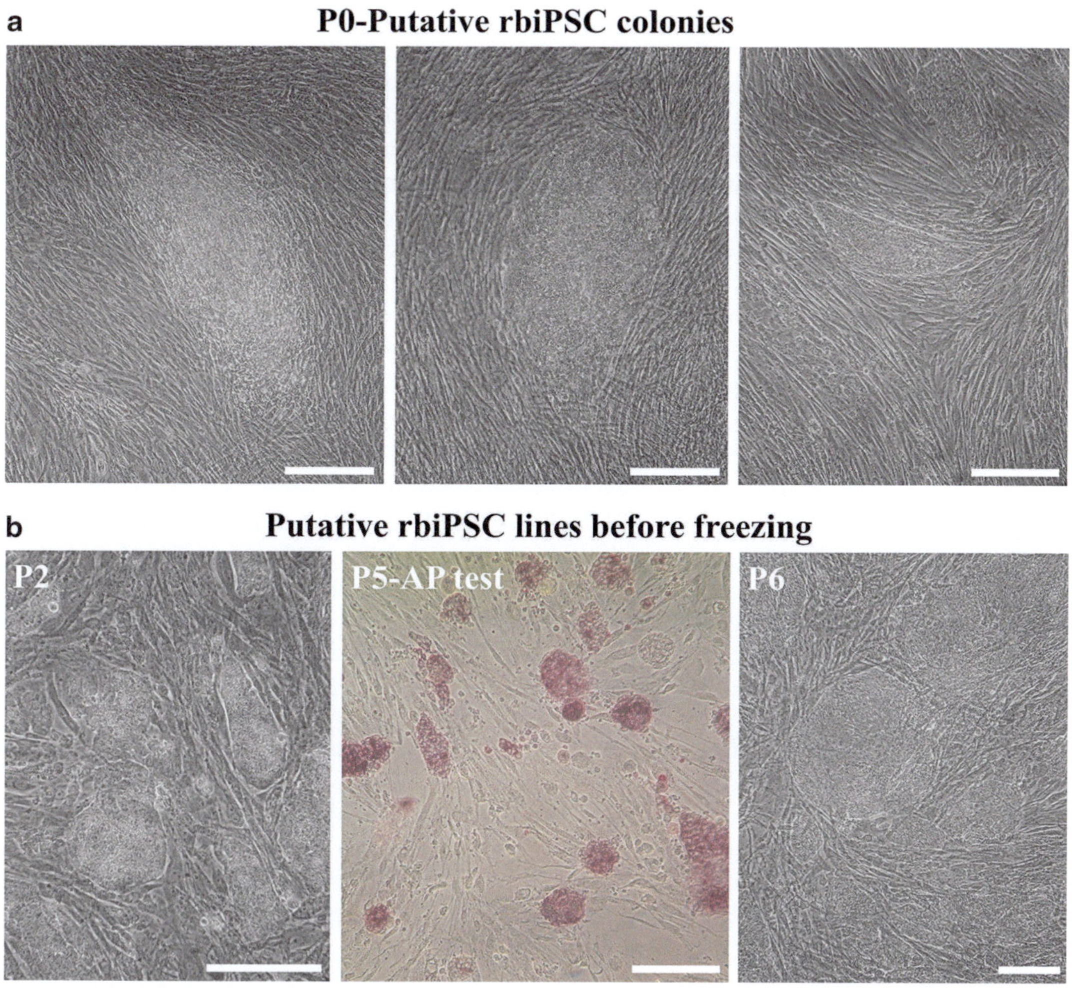

Fig. 3 Selection of putative rbiPSC clones. (**a**) Phase-contrast images of emerging rbiPSC colonies at week 3 after infection of rbFs with retroviral vectors. (**b**) Phase-contrast images of putative iPSC colonies at P2 (10–12 days after picking) and P6 (before freezing). Staining for alkaline phosphatase activity at P5. Scale bar = 50 μm

Incubate the cells at 38 °C in a humidified atmosphere of 5 % CO_2 and 5 % O_2. Change medium daily.

3. Amplify the cells that grow fast and form tight colonies with the morphology of primate pluripotent stem cells. As soon as they reach confluence, passage the cells using 1× trypsin as described in **Note 3**. Cells from one well of a 96-well culture plate must be plated onto one well of a 24-well plate (P2), then from one well of a 24-well plate into one well of a 12-well plate (P3), then from one well of 12-well plate into one well of a 6-well plate (P4), and from one well of a 6-well plate into 2 wells of a 6-well plate (P5).

4. Test the alkaline phosphate activity in the clones that maintain a stem cell morphology and high proliferation rate during the amplification period. The alkaline phosphatase activity is characteristic of highly proliferating cells and is classically used to characterize pluripotent stem cells. To determine the alkaline phosphatase activity, twice rinse one of the two wells (P5) with PBS and fix the cells by incubating with 2 mL/well of methanol for 30 min. Rinse the cells with PBS and add alkaline phosphatase substrate (2 mL/well). Incubate the cells in a dark cabinet for 15 min and then rinse the cells with water. Cells showing alkaline phosphatase activity will be colored red (Fig. 3b).
5. P5 clones showing alkaline phosphatase activity are further amplified (one well of a 6-well plate into 2 wells of a 6-well plate). The resulting P6 clones are treated as follows: cells from one well are used for freezing (*see* Section 3.11) and that from the second well are used to prepare a dry pellet. For this purpose, rinse the cells with PBS, dissociate using 1× trypsin, centrifuge the suspension for 5 min at 300 × *g*, resuspend the cells in PBS, and transfer the suspension into an Eppendorf tube. After centrifugation for 5 min at 300 × *g*, discard the supernatant and freeze the cell pellet in liquid nitrogen.

3.6 Test of Transgene and Endogenous Gene Expression by RT-PCR

1. Fully reprogrammed rbiPSCs are characterized both by the activation of the endogenous pluripotency genes *rbOct4* and *rbNanog*, and by the silencing of the reprogramming transgenes *hOCT4*, *hSOX2*, *hKLF4*, and *hc-MYC*. The expression of these genes should be carefully analyzed by semiquantitative RT-PCR using total RNA extracted from the dry rbiPSC pellet (prepared from P6 cells).
2. Isolate total RNA from dry cell pellets using the RNeasy mini kit and treat with DNAse I (follow the manufacturer's protocol). Measure the concentration of RNA in solution with a NanoDrop spectrophotometer.
3. Perform reverse transcription of the RNA. For this purpose, incubate 1 μg RNA with 200 U of M-MLV reverse transcriptase and random primer mix in a final volume of 20 μL for 5 min at 70 °C, immediately followed by incubating at 37 °C for 45 min. Dilute the cDNA solution fivefold by adding 80 μL of H_2O.
4. Prepare the PCR mix. For this purpose, mix 5 μL of cDNA, 50 mM $MgCl_2$, 20 mM dNTPs, 100 ng forward primer, 100 ng reverse primer, and 0.5 U of Euroblue Taq in a final volume of 50 μL. Perform semiquantitative PCR reactions using a thermocycler. Use the following thermal cycling conditions: an initial denaturation step at 94 °C for 5 min, followed by 35 cycles of denaturation at 94 °C for 30 s, annealing between 56 and 60 °C (depending on the primer pairs) (*see* Table 1) for 30 s, and polymerization at 72 °C for 30 s, and a final step of extension at 72 °C for 5 min.

5. Analyze the PCR products by electrophoresis on a 2 % agarose gel. Save the clones expressing *rbOct4* and the *rbNanog* (even if the clones do not show complete silencing of the human transgenes). These clones are undergoing reprogramming. Discard the clones that do not show the activation of endogenous pluripotency genes.
6. Thaw the selected clones (*see* Section 3.11) plate the cells on feeder cells, and culture in rbiPSC medium. Passage the cells once every 2 days using 1× trypsin, plate them on feeder cells at the density of 4.2×10^4 cells/cm^2, and change medium daily. Regularly monitor the expression of human transgenes in these clones until some of the clones display silencing of all four pluripotency transgenes. It could take up to 20 passages for the reprogramming to be complete.

3.7 Analysis of Pluripotency Markers by Immunofluorescence Microscopy and Flow Cytometry

1. For in situ immunolabeling, prepare glass coverslips by incubating with 0.1 % gelatin for 10 min, and then with FBS overnight. Seed the coverslips with feeder cells as described earlier. Coverslips can be placed at the bottom of 24-well plates.
2. Plate 1×10^5 rbiPSCs onto feeder cells maintained on glass coverslips. Incubate at 37 °C and 5 % CO_2 for 48 h.
3. Rinse the cells once with PBS, fix by incubating for 20 min with 2 % PFA (first add 250 μL PBS per well, then add 250 μL 4 % PFA per well), and rinse twice with PBS. Fixed cells can be store at 4 °C.
4. For the analysis of intracellular antigens, permeabilize the cells with TBS-Triton (500 μL/well) for 20 min and wash three times (10 min each) with TBS.
5. Block nonspecific binding sites with blocking solution (500 μL/well) for 1 h and incubate overnight at 4 °C with primary antibodies (OCT4, SSEA-1, SSEA-4, TRA-1-60, E-CADHERIN, N-CADHERIN, and CD90) diluted in blocking solution (Table 2).
6. Wash the cells with TBS-Tween three times for 10 min each and incubate the cells with fluorochrome-conjugated secondary antibodies (see Table 2) at room temperature for 1 h.
7. Rinse the cells with TBS-Tween three times for 10 min each and stain the nuclei by incubating the samples for 5 min with a 1-μg/ml solution of Hoechst 33342.
8. Rinse the cells four times with TBS (5 min each). Place the coverslips on glass slides with 7 μL of mounting medium and fix with DPX medium. Examine the cells under a conventional fluorescence microscope fitted with appropriate filters for Hoescht (blue), and Phycoerythrin, Rhodamine, or Alexa Fluor 555 (red) (Fig. 4a).

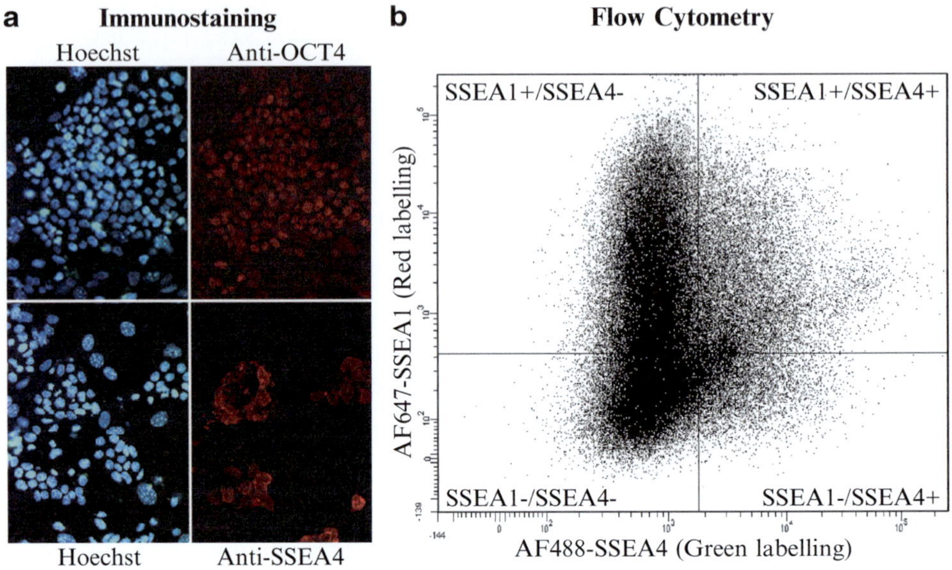

Fig. 4 Detection of pluripotency markers with specific antibodies. (**a**) Immunofluorescence labeling of rbiPSCs after immunostaining with antibodies against OCT4 (*top panel*) and SSEA-4 (*bottom panel*), and labeling of nuclei with bis-benzimide (Hoechst). (**b**) Flow cytometry analysis of rbiPSCs showing the heterogeneity in the expression of SSEA1 and SSEA4 pluripotency markers

9. For flow cytometry, dissociate the cells using 1× trypsin and label the cells in suspension using the protocol and the antibodies (SSEA-1, SSEA-4) described above.
10. Analyze the labeled cells using a flow cytometer with the help of red (633 nm) and blue (488 nm) lasers for detection of Alexa Fluor 647 or 488 labeling, respectively (Fig. 4b).

3.8 Analysis of the Expression of Pluripotency Genes by RT-qPCR

1. Extract mRNA from dry cell pellets and measure the RNA concentration. Dilute the RNA to a concentration of 0.05 ng/mL using RNAse-free water. Keep the RNA solution on ice or store at −80 °C.
2. Perform reverse transcription of the RNA using the RNA to cDNA kit. Mix 10 μL of RT buffer containing the dNTPs and the random primers with 9 μL of diluted RNA (450 ng) and 1 μL reverse transcriptase. Incubate for 5 min at 70 °C, and then for 45 min at 37 °C. Dilute the cDNA solution fivefold with water and store at −20 °C.
3. Prepare PCR reaction mixtures in microplate. For this purpose, mix 5 μL of SYBR® Green Mix, 0.44 μL of primer mix, and 3.6 μL water. Add 9 μL of the mixture to one well of a 96-well microplate. To this, add 1 μL of diluted cDNA. Perform each PCR reaction in triplicate. For each primer pair used, run a negative control in triplicate, where the reaction mixture is prepared by substituting cDNA with water. Close the microplate with an optical adhesive cover and centrifuge the plate for 1 min at 800 × *g*.

4. Run the qPCR reaction using the StepOnePlus real-time PCR system according to the manufacturer's instructions. Use 40 amplification cycles and an annealing temperature of 60 °C for the reaction. At the end of the amplification, analyze the melt-curve and verify that only the desired PCR products are formed. Using the StepOnePlus Software V2.1 (Applied Biosystems), determine the amplification efficiency for both target and reference genes from the relative values of the calibrator-normalized target gene expression. Then, normalize the expression of the target genes to those of the rabbit *Gadph* and *Tbp* (TATA-box binding protein).

3.9 Teratoma Formation

1. All material and food used for breeding the mice should be sterilized to avoid infection.
2. Teratoma formation can be induced by injecting rbiPSCs under the kidney capsule of the Severely Compromised ImmunoDeficient (SCID) mice. Inject rbiPSCs into one kidney of each mouse. Each rbiPSC line must be injected into at least two mice.
3. Preparation of rbiPSCs: dissociate the rbiPSCs using 1× trypsin, count the cells, and centrifuge the cell suspension. Resuspend 2.5×10^6 rbiPSCs in 10 μL PBS.
4. Preparation of the mouse: anesthetize the mouse by intramuscular injection of 1.4 μL/g of anesthesia mix. Allow the mouse to rest on its stomach and make a small incision (approximately 1 cm) in the skin and muscle of the back (parallel to the spinal column and at the same level as the top of the hip). Part the skin and muscle to see the kidney.
5. Inject 10 μL of the rbiPSC suspension under the kidney capsule using a glass capillary and suture the wound with screed clips. Allow the mouse to recover.
6. For the next 4–8 weeks, monitor the tumor growth by following the regular palpation of the back of the mouse. Sacrifice the mice by cervical dislocation when the manipulated kidney is approximately twice as big as the normal kidney. Surgically remove the damaged kidney.
7. Fix the kidney along with the tumor for 1 week with 4 % PFA at 4 °C.
8. Rinse the organ with PBS, embed in OCT compound, and prepare cryosections of 20 μm thickness. Mount the sections on glass slides and stain with HPS for histological analysis. The teratoma formed from pluripotent rbiPSCs will contain derivatives of the three embryonic germ layers.

3.10 Karyotyping

1. Day 1: plate a suspension of 1.5×10^7 rbiPSCs in 30 mL medium in a 175-cm^2 flask without feeder cells.
2. Day 2: treat rbiPSCs with 0.4-μg/mL colcemid solution for 4 h at 37 °C to have the cells arrested in metaphase (*see* **Note 5**).

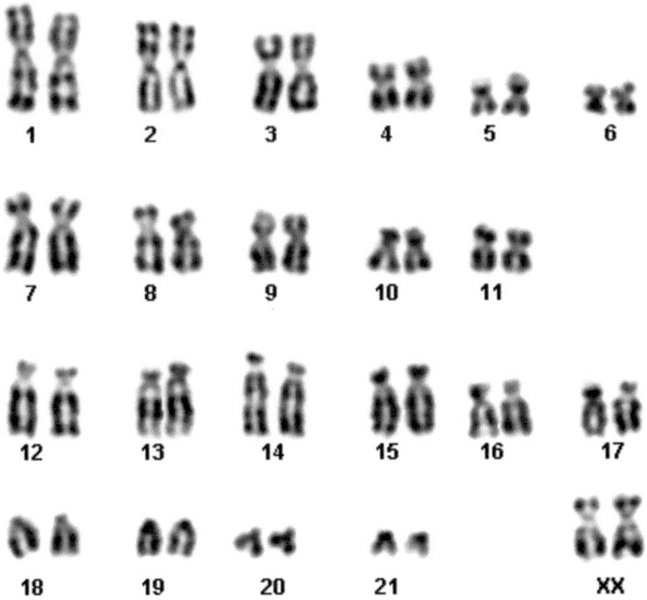

Fig. 5 G-banding karyotype of a rbiPSC line

Collect the loosely attached mitotic cells by vigorously shaking the flask. Centrifuge the cell suspension at 300 × *g* for 5 min, resuspend the pellet in 2–3 mL of warm shock solution, and incubate the cells at 37 °C for 3 min. Add 2–3 drops of cold fixation solution and centrifuge the cells again at 300 × *g* for 5 min. Resuspend the cell pellet in 200 μL of shock solution and slowly add 1–2 mL of cold fixation solution. Store the fixed cells at 4 °C.

3. Day 3: allow 1–2 drops of a cold suspension of fixed cells to fall from a height of 20–30 cm onto an ice-cold and wet glass slide. Add 4–5 drops of fixation solution to the surface of the slide. Dry the slide on a gas flame. Incubate the slides overnight in an oven at 60 °C and store the slides at room temperature.
4. Stain the chromosome spreads by incubating the slides in Giemsa staining solution for 8–10 min. Rinse the slides twice with water and allow the slides to dry. Observe the slides under a microscope fitted with a 63× objective. Count the chromosomes of at least 50 metaphase spreads.
5. For G-banding and karyotyping, digest the chromosome spread by placing the slides in a bath of 0.045 % trypsin for 75 s at 37 °C. Rinse the slides with PBS and stain the chromosomes as described before. Observe the chromosome spreads, perform pair matching, and define the chromosomal rearrangements. An example of an euploid 42XX karyotype is provided in Fig. 5.

3.11 Freezing and Thawing rbFs, MEFs or rbiPSCs

1. Dissociate the cells using 1× trypsin, count the cells, and centrifuge the suspension at 300 × *g* for 5 min. Resuspend the cells in an appropriate volume of culture medium to obtain a cell density of 8×10^6 cells/mL. Add an equal volume of cold freezing medium dropwise into the suspension. Distribute 0.5 mL each of the resulting suspension into separate Nunc cryotubes (2×10^6 cells/0.5 mL/tube). Place the tubes in a Nalgene cell-freezing container filled with isopropanol. Place the container at −80 °C for 24 h and store the cryotubes in liquid nitrogen.
2. For thawing the cells, retrieve a vial of cryopreserved cells from liquid nitrogen and place in a water bath for 2 min at 37 °C. Transfer the cell suspension into a 15-mL Falcon tube. Add 10 mL of warm culture medium dropwise into the tube. Centrifuge the cell suspension for 5 min at 300 × *g*, discard the supernatant, and resuspend the cell pellet in 3–5 mL of fresh medium. Plate the cells (2×10^6 cells) onto appropriate culture plates with or without feeder cells (1 well of 6-well plate with feeder cells for rbiPSCs and 2–3 100-mm culture dishes without feeder cells for MEFs and rbFs). Depending on the cell type, incubate the cells in two- or tri-gas incubator at 37 °C or 38 °C.
3. In contrast to MEFs and rbFs, a high proportion of rbiPSCs undergo cell death after thawing. To remove the dead cells and debris, 1 day after thawing, rinse the cells twice with PBS. Add fresh medium and culture the cells for 5–7 days before passaging, daily replacing the spent medium with fresh medium. Cells should recover after 2 or 3 passages (*see* **Note 6**) and grow as flat colonies of tightly packed cells (Fig. 6).

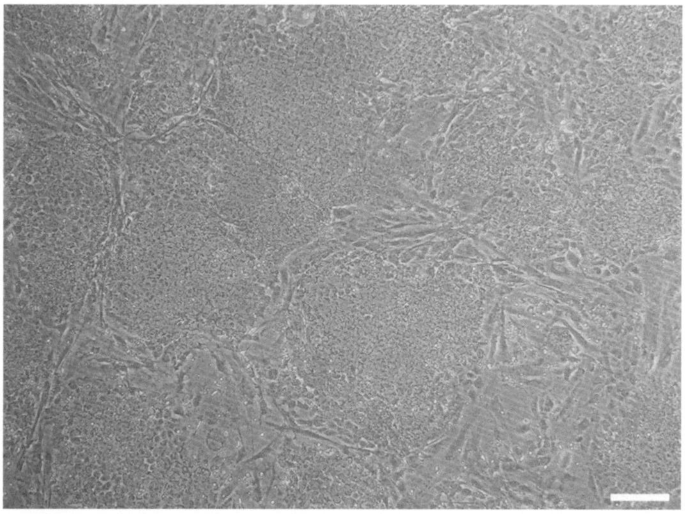

Fig. 6 Morphology of rbiPSCs. Phase-contrast image of a rbiPSC line at P30. Elongated feeder cells surround flat colonies of highly packed rbiPSCs. Scale bar = 50 μm

4 Notes

1. The pMX retroviral amphotropic particles can be produced using commercial packaging vectors such as pUMVC (Addgene, 8449) and pCMV-VSV-G (Addgene, 8454), or packaging cell line such as Platinum-A (Plat-A, Cell Biolabs Inc., RV-102).
2. Primary MEFs are prepared from 12.5-day-old OF1 mouse embryos. Collect the uterine horns from the sacrificed pregnant mouse and isolate the implanted embryos by dissecting the horns. Rinse the embryos with PBS and remove the head and the viscera with the help of a scissor. Mince the embryo bodies with a scalpel and incubate in 5 mL of 5× trypsin for 10 min at 37 °C under constant stirring. Transfer the supernatant into a 50-mL Falcon tube containing 25 mL of fibroblast culture medium to stop the enzymatic reaction. Repeat the digestion step on the embryo pieces two more times with 5 mL of 5× Trypsin. Centrifuge the dissociated cells for 10 min at 450 × *g*. Resuspend the cell pellet in an appropriate volume of fibroblast medium. Plate 10 mL of the cell suspension onto each of the 100-mm culture dishes (same number of plates as that of treating embryos). After 2–3 days, passage the confluent cultures in each plate into three new plates using 2× trypsin. Trypsinize the cells after 2–3 days of culture, count, and freeze at a density of 2×10^6 cells/0.5 mL freezing medium (90 % FBS + 10 % DMSO). Alternatively, commercial MEFs isolated from the strain CF1 could be used (ATCC, SCRC-1040).
3. Passaging cells: when the adherent cells reach confluence, they should be replated onto a fresh dish. For this purpose, remove the culture medium, rinse the cells with 3 mL PBS, incubate the cells with 1 mL of 1× or 2× trypsin (depending of the cell type) at 37 °C for 5 min, add 1 mL of culture medium to stop the enzymatic reaction, dissociate the cells by repeated pipetting, and transfer the suspension into a 15-mL Falcon tube containing 3 mL of culture medium. Count the cells using a Malassez counting chamber, centrifuge the cells at 300 × *g* for 5 min, resuspend the cells in an appropriate volume of culture medium, and plate the cells at the suitable density (variable with the cell type) on gelatin-coated culture plates with or without feeder cells.
4. The rbFs isolated from the ear skin are highly proliferating cells. Therefore, the density of rbFs plated on the feeder cells could be fewer than 5×10^4 cells/100-mm dish. If the density is higher, the proliferating non-infected rbFs may mask the emerging pre-iPSC colonies.

5. After 3 h of incubation with colcemid, every 30 min, determine the number of metaphasic rbiPSCs that appear round and refractive. To avoid the compaction of chromosome that will render the G-banding impossible, samples should be incubated for less that 5 h. The choice of incubation time should be a compromise between the number of blocked cells and the quality of metaphasic chromosomes.
6. To avoid spontaneous differentiation of rbiPSCs, it is essential to strictly follow the culture conditions described, namely the plating concentration (4.2×10^4 cells/cm^2), the daily change of medium (double the medium volume the day after the plating), and the regular passaging (once every 2 days).

Acknowledgment

This work was supported by research grants from ANR (project PLURABBIT n°PCS-09-GEM-08), COST Action (RGB-Net n°TD1101), the HyPharm Company, and the region Rhône-Alpes.

References

1. Takahashi K, Yamanaka S (2006) Induction of pluripotent stem cells from mouse embryonic and adult fibroblast cultures by defined factors. Cell 126(4):663–676
2. Takahashi K, Tanabe K, Ohnuki M, Narita M, Ichisaka T, Tomoda K, Yamanaka S (2007) Induction of pluripotent stem cells from adult human fibroblasts by defined factors. Cell 131 (5):861–872
3. Yu J, Vodyanik MA, Smuga-Otto K, Antosiewicz-Bourget J, Frane JL, Tian S et al (2007) Induced pluripotent stem cell lines derived from human somatic cells. Science 318(5858):1917–1920
4. Liu H, Zhu F, Yong J, Zhang P, Hou P, Li H et al (2008) Generation of induced pluripotent stem cells from adult rhesus monkey fibroblasts. Cell Stem Cell 3(6):587–590
5. Ezashi T, Telugu BP, Alexenko AP, Sachdev S, Sinha S, Roberts RM (2009) Derivation of induced pluripotent stem cells from pig somatic cells. Proc Natl Acad Sci U S A 106 (27):10993–10998
6. Honda A, Hirose M, Hatori M, Matoba S, Miyoshi H, Inoue K, Ogura A (2010) Generation of induced pluripotent stem cells in rabbits: potential experimental models for human regenerative medicine. J Biol Chem 285(41):31362–31369
7. Tancos Z, Nemes C, Polgar Z, Gocza E, Daniel N, Stout TA et al (2012) Generation of rabbit pluripotent stem cell lines. Theriogenology 78 (8):1774–1786
8. Osteil P, Tapponnier Y, Markossian S, Godet M, Schmaltz-Panneau B, Jouneau L et al (2013) Induced pluripotent stem cells derived from rabbits exhibit some characteristics of naive pluripotency. Biol Open 2 (6):613–628
9. Takahashi K, Okita K, Nakagawa M, Yamanaka S (2007) Induction of pluripotent stem cells from fibroblast cultures. Nat Protocol 2 (12):3081–3089
10. Sandrin V, Boson B, Salmon P, Gay W, Negre D, Le Grand R et al (2002) Lentiviral vectors pseudotyped with a modified RD114 envelope glycoprotein show increased stability in sera and augmented transduction of primary lymphocytes and CD34+ cells derived from human and nonhuman primates. Blood 100 (3):823–832
11. Mangeot PE, Duperrier K, Negre D, Boson B, Rigal D, Cosset FL, Darlix JL (2002) High levels of transduction of human dendritic

cells with optimized SIV vectors. Mol Ther 5(3):283–290

12. Tesar PJ, Chenoweth JG, Brook FA, Davies TJ, Evans EP, Mack DL et al (2007) New cell lines from mouse epiblast share defining features with human embryonic stem cells. Nature 448(7150):196–199

13. Bao S, Tang F, Li X, Hayashi K, Gillich A, Lao K, Surani MA (2009) Epigenetic reversion of post-implantation epiblast to pluripotent embryonic stem cells. Nature 461 (7268):1292–1295

Methods in Molecular Biology (2016) 1357: 173–182
DOI 10.1007/7651_2014_137

Published online: 20 November 2014

Cynomolgus Monkey Induced Pluripotent Stem Cells Generated By Using Allogeneic Genes

Nobuhiro Shimozawa

Abstract

Induced pluripotent stem (iPS) cells that are potentially similar to embryonic stem (ES) cells can be artificially established by introduction into somatic cells of the transgenes POU5F1 (also known as Oct3/4), SOX2, KLF4, and c-MYC. In cynomolgus monkeys (*Macaca fascicularis*), iPS cells generated by using these four allogeneic transgenes should be an important resource for various types of biomedical research because the use of xenogeneic transgenes may cause complications. To establish such iPS cells, cynomolgus monkey somatic cells were infected with amphotropic retroviral vectors, which were derived from Plat-A cells, containing cDNA for the cynomolgus monkey genes POU5F1, SOX2, KLF4, and c-MYC. As a result, iPS cells could be established from somatic cells from fetal liver and newborn skin of cynomolgus monkeys, similarly to the case for mouse and human somatic cells.

Keywords: Induced pluripotent stem cell, Cynomolgus monkey, Cynomolgus monkey gene, Nonhuman primate, Plat-A cell, Amphotropic retroviral vector, Cell culture, Medical research

1 Introduction

Induced pluripotent stem (iPS) cells were artificially generated from mouse somatic cells (1). These iPS cells were established by introduction of the exogenous transgenes POU5F1 (also known as Oct3/4), SOX2, KLF4, and c-MYC using retroviral vectors; they showed characteristics similar to those of embryonic stem (ES) cells, in terms of colony morphology, protein/gene expression, karyotype stability, and differentiation ability. It was also reported that mouse iPS cells differentiated into functional germ cells (2, 3). In addition, human iPS cells were successfully established (4). Nonhuman primate iPS cells were also established in rhesus monkeys (5), common marmosets (6), cynomolgus monkeys (7), and baboons (8).

However, abnormal characteristics of iPS cells have been reported (9–11), so methods for inducing more complete and simple reprogramming have been studied (12, 13). Biomedical research using nonhuman primates will be needed to examine the safety and efficacy of ES and iPS cells before they can be clinically applied in humans.

The influence that is caused by the genes from different species should be considered. Unexpected disadvantageous effects due to the use of xenogeneic transgenes, such as immune responses triggered by their expression, need to be avoided. To sidestep this problem, in order to achieve reprogramming of the cell status to one similar to ES cells, four allogeneic transgenes (cynomolgus monkey POU5F1, SOX2, KLF4, and c-MYC) were transfected to Plat-A cells to produce amphotropic retroviral vectors (14). By infecting somatic cells of cynomolgus monkeys with these retroviral vectors, cynomolgus monkey iPS cells without exogenous transgenes of an ecotropic retroviral receptor could be successfully established (15).

2 Materials

1. Phosphate buffered saline (PBS): PBS without Ca^{2+} and Mg^{2+}. Sterilize by autoclave and store at room temperature (RT).
2. Somatic cell culture (SCC) medium: Dulbecco's modified Eagle's medium (DMEM) with 10 % fetal bovine serum, 100 unit/ml penicillin, and 100 μg/ml streptomycin. For expansion of Plat-A cells, SCC medium with 1 μg/ml puromycin and 10 μg/ml blasticidin S. Store at 4 °C.
3. Trypsin/EDTA solution: 0.25 % trypsin/1 mM ethylenediaminetetraacetic acid (EDTA)·4Na solution. Aliquot and store at −20 °C.
4. Cell Strainer (mesh size: 40 μm, CORNING, Tewksbury, MA, USA).
5. Mitomycin C stock solution: 1 mg/ml mitomycin C in sterilized PBS. Aliquot and store at −80 °C.
6. CELLBANKER 1plus (Nippon Zenyaku Kogyo Co., Ltd., Fukushima, Japan). Aliquot and store at −20 °C.
7. BICELL (Nihon Freezer Co., Ltd., Tokyo, Japan). Keep at 4 °C.
8. Puromycin stock solution: 5 mg/ml puromycin in sterilized and distilled water. Aliquot and store at −20 °C.
9. Blasticidin S stock solution: 10 mg/ml blasticidin S in sterilized and distilled water. Aliquot and store at −20 °C.
10. OPTI-MEM I reduced serum medium (Life Technologies, Carlsbad, CA, USA). Store at 4 °C.
11. Fugene 6 transfection reagent (Promega, Madison, WI, USA). Store at 4 °C.

12. Retroviral vectors: pMXs retroviral vector (14) encoding cynomolgus monkey KLF4, SOX2, POU5F1 and c-MYC genes (from the JCRB Gene Bank; http://genebank.nibio.go.jp/).
13. Polybrene (Hexadimethrine Bromide) stock solution: 8 mg/ml polybrene in distilled water. Sterilize using a 0.22-μm filter and store at 4 °C.
14. Gelatin solution: 0.1 % gelatin (from porcine skin) in distilled water. Dissolve and sterilize by autoclave, and store at 4 °C.
15. Basic fibroblast growth factor (bFGF) stock solution: 10 μg/ml bFGF in PBS with 0.1 % bovine serum albumin. Sterilize using a 0.22-μm filter, aliquot and store at −20 °C.
16. GlutaMax (Life Technologies). Store at 4 °C.
17. MEM non-essential amino acids solution (NEAA) (Life Technologies). Aliquot and store at −20 °C.
18. Knockout serum replacement (KSR) (Life Technologies). Aliquot and store at −20 °C.
19. ES medium: DMEM/F12 (1:1) supplemented with 20 % KSR, 1 % GlutaMax, 0.1 mM β-mercaptoethanol, 1 % NEAA, 10 ng/ml human recombinant leukemia inhibitory factor (hLIF), 4 ng/ml bFGF, 100 unit/ml penicillin, and 100 μg/ml streptomycin. Sterilize using a 0.22-μm filter and store at 4 °C.
20. Collagenase solution: 0.1 % collagenase in Dulbecco's modified Eagle's medium (DMEM). Aliquot and store at −20 °C.
21. DAP213 solution: ES medium with 2 M dimethyl sulfoxide (DMSO), 1 M acetamide, and 3 M propylene glycol without hLIF and bFGF (16). Sterilize using a 0.22-μm filter, aliquot and store at −80 °C.

3 Methods

All experiments must conform to national regulations.

3.1 Preparation of Mouse Embryonic Fibroblast (MEF) Cells

1. Prepare a pregnant mouse at 12.5–13.5 days post-coitum. Sacrifice the mouse and disinfect the abdomen with 70 % ethanol. Remove the uterus through an abdominal incision and transfer it to a 100-mm tissue culture dish with 10 ml of PBS.
2. Take out the fetus from the uterus using sterilized scissors and tweezers. Remove its head and liver and wash it in a new 100-mm tissue culture dish with 10 ml of PBS. Transfer it to a 100-mm dish with 10 ml of SCC medium.
3. Cut the trimmed fetus into small pieces and transfer the suspension into a 50-ml centrifuge tube by passing through a 21-gauge needle. Seed the suspension in a 100-mm dish with

10 ml of SCC medium and incubate at 37 °C in 5 % CO_2 (Passage 0) (*see* **Note 1**).

4. When 80–90 % confluency is reached after 4–5 days, aspirate the medium and wash with 5 ml of PBS. Add 2 ml of trypsin/EDTA solution and incubate at 37 °C in 5 % CO_2 for 2–3 min. Add 5 ml of the SCC medium and dissociate the MEF cell monolayer into single cells by pipetting using a 5-ml plastic pipette.
5. Transfer the cell suspension into a 50-ml centrifuge tube through a Cell Strainer and centrifuge at 270 × *g* for 5 min at RT. Aspirate the supernatant and resuspend in the SCC medium. For the expansion of MEF cells, seed the cells in a new 100-mm dish at a ratio of 1:4 or 1:5 and incubate at 37 °C in 5 % CO_2 (Passage 1).
6. When 80–90 % confluency is reached after 2–3 days (Passages 2–4), repeat Steps 4 and 5.
7. To prepare feeder cells for the generation and culture of iPS cells, mitomycin C-treated MEF cells at Passages 3 and 4 are used. Add 100 μl of mitomycin C solution to a 100-mm dish with 80–90 % confluent MEF cells and incubate at 37 °C in 5 % CO_2 for 2.5–3 h.
8. Aspirate the medium and wash with 5 ml of PBS. Add 1 ml of trypsin/EDTA solution and incubate at 37 °C in 5 % CO_2 for 1–2 min. Add 5 ml of the SCC medium and dissociate the mitomycin C-treated MEF cell monolayer into single cells by pipetting using a 5-ml plastic pipette. Transfer the cell suspension into 50-ml centrifuge tubes and centrifuge at 270 × *g* for 5 min at RT.
9. Freeze MEF cells if they have not been used for culture or as feeder cells of iPS cells. Repeat Steps 4 and 5, and suspend the MEF cells at Passages 0–2 or the mitomycin C-treated MEF cells at Passages 3–4 in 2–3 ml of CELLBANKER 1plus per 100-mm dish. Transfer 1 ml of the cell suspension per cryotube and keep them in BICELL at −80 °C (*see* **Note 2**). The next day, transfer the cryotubes in liquid nitrogen (LN_2) tank.
10. For the generation of iPS cells, pick up a cryotube containing mitomycin C-treated MEF cells from the LN_2 tank and put it in a water bath at 37 °C. Transfer the suspension of MEF cells into a 15-ml centrifuge tube containing 10 ml of SCC medium and centrifuge it at 270 × *g* for 5 min at RT.
11. Aspirate the supernatant and resuspend the pellet in the SCC medium. Seed 1.2–1.6 × 10^6 cells in 10 ml of SCC medium on 100-mm gelatin-coated dishes and use within 1–2 days (*see* **Note 3**).

3.2 Preparation of Plat-A Cells

1. Pick up a cryotube containing Plat-A cells from the LN_2 tank and put it in a water bath at 37 °C. Transfer the suspension of Plat-A cells into a 15-ml centrifuge tube containing 10 ml of SCC medium and centrifuge it at 270 × *g* for 5 min at RT.
2. Aspirate the supernatant and resuspend the pellet in the SCC medium with puromycin and blasticidin S. Seed the suspension in 100-mm dishes and incubate at 37 °C in 5 % CO_2.
3. When 80–90 % confluency is reached, aspirate the medium and wash with 5 ml of PBS (*see* **Note 4**). Add 1 ml of trypsin/EDTA solution and incubate at 37 °C in 5 % CO_2 for 1 min.
4. Add 5 ml of the SCC medium and dissociate the cell monolayer into single cells by pipetting using a 5-ml plastic pipette. Transfer the cell suspension into a 50-ml centrifuge tube and centrifuge at 270 × *g* for 5 min at RT.
5. Aspirate the supernatant and, for the expansion of Plat-A cells, seed the cells in a new 100-mm dish at a ratio of 1:5 and incubate at 37 °C in 5 % CO_2.
6. Freeze Plat-A cells if they have not been used for expansion or the production of retroviral vectors. Repeat Steps 3 and 4 and resuspend the pellet in 3 ml of CELLBANKER 1plus per 100-mm dish. Transfer 1 ml of the cell suspension per cryotube and keep them in BICELL at −80 °C (*see* **Note 2**). The next day, transfer the cryotubes in LN_2 tank.

3.3 Preparation of Cynomolgus Monkey Somatic Cells

1. Collect some fetal liver or newborn skin tissue. Mince it thoroughly in a little SCC medium in the lid of a 60-mm dish using scissors and transfer the minced tissue pieces in 10 ml of SCC medium in a 100-mm dish. Incubate at 37 °C in 5 % CO_2 and change the medium after 5–7 days.
2. Change the medium two times a week. When 80–90 % confluency is reached, aspirate the medium and wash with 5 ml of PBS. Add 2 ml of trypsin/EDTA solution and incubate at 37 °C in 5 % CO_2 for 5 min.
3. Add 5 ml of SCC medium and dissociate the cell monolayer into single cells by pipetting using a 5-ml plastic pipette. Transfer the cell suspension into a 50-ml centrifuge tube through a Cell Strainer and centrifuge at 270 × *g* for 5 min at RT.
4. Aspirate the supernatant and resuspend in SCC medium. Seed the suspension in 10 ml of SCC medium in a new 100-mm dish at a ratio of 1:3 and incubate at 37 °C in 5 % CO_2.
5. Repeat Steps 2 and 3, except the use of the Cell Strainer.
6. Aspirate the supernatant and proceed to Step 7 or 8.
7. For freezing, suspend the pellet in 2–3 ml of CELLBANKER 1plus per 100-mm dish. Transfer 1 ml of the cell suspension per cryotube and keep them in BICELL at −80 °C (*see* **Note 2**). The next day, transfer the cryotubes in LN_2 tank.

8. For the generation of iPS cells, suspend the pellet in SCC medium, seed 1.5×10^5 cells in 5 ml of medium in a 60-mm dish and incubate at 37 °C in 5 % CO_2. The next day, use for infection with amphotropic retroviral vectors.

3.4 Production of Retroviral Vectors

1. Pick up a tube containing Plat-A cells from the LN_2 tank and put it into a water bath at 37 °C. Transfer the cell suspension into a 15-ml centrifuge tube containing 8 ml of SCC medium and centrifuge at $270 \times g$ for 5 min at RT.
2. Aspirate the supernatant and resuspend the pellet in the SCC medium. Seed 1.5×10^6 cells in 5 ml of SCC medium in a 60-mm dish and incubate at 37 °C in 5 % CO_2 for 24 h. Prepare four 60-mm dishes.
3. Transfer 100 μl of OPTI-MEM I reduced serum medium into a 1.5-ml sampling tube, add 9 μl of Fugene 6 transfection reagent to it and mix by tapping gently. Prepare four tubes and keep at RT for 5 min.
4. Add 3 μg of pMXs plasmid, Oct3/4, Sox2, Klf4, or c-Myc, to the solution, mix by tapping gently, and keep at RT for 15 min (*see* **Note 5**).
5. Add the mixture gently into the four dishes used for culturing Plat-A cells and incubate at 37 °C in 5 % CO_2 for 24 h (*see* **Note 6**).
6. Change to fresh SCC medium gently and incubate at 37 °C in 5 % CO_2 for 24 h (*see* **Note 7**).
7. Collect the supernatants with viruses from the four dishes into a 15-ml tube through a 0.45-μm filter. Add 4 μl of polybrene stock solution to the virus supernatants and mix gently.

3.5 Generation of iPS Cells

1. Aspirate the medium used for culturing the cells from cynomolgus fetal liver or newborn skin (Step 8 of Section 3.3), transfer 2 ml of the virus supernatants containing polybrene (Step 7 of Section 3.4) and incubate at 37 °C in 5 % CO_2 for 24 h.
2. Aspirate the virus supernatants, change to fresh SCC medium and incubate at 37 °C in 5 % CO_2. Change the medium after 2–3 days.
3. Seven days after infection, repeat Steps 5 and 6 of Section 3.3 to collect the infected cells. Seed $1.5–5 \times 10^5$ infected cells in 10 ml of SC medium per 100-mm dish prepared in Step 11 of Section 3.1 and incubate at 37 °C in 5 % CO_2 overnight.
4. Aspirate the medium and wash with 5 ml of PBS. Transfer 10 ml of ES medium and incubate at 37 °C in 5 % CO_2. Change the medium every other day. After approximately 3 weeks of seeding of infected cells, ES-like colonies appear and should be passaged (Fig. 1a, b) (*see* **Note 8**).
5. To passage ES-like colonies, prepare 35-mm dishes according to Step 11 of Section 3.1 and use within 1–2 days (*see* **Note 9**).

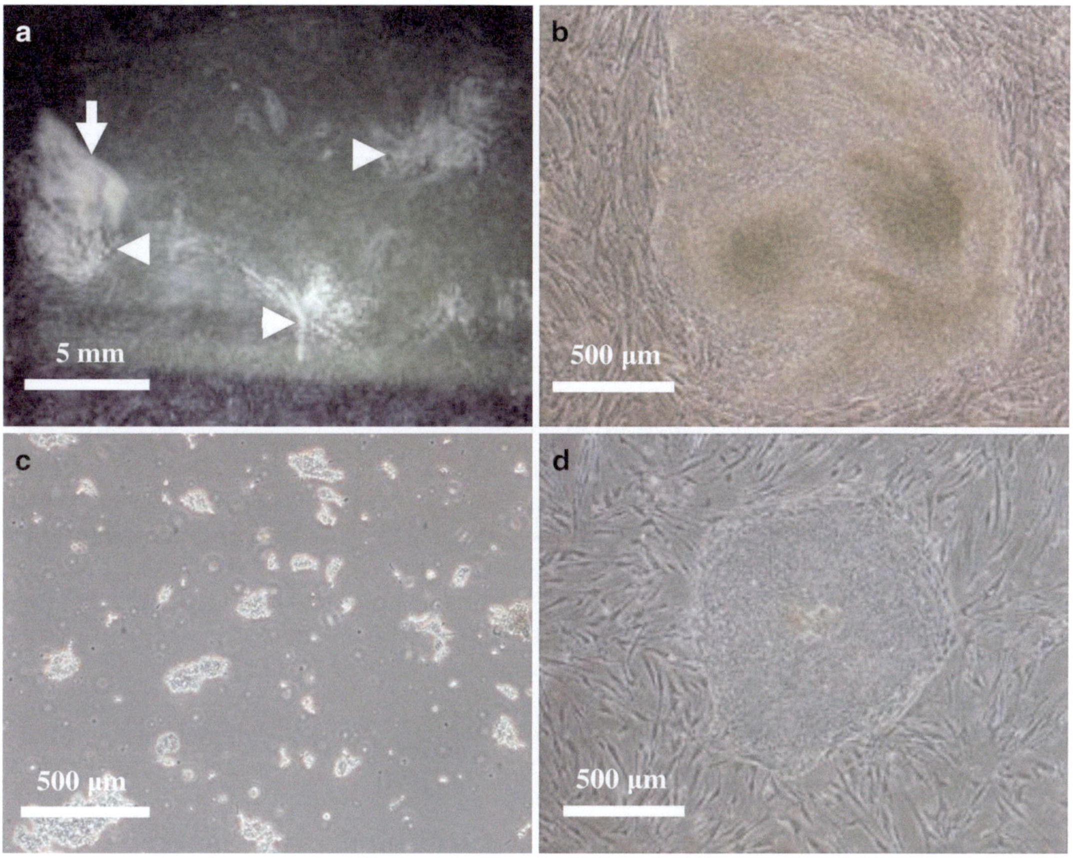

Fig. 1 Generation and passage of iPS cells. (**a**) The *arrow* and *arrowheads* indicate the iPS cell colony and the non-iPS cell colonies, respectively. The iPS cell colony consists of cells aggregating more densely than in the non-iPS cell colonies (**b**) The iPS cell colony consists of cells with large nuclei, similar to ES cell colony, having a flat morphology. (**c**) The detached iPS cell colonies that divided into small clusters by pipetting using a 1,000-μl tip or a 5-ml plastic pipette. The attachment of the iPS cell colony to the culture dish is weakened by treatment with collagenase and the edge is turned up. Such iPS colonies are easily detached from the bottom of the dish by pipetting using a 200-μl tip or a 5-ml plastic pipette. (**d**) An iPS cell colony undergoing stable passaging

3.6 Passage and Culture of iPS Cells

1. Trace the edge of an ES-like colony with a 26-gauge needle. Detach the colony from the bottom of the dish by pipetting using a 200-μl tip. Transfer the colony with a little medium using a 200-μl tip to an empty 35-mm dish and divide the colony into small clusters by pipetting using a 200-μl tip.
2. Transfer all of the clusters in 2 ml of ES medium into the 35-mm dish prepared in Step 5 of Section 3.5 and incubate at 37 °C in 5 % CO_2 (Passage 1). Change the medium every day. Passage the iPS cells after approximately 5–7 days.

3. Aspirate the medium and wash with 2 ml of PBS. Transfer 500 μl of collagenase solution and incubate at 37 °C in 5 % CO_2 for 3–5 min.
4. Confirm the detached edge of the colonies under an inverted microscope. Aspirate the collagenase solution, add 2 ml of the ES medium and detach all of the colonies from the bottom of the dish by pipetting using a 200-μl tip under a stereomicroscope. Collect the detached colonies into a 15-ml centrifuge tube and centrifuge at 190 × *g* for 3 min at RT.
5. Aspirate the supernatant, add 500 μl of the ES medium and divide the colonies into small clusters by pipetting using a 1,000-μl tip. Transfer the suspension into the 35-mm dish prepared in Step 5 of Section 3.5 (Passage 2) (*see* **Note 10**).
6. When the culture of iPS cells is stable, these cells can be expanded at a ratio of 1:5–10. Repeat Steps 3–5, or when a 60-mm dish is used, treat with 1 ml of collagenase solution at 37 °C in 5 % CO_2 for 3–5 min and add 4 ml of the ES medium into the dish. Detach the colonies, divide them into small clusters by pipetting using a 5-ml plastic pipette, and passage the iPS cells using 60-mm dishes (Fig. 1c, d).

3.7 Vitrification and Thawing of iPS Cells

1. Put empty cryotubes and DAP213 solution on ice. Transfer the LN_2 into a Dewar flask. Repeat Steps 3 and 4 of Section 3.6 to collect the iPS cells.
2. Aspirate the supernatant and resuspend the pellet in 400–600 μl of DAP213 solution by gentle pipetting using a 1,000-μl tip. Transfer 200 μl of the suspension per cryotube on ice into the LN_2 in the Dewar flask immediately (*see* **Note 11**).
3. For thawing, put a 15-ml centrifuge tube containing 10 ml of ES medium in an incubator at 37 °C in advance.
4. Pick up the cryotube containing iPS cells from the LN_2 tank. Take out the 15-ml centrifuge tube containing ES medium, add 1,000 μl of the ES medium into the cryotube and suspend by rapid pipetting using a 1,000-μl tip (*see* **Note 12**). Transfer the suspension into the 15-ml centrifuge tube and centrifuge at 190 × *g* for 3 min at RT.
5. Aspirate the supernatant and resuspend the pellet in ES medium. Seed the suspension in 5 ml of ES medium in 60-mm dishes prepared according to Step 11 of Section 3.1 at a ratio of 1:1 or 1:2 and incubate at 37 °C in 5 % CO_2.

4 Notes

1. If the number of fetuses is ten, aliquot the suspension of primary MEF cells into ten dishes.
2. When resuspending MEF cells in CELLBANKER 1plus, transfer the cell suspension into a cryotube and put it in BICELL containing the cryotubes at −80 °C as soon as possible.
3. When 60-mm or 35-mm dishes are used for the culture of iPS cells, seed 5–7 × 10^5 and 2.5–3.5 × 10^5 MEF cells, respectively.
4. Because Plat-A cells are easy to detach from dishes, gradually pour PBS into the dishes and wash the cells.
5. If introduction of the transgenes into the somatic cells has been confirmed, pMX containing cDNA of GFP and a dish with Plat-A cells should also be added.
6. Because Plat-A cells are easy to detach from dishes, gradually add the mixture to the dishes.
7. Because Plat-A cells are easy to detach from dishes, gradually pour SCC medium into the dishes.
8. ES-like colonies have a flat morphology, consisting of cells with large nuclei.
9. A 35-mm dish should be used rather than a six-well plate because only dishes used to culture iPS cells for passaging should be taken out of the incubator; this can thus be avoided for iPS cells that are not for passaging.
10. If the number of colonies is five or less and more than five, 35-mm and 60-mm dishes should be used for passaging, respectively.
11. The process from resuspending in DAP 213 solution to putting into the LN_2 should be conducted within 30 s.
12. Discard LN_2 if any remains in the cryotube. Within 5 s, add ES medium at 37 °C to the cryotube and suspend by rapid pipetting so as to avoid recrystallization.

Acknowledgements

I thank Ryoichi Ono, Manami Shimada, Hiroaki Shibata, Ichiro Takahashi, Hiroyasu Inada, Tatsuyuki Takada, Tetsuya Nosaka, and Yasuhiro Yasutomi for this experiment. I am also grateful to Prof. Toshio Kitamura for providing pMXs retroviral vectors and Plat-A cells.

References

1. Takahashi K, Yamanaka S (2006) Induction of pluripotent stem cells from mouse embryonic and adult fibroblast cultures by defined factors. Cell 126:663–676
2. Okita K, Ichisaka T, Yamanaka S (2007) Generation of germ-line-competent induced pluripotent stem cells. Nature 448:313–317
3. Nakaki F, Hayashi K, Ohta H, Kurimoto K, Yabuta Y, Saitou M (2013) Induction of mouse germ-cell fate by transcription factors *in vitro*. Nature 501:222–226
4. Takahashi K, Tanabe K, Ohnuki M, Narita M, Ichisaka T, Tomoda K et al (2007) Induction of pluripotent stem cells from adult human fibroblasts by defined factors. Cell 131:861–872
5. Liu H, Zhu F, Yong J, Zhang P, Hou P, Li H et al (2008) Generation of induced pluripotent stem cells from adult rhesus monkey fibroblasts. Cell Stem Cell 3:587–590
6. Tomioka I, Maeda T, Shimada H, Kawai K, Okada Y, Igarashi H et al (2010) Generating induced pluripotent stem cells from common marmoset (*Callithrix jacchus*) fetal liver cells using defined factors, including Lin28. Genes Cells 15:959–969
7. Okahara-Narita J, Umeda R, Nakamura S, Mori T, Noce T, Torii R (2012) Induction of pluripotent stem cells from fetal and adult cynomolgus monkey fibroblasts using four human transcription factors. Primates 53:205–213
8. Navara CS, Hornecker J, Grow D, Chaudhari S, Hornsby PJ, Ichida JK et al (2013) Derivation of induced pluripotent stem cells from the Baboon: a nonhuman primate model for preclinical testing of stem cell therapies. Cell Reprogram 15:495–502
9. Kim K, Doi A, Wen B, Ng K, Zhao R, Cahan P et al (2010) Epigenetic memory in induced pluripotent stem cells. Nature 467:285–290
10. Lister R, Pelizzola M, Kida YS, Hawkins RD, Nery JR, Hon G et al (2011) Hotspots of aberrant epigenetic reprogramming in human induced pluripotent stem cells. Nature 471:68–73
11. Nishino K, Toyoda M, Yamazaki-Inoue M, Fukawatase Y, Chikazawa E, Sakaguchi H et al (2011) DNA methylation dynamics in human induced pluripotent stem cells over time. PLoS Genet 7:e1002085
12. Maekawa M, Yamaguchi K, Nakamura T, Shibukawa R, Kodanaka I, Ichisaka T et al (2011) Direct reprogramming of somatic cells is promoted by maternal transcription factor Glis1. Nature 474:225–229
13. Hou P, Li Y, Zhang X, Liu C, Guan J, Li H et al (2013) Pluripotent stem cells induced from mouse somatic cells by small-molecule compounds. Science 341:651–654
14. Kitamura T, Koshino Y, Shibata F, Oki T, Nakajima H, Nosaka T et al (2003) Retrovirus-mediated gene transfer and expression cloning: powerful tools in functional genomics. Exp Hematol 31:1007–1014
15. Shimozawa N, Ono R, Shimada M, Shibata H, Takahashi I, Inada H et al (2013) Cynomolgus monkey induced pluripotent stem cells established by using exogenous genes derived from the same monkey species. Differentiation 85:131–139
16. Fujioka T, Yasuchika K, Nakamura Y, Nakatsuji N, Suemori H (2004) A simple and efficient cryopreservation method for primate embryonic stem cells. Int J Dev Biol 48:1149–1154

Methods in Molecular Biology (2016) 1357: 183–193
DOI 10.1007/7651_2014_159

Published online: 26 December 2014

Induced Pluripotent Stem Cells from Nonhuman Primates

Anuja Mishra, Zhifang Qiu, Steven L. Farnsworth, Jacob J. Hemmi, Miao Li, Alexander V. Pickering, and Peter J. Hornsby

Abstract

Induced pluripotent stem cells from nonhuman primates (NHPs) have unique roles in cell biology and regenerative medicine. Because of the relatedness of NHPs to humans, NHP iPS cells can serve as a source of differentiated derivatives that can be used to address important questions in the comparative biology of primates. Additionally, when used as a source of cells for regenerative medicine, NHP iPS cells serve an invaluable role in translational experiments in cell therapy. Reprogramming of NHP somatic cells requires the same conditions as previously established for human cells. However, throughout the process, a variety of modifications to the human cell protocols must be made to accommodate significant species differences.

Keywords: Induced pluripotent stem cells, Nonhuman primates, Marmoset, Reprogramming, Retroviruses, Teratoma

1 Introduction

Following the generation of human induced pluripotent stem cells by Yamanaka's group [1, 2], the first nonhuman primate species to be reprogrammed was the rhesus macaque (*Macaca mulatta*) [3]. Following this, the second nonhuman primate was the common marmoset (*Callithrix jacchus*) by our group, published in 2010 [4]. Another group subsequently published a somewhat different method for the generation of marmoset iPS cells [5]. This chapter focuses on the marmoset, as the species with which we are most familiar. We have used essentially the same techniques to generate and grow iPS cells from the chimpanzee (*Pan troglodytes*) and other primates.

Based on the genetic relatedness of humans and these various NHP species, methods developed for human cells can generally be used for NHPs. However, throughout the process of generation, growth, and differentiation of NHP iPS cells, we have found it necessary to adjust the conditions previously found suitable for human cells to these other species. Some advances described for human cells such as the use of TeSR1 medium—the first defined medium described for human pluripotent cells [6]—have not been found suitable for marmoset iPS cells. On the other hand,

E8 medium, essentially a simpler version of TeSR1 [7], has worked well in our hands for NHP iPS cells. The necessity of adapting each method developed for human pluripotent cells to each NHP species may give the impression that the cells are difficult to derive and maintain. However, when optimized conditions are used, we find that NHP iPS cells are no more difficult to handle than human iPS cells, reflecting the fact that most attention has been paid to human cells, and therefore, unsurprisingly, protocols have been fine-tuned for the latter species.

2 Materials

Maintain sterility for all components. Antibiotics (100 μg/ml penicillin and 50 μg/ml gentamicin) are added to all components that come into contact with cells to prevent bacterial contamination (**Note 1**).

2.1 Cell Culture Medium

1. Dulbecco's Modified Eagle's Medium (DMEM) with 10 % Cosmic Calf Serum (CCS; Thermo Hyclone catalog number SH30087.03HI): used for skin fibroblast culture and for the first stage of reprogramming of fibroblasts to iPS cells (**Note 2**).
2. E8 medium [7]: used for all routine growth of iPS cells, as well as expansion of newly derived clones of cells; made up from components as described [7] or available from commercial sources. We prepare it from a basal medium comprising DMEM/F12 (Sigma catalog number D8437). Many NHP iPS cell clones require the addition of 10 % fetal bovine serum (FBS) to E8 (**Note 3**). A batch of the medium can be divided into 50 ml tubes to be stored at −20 °C; thaw out just before use.
3. DMEM/F12 with Knockout Serum Replacement (KSR): a medium comprising DMEM/F12, 20 % KSR (Life Technologies catalog number 10828-028), and 20 ng/ml FGF2, with the addition of sodium bicarbonate (25 ml of 18 mg/ml $NaHCO_3$ added to 100 ml medium) to avoid excessive acidity. We use this medium in the early expansion of reprogrammed cells (*see* Section 3.2).
4. mFreSR (STEMCELL Technologies Inc., Vancouver, Canada): used for cell cryopreservation. Aliqout and store at −20 °C until needed.
5. ROCK inhibitor Y-27632: add to the medium for iPS cells. The compound is made up at 10 mM in water and stored frozen in aliquots.

2.2 Enzymes

1. Crude collagenase (Sigma Type I) for preparation of fibroblasts from skin.
2. Accutase (a mix of enzymes with proteolytic and collagenolytic activities) available from several commercial sources as a pre-made solution: use for subculture of pluripotent cells. Store at −20 °C in 15 ml aliquots and thaw just before use.

2.3 Materials for Reprogramming

1. Use the reprogramming factors (Oct4, Sox2, Klf4, c-Myc) expressed by retroviruses in the pMXs vector [2]. A suitable preparation of the mix of the four viruses can be prepared in-house [8] by transfection of the retroviral plasmids into a suitable packaging cell lines such as Plat-A cells [9] or can be purchased as a ready-to-use mix (Salk Institute, Gene Transfer, Targeting and Therapeutics Core) (**Note 4**).
2. The supernatant from the packaging cells should be filtered through a 45 μm syringe filter and then can be aliquoted and stored at −80 °C. Refreezing aliquots is not advisable.
3. Valproic acid (VPA) is used as an epigenetic modifier to enhance the rate of reprogramming. It is made up as a 100 mM stock in water and added to cells at 1 mM [10].

2.4 Extracellular Matrix Components

1. For routine feeder-free culture, culture dishes can be coated with Matrigel (BD Biosciences catalog number 354234). When received, Matrigel should be thawed once to 4 °C and aliquoted in 300 μl aliquots. Store at −20 °C.
2. Dishes, pipettes, medium, and other items should be pre-cooled to 4 °C before starting the protocol.
3. Mix 25 ml DMEM/F12 with 300 μl Matrigel using a polyethylene transfer pipette (Fisher Scientific, catalog number 13-711-20).
4. Pipette 2 ml of the mix into each 35 mm dish. At this point, the dishes and contents can be allowed to warm to room temperature.
5. Place the dishes on a rotating shaker for 1 h.
6. Store dishes at 4 °C and use within 1 week.

2.5 Cells for Preparing Feeder Layers

1. Mitotically inactivated mouse embryo fibroblasts (**Note 5**) can be prepared from mouse embryos in-house or can be conveniently obtained commercially (e.g., mitomycin inactivated, GlobalStem Inc., catalog number GSC-6001M). Cryopreserved cells are stored in liquid nitrogen until needed.
2. Cells are thawed at 37 °C and then plated on Matrigel-coated dishes in DMEM/CCS. Dishes are incubated at 37 °C. They can be used the following day or up to 3–4 days after plating.

3 Methods

All procedures should be carried out under BSL2 conditions (**Note 6**).

3.1 Preparation of NHP Skin Fibroblasts

1. Obtain a skin sample from the animal. Include the subcutaneous tissue whenever possible. Once excised from the animal, the tissue should be kept on ice in a suitable container, but it is not necessary to add medium or buffer. Small samples such as punch biopsies from living animals can be successfully used. Animals that have died of natural causes can also yield skin for successful culture, if the skin is obtained within a few hours of death. Skin samples can be shipped on ice and/or stored at 4 °C for several days and still yield a large number of viable cells.
2. Dissolve collagenase at 5 mg/ml in DMEM/CCS. Add the collagenase solution to the skin sample in a dish. Using scalpel and forceps, cut the skin into small fragments (1–2 mm). Incubate the fragments in the collagenase at 37 °C in an incubator. After 1 h, flush the fragments through a polyethylene transfer pipette. Repeat every hour, until the tissue has dissociated to single cells and small clumps of cells. This may take several hours.
3. When the tissue has dissociated, centrifuge the single cells and clumps and resuspend them in DMEM/CCS. Plate the cells in 35 mm, 6 cm, or larger dishes as needed. Change the medium the next day and every 2 days thereafter.
4. Cells can be infected with reprogramming viruses as soon as they start dividing rapidly. Some cells should be continued in standard fibroblast culture and then cryopreserved in 5 % DMSO for optional later repeat of reprogramming.

3.2 Reprogramming Using Retroviruses Encoding Oct4, Sox2, Klf4, and c-Myc

1. Grow skin fibroblasts as described above and ensure that the cells are dividing at a maximal rate (**Note 7**).
2. Subculture the cells into wells of a 6-well culture plate. The optimal cell density should be determined for each population of fibroblasts, but typically about 60–70 % cell confluency is optimal.
3. Thaw a suitable aliquot of frozen viral supernatant, containing all four viruses, with warm water, not allowing the supernatant to exceed room temperature. For each well use 400 μl of virus mix plus 2 ml fibroblast culture medium. Add 8 μg/ml Polybrene to the mix [8]. Replace the culture medium with the virus mix (**Note 8**).
4. Allow the cells to equilibrate in the incubator for 5 min. Wrap the plate in Parafilm and then transfer to the centrifuge.

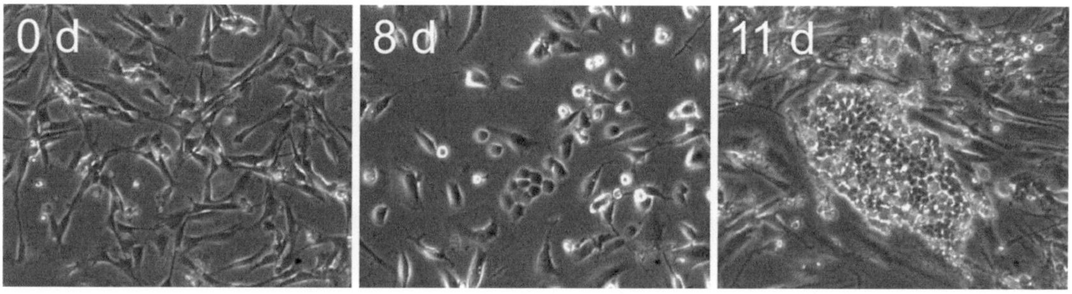

Fig. 1 Derivation of marmoset iPS cells from skin fibroblasts. (**0 d**) Skin fibroblasts (adult marmoset) growing in culture before infection with reprogramming retroviruses. (**8 d**) About 8 days after infection, small numbers of cells with an altered morphology (epithelial rather than fibroblastic) are noted, if reprogramming is proceeding successfully. (**11 d**) About 11 days after infection, distinct colonies of cells that have the appearance of iPS cells are noted; cells are small, rapidly dividing, with prominent nuclei

5. Using a suitable plate carrier (e.g., Beckman Coulter Inc., SX4750 Rotor), centrifuge the plate at 1,200 × *g* for 1 h at 15 °C [8].
6. Remove the Parafilm and place the plate in the incubator. After 3 h, add extra DMEM/CCS to the virus mix to fill the well.
7. After 24 h, repeat the process of infection (steps 3–6).
8. After a further 24 h, replace the medium with DMEM/CCS plus 1 mM VPA. If the cells become confluent, subculture them at a ratio typical for fibroblasts (e.g., 1:4).
9. After 6 days in these conditions, subculture the cells using trypsin at a ratio of 1:4. Plate the cells onto a mouse embryo fibroblast feeder layer in DMEM/F12 with 20 % KSR (**Note 9**).
10. Over the next 1–2 weeks, colonies of cells of altered appearance should become apparent (Fig. 1). The change in morphology is initially the result of a mesenchymal to epithelial transition [11]. As the colonies develop, they adopt an appearance characteristic of iPS cells (small, rapidly dividing cells with prominent nuclei). Colonies of NHP iPS cells are generally flat and do not adopt the domed appearance of mouse pluripotent cells.
11. Once the colonies have the appearance of typical iPS cells, change the medium to E8 with 10 % FBS and 10 μM ROCK inhibitor.
12. When colonies have developed to a size of several hundred cells, they are isolated by detaching them with Accutase. Flood the dish with Accutase. Wait until cells in the colonies round up. Under a phase contrast microscope, aspirate the colony using a fine-tip polyethylene transfer pipette (Sarstedt, catalog number 86.1180).
13. Plate the cells in E8 with 10 % FBS and 10μM Rock inhibitor on a mouse embryo fibroblast feeder layer in a small well (e.g., 48-well plate).

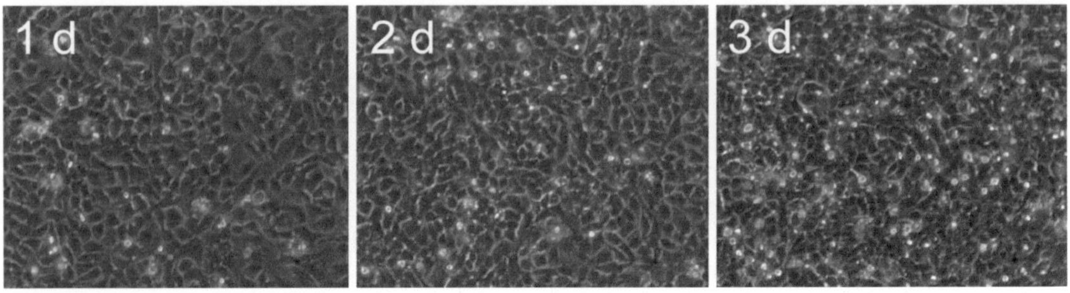

Fig. 2 Growth of marmoset iPS cells under feeder-free mass culture conditions. Cells of clone #15 (4) were plated on Matrigel-coated dishes in E8 medium with 10 % FBS. The figure shows the appearance of the cells at 1–3 days after plating as they achieve a high density again. As the cells grow to high density, it is typical to observe rounded cells, some of which detach or adhere to the surface of the monolayer. The medium also becomes acidic, even though it is changed every 24 h. The level of acidity is a useful indication that the cells should be passaged again

14. As the clone grows, transfer to successively larger wells and dishes.
15. Cryopreserve the clone as soon as it comprises sufficient cells.

3.3 Routine Maintenance of iPS Cells

1. NHP iPS cells can be grown as mass cultures [12] using E8 and Matrigel-coated dishes (Fig. 2). This is a much simpler protocol than the traditional colony-based growth on feeder layers. Use E8 medium and 10 μM ROCK inhibitor, with the addition of 10 % FBS as necessary (**Note 3**).
2. Change the medium every day. The pH of the medium gives a good indication of when to subculture; do not let the medium become too acidic, although high cell density with some accompanying acidity is beneficial for cell growth.
3. For subculturing, aspirate the medium, wash the monolayer of cells once with Accutase, add 1 ml more Accutase to the cells, and place the dish in the incubator. When the cells begin to round up but are not yet detaching, remove the Accutase from the monolayer and flush the cells off the dish into fresh medium using a polyethylene transfer pipette (**Note 10**).
4. Plate 1.5 ml cell suspension per 35 mm dish and then change the next day with 2 ml medium. Generally, a split ratio of 1:2 is suitable. Using this split ratio, NHP iPS cells can often be grown with a subculture every 2 days.
5. Clones of iPS cells should be cryopreserved when the clone is first isolated and again at early passages (**Note 11**). Remove the cells from the plate using Accutase, as for subculturing. Centrifuge the cells at 156 × *g* for 3 min and then resuspend in 1.8 ml mFreSR per vial. Place the vials in a CoolCell (BioCision) at −80 °C overnight, and then transfer to liquid nitrogen for long-term storage.

6. At early passages, perform basic characterization of the clone, including expression of pluripotency genes (particularly *NANOG* and *OCT4/POU5F1*), proper silencing of the retroviral genes, and karyotyping [4].
7. To recover iPS cells from liquid nitrogen storage, place the vial in 37 °C water and thaw as rapidly as possible [13]. Add 1 ml warm medium and transfer to a 15 ml tube. After 2 min, add 2 ml more medium; after a further 2 min, add 4 ml medium; and then after 2 more minutes, centrifuge the cells at 156 × *g* for 3 min. Add 1.5 ml medium to the cell pellet for a 35 mm dish and transfer to the incubator.

3.4 Investigation of Pluripotency by Teratoma Formation

This section describes a protocol for assessment of the differentiation potential of pluripotent cells by teratoma formation (**Note 12**):

1. Various types of immunodeficient mice have been used successfully for the formation of teratomas. Following earlier demonstrations that more profoundly immunodeficient mice are superior for efficiency and rapidity of teratoma formation [14], we use *Rag2*$^{-/-}$, *Il2rg*$^{-/-}$ mice (available from Taconic, model 4111). These animals must be housed under conditions suitable for immunodeficient mice, the most important of which is the use of microisolator cages that provide a cage-level barrier against exposure to pathogens. With care not to expose these mice to pathogens derived from conventional mice, this strain can be housed without evidence of bacterial/viral disease under relatively convenient housing conditions.
2. Prepare cells that are being tested for teratoma formation as for routine subculture as described above. Produce a pellet of at least 10^6 cells. Cover the pellet with culture medium in a 1.6 ml tube and transport on ice for injection into the mouse.
3. Remove the medium from the pellet and resuspend in 20 μl of a mix of 50 % Matrigel and 50 % DMEM/F12 (*see* Section 2.4) kept on ice until used.
4. Anesthetize the mouse with Avertin [15] or a suitable alternative.
5. After the cells are mixed into the 50 % Matrigel, draw the cell suspension up into a cold 50 μl glass syringe (Hamilton) fitted with a suitable needle for subcutaneous injection (**Note 13**).
6. Inject the cell suspension subcutaneously; an ideal location is the skin on the head just caudal to the external ear. This location provides nearby blood vessels for growth support and is in a location that cannot be reached by the mouse during development of the teratoma.
7. After teratomas can be felt under the skin, euthanize the mice and excise the tumors for histological or other processing (Fig. 3).

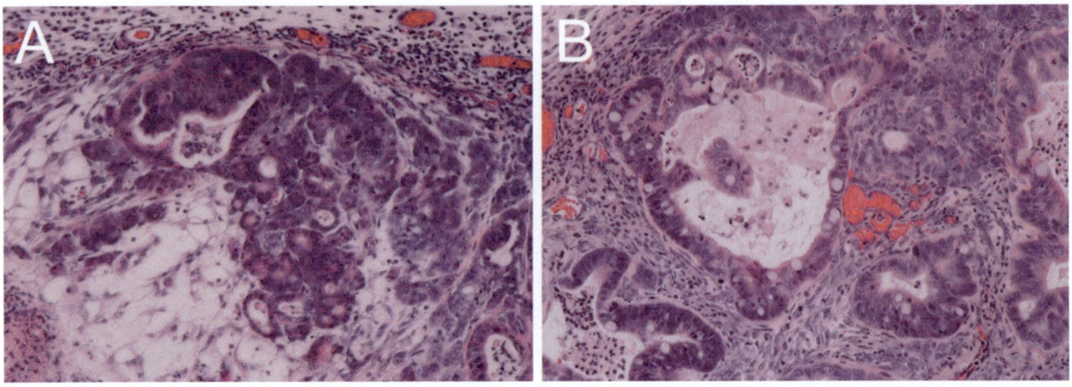

Fig. 3 Histological appearance of teratomas formed from a marmoset iPS cell line (B8; *see* ref. (4)). (**a**, **b**) Shows examples of various tissue structures within teratomas. Staining with tissue-specific antibodies shows that tissues within teratomas originate from all three germ layers (4). Hematoxylin and eosin-stained sections from conventionally fixed and paraffin-embedded tissue samples

4 Notes

1. The combination of penicillin and gentamicin is very effective against bacterial contamination and is helpful when initiating cultures from potentially nonsterile skin samples. We have tested several antimycotics for prevention of yeast contamination and have not found any that do not interfere with the growth of fibroblasts or iPS cells. Contamination by yeast is an occasional problem with primary cultures, and if it occurs contaminated cultures should be discarded as soon as the growth of yeast is noted.
2. Cosmic Calf Serum is a commercially available substitute for fetal bovine serum. It supports good growth of fibroblasts and many other cells.
3. Some marmoset iPS cell clones can be grown in E8 with ROCK inhibitor without further additions. Other clones grow much better with the addition of 10 % FBS. We use "ES-qualified" FBS (GlobalStem). In addition to E8, some other defined media such as Pluriton (Stemgent) may be used for marmoset iPS cells [16]. E8 is preferred, because it can readily be made up from defined components, enabling it to be modified as needed.
4. Although many protocols for reprogramming have been published, in our hands the original pMXs-based retroviral vectors have been the most reliable. We have also used a polycistronic retroviral vector successfully for marmoset cells [16], but this has proven to be much less efficient than the mix of four

retroviruses. We also found that a Sendai virus-based kit ("CytoTune," developed by DNAVEC Corporation and available from Life Technologies) can be used successfully for chimpanzee cells, but currently we have not been able to use this method to reprogram marmoset cells. Interestingly, this kit has been reported to be able to generate induced neural progenitor cells, but not iPS cells, using rhesus macaque fibroblasts [17]. Another study showed that cells that produced dysgerminomas were generated by the introduction of Oct4/Sox2/Klf4/c-Myc reprogramming factors in marmoset cells using lentiviral vectors [18]. Clearly it is desirable to develop robust protocols that use nonintegrating vectors or purely chemical methods for reliable reprogramming of NHP cells.

5. The use of E8 medium, with the addition of FBS when needed, and Matrigel-coated plates makes the routine mass culture of NHP iPS cells feasible without the use of traditional feeder layers. However, when early reprogrammed clones are being expanded, we find that mouse fibroblast feeder layers are required for successful expansion of clones.
6. Refer to your institution's specific guidelines for the handling of materials from NHPs. Generally, work with NHP materials, as well as retroviruses, should be under BSL2 conditions [19].
7. The process of infecting cells with retroviruses is simple but can be of low efficiency. Cells must be dividing to be infected, and the use of a positively charged polymer with low speed centrifugation of the culture plate ("spinoculation") can greatly increase infection rates [8]. A high infection rate is critical because a cell must be infected with all four retroviruses to undergo reprogramming.
8. The amount of virus that is required for successful reprogramming is difficult to predict in advance of the experiment. If possible, prepare batches of virus that are large enough to permit multiple reprogramming experiments. As necessary, adjust the amount of virus used per experiment to permit reprogramming for that particular somatic cell culture. In a successful experiment, large numbers of iPS cell clones will be obtained, more than what is typically needed. If the number of clones obtained is very low, or zero, attempt to increase the virus as much as possible so that some successful reprogramming occurs.
9. We use this medium at this stage of the reprogramming because it is permissive for the growth of pluripotent cells, while not encouraging the growth of fibroblasts. Therefore, the culture can be maintained for 7–14 days without subculturing, as the iPS cell colonies appear and the fibroblasts slowly degenerate. Although E8 is better for the long-term maintenance of NHP iPS cells, it does also permit growth of fibroblasts.

10. Although conventionally cells are resuspended in fresh medium and centrifuged before replating, this step is unnecessary for efficient subculturing of pluripotent cells, as noted previously for an EDTA-based protocol [20]. Timing of Accutase treatment is critical and must be customized for each cell line. Under these conditions, it is simple to aspirate the Accutase without disturbing the cell monolayer and then to flush the cells off into fresh medium, which also serves to neutralize the enzymes in Accutase.
11. We have not extensively determined the importance of passage number on the properties of NHP iPS cells. However, it is always advisable to use cells at a known passage and usually at as low a passage number as practical.
12. The teratoma assay, as an assay for pluripotency, has been criticized on several grounds [21], yet remains an important test for NHP iPS cells. For human cells, patterns of gene expression under relatively simple differentiation conditions (e.g., embryoid body formation) have been defined adequately such that a newly derived cell line can be verified to be a genuine iPS cell line without the teratoma assay. However, for many other species, similar assays have not yet been developed. In those species the teratoma assay remains the "gold standard" for pluripotency determination.
13. The needle should be carefully chosen to permit the injection of a gel (Matrigel plus cells) efficiently under the skin. We use a custom 22 gauge needle (Hamilton Company) that is short (0.5 in.) and has a 45° beveled tip.

Acknowledgments

This work was supported by VA grant I01BX001454, NIH grants R21 AG033286 and R03 AG045481, and by grants from the Owens Medical Research Foundation and the Ted Nash Long Life Foundation.

References

1. Takahashi K, Yamanaka S (2006) Induction of pluripotent stem cells from mouse embryonic and adult fibroblast cultures by defined factors. Cell 126:663–676
2. Takahashi K, Tanabe K, Ohnuki M, Narita M, Ichisaka T, Tomoda K, Yamanaka S (2007) Induction of pluripotent stem cells from adult human fibroblasts by defined factors. Cell 131:861–872
3. Liu H, Zhu F, Yong J, Zhang P, Hou P, Li H, Jiang W, Cai J, Liu M, Cui K, Qu X, Xiang T, Lu D, Chi X, Gao G, Ji W, Ding M, Deng H (2008) Generation of induced pluripotent stem cells from adult rhesus monkey fibroblasts. Cell Stem Cell 3:587–590
4. Wu Y, Zhang Y, Mishra A, Tardif SD, Hornsby PJ (2010) Generation of induced pluripotent stem cells from newborn marmoset skin fibroblasts. Stem Cell Res 4:180–188
5. Tomioka I, Maeda T, Shimada H, Kawai K, Okada Y, Igarashi H, Oiwa R, Iwasaki T, Aoki M, Kimura T, Shiozawa S, Shinohara H, Suemizu H, Sasaki E, Okano H (2010) Generating induced pluripotent stem cells from common

marmoset (*Callithrix jacchus*) fetal liver cells using defined factors, including Lin28. Genes Cells 15:959–969

6. Ludwig TE, Levenstein ME, Jones JM, Berggren WT, Mitchen ER, Frane JL, Crandall LJ, Daigh CA, Conard KR, Piekarczyk MS, Llanas RA, Thomson JA (2006) Derivation of human embryonic stem cells in defined conditions. Nat Biotechnol 24:185–187
7. Chen G, Gulbranson DR, Hou Z, Bolin JM, Ruotti V, Probasco MD, Smuga-Otto K, Howden SE, Diol NR, Propson NE, Wagner R, Lee GO, Antosiewicz-Bourget J, Teng JM, Thomson JA (2011) Chemically defined conditions for human iPSC derivation and culture. Nat Methods 8:424–429
8. Wu Y, Melton DW, Zhang Y, Hornsby PJ (2009) Improved coinfection with amphotropic pseudotyped retroviral vectors. J Biomed Biotechnol 2009:901079
9. Morita S, Kojima T, Kitamura T (2000) Plat-E: an efficient and stable system for transient packaging of retroviruses. Gene Ther 7:1063–1066
10. Huangfu D, Osafune K, Maehr R, Guo W, Eijkelenboom A, Chen S, Muhlestein W, Melton DA (2008) Induction of pluripotent stem cells from primary human fibroblasts with only Oct4 and Sox2. Nat Biotechnol 26:1269–1275
11. Polo JM, Hochedlinger K (2010) When fibroblasts MET iPSCs. Cell Stem Cell 7:5–6
12. Chen KG, Mallon BS, Hamilton RS, Kozhich OA, Park K, Hoeppner DJ, Robey PG, McKay RD (2012) Non-colony type monolayer culture of human embryonic stem cells. Stem Cell Res 9:237–248
13. Valbuena D, Sanchez-Luengo S, Galan A, Sanchez E, Gomez E, Poo ME, Ruiz V, Genbacev O, Krtolica A, Pellicer A, Moreno R, Simon C (2008) Efficient method for slow cryopreservation of human embryonic stem cells in xeno-free conditions. Reprod Biomed Online 17:127–135
14. Koch CA, Jordan CE, Platt JL (2006) Complement-dependent control of teratoma formation by embryonic stem cells. J Immunol 177:4803–4809
15. Papaioannou VE, Fox JG (1993) Efficacy of tribromoethanol anesthesia in mice. Lab Anim Sci 43:189–192
16. Wu Y, Mishra A, Qiu Z, Farnsworth S, Tardif SD, Hornsby PJ (2012) Nonhuman primate induced pluripotent stem cells in regenerative medicine. Stem Cells Int 2012:767195
17. Lu J, Liu H, Huang CT, Chen H, Du Z, Liu Y, Sherafat MA, Zhang SC (2013) Generation of integration-free and region-specific neural progenitors from primate fibroblasts. Cell Rep 3:1580–1591
18. Yamaguchi S, Marumoto T, Nii T, Kawano H, Liao J, Nagai Y, Okada M, Takahashi A, Inoue H, Sasaki E, Fujii H, Okano S, Ebise H, Sato T, Suyama M, Okano H, Miura Y, Tani K (2014) Characterization of common marmoset dysgerminoma-like tumor induced by the lentiviral expression of reprogramming factors. Cancer Sci 105:402–408
19. Adams SR, Muchmore E, Richardson JH (1995) Biosafety. In: Bennett BT, Abee CR, Henrickson RV (eds) Nonhuman primates in biomedical research: biology and management. Academic Press, San Diego, pp 377–420
20. Beers J, Gulbranson DR, George N, Siniscalchi LI, Jones J, Thomson JA, Chen G (2012) Passaging and colony expansion of human pluripotent stem cells by enzyme-free dissociation in chemically defined culture conditions. Nat Protoc 7:2029–2040
21. Dolgin E (2010) Putting stem cells to the test. Nat Med 16:1354–1357

Methods in Molecular Biology (2016) 1357: 195–220
DOI 10.1007/7651_2015_279

Published online: 04 July 2015

Computational Biology Methods for Characterization of Pluripotent Cells

Marcos J. Araúzo-Bravo

Abstract

Pluripotent cells are a powerful tool for regenerative medicine and drug discovery. Several techniques have been developed to induce pluripotency, or to extract pluripotent cells from different tissues and biological fluids. However, the characterization of pluripotency requires tedious, expensive, time-consuming, and not always reliable wet-lab experiments; thus, an easy, standard quality-control protocol of pluripotency assessment remains to be established. Here to help comes the use of high-throughput techniques, and in particular, the employment of gene expression microarrays, which has become a complementary technique for cellular characterization. Research has shown that the transcriptomics comparison with an Embryonic Stem Cell (ESC) of reference is a good approach to assess the pluripotency. Under the premise that the best protocol is a computer software source code, here I propose and explain line by line a software protocol coded in R-Bioconductor for pluripotency assessment based on the comparison of transcriptomics data of pluripotent cells with an ESC of reference. I provide advice for experimental design, warning about possible pitfalls, and guides for results interpretation.

Keywords: Cellular reprogramming, iPSC, Computational biology, Transcriptomics, R, Bioconductor, Hierarchical clustering, Principal component analysis, Scatterplots

1 Introduction

The conversion of unipotent somatic cells into pluripotent cells by cellular reprogramming has wide opened the doors of research in regenerative medicine and drug screening applications. Such conversion was made possible by the pioneer work of Yamanaka [1, 2] using a reprogramming cocktail of four factors (POUF5F1/OCT4, SOX2, KLF4, and C-MYC), and Thomson [3], using OCT4, SOX2, NANOG, and LIN28, generating the so-called induced pluripotent stem cells (iPSCs). However, these approaches suffer from drawbacks, such as the use of oncogenes like C-MYC, they are low efficient, and introduce foreign genomic integrations in the host cells. To solve these problems, a great variety of methods have been developed like generating iPSCs with fewer factors [4–6], or replacing factors by other molecules to circumvent the use of oncogenes or to avoid genome integrations [7–12]. A method has even been reported to generate pluripotent cells from specific origin without reprogramming factors but by manipulating the culture conditions [13]. To improve the reprogramming efficiency, techniques have been developed based on the introduction of additional factors [14], or on the use of other molecules [15–18]. Thus, though multiple ways to generate pluripotent cells have been established and more are continuously appearing, an easy characterization to check the pluripotency of the cells has not been fully established yet. There are several in vitro and in vivo criteria which can be used to evaluate the quality of mouse iPSCs: (1) Expression of pluripotent markers such as OCT4, SOX2, NANOG, and others. (2) In vivo teratoma formation or in vitro embryo body formation. (3) Chimera generation ability. (4) Germline contribution. (5) Germline transmission that is the most determinant criterion, especially through tetraploid complementation assays [19]. The latter is considered as the "golden standard" for pluripotency characterization. However, with the exception of the gene expression characterization, the other techniques are cumbersome and time-consuming. A good characterization of human iPSCs is even more difficult, since due to technical and ethical reasons, less assays are available for checking the pluripotency of human cells. Only the evaluation of pluripotent marker expression and teratoma formation are the available criteria for characterizing human cells pluripotency. Thus, according to guidelines for cellular pluripotency, issued by the U.S. National Institutes of Health (NIH), both teratoma formation assays in immunodeficient mice and transcriptomics analysis are required for ascertaining the pluripotent nature of human cells [20]. However, it has been shown that not fully reprogrammed human iPSCs lines are able to form teratomas, suggesting that teratoma formation cannot be the ultimate test to judge the quality of human iPSCs [21]. Therefore, in the human iPSC case, transcriptomics analysis remains the most reliable tool to assess the pluripotency of the reprogrammed cells. In a typical reprogramming experiment, the transcriptomics profiles of the initial non-pluripotent population are compared against the profile of the reprogrammed population using as a positive control an ESC population. The degree to which the transcriptomics profile of the reprogrammed population departs from the original population to reach the profile of the fully pluripotent ESC control is an indication of the level of pluripotency achieved during the reprogramming process. However, even though this assessment seems easy, there are controversies about the characterization of some human pluripotent cells [22–24]. Differences between ESCs and reprogrammed cells have been reported from transcription [25] and epigenetic [26, 27] profiles, while some work [28, 29] showed that there are only a few gene expression differences between iPSCs and ESCs.

Because of ethical and technical reasons [30], it is more feasible to work with pluripotent cells from mouse, rather than from human. Human ESCs (hESCs) are similarly pluripotent as mouse ESCs (mESCs) but they are also unique in many aspects. In contrast to mESCs, hESCs form flat and larger colonies, divide slower, and rely fundamentally on Activin/TGF-β signaling for self-renewal [31]. The subsequent isolation of a unique pluripotent stem cell population from post-implantation mouse epiblast-derived stem cells (mEpiSCs) [32, 33] hinted that hESCs may exist in a primed pluripotent state, i.e., already poised for early differentiation [34]. mEpiSCs resemble hESCs in terms of appearance, clonogenicity, culture requirements, molecular markers, gene expression, signaling responses, and differentiation behavior [35]. Nevertheless, mEpiSCs, unlike mESCs, rarely incorporate into the blastocyst inner cell mass (ICM) and are incompetent to contribute to chimeras [35], suggesting that they exist in a functionally distinct pluripotent state from the naïve [34] ground state. Thus, the ground state naïve pluripotency is established in the epiblast and may be captured in the in vitro ESCs. After implantation, the epiblast becomes primed for lineage specification and commitment. Thus, mEpiSCs are the in vitro counterpart of primed epiblast [34], in cell cultures rodent ESCs represent the ground state naïve pluripotency, and rodent EpiSCs and primate ESCs represent the primed state. Therefore, mEpiSCs are often used as a model to study hESCs [36], and we have also applied the protocol for pluripotent assessment for the mEpiSCs [37, 38].

The information collected by transcriptomics analysis is a valuable troubleshooting tool to search for possible causes of errors in case something has gone wrong during the reprogramming process, and to find possible somatic memory events. In this chapter I will describe a computational biology protocol that evaluates whether a population is pluripotent, from the transcriptomics experimental design to the data interpretation. Thus the reader can learn to perform the pluripotency assessment, interpret the results, and become familiar with typical pitfalls that can ruin the results of potentially good projects.

2 Materials

The first stages (from Section 3.3 to Section 3.5) of transcriptomics processing depend on the platform type used to run the experiment. Initially I will describe a reprogramming experiment performed for the most commonly used microarray platform, the Affymetrix one. Free available data downloaded from the public Gene Expression Omnibus (GEO) database at the National Center for Biotechnology Information (NCBI) in the USA (http://www.ncbi.nlm.nih.gov/geo/) is used to explain the protocol.

There are several ways of searching for information in the GEO database.

- If the ID of the record is known in advance, it is possible to access it directly from the GEO portal http://www.ncbi.nlm.nih.gov/geo/, and write the ID in the search form of the browser (*see* **Note 1**).
- If the ID record is not known, it is possible to search for it through the NCBI portal http://www.ncbi.nlm.nih.gov/. In such case, the user has to write a keyword in the "search" form, click on the GEO DataSets icon, and browse the list of results.

We will illustrate the pluripotency assessment protocol with the transcriptomics data from iPSCs derived from adult Neural Stem Cells (NSCs) using Pou4f1/Oct4 and Klf4 as reprogramming factors deposited in the GEO database [5]. Applying the aforementioned method for searching GEO data writing "Pluripotent stem cells induced from adult neural stem cells by reprogramming with two factors" in the search form of the http://www.ncbi.nlm.nih.gov/ server, and browsing the results for the GEO record with the title "Pluripotent stem cells induced from adult neural stem cells by reprogramming with two factors" that is the title of the associated article [5], the user will find the record: GSE10806. The relevant information about such record for applying the pluripotency check protocol is:

1. Series: GSE10806.
2. Platform: GPL1261 [Mouse4302_2] Affymetrix Mouse Genome 430 2.0 Array.
3. Samples:

   GSM272753 Embryonic Stem cells sample 1.

   GSM272836 Embryonic Stem cells sample 2.

   GSM272837 Embryonic Stem cells sample 3.

   GSM272839 Induced pluripotent stem (iPS) cells (Oct4, Klf4) sample 2.

   GSM272846 Induced pluripotent stem (iPS) cells (Oct4, Klf4) sample 3.

   GSM272847 Neural stem cells (NSC) sample 2.

   GSM272848 Neural Stem cells (NSC) sample 3.

   GSM272890 Induced pluripotent stem (iPS) cells (Oct4, Klf4) sample 1.

   GSM279200 Induced pluripotent stem (iPS) cells (Oct4, Sox2, c-Myc, Klf4) sample 1.

   GSM279201 Induced pluripotent stem (iPS) cells (Oct4, Sox2, c-Myc, Klf4) sample 2.

   GSM279202 Induced pluripotent stem (iPS) cells (Oct4, Sox2, c-Myc, Klf4) sample 3.

The data that we will use to illustrate the proposed protocol are the two records (GSM272847, GSM272848) corresponding with the transcriptomics information of the mouse NSCs from which the iPSCs were derived; the three records (GSM272890, GSM272839, GSM272846) corresponding with the transcriptomics information of the mouse induced with two factors (Oct4, Klf4) pluripotent stem cells (iPSCs); and the three records (GSM272853, GSM272836, GSM272837) corresponding with the transcriptomics information of the ESCs of reference. To obtain the data necessary for further analysis:

1. Click on the GSE10806.
2. Go to the end of the page ("Supplementary file" table) and click "ftp" on the "Download" column.
3. Click "Save File" to save the GSE10806_RAW.tar file.

Once the data has been downloaded, set the working directory:

1. Create a directory, for example, with the name `Pluripotency`.
2. Create a subdirectory (e.g., with the name CEL) inside the directory to unpack the CEL files.
3. Move the GSE10806_RAW.tar file to the `/Pluripotency/CEL` directory.[1]
4. "Untar" the GSE10806_RAW.tar file. To perform such task depending on the user software installation, click on the file name or open a terminal, and go to the folder that contains the file and type:

```
tar xvf GSE10806_RAW.tar
```

5. This will produce a list of 11 CEL.gz files (GSM272753.CEL.gz, GSM272836.CEL.gz, GSM272837.CEL.gz, GSM272839.CEL.gz, GSM272846.CEL.gz, GSM272847.CEL.gz, GSM272848.CEL.gz, GSM272890.CEL.gz, GSM279200.CEL.gz, GSM279201.CEL.gz, GSM279202.CEL.gz). "Unzip" the CEL.gz files. To perform such task depending on the software installation the user can click on each file name or open a terminal, and go to the `/Pluripotency/CEL`, and type:

```
unzip *.gz
```

In this example, for brevity initially we will not use the three files associated with the four factor iPSCs.

[1] The directory separator is / or \ for Linux or Windows operating system, respectively.

3 Methods[2]

The data processing will be performed with the software R that is a free dialect implementation (part of the GNU project) of the statistics commercial programming language S. The user can download precompiled binary distributions of the base system and contributed packages for Linux (Debian, Ubuntu, Fedora), Mac OS X, and Windows operating systems from the Comprehensive R Archive Repository (CRAN) at http://cran.r-project.org/ following the instructions of the link. R is part of many Linux distributions; therefore, the user should check whether R is already installed.

3.1 Install Bioconductor

R includes packages for all types of statistical analysis. As a tool for bioinformatics data analysis, we will use Bioconductor (http://www.bioconductor.org/) running on R. To install Bioconductor

1. Open a terminal and launch R typing

```
R
```

   In case of Linux distributions, such as Ubuntu, it is advisable when running R for installing Bioconductor to have superuser rights. In such case, type on the terminal `sudo R`, and write your password when the system asks for it.
2. To install Bioconductor type

```
source("http://bioconductor.org/biocLite.R")
biocLite()
```

   This command installs a selection of core Bioconductor packages: affy, affydata, affyPLM, affyQCReport, annaffy, annotate, Biobase, biomaRt, Biostrings, DynDoc, gcrma, genefilter, geneplotter, GenomicRanges, hgu95av2.db, limma, marray, multtest, vsn, and xtable. After downloading and installing these packages, the script prints "Installation complete" and "TRUE".
3. To install a particular package, e.g., `limma`, type in an R command window:

```
source("http://bioconductor.org/biocLite.R")
biocLite("limma")
```

[2] The commands to be input in the computer console will appear in this chapter inside boxes.

Be aware that the R interpreter is case sensitive. It distinguishes between uppercase and lowercase characters, and sometimes, depending on the percentage of the free working memory, this installation method does not work and it is necessary to restart R before running the previous commands. To quick R write:

```
q()
```

Another option to install a package in Bioconductor is to run

```
install.packages("limma")
```

In such case a window appears, in which it is necessary to click on the CRAN mirror from which one wishes to download the package (usually, the mirror closer geographically could be faster).

3.2 Design of the Transcriptomics Experiment to Assess Pluripotency

To design an experiment to assess the pluripotency of a cell population based on transcriptomics analysis, it is necessary to decide in advance:

- The microarray platform to be used: Very often such decision is constrained by the infrastructure available in the researcher's laboratory or in the service provider. In the example used to illustrate the present protocol, Affymetrix Mouse Genome 430 2.0 Arrays have been used.
- The negative and positive controls: The best population for negative control is the population from which the pluripotent cells have been derived. In this example, the negative control are the Neural Stem Cells (NSCs), while the best population for the positive control are the ESCs.
- Type of replicates. Once the user is confident with the performance of the transcriptomics infrastructure, it is much better to use biological instead of technical replicates.
- The number of replicates: For each population (reprogrammed, positive and negative control) use at least three biological replicates. In case of limited economic resources, give priority to the reprogrammed populations, then to the positive population, and finally to the negative one. Anyway, be aware that an initial economical setting trying to save replicates could turn into a much more expensive experiment if something goes wrong (*see* **Note 2**). In this example, three replicates have been used initially hybridized for the negative control population but the hybridization of one went wrong and there remained only two replicates (GSM272847, GSM272848) for the negative control, three replicates for the reprogrammed population (GSM272890, GSM272839, GSM272846), and three for the positive control (GSM272753, GSM272836, GSM272837).

3.3 Read the Data

To read the data into R, open a console, go to the working directory in which you wish to process the data (in this example: Pluripotency), run R (typing R on the console), load the Bioconductor libraries and read the data. These steps are implemented by typing the following instructions:

```
library(affy)
  DirCEL="./CEL/"
FilNam=c("GSM272847.CEL","GSM272848.CEL",
"GSM272890.CEL","GSM272839.CEL","GSM272846.CEL",
"GSM272753.CEL","GSM272836.CEL","GSM272837.CEL")
NumSam=length(FilNam)
for (i in c(1:NumSam))
{FilNam[i]=paste(DirCEL,FilNam[i],sep="")}
FulFilNam=FilNam
ab=read.affybatch(filenames=FulFilNam)
```

The library `affy` is the Bioconductor library to process Affymetrix arrays. In case of other arrays or Next Generation Sequencing (NGS) data, other libraries will be necessary. The variable `DirCEL` is the name of the folder directory under the present working directory in which the CEL files of the Affymetrix arrays have been stored.

The function `c` concatenates the names of the CEL files, saving them in the list `FilNam`.

The command `NumSam=length(FilNam)` calculates and saves into the variable `NumSam` the number of arrays (in this case 8).

The loop `for (...) {...}` goes through the `NumSam` CEL files and appends the CEL directory name to the CEL file names.

The operator `[i]` returns a single file name of the list of names `FilNam`.

The command `ab=read.affybatch(filenames=FulFil)` reads all the CEL files and puts the data into a Bioconductor object `ab`.

3.4 Inspect the Raw Data

It is wise to have a glance at the microarrays images before going further with the processing. Experimental problems (such as slides scratches, bubbles, or unbalanced regions) might be revealed and subsequent action concerning the data could be taken (*see* **Note 3**). After reading the data, the user can write in the R terminal.

```
image(ab[,i])
```

where `i` is the index of the array the user wants to inspect. The command `image(ab)` shows all the images one by one asking to hit the < return > key to see the next plot. Be aware that this procedure could be time-consuming depending on the number of processed arrays.

3.5 Normalize the Data

Before comparing the gene expression values from samples (even if they came from exactly the same experimental conditions), it is necessary to normalize them to put all on the same scale (in terms not only of range but also of mean and standard deviation). There exist several normalization techniques to perform such task, and some of them are platform dependent (*see* **Note 4**). In this example, we use the Robust Multi-array Analysis (RMA) normalization from Irizarry [39].

```
eset = rma(ab)
```

The function `rma` from the `affy` package computes the RMA normalization and puts the results in the expression set object `eset`. The expression measurements in `eset` are in $\log_2$ scale; thus it is not necessary to perform any further variance stabilization step and this is different from most of the other normalization methods (*see* **Note 5**).

3.6 Define the Indexes of the replicates of Each Population

The list of data file names `FilNam=c` provided in Section 3.3 does not have any phenotypic information defining to which population type corresponds each file. Each population has a collection of replicates, in this example, two for the NSCs of origin, three for the reprogrammed two factor iPSCs, and three positive for control ESCs. The way to define the association between the data files and their population identity is by following the instructions:

```
Ind=c(1,1,2,2,2,3,3,3)
IndOri=which(Ind==1)
IndiPSC=which(Ind==2)
IndESC=which(Ind==3)
```

The function `c` concatenates the indexes of the CEL files in the order previously provided in the list `FilNam`, where the index `1` is assigned to the NSC samples (using the function `IndOr-i=which(Ind==1)`), the index `2` to the iPSCs, and the index `3` to the ESCs.

3.7 Perform the Principal Component Analysis (PCA)

The Principal component analysis (PCA) involves a mathematical procedure that transforms a number of (possibly) correlated variables into a (smaller) number of uncorrelated variables called Principal Components (PCs) with the objective of reducing the data dimensionality. The first principal component accounts for as much of the information (in terms of variability) in the data as possible, and each successive component accounts for as much of the remaining information as possible. Thus the PCA is a way to represent the very high dimensional space of each transcriptomics sample (a space of more than 20,000 dimensions, one for each transcript) into a much reduced dimensional space through an optimal rotation and projection preserving as much information as possible. This is performed by writing in the R console.

```
library(limma)
pca=prcomp(eset,scale=TRUE)
Rot=pca$rotation
```

With the `prcomp` command the initial expression values in the object `eset` are transformed to the `pca` values. `prcomp` needs the "limma" package to be loaded with the function `library (limma)` to understand properly the object `eset`. The operator `$` extracts the rotation matrix from the `pca` object. The signs of the columns of the rotation matrix `Rot` generated by `prcomp` are arbitrary, and so may differ between different builds of R. Once we have the PCs, we can represent them into two- and tridimensional spaces.

3.8 Draw the Bidimensional Principal Component Analysis (PCA)

First, we plot the 2D plot by writing in the R console:

```
plot(Rot[,1],Rot[,2],xlab="PC1",ylab="PC2")
```

The function `plot` draws the first PC (that is first column of the rotation matrix `Rot[,1]`) against the second PC (that is second column of the rotation matrix `Rot[,2]`). To write the sample labels, we should calculate the mean of the replicates in the PC space by typing in the R console:

```
RotT=t(Rot)
MeaPCOri=rowMeans(RotT[,IndOri])
MeaPCiPSC=rowMeans(RotT[,IndiPSC])
MeaPCESC=rowMeans(RotT[,IndESC])
```

The function `t` transposes the rotation matrix `Rot` to facilitate the calculation of the means of the PCs of each population type using the function `rowMeans`. Then, we place the population labels `Ori` for the original population from where the IPSCs are derived, `iPSC` for the iPSC population and `ESC` for the ESC population.

```
X=c(MeaPCOri[1],MeaPCiPSC[1],MeaPCESC[1])
Y=c(MeaPCOri[2],MeaPCiPSC[2],MeaPCESC[2])
MeaLabels=c("Ori","iPSC","ESC")
text(X,Y,labels=MeaLabels)
```

The function `c` concatenates the three population names and stores them in the list of labels `MeaLabels`. Such labels are placed on plots over the PC coordinates defined by the vectors `X` and `Y`. Finally, we color the samples:

```
points(RotT[1,IndOri],RotT[2,IndOri],col="red",pch=21)
points(RotT[1,IndiPSC],RotT[2,IndiPSC],col="blue",pch=21)
points(RotT[1,IndESC],RotT[2,IndESC],col="green",pch=21)
```

The parameter `col` defines the color of the point and the parameter `pch` the "plotting character", i.e., the symbol to be drawn in the point position, `pch=21` codes for circle. The pairwise bidimensional PCA plots produced by these two blocks of instruction are depicted in Fig. 1a. The PCA information is complementary to the one provided by the hierarchical clustering (*see* Section 3.9) and is a good validation for both methods that their results behave in a coherent way, but this is not always the case. In theory the closer the points representing the reprogrammed population are to the points representing the ESCs, the more similar are both populations and we can consider the reprogrammed population as more pluripotent (*see* **Note 6**) but to use only PCA as a method to assess pluripotency can have several pitfalls (*see* **Note 7**). A PCA with an additional dimension (3D PCA) is usually not a good tool to assess the pluripotency (*see* **Note 8**).

3.9 Perform the Hierarchical Clustering of Samples

To see how the different replicates cluster together

```
Dat=exprs(eset)
DatT=t(Dat)
Dis=as.dist(1-cor(Dat, method="pearson"))
Dis=dist(DatT)
Hie=hclust(Dis)
labels=c("Origen1","Origen2",
"iPSC","iPSC2","iPSC3",
"ESC1","ESC2","ESC3")
plot(Hie,labels=labels)
```

The command `Dat=exprs(eset)` extracts the numerical expression values from the expression set object `eset` and saves them in the matrix `Dat`. The function `t` calculates the `Dat` matrix transpose. The command `Dis=as.dist(1-cor(Dat, method="pearson"))` calculates the distances among all the samples using the "one minus Pearson correlation metric". More metrics are possible but it is advisable to use this one (*see* **Note 9**).

The function `hclust` performs the hierarchical clustering.

The function `c` concatenates the names of the samples that we want to appear as labels of the cluster branches, saving them in the list `labels` to be used by the function `plot` that draws the clustering as shown in Fig. 1b.

A necessary but not sufficient condition for a pluripotent population to be well reprogrammed is that all its replicates cluster within the positive control ESCs replicates and not within the cluster of the original populations from which they were derived. According to an ideal scenario, the pluripotent populations have to be so close to the ESC population that their replicates intermingle.

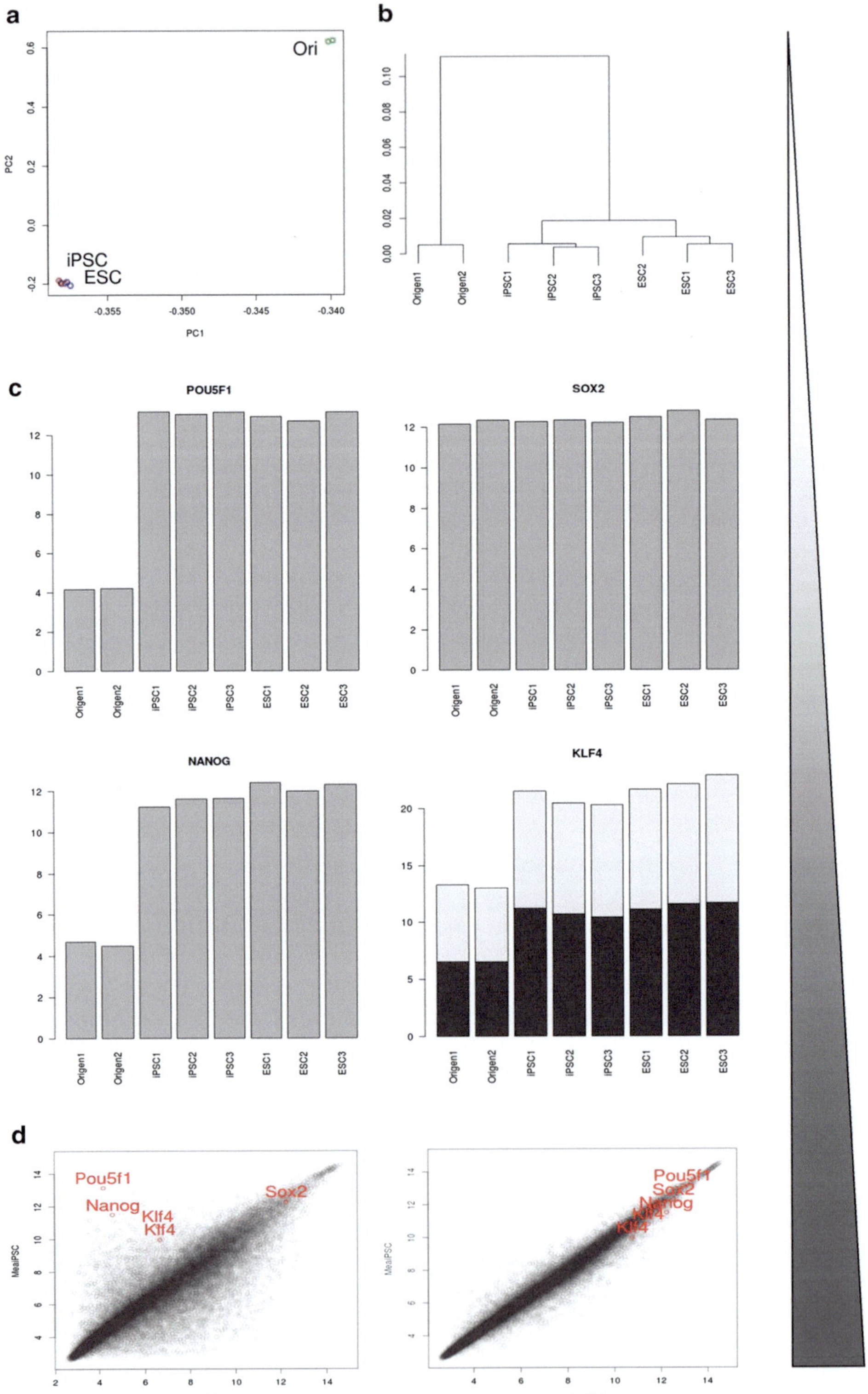

Fig. 1 Graphical evaluation of the quality of the reprogrammed cells. (**a**) Bidimensional Principal Component Analysis (PCA) of samples. (**b**) Hierarchical clustering of samples. (**c**) *Bar plots* of the expression for all the

3.10 Annotate the Arrays

To know which probe sets in the arrays target each gene, it is necessary to annotate the arrays with genetic information using a metadata package from Bioconductor. To find which metadata package corresponds to the used data stored in the expression set variable `eset`, simply write in the R command line:

```
annotation(eset)
```

In this example the function output is `"mouse4302"`. For most platforms the character returned by the function `annotation` is the name of the annotation package that we need to install, followed by a `.db` extension. Thus, in our case it is necessary to download the annotation package `mouse4302.db` using the following commands in the R terminal:

```
source("http://bioconductor.org/biocLite.R")
biocLite("mouse4302.db")
```

Once we have downloaded the `mouse4302.db` package in R, we can load its libraries in the current R session using:

```
library(mouse4302.db)
library(mouse4302cdf)
```

The `mouse4302cdf` is the R environment containing the structures of the `mouse4302` Chip Definition File (CDF). The CDF file contains the information of the mapping of the chip probes targeting each gene.

3.11 Extract the Official Gene Symbols

Put in a list (named for example `SYMBOL`) the official gene symbols of the arrays using the command in the R terminal.

```
SYMBOL=sapply(contents(mouse4302SYMBOL),paste,col-
lapse=", ")
```

The `contents` method retrieves all the values in the environment `mouse4302SYMBOL`, and the method `sapply` uses the function `paste` over all the elements `mouse4302SYMBOL` to concatenate the symbol names of all the genes into a vector separated by commas.

Fig. 1 (continued) replicates in each population of all the probe sets of a selection of pluripotent markers. The overlapping bars in different gray tones correspond to the cases (KLF4), in which a gene is targeted by more than a probe set. (**d**) Pairwise scatterplots of the mean values of iPSCs versus the population of origin (*left*) and versus the ESCs (*right*). The positions of the selected pluripotent markers are overimposed in *red*. The *triangle to the right* symbolizes the degree of pluripotency assessment associated with the corresponding stage of the computational analysis. The *darker the triangle*, the more useful the corresponding technique to address the pluripotency is; thus the PCA and the hierarchical clustering are the least informative techniques and the pairwise scatterplot of the mean values of iPSCs versus the ESCs is the most informative ones

3.12 Visualize the Expression of Some Pluripotent Markers

To have an initial indication of the quality of pluripotent cells, it is advisable to visualize the expression of some pluripotent markers such as *POU5F1/OCT4*, *SOX2*, *NANOG*, and *KLF4*. To find the indexes of the probe sets targeting such genes, we use the function `which`:

```
IndPOU5F1=which(SYMBOL=="Pou5f1")
IndSOX2=which(SYMBOL=="Sox2")
IndNANOG=which(SYMBOL=="Nanog")
IndKLF4=which(SYMBOL=="Klf4")
```

Be aware that in the annotation packages, the mouse gene symbols start with a first uppercase character, and the rest of the characters are lowercase, whereas in the human case all the characters of the gene symbol are in uppercase (thus, for human arrays the user has to write after `SYMBOL==` the gene names with all characters in uppercase).

The vectors of the indexes of all the probe sets targeting each gene are stored in the variables (`IndPOU5F1`, `IndSOX2`, `IndNANOG`, `IndKLF4`). To visualize the expression of all the probe sets targeting each maker, use the function `barplot` by writing in the R terminal:

```
library(lumi)
Dat=exprs(eset)
barplot(Dat[IndPOU5F1,],names.arg=labels,main="POU5F1",
las=2)
barplot(Dat[IndSOX2,],names.arg=labels,main="SOX2",las=2)
barplot(Dat[IndNANOG,],names.arg=labels,main="NANOG",las=2)
barplot(Dat[IndKLF4,],names.arg=labels,main="KLF4",las=2)
```

The argument `las=2` rotates the sample names 90° to have space to write all of them.

Since such function `barplot` belongs to the library `lumi`, first it is necessary to load such library. The command `Dat=exprs(eset)` is performed to extract the numerical expression values from the expression set object `eset`. The produced barplots are depicted in Fig. 1c.

3.13 Calculate the Mean Values of Each Population

Each population is represented by a collection of replicates. Once we observed in the hierarchical clustering that the replicates of each population cluster together (Fig. 1b), we can take the averages of the replicates of such population as a representative of each of the three populations. Once the indexes of each population are defined in Section 3.6, the means are calculated using the function `rowMeans` before extracting the numerical expression information of the expression set object `eset` using the function `exprs`. Such calculation is performed by writing in the R console:

```
MeaOri=rowMeans(exprs(eset)[,IndOri])
MeaiPSC=rowMeans(exprs(eset)[,IndiPSC])
MeaESC=rowMeans(exprs(eset)[,IndESC])
```

3.14 Calculate the Correlation Between the Mean Representatives of Each Population

To quantify how similar are the different populations and how much the reprogrammed population departs from its original state towards the target pluripotent state represented by the ESCs, we use the Pearson correlation, but this time applied on the mean representative of each population `{MeaOri, MeaiPSC, MeaESC}`. Such calculation is performed by writing in the R console:

```
CoriPSCOri=cor(MeaiPSC,MeaOri,method="pearson")
CoriPSCESC=cor(MeaiPSC,MeaESC,method="pearson")
```

In the current example we obtain for the correlation between the reprogrammed population and the population of origin `CoriPSCOri=0.8956957`, and between the reprogrammed population and ESCs a correlation `CoriPSCESC=0.9864078`. We consider that two transcriptomics profiles obtained by microarrays are very similar if their correlation is higher than 0.98 (*see* **Note 10**); thus, in this example we can consider that the reprogrammed population is similar to the ESC of reference. Such analysis has been performed with the two factor reprogrammed iPSCs, but it is very simple to repeat the whole analysis with the four factor iPSCs (*see* **Note 11**).

3.15 Draw the Pairwise Scatterplots

The best way to visualize the extent to which the reprogrammed population departs from the original population (negative control) from which it has been derived to approach the target pluripotent state (positive control) represented by the ESCs is to depict in a pairwise fashion all the probes of the mean of each population (`MeaOri`, `MeaPlu`, `MeaESC`) calculated in Section 3.13 corresponding to the NSCs, two factor iPSCs, and ESCs, respectively. It is also very useful to overimpose in such plots the position of the pluripotent markers. To facilitate the use of such markers, it is handy to collect them into a list using the R function `list`:

```
IndPluMar=list(IndPOU5F1,IndSOX2,IndNANOG,IndKLF4)
NumPluMar=length(IndPluMar)
```

Thus we can use across the rest of the protocol the variable IndPluMar storing the indexes of all the markers. The function `length` calculates the number of pluripotent markers selected. This approach allows the use of a longer or shorter number of pluripotent markers depending on the graphical space available. Then, the scatterplots for the iPSCs versus the NSCs are depicted using the `plot` function as follows:

```
plot(MeaOri,MeaiPSC,lwd=0.1)
for (i in c(1:NumPluMar))
{
  Ind=IndPluMar[[i]]
  Name=SYMBOL[Ind]
  Name=Name[1]
text(MeaOri[Ind],MeaiPSC[Ind],col="red",pos=3,off-
set=0.5,labels=Name)
points(MeaOri[Ind],MeaiPSC[Ind],col="red")
}
```

The `plot` parameter `lwd` specifies the line width for lines appearing in the legend. Once the plot is drawn, the loop `for(.){.}` goes through the `NumPluMar` pluripotent markers to mark their position in red and overimpose their names. The parameter `pos=3` is set to write the gene names with the `text` function over[3] the probe set coordinates and the parameter `offset=0.5` to do it with an offset of 0.5. The function `points` draws a circle in the position of each probe set, while the variable `col="red"` indicates that such circle has to be red. The `[[.]]` operator returns the object type stored in `IndPluMar`, while `[.]` returns a single component of a list.

Analogously, the scatterplots for the iPSCs versus the ESCs are depicted using the `plot` function as follows:

```
plot(MeaESC,MeaiPSC,lwd=0.1)
for (i in c(1:NumPluMar))
{
  Ind=IndPluMar[[i]]
  Name=SYMBOL[Ind]
  Name=Name[1]
text(MeaESC[Ind],MeaiPSC[Ind],col="red",pos=3,off-
set=0.5,labels=Name)
    points(MeaESC[Ind],MeaiPSC[Ind],col="red")
}
```

The pairwise scatterplots produced by these two blocks of instructions are depicted in Fig. 1d. The more similar are two populations, the less scattered and closer to the diagonal have to be the dots representing the expression values of each transcript in the two depicted populations (*see* **Note 12**).

[3] The parameter pos is a position specifier for the text. If specified this overrides any adj value given. Values of 1, 2, 3, and 4, respectively, indicate positions below, to the left of, above, and to the right of the specified coordinates.

3.16 Final Assessment of Pluripotency Based on Computational Analysis of Transcriptomics Data

Using only one of the described methods to assess the pluripotency is unreliable and it is advisable to use the battery of all the methods described here and that constitute altogether the pluripotent assessment protocol, since they complement each other depicting several views of the data from a different angle that work in synergy.

1. Thus, the bidimensional PCA of Section 3.8 provides a first glance of the relative distances between the different samples in a space of reduced number of dimensions, in which the results appear usually appealing configurations but they have to be considered in a cautious way (*see* **Notes 6–8**).
2. The hierarchical cluster of samples of Section 3.9 is important to inspect for non-replication problems. In a good scenario, the reprogrammed samples should not cluster with the original population but with the ESCs. In the best scenario, the reprogrammed samples should form a unique cluster with the ESCs, intermingling their replicates. The hierarchical clustering provides a numerical distance scale based on the "one minus the Pearson correlation" metric; however the projection of the hierarchy nodes (in which the tree branches join) over such scale provides a measurement of the correlation between the virtual populations represented by the sample branches converging in such nodes; thus it is difficult to provide a pluripotency metric based on the hierarchical clustering.
3. Once such clustering shows a good replication behavior, we can confidently calculate the transcriptomics representative of each population using the replicates means (*see* Section 3.13). Based on such representative we can calculate the Pearson correlation ρ among the samples in Section 3.14. If we find a correlation $\rho > 0.98$ between the reprogrammed samples and the ESCs, we can start to think that both populations are very similar at transcriptomics level.
4. Even though a good correlation between reprogrammed samples and ESCs is a good indicator of pluripotency, it is very important to see whether the common pluripotency markers (*POU5F1/OCT4*, *SOX2*, *NANOG*, *KLF4*, ...) have achieved expression values in the reprogrammed population similar to those of the ESCs. This can be observed with the barplots of Section 3.12 and with the scatterplots of Section 3.15.
5. Once, following the clustering, we are sure that we can use the replicate means as population representatives, and we have assessed the pluripotency by the main pluripotency markers, we can fuse such information in the scatterplots of Section 3.15. The scatterplot between the reprogrammed representative and the ESC representative provides a good glimpse of the reprogramming process achievements: In one shot, they show all the data transcripts, depicting how similar the reprogrammed and

the ESC populations are, and how similar the main pluripotency markers in both populations are. Additionally, the scattering along the diagonal, although not providing direct information of the pluripotency, could inform us about potential normalization problems.

This protocol has been designed to be simple but simultaneously very flexible. Only changing very few code lines, it can be adapted to assess other experiments (*see* **Note 13**) and with minimal graphical requirements (that are operating system dependent), to run in different operation systems. A more graphics-oriented version of the protocol has been implemented in the commercial numerical computing environment MATLAB (MathWorks) and successfully applied by this author in multiple publications for reprogramming studies of mouse [5, 13, 14, 37–44] and human [6, 22, 24, 45].

Be aware that the transcriptomics information does not represent the whole cellular state of a population (*see* **Note 14**). However, the whole protocol described here for transcriptomics data can be performed with other –omics data, changing accordingly the preprocessing Sections 3.3–3.5.

4 Notes

1. Different types of records in the GEO database exist, each of them starting with a different ID:
   - GSMxxxx Samples: Describes the conditions under which an individual sample was handled, the manipulations it underwent, and the abundance measurement of each element derived from it. Each Sample record is assigned a unique and stable GEO accession number (GSMxxxx). A Sample entity must reference only one platform and may be included in multiple series.
   - GSExxxx Series: Are original submitter-supplied record that summarizes a study. A Series record links together a group of related Samples and provides a focal point and description of the whole study. Series records may also contain tables describing extracted data, summary conclusions, or analyses.
   - GDSxxxx DataSets: Are collections of biologically and statistically comparable samples processed using the same platform. They are created by the GEO staff. Samples within a DataSet refer to the same Platform, i.e., they share a common set of array elements. Value measurements for each Sample within a DataSet are assumed to be calculated in an equivalent manner, i.e., considerations

such as background processing and normalization are consistent across the DataSet. Information reflecting experimental factors is provided through DataSet subsets.

- GPLxxxx Platforms. A Platform record is composed of a summary description of the array or sequencer and, for array-based Platforms, a data table defining the array template.

2. If only two replicates of a population are used and in case that one fails, it is not very feasible to determine the wrong sample. And to repeat the experiment with another replicate could result (due to batch effects) that the new replicate does not cluster well with the previous ones. Thus, the initial cost avoidance of performing three transcriptomics assays per population (saving one additional transcriptomics assay for doing two instead of three) incurs finally in the cost of five assays.
3. Since the probes in an array targeting a gene are distributed across all the slide surface of the array, a small scratch or some bubbles usually affect only very few of the multiple probes targeting each gene. Thus, the effect of such defects will be diminished during the preprocessing stages. Also, the effect of linearly unbalanced regions, due to slide displacement or non-uniform scanning, is usually corrected by the normalization stage. The bigger issue arises in case of substantial nonuniform image intensity distribution or regions with different background spread across significant areas of the slide surface; therefore I would advise discarding slides in which such phenomenon appears.
4. The typical normalization methods for Affymetrix arrays are MAS 5.0 (of the Affymetrix Microarray Suit), RMA and GCRMA (an RMA method that accounts for the GC content of the probes), provided in the functions `mas5`, `rma` and `gcrma` of the `affy` library and the Probe Logarithmic Error Intensity Estimate (PLIER) method provided in the `plier` library. Additionally, the Affymetrix arrays allow to assess statistically whether a gene has been expressed or not by providing a ternary label {A,M,P} for the absent (A) gene (not expressed), marginally (M) expressed gene, and present (P) gene (expressed), respectively.

   For Illumina arrays, it is possible to use a combination of RMA for background corrections and a Quantile normalization implemented in the `lumi` package.

   For single color Agilent arrays, a simple approach is to use a Quantile normalization that can be performed with the following command from the `limma` package:

```
library(limma)
exprSet.quantile = normalizeQuantiles(eset)
```

For two color Agilent arrays, a possible approach is to first normalize the two colors of each array using a within-array normalization based on the Loess "(LOcal regrESSion)" regression and then use a Quantile normalization for the inter-array normalization, both implemented in the `limma` package.

5. The Affymetrix-associated normalization methods RMA and GCRMA provide the normalized expression values in $\log_2$ scale. This differs from the other normalization methods from other transcriptomics platforms that return the normalized values in an untransformed scale, for which to stabilize the variance it is necessary to perform an additional step, for example, with a $\log_2$ scale transformation.
6. The pluripotent assessment based on PCAs is hindered by several features that do not allow us to trust totally the pluripotency results based only on PCA, and to use PCA only as an additional tool to study pluripotency. Such features are (1) Loss of information in the PCA: the PC used in a 2D or 3D PCA represents only a percentage (very often not higher than 60 %) of the transcriptomics information (the distribution of such information can be analyzed with Pareto charts). (2) The distances between points in the PCA plots are only relative distances; then they depend on the combination of the transcriptomics values of all the samples represented in the PCA plot. (3) The distance in each coordinate is not equivalent, the first PC is more informative than the second PC, and thus the distances along the first PC are more important than the distances along the second PC. Thus, we cannot establish a threshold to measure the pluripotency based on the relative distances between the reprogrammed samples and the ESCs in the PCA plots. Additionally, the PCA is more informative as when to visualize the relative distance between multiple samples. In the case of two samples (i.e., because lack of negative control) and not replicates, to use PCA has no sense since the relative distance between two samples is arbitrary.
7. To use only PCA as a method to assess pluripotency can lead to the error to decide that a cell is pluripotent when it is not the case. Such pluripotency assessment was one of the pitfalls of the claimed pluripotent human adult Germ Stem Cells (haGSCs) [46].

After applying this computational pluripotency assessment protocol, we demonstrated that indeed the haGSCs were not pluripotent cells [42]. The reader can replicate such demonstration with this protocol, downloading the [46] associated data from the GEO database (GSE11350) and setting the following variables in the protocol:

In Section 3.3.

Set the names of the CEL files:

```
FilNam=c("GSM282008.CEL","GSM282012.CEL","GSM282013.
CEL",
"GSM282014.CEL","GSM282015.CEL","GSM282016.CEL",
"GSM282017.CEL","GSM282018.CEL","GSM282019.CEL",
"GSM282009.CEL","GSM282010.CEL","GSM282011.CEL")
```

In Section 3.9.

Set the labels of the replicates:

```
labels=c("Origen1","Origen2","Origen3",          # hGSC
         "Reprogram1","Reprogram2","Reprogram3", # haGSC
         "Reprogram4","Reprogram5","Reprogram6",#
         "ESC1","ESC2","ESC3")                    # ESC
```

In Section 3.10.

Set the annotation for the Affymetrix Human Genome U133 Plus 2.0 Arrays:

```
library(hgu133plus2.db)
library(hgu133plus2cdf)
```

In Section 3.12.

Eliminate the index to POU5F1 since such gene is not included in the hgu133plus2 annotation package. One alternative to include such important pluripotent marker is to use the biomaRt annotation package.

In Section 3.13.

Set the indexes of the replicates of each population type:

```
Ind=c(1,1,1,2,2,2,2,2,2,3,3,3)
```

8. The 3D PCAs usually look appealing, but due to the projection distortions of representing 3D objects in the 2D space of a paper or a slide, it could arise visual effects such as overimposed samples in the perpendicular to the visualization surface projection that makes samples appear closer than they are in reality, resulting obstructions in the result interpretation. If the reader wishes to perform a tridimensional PCA, the package `rgl` is

required to draw the 3D graphs; and to install it, it is necessary to have in the operating system the Mesa 3D graphics library. Mesa is an open-source implementation of the OpenGL specification—a system for rendering interactive 3D graphics. The following Mesa libraries: `libglu1-mesa-de`, `libgl1-mesa-dev` and `mesa-common-dev` packages are required. In the case of a Linux system, to install Mesa, open a system terminal and type:

```
sudo apt-get install libglu1-mesa-dev
```

Input the password and all the necessary Mesa packages will be installed. Then, to install the `rgl` library, type in the R console:

```
source("http://bioconductor.org/biocLite.R")
biocLite("rgl")
```

Now the system is ready to draw 3D PCAs. Type the following instructions in an R console:

```
comp=c(1,2,3)
library(rgl)
rgl.clear()
bg3d("white") # Set to white the background color
comp=c(1,2,3)
plot3d(Rot[,comp[1]],Rot[,comp[2]],Rot[,comp[3]],
    col=rainbow(3),
    size=2,
    type='s',
    xlab=colnames(Rot[comp[1]],
    ylab=colnames(Rot[comp[2]],
    zlab=colnames(Rot[comp[3]],
    top=T)
# Write the labels
Z=c(MeaPCOri[3],MeaPCiPSC[3],MeaPCESC[3])
text3d(X,Y,Z,text=MeaLabels,adj=1.5)
```

A way to avoid the overlapping issue (at least for slide presentations) is to generate movies with rotations in the three possible orthogonal axes of the 3D PCA representation.

9. The simplest way to calculate the distance between the samples is with the R function `Dis=dist(DatT)` that uses the Euclidean metric, but such metric is not very proper to calculate distances between the samples in the very high dimensional transcriptomics space where gene expression values with high dynamic range dominate the genes with low dynamic range.
10. Although a Pearson correlation ρ higher than 0.85 seems to be very high in a lot of applications, it is wrong to consider such

threshold of similarity sufficient to assume the similarity of two transcriptomic profiles obtained by microarrays. The idea of good correlation $\rho > 0.85$ came from analysis of Gaussian distributed data, but the transcriptomic profiles are far from following such distributions. They have a high percentage of non- or very low-expressed genes that displace the transcriptomics distribution to the low expression range in a nonsymmetric manner. Also, independently of the origin of the samples, all transcriptomics samples share a set of very highly expressed housekeeping genes that artificially shrink the distance between otherwise very dissimilar populations. Thus, even for very distinct cellular populations, it is very common to observe transcriptomics profiles with Pearson correlations ρ higher than 0.85.

11. To repeat the whole protocol for pluripotency assessment for the four factor iPSCs, the only part to be modified in the software is the list with the CEL files of the presented example by substituting the three CEL files of the two factor iPSCs with the three CEL files of the four factor iPSCs; thus use the following list in Section 3.3.

```
FilNam=c("GSM272847.CEL","GSM272848.CEL",
"GSM279200.CEL","GSM279201.CEL","GSM279202.CEL",
"GSM272753.CEL","GSM272836.CEL","GSM272837.CEL")
```

After running again the normalization, it can be found that the Pearson correlation between the four factor iPSCs and the population of origin is `CoriPSCOri=0.8914526`, and between the four factor iPSCs and ESCs is `CoriPSCESC=0.9808192`. These values are very similar but slightly worse than the obtained for the two factors reprogramming (`CoriPSCOri=0.8956957` and `CoriPSCESC=0.9864078`). This example illustrates how using more reprogramming factors is not always a guarantee for improving the reprogramming process, and why it is crucial to use a good pluripotency assessment protocol and not rely on preconceived assumptions on the not-fully-understood-yet reprogramming mechanism.

12. Independently of the degree of similarity between two transcriptomics profiles, their pairwise scatterplots should always scatter along the diagonal. Otherwise, we can suspect that some problem has occurred during the data handling, very probably due to an improper normalization.

13. To assess the pluripotency of other experiments based on Affymetrix transcriptomics data, it is enough to change the following lines of the protocol.
    - In Section 3.3.
      List of CEL file names: `FilNam=c(.)`, substituting the dot by the desired CEL file names.
    - In Section 3.9.
      Set the labels of the replicates: `labels=c("Origen1",...)`.
    - In Section 3.10.
      Set the annotation obtained with the command `annotation(eset)` that provides an annotation label: `AnnotationLabel`, to be used to load the corresponding annotation libraries: `library(AnnotationLabel.db)` and `library(AnnotationLabelcdf)`.
    - In Section 3.13.
      List of the numerical indexes of replicates for the three types of populations (Origen=1, Pluripotent=2, ESC=3): `Ind=c(1,...,2,...3,...)`.
14. Reprogrammed populations with high similarities at transcriptomics level with ESCs have the same developmental potential as the ESCs, i.e., in the mouse case the reprogrammed cells have the capability of raising chimeras, that is the "golden standard" of pluripotency. Anyway, such information does not mean that the reprogrammed populations are "identical" but only "similar" to the ESCs of reference, since a reprogramming somatic memory exists at epigenetic level, where differences between reprogrammed cells in relation to ESCs can be observed for example with DNA methylomics analysis [47].

Acknowledgments

This work has been supported by the Basque Country Foundation of Science (Ikerbasque).

References

1. Takahashi K, Yamanaka S (2006) Induction of pluripotent stem cells from mouse embryonic and adult fibroblast cultures by defined factors. Cell 126(4):663–676
2. Okita K, Ichisaka T, Yamanaka S (2007) Generation of germline-competent induced pluripotent stem cells. Nature 448(7151):313–317
3. Yu J, Vodyanik MA, Smuga-Otto K, Antosiewicz-Bourget J, Frane JL, Tian S, Nie J, Jonsdottir GA, Ruotti V, Stewart R et al (2007) Induced pluripotent stem cell lines derived from human somatic cells. Science 318(5858):1917–1920
4. Kang L, Wu T, Tao Y, Yuan Y, He J, Zhang Y, Luo T, Kou Z, Gao S (2010) Viable mice produced from three-factor induced pluripotent stem (iPS) cells through tetraploid complementation. Cell Res 21(3):546–549

5. Kim JB, Zaehres H, Wu G, Gentile L, Ko K, Sebastiano V, Araúzo-Bravo MJ, Ruau D, Han DW, Zenke M, Schöler HR (2008) Pluripotent stem cells induced from adult neural stem cells by reprogramming with two factors. Nature 454(7204):646–650
6. Kim JB, Greber B, Araúzo-Bravo MJ, Meyer J, Park KI, Zaehres H, Schöler HR (2009) Direct reprogramming of human neural stem cells by OCT4. Nature 461(7264):649–653
7. Feng B, Jiang J, Kraus P, Ng JH, Heng JCD, Chan YS, Yaw LP, Zhang W, Loh YH, Han J et al (2009) Reprogramming of fibroblasts into induced pluripotent stem cells with orphan nuclear receptor Esrrb. Nat Cell Biol 11 (2):197–203
8. Chen J, Liu J, Yang J, Chen Y, Chen J, Ni S, Song H, Zeng L, Ding K, Pei D (2010) Bmps functionally replace Klf4 and support efficient reprogramming of mouse fibroblasts by Oct4 alone. Cell Res 21(1):205–212
9. Moon JH, Heo JS, Kim JS, Jun EK, Lee JH, Kim A, Kim J, Whang KY, Kang YK, Yeo S et al (2011) Reprogramming fibroblasts into induced pluripotent stem cells with Bmi1. Cell Res 21(9):1305–1315
10. Maekawa M, Yamaguchi K, Nakamura T, Shibukawa R, Kodanaka I, Ichisaka T, Kawamura Y, Mochizuki H, Goshima N, Yamanaka S (2011) Direct reprogramming of somatic cells is promoted by maternal transcription factor Glis1. Nature 474(7350):225–229
11. Heng JCD, Feng B, Han J, Jiang J, Kraus P, Ng JH, Orlov YL, Huss M, Yang L, Lufkin T et al (2010) The nuclear receptor Nr5a2 can replace Oct4 in the reprogramming of murine somatic cells to pluripotent cells. Cell Stem Cell 6 (2):167–174
12. Ichida JK, Blanchard J, Lam K, Son EY, Chung JE, Egli D, Loh KM et al (2009) A small-molecule inhibitor of TGF-β signaling replaces Sox2 in reprogramming by inducing Nanog. Cell Stem Cell 5(5):491–503
13. Ko K, Tapia N, Wu G, Kim JB, Araúzo-Bravo MJ, Sasse P, Glaser T, Ruau D, Han DW, Greber B, Hausdölrfer K, Sebastiano S, Stehling M, Fleischmann BK, Brülstle O, Zenke M, Schöller HR (2009) Induction of pluripotency in adult unipotent germline stem cells. Cell Stem Cell 5(1):87–96
14. Singhal N, Graumann J, Wu G, Araúzo-Bravo MJ, Han DW, Greber B, Gentile L, Mann M, Schöler HR (2010) Chromatin-remodeling components of the Baf complex facilitate reprogramming. Cell 141(6):943–955
15. Esteban MA, Wang T, Qin B, Yang J, Qin D, Cai J, Li W, Weng Z, Chen J, Ni S et al (2010) Vitamin c enhances the generation of mouse and human induced pluripotent stem cells. Cell Stem Cell 6(1):71–79
16. Liang G, He J, Zhang Y (2012) Kdm2b promotes induced pluripotent stem cell generation by facilitating gene activation early in reprogramming. Nat Cell Biol 14(5):457–466
17. Han J, Yuan P, Yang H, Zhang J, Soh BS, Li P, Lim SL, Cao S, Tay J, Orlov YL et al (2010) Tbx3 improves the germ-line competency of induced pluripotent stem cells. Nature 463 (7284):1096–1100
18. Hirata T, Amano T, Nakatake Y, Amano M, Piao Y, Hoang HG, Ko MSH (2012) Zscan4 transiently reactivates early embryonic genes during the generation of induced pluripotent stem cells. Sci Rep 2
19. Zhao X, Li W, Lv Z, Liu L, Tong M, Hai T, Hao J, Guo C, Ma Q, Wang L et al (2009) ips cells produce viable mice through tetraploid complementation. Nature 461(7260):86–90
20. In stem cell information. http://stemcells.nih.gov/research/registry/pluripotent_faq/. Accessed September 13, 2012
21. Chan EM, Ratanasirintrawoot S, Park IH, Manos PD, Loh YH, Huo H, Miller JD, Hartung O, Rho J, Ince TA et al (2009) Live cell imaging distinguishes bona fide human iPS cells from partially reprogrammed cells. Nat Biotechnol 27(11):1033–1037
22. Ko K, Araúzo-Bravo MJ, Tapia N, Kim J, Lin Q, Bernemann C, Han DW, Gentile L, Reinhardt P, Greber B, Schneider RK, Kliesch S, Zenke M, Schöller HR (2010) Human adult germline stem cells in question. Nature 465 (7301):E1-E1
23. Tapia N, Araúzo-Bravo MJ, Ko K, Schöler HR (2011) Concise review: challenging the pluripotency of human testis-derived esc-like cells. Stem Cells 29(8):1165–1169
24. Ko K, Reinhardt P, Tapia N, Schneider RK, Araúzo-Bravo MJ, Han DW, Greber B, Kim J, Kliesch S, Zenke M, Schöler HR (2011) Brief report: evaluating the potential of putative pluripotent cells derived from human testis. Stem Cells 29(8):1304–1309
25. Chin MH, Mason MJ, Xie W, Volinia S, Singer M, Peterson C, Ambartsumyan G, Aimiuwu O, Richter L, Zhang J et al (2009) Induced pluripotent stem cells and embryonic stem cells are distinguished by gene expression signatures. Cell Stem Cell 5(1):111–123
26. Bock C, Kiskinis E, Verstappen G, Gu H, Boulting G, Smith ZD (2011) Reference maps of human es and ips cell variation enable high-throughput characterization of pluripotent cell lines. Cell 144(3):439–452

27. Ruiz S, Diep D, Gore A, Panopoulos AD, Montserrat N, Plongthongkum N, Kumar S, Fung HL, Giorgetti A, Bilic J, Batchelder EM, Zaehres H, Kan NG, Schöler HR, Mercola M, Zhang K, Izpisua Belmonte JC (2012) Identification of a specific reprogramming-associated epigenetic signature in human induced pluripotent stem cells. Proc Natl Acad Sci U S A 109:16196–16201
28. Guenther MG, Frampton GM, Soldner F, Hockemeyer D, Mitalipova M, Jaenisch R, Young RA (2010) Chromatin structure and gene expression programs of human embryonic and induced pluripotent stem cells. Cell Stem Cell 7(2):249–257
29. Chin MH, Pellegrini M, Plath K, Lowry WE (2010) Molecular analyses of human induced pluripotent stem cells and embryonic stem cells. Cell Stem Cell 7(2):263–269
30. Thomson JA, Itskovitz-Eldor J, Shapiro SS, Waknitz MA, Swiergiel JJ, Marshall VS, Jones JM (1998) Embryonic stem cell lines derived from human blastocysts. Science 282(5391): 1145–1147
31. Smith AG (2001) Embryo-derived stem cells: of mice and men. Annu Rev Cell Dev Biol 17:435–462
32. Brons I, Smithers L, Trotter M, Gunn PR, Sun B, de Sousa S, Howlett S, Clarkson A, Richter LA, Pedersen R, Vallier L (2007) Derivation of pluripotent epiblast stem cells from mammalian embryos. Nature 448:191–195
33. Tesar PJ, Chenoweth JG, Brook FA, Davies TJ, Evans EP, Mack DL, Gardner RL, McKay RDG (2007) New cell lines from mouse epiblast share defining features with human embryonic stem cells. Nature 448(7150): 196–199
34. Nichols J, Smith AG (2009) Naive and primed pluripotent states. Cell Stem Cell 4:487–492
35. Guo G, Yang J, Nichols J, Hall JS, Eyres I, Manseld W, Smith A (2009) Klf4 reverts developmentally programmed restriction of ground state pluripotency. Development 136(7): 1063–1069
36. Vallier L, Touboul T, Chng Z, Brimpari M, Hannan N, Millan E, Smithers LE, Trotter M, Rugg-Gunn P, Weber A, Pedersen RA (2009) Early cell fate decisions of human embryonic stem cells and mouse epiblast stem cells are controlled by the same signalling pathways. PLoS One 4(6):e6082
37. Han DW, Tapia N, Joo JY, Greber B, Araúzo-Bravo MJ, Bernemann C, Ko K, Wu G, Stehling M, Do JT, Schöler HR (2010) Epiblast stem cell subpopulations represent mouse embryos of distinct pregastrulation stages. Cell 143(4):617–627
38. Han DW, Greber B, Wu G, Tapia N, Araúzo-Bravo MJ, Ko K, Bernemann C, Stehling M, Schöler HR (2011) Direct reprogramming of fibroblasts into epiblast stem cell. Nat Cell Biol 13(1):66–71
39. Irizarry RA, Bolstad BM, Collin F, Cope LM, Hobbs B, Speed TP (2003) Summaries of affymetrix GeneChip probe level data. Nucleic Acids Res 31(4), e15
40. Kim JB, Sebastiano V, Wu G, Araúzo-Bravo MJ, Sasse P, Gentile L, Ko K, Ruau D, Ehrich M, van den Boom D, Meyer J, Hübner K, Bernemann C, Ortmeier C, Zenke M, Fleischmann BK, Zaehres H, Schöler HR (2009) Oct4-induced pluripotency in adult neural stem cells. Cell 136(3):411–419
41. Ko K, Wu G, Araúzo-Bravo MJ, Kim J, Francine J, Greber B, Mühlisch J, Joo JY, Sabour D, Frühwald MC, Tapia N, Schöler HR (2011) Autologous pluripotent stem cells generated from adult mouse testicular biopsy. Stem Cell Rev 8(2):435–444
42. Ko K, Araúzo-Bravo MJ, Kim J, Stehling M, Schöler HR (2010) Conversion of adult mouse unipotent germline stem cells into pluripotent stem cells. Nat Protoc 5(5):921–928
43. Do JT, Joo JY, Han DW, Araúzo-Bravo MJ, Kim MJ, Greber B, Zaehres H, Sobek-Klocke I, Chung HM, Schöler HR (2009) Generation of parthenogenetic iPS cells from parthenogenetic neural stem cells. Stem Cells 27(12): 2962–2968
44. Kim JB, Zaehres H, Araúzo-Bravo MJ, Schöler HR (2009) Generation of induced pluripotent stem cells from neural stem cells. Nat Protoc 4 (10):1464–1470
45. Zaehres H, Kögler G, Araúzo-Bravo MJ, Bleidissel M, Santourlidis S, Weinhold S, Greber B, Kim JB, Buchheiser A, Liedtke S, Graffmann N, Zhao X, Meyer J, Reinhardt P, Burr B, Waclawczyk S, Ortmeier C, Uhrberg M, Schöler HR, Cantz T, Wernet P (2010) Induction of pluripotency in human cord blood unrestricted somatic stem cells. Exp Hematol 38 (9):809–818
46. Conrad S, Renninger M, Hennenlotter J, Wiesner T, Just L, Bonin M, Aicher W, Bühring HJ, Mattheus U, Mack A, Wagner HJ, Minger S, Matzkies M, Reppel M, Hescheler J, Sievert KD, Stenz A, Skutella T (2008) Generation of pluripotent stem cells from adult human testis. Nature 456:344–349
47. Luu PL, Schöler HR, Araúzo-Bravo MJ (2013) Disclosing the crosstalk among DNA methylation, transcription factors, and histone marks in human pluripotent cells through discovery of DNA methylation motifs. Genome Res 23 (12):2013–2029

Methods in Molecular Biology (2016) 1357: 221–229
DOI 10.1007/7651_2015_243

Published online: 12 April 2015

cGMP-Compliant Expansion of Human iPSC Cultures as Adherent Monolayers

Ann M. Parr, Patrick J. Walsh, Vincent Truong, and James R. Dutton

Abstract

Therapeutic uses of cells differentiated from human pluripotent stem cells (hPSCs), either embryonic stem (ES) cells or induced pluripotent stem cells (iPSCs), are now being tested in clinical trials, and it is likely that this will lead to increased commercial interest in the clinical translation of promising hPSC research. Recent technical advances in the use of defined media and culture substrates have significantly improved both the simplicity and predictability of growing hPSCs, allowing a much more straightforward application of current good manufacturing practices (cGMP) to the culture of these cells. In addition, the adoption of cGMP-compliant techniques in research environments will both improve the replication of results and make the transition of promising investigations to the commercial sector significantly less cumbersome. However, passaging methods for hPSCs are inherently unpredictable and rely on operator experience and expertise. This is problematic for the cell manufacturing process where operator time and process predictability are often determining cost drivers. We have adopted a human iPSC system using defined media and a recombinant substrate that employs cell dissociation with a hypertonic citrate solution which eliminates variability during hPSC cell expansion and provides a simple cGMP-compliant technique for hiPSC cultivation that is appropriate in both research and commercial applications.

Keywords: Pluripotent stem cells, iPS cell culture, Cell manufacturing

1 Introduction

Since their introduction, it has been widely anticipated that human induced pluripotent cells will be utilized to generate differentiated cells for clinical therapies. For clinical translation, the iPS cell lines must be derived, cultured, and differentiated using protocols compliant with current good manufacturing practices.

This has led to the rapid development of various fully defined media and culture substrate combinations for the culture of undifferentiated hiPSCs. However, options for passaging hiPSCs suitable in an ideal manufacturing setting remain limited.

The methods first described for hPSC subculture involved the maintenance of colonies grown atop mitotically inactivated stromal support layers, often irradiated or mitomycin C treated embryonic mouse fibroblasts. Propagation of these cultures requires manual dissection of colonies into smaller fragments that would then be collected and seeded onto new stroma. As human pluripotent stem cell culture systems have become more refined with respect to their composition, the manual element of propagation has persisted in the form of enzymatic digestion and scraping, or chelation and forceful, direct trituration of colonies from culture surfaces. These passaging methods make the translation of research-grade, benchtop processes to clinical-grade, clean room manufacturing difficult.

An ideal passaging system for iPS cell line derivation and propagation uses fully defined reagents and can be employed in both an open plate format and fully closed systems. There should be no physical, direct contact with the cells to release them from the substrate. Culture manipulations should be simple and streamlined to reduce error and minimize clean room operator involvement. The system should also be scalable to address a variety of culture platforms, surface areas, and cell numbers. We describe here a method for passaging hPSCs within the context of a cGMP-compatible culture system that allows for closed-system manipulation of cultures for the purposes of maintenance and expansion, similar to common trypsinization techniques used for primary stroma and adherent cell lines.

cGMP-compliant manufacture of human induced pluripotent stem cell lines is still evolving and many questions, which fall outside the scope of this chapter, have yet to be answered with regard to reprogramming technologies and optimal cell sources. Assuming investigators have acquired informed consent for tissue donation and have successfully reprogrammed the chosen somatic cells to a pluripotent fate using a vector system suitable to the task, and isolated colonies (if necessary) into small flasks; this chapter describes the flask-based maintenance and expansion aspect of the cell culture process that can be made part of a cGMP compliant manufacturing standard operating procedure.

Strategies amenable to flask-based culture of hiPSCs in cGMP-compatible conditions are limited. While alternatives exist not described here, we have found success with a culture system comprised of *Essential 8* [1] medium and a recombinant human vitronectin matrix. Colonies are dissociated as large fragments using a modified hypertonic citrate formulation [2], which allows for colonies to be dislodged with only medium turbulence following a brief incubation period. We have optimized the citrate formulation specific to the culture environment. Optimal results are obtained with a 15 mM citrate solution and osmolality adjusted to fall between 625 and 650 mOsmol/kg with potassium chloride in water.

The process as described here applies to passage-two or later established iPSC lines, but the technology could be scaled down and applied to individual passage-zero colonies if they have been derived clonally by physical sequestration in small-scale cultures before expanding into T25 flasks between passages one and two (*see* **note 1**). The process is likewise described as open and flask-based, but could be easily closed and performed outside a biosafety cabinet with the use of appropriate tubing sets, feed bags, sterile connectors, and welds.

2 Materials

2.1 Reagents

1. Sodium citrate (Sigma W302600).
2. Potassium chloride (Sigma P9333).
3. Cell culture-grade water (Life Technologies A1287303).
4. DPBS (Life Technologies 14040-133).
5. Recombinant Human Vitronectin (Peprotech AF-140-09).
6. *Essential 8* Medium (Life Technologies A1517001).
7. DMEM F/12 (Life Technologies 11039-021).
8. T-25 flasks (Thermo Fisher Scientific 1012628).
9. 15 mL conical tubes (Thermo Fisher 0553859A).
10. 50 mL conical tubes (Thermo Fisher 0553860).
11. 250 mL bottle top filters (0.2 μm) (Denville Scientific F0706).

2.2 Formulating Hypertonic Citrate

1. Weigh 4.4 g sodium citrate.
2. Weigh 25 g potassium chloride.
3. Combine sodium citrate and potassium chloride and dissolve into 1 L cell culture-grade water.
4. Assess osmolality with a properly calibrated osmometer (Vapro 5600 Osmometer, Wescor) and adjust osmolality if necessary to ~640 mOsmol/kg.
   - Increase osmolality by adding 400 mM potassium chloride.
   - Decrease osmolality by adding 15 mM sodium citrate.
5. Sterilize using 250 mL 0.2 μm bottle top vacuum filters and store at room temperature. The prepared chelation reagent is stable for over 1 year.

2.3 Preparing Recombinant Vitronectin Solution (500 μg/mL Stock)

1. Label 20 sterilized 0.6 mL microcentrifuge tubes.
2. Microcentrifuge 1 mg vial of lyophilized vitronectin (pulse to 10,000 × *g*).
3. Inside biosafety cabinet, dissolve 1 mg vitronectin into 2 mL water.

4. Aliquot 100 μL vitronectin stock solution into each tube (*see* **note 2**).
5. Freeze stock solution aliquots −20 °C to −80 °C (*see* **note 3**).

2.4 Preparing Essential 8 Medium

1. Thaw 10 mL Essential 8 Supplement overnight at 4 °C.
2. Bring *Essential* 8 basal medium and supplement into biosafety cabinet.
3. Discard 10 mL of the 500 mL Essential 8 basal medium.
4. Combine the 10 mL Essential 8 Supplement with the remaining basal medium.
5. Protect the completed Essential 8 media from light and store at 4 °C (*see* **note 4**).

3 Methods

3.1 Maintenance of Adherent Monolayer iPSC Cultures

1. In our experience, hiPSC cultures maintained as described here using *Essential* 8 medium on a recombinant vitronectin substrate require passage every 3–4 days regardless of confluence. A properly seeded culture should be roughly 40 % confluent 24 h following passage. At this seeding density, cultures should achieve 70 % confluence by days 3–4 (*see* **note 5**).
2. In the biosafety cabinet remove an aliquot of Essential 8 Medium and allow it to reach room temperature.
3. Remove the culture vessel from the incubator and place it in the biosafety cabinet. Aspirate and discard spent medium from flask.
4. Add 5 mL *Essential* 8 Medium being careful not to pipette directly onto the cells.

   Inspect the cell culture before returning the flask to a 37 °C, 5 % CO_2 incubator (*see* Section 4).
5. Repeat medium exchange daily, inspecting the cells each day to gauge cell morphology and colony confluence to determine the culture readiness for passage.

3.2 Coating Tissue Culture Flasks with Recombinant Vitronectin

1. In the biosafety cabinet thaw and add one 100 μL aliquot of (500 μg/mL) stock vitronectin into 10 mL DPBS (5 μg/mL final concentration) in a sterile 15 mL conical tube.
2. Mix and pipette 2.5 mL dilute vitronectin coating into each T25 flask and rock the flask to ensure consistent coverage of the culture area.
3. Incubate the flask for a minimum of 1 h at room temperature (*see* **note 6**).

3.3 Dissociation of Adherent Monolayer iPSC Cultures Without Direct Colony Contact by the Operator and Minimal Shear Stress

1. This method is designed to dissociate the adherent monolayer into appropriately sized colony pieces (*see* Section 3.4) that balance cell survival and culture expansion with minimal physical intervention. The procedure reduces culture variation by eliminating steps dependent on operator skill and experience.
2. Remove cell culture flask from the incubator and place inside biosafety cabinet. Uncap flask and aspirate culture medium.
3. Carefully coat the cell monolayer with 2.5 mL hypertonic citrate. Rock the flask to ensure an even distribution of reagent over all the culture then carefully aspirate the chelation solution from the cells.
4. Carefully coat the cell monolayer with a fresh 2.5 mL of hypertonic citrate solution. Again, rock the flask to ensure even coverage of the cell monolayer.
5. Return the flask to the incubator and incubate vessel for 6 min at 37 °C.
6. Remove culture from incubator and observe colonies using an inverted microscope (*see* **note 7**).
7. While keeping flask flat and level, vigorously swirl the citrate in a circular motion for approximately 15 s to dislodge colonies from the vitronectin matrix.
8. If few or no detaching colonies can be observed, return the flask to the incubator and continue to incubate the culture at 37 °C for an additional 2 min before observing the colonies again.
9. Monitor for detachment every 2 min until the majority of colonies can be seen floating in suspension following the vigorous swirling (*see* **note 8**).

3.4 Quenching the Chelation Reaction and Collecting the Dissociated Colonies

1. Once the colonies have detached sufficiently, return the flask to the biosafety cabinet, uncap the flask, and add 10 mL DMEM F/12 (*see* **notes 9** and **10**).
2. Prop the flask on its short edge opposite its opening at a 45° angle for 5 min. The colonies will be observed to gravity-settle into the corner of the flask if dissociated into fragments comprised of 50–200 cells (*see* **note 11**).
3. Carefully collect and dispose of the supernatant, being careful not to disturb the settled aggregate layer deposited in the corner of the flask.
4. Approximately 1 mL of aggregate layer and medium will be settled at the bottom of the flask. Add 3 mL of Essential 8 medium to the settled aggregate layer (*see* **note 12**).

3.5 Distribution and Expansion of Dissociated Colonies

1. Aspirate vitronectin coating from new flasks.
2. Add 4 mL Essential 8 Medium to each new flask.
3. Gently shake the cell suspension in the flask being passaged to evenly suspend the colony fragments. At this stage there is no requirement for additional pipetting to break up the colonies (*see* **note 13**).
4. Gently collect the dissociated colonies in a pipette and distribute 1 mL of the dissociated culture into each new flask.
5. Cap the flasks and evenly distribute the colonies by shaking back and forth or in a figure of 8 motion (*see* **note 14**).
6. Portions of unused dissociated colony suspension can be banked or taken for culture characterization (e.g., Karyotype, analysis, mRNA or protein expression determination) or used to initiate differentiation protocols.

4 Expected Results

Figure 1 illustrates the daily expected morphology and confluence progression seen when adopting the methods described 15 min after passage the colony pieces should be adhering to the coated flask surface and be evenly distributed. 24 h after passage the colonies are slightly domed in appearance and possess phase bright edges as can be observed in Fig. 1a. 48 h after passage the colonies have adopted a much flatter morphology and may begin to merge (*see* Fig. 1b). 3 days after passage the culture will be approaching readiness for passage with large flat colonies as shown in Fig. 1c. Cultures reaching 70–80 % confluency are ready for passage. In our experience, a small degree of cell differentiation characterized by patches of flattened cells should be expected and tolerated (generally 5 % of the culture area or less). The differentiated cells are essentially eliminated during the chelation passaging technique.

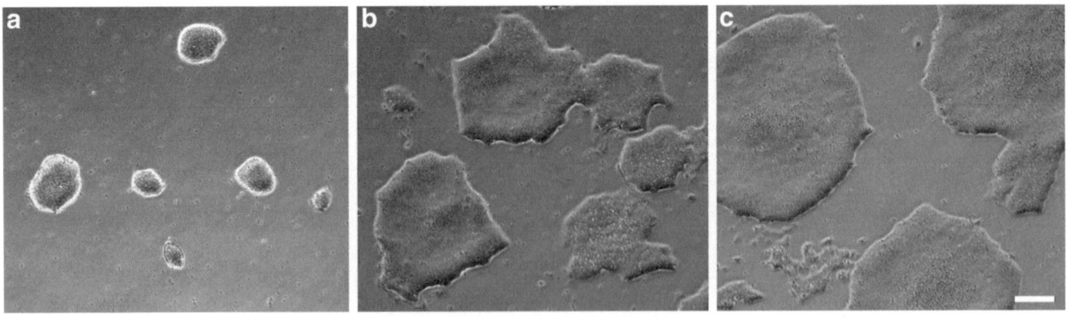

Fig. 1 Adherent hiPSC culture using Essential 8 media, vitronectin substrate, and hypertonic citrate passaging. (**a**) hiPSC colonies 24 h after passage. (**b**) Colonies 48 h after passage at 50 % confluency. (**c**) Colonies 72 h after passage at 70 % confluency. Scale bar 250 μm

5 Notes

1. We have also used citrate passaging to rapidly establish iPSC cultures from single picked colonies transferred quickly into flask-based culture for expansion and banking. Individual colonies are manually pipetted as colony pieces from the reprogramming plate and cultured in vitronectin-coated 4.8 cm^2 wells. Colony density is increased by citrate passage into a new single 4.8 cm^2 well. When this culture achieves 70–80 % confluency the colonies can be transferred using citrate passaging to a T-25 flask and the culture continued in the flask format as described above. This simple technique quickly establishes the new iPSC lines without manual intervention after the initial colony selection. In addition there is a reduction in the amount of operations in an open plate format, reducing the risk of compromising sterility.
2. Each 100 μL aliquot is sufficient to coat 100 cm^2 culture area, or 4 T25 flasks.
3. Reconstituted stock vitronectin solution is stable for 3 months at −20 °C.
4. Do not repeatedly warm whole bottles of completed media to room temperature and then return them to 4 °C. In the biosafety cabinet remove a sufficient aliquot of media to complete the task and warm that room temperature and return the unused media to the fridge without warming. Completed Essential 8 medium is stable at 4 °C for 2 weeks.
5. hiPSC growth rates may vary slightly between individual cell lines, but expansion from 70 % confluence can generally be maintained with a volume split ratio of 1:3 to 1:4.
6. Several conditions are acceptable for applying a vitronectin coating to flaskware. In addition to coating for 1 h at room temperature, 1 h incubation at 37 °C or 4 °C incubation overnight have also been successful. Coated flasks can also be aspirated dry and stored −20 °C for use >1 month later. The volume of coating can be scaled appropriately for larger culture areas (*see* Table 1).
7. Colonies can be observed macroscopically to detach into the citrate as early as 6 min post-incubation. Required incubation times tend to be confluence-based. Cultures near 50 % confluence may dissociate within 6 min, while cultures greater than 80 % confluent may require up to 15 min incubation.
8. In our experience a failure of the colonies to detach even after a 15 min incubation period is generally attributable to a failure of the chelation reagent due to incorrect formulation. Reformulate new chelation reagent paying careful attention to the

Table 1
Scaling up the use of hypertonic citrate passaging reagent

Culture vessel	cm^2 per chamber	Volume required (mL)
12-well plate	4.8	0.5
6-well plate	9.6	1
T-25 flask	25	2.5
T-75 flask	75	7.5
T-150 flask	150	15
T-225 flask	225	22.5
CS-1 cell stack	636	63.6

Volumes of citrate passaging reagent (mL) required to passage hiPSCs on adherent substrates in culture formats of increasing area (in cm^2)

osmolarity of the reagent before repeating the passage process. Cells treated with citrate reagent that do not detach after 15 min of treatment should be discarded as they can rarely be returned to routine culture.

9. This "quench volume" inhibits the chelation activity of the citrate and equilibrates the culture to a physiological osmolality.
10. Complete Essential 8 medium can also be used for this step but as this medium is then discarded it is generally more cost-effective to use DMEM F/12 medium for quenching.
11. Alternatively, cell suspensions can be collected into 15 or 50 mL sterile conical tubes and centrifuged at 100 rcf for 3 min to pellet colonies. Remove the media and during resuspension be careful to pipet gently or simply shake the tube to dislodge and resuspend the ensuing pellet while avoiding excess shear.
12. Addition of 3 mL Essential 8 medium gives 4 mL cell suspension that can be easily divided among 4 flasks to perform a 1:4 volume split.
13. Citrate treatment will compromise colony integrity. Avoid excess shear stress while pipetting colonies to prevent extensive colony fragmentation. Single cells and the smallest colony fragments will not adhere, resulting in poor passaging yield and eventual culture loss or establishment of a culture with cells containing chromosomal aberrations.
14. The majority of colonies can be seen to rapidly adhere within 15 min. It is important to evenly distribute the colony suspension as soon as possible to avoid heterogeneous plating density on the culture substrate.

Acknowledgements

The authors acknowledge support from NIH grant U01HL100407 and the University of Minnesota Foundation.

References

1. Chen G et al (2011) Chemically defined conditions for human iPSC derivation and culture. Nat Methods 8(5):424–429
2. Nie Y et al (2014) Scalable passaging of adherent human pluripotent stem cells. PLoS One 9(1): e88012

Methods in Molecular Biology (2016) 1357: 231–267
DOI 10.1007/7651_2014_156

Published online: 03 February 2015

Analysis of the Mitochondrial DNA and Its Replicative Capacity in Induced Pluripotent Stem Cells

Gael Cagnone, Vijesh Vaghjiani, William Lee, Claire Sun, Jacqueline Johnson, Ka-Yu Yeung, and Justin C. St. John

Abstract

The mitochondrial genome resides in the mitochondrion of nearly all mammalian cells. It is important for energy production as it encodes 13 of the key subunits of the electron transfer chain, which generates the vast majority of cellular ATP through the process of oxidative phosphorylation. As cells establish pluripotency, they regulate their mtDNA copy number so that they possess few copies but sufficient that they can be replicated to match the differentiated cell-specific requirements for ATP derived through oxidative phosphorylation. However, the failure to strictly regulate this process prevents pluripotent cells from differentiating. We describe a series of protocols that analyze mtDNA copy number, DNA methylation within the nuclear-encoded mtDNA-specific polymerase, and gene expression of the other factors that drive replication of the mitochondrial genome. We demonstrate how to measure ATP-generating capacity through oxygen respiratory capacity and total cellular ATP and lactate levels. Finally, we also describe how to detect mtDNA variants in pluripotent and differentiating cells using next-generation sequencing protocols and how the variants can be confirmed by high-resolution melt analysis.

Keywords: Mitochondrial DNA, Variants, Next-generation sequencing, Pluripotent cells, Differentiation, Transcription and replication, Oxygen consumption, ATP

1 Introduction

The mitochondrial genome is 16.6 kb in size and encodes 13 of the subunits of the electron transfer chain, which is the intramitochondrial apparatus that generates the majority of cellular ATP through the biochemical process of oxidative phosphorylation (OXPHOS) (1). Mitochondrial DNA (mtDNA) also encodes 22 tRNAs and 2 rRNAs and has one noncoding region, known as the D-loop. This is the mitochondrial genome's control region, where the nuclear-encoded transcription and replication factors that translocate to the mitochondrion functionally interact with mtDNA to initiate the first transcription and then replication of this bacterial-originating genome (2). The D-loop also has two hypervariable regions, which are characterized by sequences specific to maternal relatives and demonstrate the maternal inheritance of mtDNA (3).

Embryonic and induced pluripotent stem cells are excellent models to study and understand mitochondrial transcription and replication events during early development (4). mtDNA is present in large numbers in the egg at the time of fertilization (5). Once the egg has been fertilized, the number of mtDNA copies per cell is significantly diluted out as embryonic cells progress through preimplantation development to the blastocyst stage (5). At the blastocyst stage, mtDNA replication is restricted to the trophoblast cells. These are the first cells of the embryo to undergo differentiation. The inner cell mass cells, the cells that give rise to the fetus and embryonic stem cells, continue to dilute out their mtDNA copy number until levels are between 200 and 400 copies per cell (6). This is a significant developmental step, which establishes the *mtDNA set point* (6, 7).

The mtDNA set point in embryonic stem cells is defined as the state where all undifferentiated cells are under the control of the OCT4-SOX2-NANOG network and possess few copies of mtDNA. This promotes their ability to proliferate and remain pluripotent but, at the same time, maintain the potential to undergo differentiation and expand their mtDNA copy number. Consequently, the mtDNA set point ensures that all cells can expand their mtDNA copy number to meet the metabolic demands of the specialized cell that they fully differentiate into. This allows heart and muscle precursor cells to propagate high numbers of mtDNA copy to meet their high demands for OXPHOS-derived ATP, while blood and other cells from the spleen have fewer copies as their metabolic requirements are primarily fueled through glycolysis (6). The increases in mtDNA copy number take place in a synchronous manner with the expression of key genes during development. For example, during astrocyte differentiation, mtDNA copy number will remain low as cells continue to express NCAM, NESTIN, MUSASHI1, and CD133. However, as cells start to express endpoint markers, such as GFAP, they increase mtDNA copy number significantly (7).

Assessment of mtDNA copy number is also essential to determine whether a somatic cell that has been reprogrammed to a pluripotent state, i.e., induced pluripotent stem cells, has regulated their mtDNA copy number to low levels as part of the induction of dedifferentiation (8). It is equally important to determine whether induced pluripotent stem cells have the potential to increase their mtDNA copy number in a cell-specific manner during differentiation as a means of assessing their capability to fully differentiate.

mtDNA copy number is regulated by nuclear-encoded transcription and replication factors that translocate to the mitochondrion to drive firstly transcription and then replication (2). The key players in mitochondrial replication are mitochondrial transcription factor A (TFAM); the mitochondrial DNA-specific helicase, TWINKLE; the mitochondrial single-stranded binding protein

(MTSSB); and the mtDNA-specific polymerase, DNA polymerase gamma (POLG). POLG consists of two subunits, one is the accessory subunit known as POLGB, and the other is the catalytic subunit POLGA. One key developmental process is the regulation of POLGA expression during differentiation and development, which is mediated by DNA methylation, again in a cell-specific manner (6). The levels of DNA methylation during differentiation can be assessed by measuring the levels of 5-methylcytosine (5-mC) and 5-hydroxymethylcytosine (5-hmC), which are indicative of active and transient DNA methylation (9).

It is becoming increasingly important to measure the levels of mutations, which are present in stem cells that have been derived from patients carrying a mitochondrial disease, and define the selection of background mtDNA variants that are preferentially amplified during the induction of pluripotency. This could be performed using next-generation sequencing protocols, such as those used in conjunction with the Ion Torrent PGM or Illumina platforms.

Here, we describe a series of essential assays that define mtDNA replication, copy number, and ATP-generating capacity, which can be determined through measurement of oxygen consumption rates with an OROBOROS Oxygraph-2k, and total ATP and lactate assays. mtDNA copy number can be assessed through real-time PCR, while the most appropriate mechanism for measuring 5-mC and 5-hmC activities is through the use of immunoprecipitation of methylated DNA (MeDIP). When assessing the interactions of the nuclear transcription and replication factors that drive transcription and replication of the mitochondrial genome, it is essential to assess these genes in terms of their levels of expression. This can be effectively performed by using a real-time RT-PCR assay and comparing the levels of expression to known housekeeping genes. We favor the use of the Ion Torrent PGM with the CLC Genomics Workbench bioinformatics package for detecting mtDNA variants. Our protocols are sensitive to very low levels (approx. 3 %), which will provide an overall determination of the level of variants present. The variants can then be verified using high-resolution melt (HRM) analysis.

2 Materials

2.1 Culture of Human-Induced Pluripotent Stem Cells

1. Dulbecco's Modified Eagle Medium/Nutrient Mixture F12 with GlutaMAX (DMEM/F12) (Life Technologies; cat. no. 10565-018).
2. KnockOut serum replacement (Life Technologies; cat. no. 10828-028).

3. MEM nonessential amino acids (100×) (Life Technologies; cat. no. 11140-050).
4. 55 mM β-mercaptoethanol (Life Technologies; cat. no. 21985-023).
5. Dulbecco's phosphate-buffered saline (PBS) (Life Technologies; cat. no. 14190-144).
6. 10 μg/mL basic fibroblast growth factor (bFGF) solution (bFGF powder; Millipore; cat. no. GF003, dissolved in PBS).
7. Serum replacement (SR) medium: For 250 mL, combine 200 mL of DMEM/F12 with GlutaMAX, 50 mL KnockOut serum replacement, 2.5 mL MEM nonessential amino acids, and 450 μL β-mercaptoethanol and filter sterilize to 0.22 μm. The medium is supplemented with bFGF to a final concentration of 10 ng/mL just prior to use.
8. 4 mg/mL collagenase I solution (collagenase I powder; Worthington; cat. no. 4196, dissolved in DMEM/F12 medium).
9. 10 mg/mL dispase solution (dispase powder; Life Technologies; cat. no. 17105-041, dissolved in SR medium).
10. 0.1 % gelatin in sterile water (Sigma-Aldrich; cat. no. G1890).
11. 60-mm culture dishes (Falcon; cat. no. 353037).
12. 100-mm culture dishes (Falcon; cat. no. 353003).
13. 25-cm^2 culture flasks (Falcon; cat. no. 353108).
14. 15-mL sterile conical tube (Corning; cat. no. 430791).
15. Sterile Eppendorf tube (Axygen; cat. no. MCT175C).
16. 6-well ultralow attachment plates (Corning; cat. no. 3471).

2.2 Oxygen Consumption Measurement

1. Polarographic oxygen sensors (the Oxygraph-2K, OROBOROS INSTRUMENTS Corp., Innsbruck, Austria) with 10- and 50-μL syringes.
2. Hank's balanced buffer solution (Life Technologies, #14025076) and sodium hydrosulfite (Sigma-Aldrich; cat. no. 157953).
3. Reagents: oligomycin (Omy, 5 mg/mL, #75351), carbonyl cyanide *p*-(trifluoro-methoxy) phenyl-hydrazone (FCCP, 10 μM, #C2920), antimycin A (Ama, 5 mM, #A0274), and rotenone (Rot, 1 mM, #R8875). Reagents can be purchased from Sigma-Aldrich and are dissolved in 100 % EtOH. Store reagents at −20 °C.

2.3 ATP Assay

1. Luminescence plate reader. FLUOstar OPTIMA (BMG LABTECH, Offenburg, Germany) may be used to record luminescence.

2. 96-well white plate (Bio-Rad, #HSP-9601).
3. ATPlite luminescence assay system (PerkinElmer, #6016941). This kit contains cell lysis solution, substrate buffer solution, ATP standard, substrate solution (luciferase/luciferin). Store kit at 4 °C.

2.4 Lactate Assay

1. Absorbance plate reader. FLUOstar OPTIMA (BMG LABTECH, Offenburg, Germany) may be used to record absorbance.
2. 96-well transparent plate (U-shaped well) (Thermo Scientific, #262162).
3. Lactate colorimetric assay kit. Lactate Colorimetric Assay Kit II (BioVision, #K627-100) may be used. This kit contains L(+)-lactate standards, lactate enzyme mix, lactate substrate mix, and lactate assay buffer. Store kit at −20 °C.

2.5 mtDNA Copy Number

1. BIOTAQ™ DNA polymerase kit (Bioline; cat. no. BIO-21040).
2. dNTP (Bioline; cat. no. BIO-39028).
3. Primers (Sigma-Aldrich) (*see* primer, Table 1).
4. QIAquick Gel Extraction Kit (Qiagen; cat. no. 28704).

2.6 Immunoprecipitation of Methylated DNA (MeDIP)

1. Covaris Adaptive Focused Acoustics (AFA™) S220 system including Covaris SonoLab 7 software (Applied Biosystems; cat. no. 4465653) with Covaris milliTUBE (Covaris, cat. no. 520056).
2. Agarose gel is made with agarose (Bioline; cat. no. BIO-41025) with 1× TE buffer.
3. 10× IP buffer consists of 100 mM sodium phosphate (pH 7.0; Sigma-Aldrich; cat. no. 342483), 1.4 M NaCl (Sigma-Aldrich; cat. no. S5886), and 0.5 % Triton X-100 (Sigma-Aldrich; cat. no. T8532).
4. Anti-5mC antibody (Active Motif; cat. no. 39649) and anti-5hmC antibody (Active Motif; cat. no. 39999).
5. Dynabeads® Protein G (Life Technologies; cat. no. 10009D).
6. PBS/BSA buffer is made with 0.1 % BSA (Sigma-Aldrich; cat. no. A8806) in 1× PBS (Life Technologies; cat. no. 14190-144).
7. Magnetic rack (Invitrogen; cat. no. 123-21D).
8. 10× proteinase K digestion buffer consists of 50 mM Tris–HCl (pH 8.0; Merck; cat. no. 1082191000), 10 mM EDTA (pH 8.0; Sigma-Aldrich; cat. no. ED), and 1.0 % SDS (Sigma-Aldrich; cat. no. L6026).
9. Proteinase K (Bioline; cat. no. BIO-37037).
10. QIAquick PCR purification kit (Qiagen; cat. no. 28104).

Table 1
Primer sequences for gene expression and copy number and their melting temperature

Gene	Forward primers	Reverse primers	Product size (bp)	Annealing temp (°C)
β-ACTIN	CAAAACCTAACTTGCGCAGA	TTTTAGGATGGCAAGGGACT	261	53
MTSSB	GAAGCCATGTTTCGAAGACCTG	CTGATATTCTGTGCCATGTTGTC	271	57
POLGA	CACACCTAAACTCATGGCAC	GTCCACGTCGTTGTAAGGTC	436	56
POLGB	GTTTGCCATGAGTCCATCTAAC	CTCTGTCAGCTGGAAAGAATC	280	56
TFAM	ATTGGGGTCGGGTCACTGCCTCA	TACCTGCCACTCCGCCCTATAAGC	361	63
TWINKLE	GCACAAGTCCATCGTATCTTTC	CATACTCACTGATGAATGTCGTC	197	56
β-GLOBIN	CAACTTCATCCACGTTCACC	GAAGAGCCAAGGACAGGTAC	268	57
mtDNA	CGAAAGGACAAGAGAAATAAGG	CTGTAAAGTTTTAAGTTTTATGCG	152	53

2.7 Analysis of Gene Expression of mtDNA Replication Factors and mtDNA-Encoded Genes

2.7.1 cDNA Synthesis

1. QIAshredder columns (Qiagen; cat. no. 79654).
2. RNeasy Plus Mini Kit (Qiagen; cat. no. 74104).
3. DNase I kit (Qiagen; cat. no. 79254).
4. Molecular grade ethanol (Sigma-Aldrich; cat. no. E7023).
5. SuperScript III cDNA synthesis kit (Life Technologies; cat. no. 18080-051).

2.7.2 Real-Time Quantitative PCR

1. SensiMix Hi-ROX (Bioline; cat. no. QT605-20).
2. Gene-specific primers (Sigma-Aldrich) (*see* primer, Table 1).

2.8 Analysis of mtDNA Variants by Next-Generation Sequencing

2.8.1 Long PCR

1. Platinum *Taq* high-fidelity DNA polymerase kit (Invitrogen/ Life Technologies; cat. no. 11304-029). The kit is composed of:
   (a) Platinum *Taq* DNA polymerase high fidelity (5 U/μL)
   (b) 10× high-fidelity PCR buffer
   (c) 50 mM $MgSO_4$
2. 10 mM each dNTP mix (Bioline).
3. Forward primer 1: 5′-GACGGGCTCACATCACCCCATAA-3′ (10 μM, Sigma-Aldrich).
4. Reverse primer 1: 5′-GCGTACGGCCAGGGCTATTGGT-3′ (10 μM, Sigma-Aldrich).
5. Forward primer 2: 5′-GCCACAACTAACCTCCTCG-GACTCCT-3′ (10 μM, Sigma-Aldrich).
6. Reverse primer 2: 5′-GGTGGCTGGCACGAAATTGACC-3′ (10 μM, Sigma-Aldrich).
7. Autoclaved sterilized ddH_2O.
8. 0.2-mL sterile polypropylene PCR tubes (Axygen/Corning; cat. no. 6571).

2.8.2 Purification of PCR Products

1. QIAquick PCR purification kit (Qiagen; cat. no. 28104). The kit is composed of:
   (a) QIAquick spin columns
   (b) Buffer PB
   (c) Buffer PE
   (d) Buffer EB
   (e) pH Indicator I
   (f) Collection tubes (2 mL)
   (g) Loading dye
2. 3 M sodium acetate, pH 5.0 (Sigma-Aldrich; cat. no. S2889).
3. Autoclaved sterilized ddH_2O.
4. 1.7-mL sterile polypropylene Snaplock microcentrifuge tubes (Axygen/Corning; cat. no. MCT-175-C).

2.8.3 Quantification of Double-Stranded DNA

1. Qubit 2.0 Fluorometer (Invitrogen/Life Technologies; cat. no. Q32866).
2. Qubit dsDNA HS Assay Kit (Invitrogen/Life Technologies; cat. no. Q32851). The kit is composed of:
   (a) Qubit dsDNA HS reagent (200× concentrate in DMSO)
   (b) Qubit dsDNA HS buffer
   (c) Qubit dsDNA HS Standard #1 (0 ng/μL in TE buffer)
   (d) Qubit dsDNA HS Standard #2 (10 ng/μL in TE buffer)
3. Qubit assay tubes (Invitrogen/Life Technologies; cat. no. Q32856)

2.8.4 Shearing of Long PCR Amplicons

1. Covaris Adaptive Focused Acoustics (AFA™) S220 system including Covaris SonoLab 7 software (Applied Biosystems; cat. no. 4465653).
2. Covaris microTUBE (Covaris; cat. no. 520096).

2.8.5 Sample Purification

1. Agencourt AMPure XP kit (Beckman Coulter: cat. no. A63881), which is composed of Agencourt AMPure XP reagent beads.
2. DynaMag-2 magnetic rack (Invitrogen; cat. no. 123-21D).
3. Low TE buffer (Ion Fragment Library Kit).
4. 1.5-mL LoBind tubes (Eppendorf; cat. no. 022431021).

2.8.6 Size Selection of the DNA Library

1. E-Gel iBase and E-Gel Safe Imager Combo Kit (Invitrogen; cat. no. G6465).
2. E-Gel SizeSelect 2 % agarose gel (Invitrogen; cat. no. G6610-02).
3. Low TE buffer (Ion Fragment Library Kit).
4. 100-bp DNA ladder (Invitrogen; cat. no. 15628-019).

2.8.7 Quantification of the DNA Libraries

1. Agilent 2100 Bioanalyzer, which includes the Agilent 2100 Expert Software (Agilent Technologies; cat. no. G2947CA). This comprises:
   (a) Chip-priming station
   (b) IKA vortex mixer
2. Agilent High Sensitivity DNA Kit (Agilent Technologies; cat. no. 5067-4626), which is composed of:
   (a) High sensitivity DNA chips
   (b) Electrode cleaner
   (c) Syringe
   (d) Spin filters
   (e) High sensitivity DNA ladder
   (f) High sensitivity DNA markers 35/10380 bp
   (g) High sensitivity DNA gel matrix

2.8.8 DNA Template Amplification Through Emulsion PCR

This includes materials for the preparation of the IKA DT-20 solution and the Ion Sphere Particles, the emulsion, emulsion PCR, and recovery of the Ion Sphere Particles.

1. Ion Xpress Template Kit (Invitrogen/Life Technologies; cat. no. 4466457). This is composed of:
   (a) Emulsion oil
   (b) Ion Sphere wash solution
2. Ion Template Reagents Kit (Invitrogen/Life Technologies; cat. no. 4466462). This is composed of:
   (a) Polymerase
   (b) Amplification buffer
   (c) $MgCl_2$ solution
   (d) dNTPs
   (e) Thermostable pyrophosphatase (TAP)
   (f) Clonal amplification primer mix
   (g) Enrichment primer
   (h) Ion Sphere Particles
3. Ion Template Solutions Kit (Invitrogen/Life Technologies; cat. no. 4466463). This comprises:
   (a) Recovery solution
   (b) Wash solution
4. Eppendorf Combitips Plus, 10 mL, sterile (Eppendorf; cat. no. 022496123).
5. IKA ULTRA-TURRAX Tube Drive for the PGM System (Ion Torrent; cat. no. 4464747).
6. Repeater Plus pipette (Eppendorf; cat. no. 022260201).
7. MicroAmp Optical 96-Well Reaction Plates (Applied Biosystems; cat. no. N8010560).
8. MicroAmp Optical 8-Cap Strips (Applied Biosystems; cat. no. N8010535).
9. GeneAmp 96-well PCR System 9700 (Applied Biosystems; cat. no. N8050200).
10. 1-Butanol (molecular biology grade, $\geq$99 %, Sigma-Aldrich; cat. no. B7906).
11. PVC basin (VWR; cat. no. 4415129).
12. 1.5-mL LoBind tubes (Eppendorf; cat. no. 022431021).

2.8.9 Enrichment of Template-Positive Ion Sphere Particles

1. Tween-20 (Sigma-Aldrich; cat. no. P9416).
2. 1 M NaOH (Sigma-Aldrich; cat. no. S5881).
3. DynaMag-2 Magnet (Invitrogen; cat. no. 123-21D).

4. Dynabeads MyOne Streptavidin C1 magnetic beads (Invitrogen; cat. no. 650.01).
5. 1.5-mL LoBind tubes (Eppendorf; cat. no. 022431021).
6. Rotating wheel.
7. Wash solution (from the Ion Template Solutions Kit).

2.8.10 Assessment of the Quality of the Enriched Ion Sphere Particles

1. Qubit 2.0 Fluorometer (Invitrogen; cat. no. Q32866).
2. Qubit assay tubes (Invitrogen; cat. no. Q32856).
3. 0.2-mL RNase-free PCR tubes (Ambion; cat. no. AM12225).
4. 1.5-mL DNA LoBind tubes (Eppendorf; cat. no. 022431021).
5. Qubit 2.0 Easy Calculator Microsoft Excel Spreadsheet (download online).
6. Buffers:
   (a) 1× TEX (10 mM Tris, 1 mM EDTA, 0.01 % Triton X-100, pH 8)
   (b) 1× SSPE (diluted from a 20× stock, Ambion; cat. no. AM9767)
7. Oligonucleotides, HPLC purified, 250-nM synthesis scale (IDT):
   (a) B′-FAM 5′-CTGAGACTGCCAAGGCACACAGGGGATAGG-3′
   (b) A-CY5 5′-CCATCTCATCCCTGCGTGTCTCCGACTCAG-3′

2.8.11 Generation of the DNA Libraries for Ion Torrent PGM Sequencing

This comprises materials for end repair of the sheared DNA, ligation of adaptors, and amplification of the 200-bp DNA libraries:

1. The Ion Fragment Library Kit (cat. no. 4466464). This comprises the:
   (a) 5× end repair buffer
   (b) End repair enzyme
   (c) 10× ligase buffer
   (d) DNA ligase
   (e) Adapters
   (f) Platinum PCR SuperMix high fidelity
   (g) Library Amplification Primer Mix
   (h) Low TE buffer

2.8.12 Sequencing of Libraries with the Ion Torrent PGM

This includes the materials required for initialization of the Ion Torrent PGM sequencer and loading Ion Sphere Particles onto the Ion Chip:

1. The Ion Sequencing Kit (Invitrogen/Life Technologies; cat. no. 4466456). This is composed of:

   Ion PGM Supplies Kit (cat. no. 4466458), composed of:

   (a) Wash bottle sippers
   (b) Conical tube sippers
   (c) Conical tubes and labels
   (d) Wash 1 bottle (250 mL) and label
   (e) Wash 2 bottle (2 L) and label
   (f) Wash 3 bottle (250 mL) and label
   (g) Ion Sequencing Reagents Kit (cat. no. 4466459), composed of:
   (h) dGTP
   (i) dCTP
   (j) dATP
   (k) dTTP
   (l) Sequencing polymerase
   (m) Sequencing primer
   (n) Control Ion Sphere Particles
   (o) Ion PGM reagents kit (cat. no. 4466460). This is composed of:
       - W2 solution
       - Annealing buffer
       - PGM cleaning tablet
       - W3 solution
2. Industrial Grade Liquid Argon, 160-L cylinder (230 PSI) (Airgas; cat. no. AR160LT230).
3. Multistage (dual-stage) gas regulator (0–50 PSI, 2–3 bar output) (VWR International; cat. no. 55850-422).
4. Magnetic stirrer base MS-3000 (Fisher Biotec; cat. no. BS-010301-ABF).
5. 4-cm magnetic stirrer bar (ProSciTech; cat. no. L35-4008).
6. Glycerol, molecular biology grade (Sigma-Aldrich; cat. no. G5516).
7. Sonication water bath and floating PCR tube holder (Branson; cat no: 952-118).

Analysis of Sequence Outputs

CLC Genomics Workbench 7 (www.clcbio.com).

2.9 High-Resolution Melting Analysis

1. PCR thermocycler. MultiGene™ OptiMax Thermal Cycler (Labnet) may be used.
2. Light scanner with high-resolution melting analysis software. LightScanner System and Call-IT 2.0 software (Idaho Technologies, Salt Lake City, Utah) may be used.
3. 96-well white and black plate. Hard-Shell® Low-Profile Thin-Wall 96-Well Skirted PCR plates (Bio-Rad, Hercules; CA #HSP-9665) may be used.
4. Mineral oil (Sigma-Aldrich).
5. High-resolution melting enzyme mix. HRM Mix (TrendBio) may be used.
6. PCR primers (10 μM).
7. Optical sealing tape. Optical sealing tape (Bio-Rad, Hercules, CA #223-9444) may be used.

3 Methods

3.1 Manual Passage of Human-Induced Pluripotent Stem (iPS) Cells (See Note 1)

1. Under the dissecting microscope, assess colonies for differentiation and determine the most undifferentiated colonies for transfer (*see* **Note 2**).
2. Change medium on all preprepared MEF feeder plates to SR medium (*see* **Notes 3** and **4**).
3. From colony plates, choose areas of undifferentiated cells only for transfer by selectively cutting the colony into small pieces to generate small clumps of undifferentiated cells. This can be done with a pulled glass capillary or a 30-G needle.
4. Detach the pieces from the plate by gently teasing them up with the end of a capillary, needle, or pipette.
5. Using a P20 pipette, transfer 6–9 colony pieces onto each prepared feeder plate, and place them carefully on plate to avoid damage to the feeder layer.
6. Carefully transfer plate to 37 °C incubator so colony pieces do not move around and leave overnight to allow cells to attach.
7. Medium change plates daily until ready for next passage, usually 6–7 days.

3.2 Transition from Manual to Enzymatic Passage

1. To set up an enzymatic culture from manual passage dishes, cut colonies as per manual passage method, choosing only undifferentiated areas of colonies (*see* **Notes 5** and **6**).
2. Transfer colony pieces to a sterile Eppendorf containing 400 μL of serum replacement (SR) medium.
3. Use a 200 μL pipette to gently pipette up and down until colony pieces have broken up to a relatively homogeneous

suspension. Transfer suspension to a 25-cm^2 flask (previously seeded with MEFs) containing SR medium + 10 ng/mL bFGF (*see* **Note 7**).

4. Medium change flask daily and passage every 4–7 days as necessary.

3.3 Enzymatic Passage of iPS Cells

1. Remove medium from culture flask.
2. Add collagenase I solution to flask (e.g., 2 mL for 25-cm^2 flask).
3. Incubate flask at 37 °C for 10–15 min, undisturbed, if using freshly made collagenase or 20 min if using thawed collagenase.
4. During incubation, prepare new flasks by adding SR + 10 ng/mL bFGF medium to flasks previously seeded with MEFs.
5. Tap flask to dislodge cells and add sufficient SR + bFGF medium to dilute collagenase solution (*see* **Note 8**). Wash medium over surface of flask until majority of cells are removed, some differentiated cells and feeder cells may remain attached to the surface of the flask.
6. Transfer cell suspension to a 15-mL conical tube and centrifuge for 3 min at 150 × *g*.
7. Aspirate medium and gently resuspend pellet in 1 mL of SR + bFGF medium using a 1,000-μL tip; add further SR + bFGF medium and gently resuspend until a homogeneous suspension is achieved.
8. For continuous cultures, inoculate new flasks with desired dilution of cell suspension (*see* **Note 9**).
9. Observe cells and medium change daily. Passage when confluent, usually every 4–7 days.

3.4 Spontaneous Differentiation of iPS

1. Wash iPS plates with PBS and then incubate with dispase solution for 10 min at 37 °C to allow intact colonies to become detached.
2. Add SR medium, without bFGF, and transfer colonies to a 15-mL tube; centrifuge for 3 min at 150 × *g*.
3. Remove supernatant and then gently resuspend in SR medium, without bFGF, until a homogeneous suspension is achieved.
4. Transfer suspension to ultralow attachment plates to induce the formation of embryoid bodies (EBs).
5. Culture EBs in suspension for 7 days then transfer to 0.1 % gelatin-coated tissue culture plates to let attachment and spontaneous differentiation occur.
6. Change SR medium every 3–4 days until differentiation is complete.

3.5 To Measure Oxygen Consumption in iPS Cells

Measuring change in O_2 concentration (oxygen flux) during incubation with intact living cells can inform on the oxidative phosphorylation (OXPHOS) state in the mitochondria. Generally, mitochondria from embryonic as well as pluripotent stem cells show low oxygen consumption rate (OCR), and the utilization of OXPHOS inhibitors as well as uncouplers of the proton gradient is required to assess the respiration capacity of the cells:

1. Power on the Oxygraph-2K (O2K) and initiate the OROBOROS DataLab software (*see* **Note 10**).
2. Wash chambers with water, EtOH 80 %, and EtOH 100 % and then water before starting calibration of the oxygen sensors.
3. Fill the O2K chambers with 5 % sodium hydrosulfite and wait for the oxygen concentration to be minimal (~0 %). In the calibration parameters, assign 0 % oxygen to each oxygen sensor.
4. Wash chambers with water, EtOH 80 %, and EtOH 100 % and then water before filling up with HBSS. Wait for oxygen concentration to stabilize, and then assign 100 % oxygen to each sensor.
5. Close completely each O2K chamber (no air bubble) and inject up to 50 μL of cell suspension per chamber. Before injection, cells are washed, counted, and resuspended in HBSS. For effective oxygen consumption measurement, inject at least half a million cells per chamber (*see* **Note 11**). As an experimental parameter, normalize oxygen flux to the cell number corresponding to cell count (*see* **Note 12**).
6. Five minutes after injection, record the OCR in each chamber. This measure indicates the basal rate ("routine") of OXPHOS.
7. Omy is an inhibitor of ATP synthase complex V. Inject 2 μL Omy to each chamber and, after stabilization, record the OCR (*see* **Note 13**). This indicates the non-phosphorylating rate of respiration. By inhibiting complex V, the proton gradient becomes maximal and oxygen flux reflects proton leak (mainly) as well as other oxidative reactions in the cells.
8. FCCP is an uncoupler of the proton gradient. By adding successive volumes of 2 μL FCCP, titrate to obtain a maximal OCR (maximal stimulated respiratory capacity). Uncoupler titration must be performed carefully as over-titration will inhibit OCR (*see* **Note 14**). Maximal OCR under uncoupling conditions reflects the maximal activity of the electron transport chain (ETC).
9. Rot and Ama inhibit OXPHOS through complexes I and III. Add 1 μL Ama and 1 μL Rot and record OCR. This measure corresponds to the OXPHOS state under maximal inhibition and reflects residual oxygen consumption (ROX) in the cells. The ROX level is used as a background value of non-OXPHOS activity that must be subtracted to previously recorded OCRs.

10. After definitive OXPHOS inhibition, the experiment is finished and each chamber requires thorough washing (*see* **Note 15**). Rinse the chamber with water, EtOH 80 %, and EtOH 100 %. Adding spare cells to each chamber is effective to eliminate excess of inhibitors.

Interpretations of cellular respiration are based on corrected OCR measurements for basal (6), non-phosphorylating (7), and uncoupled (8) OXPHOS states. Uncoupling control ratio (uncoupled/basal) is an index of the ETC reserve capacity. It decreases under high ATP demand or because of OXPHOS dysfunction.

3.6 To Measure ATP Content in iPS Cells

ATP is a marker of cell viability, and ATP monitoring can be used to assess the level of metabolic quiescence in iPS cells and the proliferative impact of a wide range of drugs or biological compounds. ATP content can be monitored on the basis of the production of light caused by the reaction of ATP with added luciferase and D-luciferin (firefly enzymes). ATPlite luminescence assay is sensitive down to 5 cells in a 100-μL medium (*see* **Note 16**), but a higher cell number is recommended to obtain consistent luminescence. Moreover, luminescence can be acquired without the need of injectors as the signal has a half-life of greater than 5 h:

1. Harvest, count, and resuspend cell in 100 μL of culture medium, and then pipette the 100 μL cell suspension into the wells of a 96-well plate.
2. In parallel, add 100 μL of culture medium without cells in separate wells of the same plate that is used for the experimental samples. These wells will contain a serial dilution of ATP standards (from 10^{-4} M down to blank) (*see* **Note 17**).
3. Add 50 μL of lysis solution to every well and shake the plate for 5 min in an orbital shaker at 700 rpm. This lyses the cells and stabilizes the ATP.
4. In wells containing culture medium only, add the serial dilution of 10 μL ATP standards and shake the plate for 5 min in an orbital shaker at 700 rpm.
5. Add 50 μL substrate solution to each well and shake the plate for 5 min in an orbital shaker at 700 rpm.
6. Cover the plate in foil for 10 min to prevent access to light and measure the luminescence using a plate reader set for luminescence acquisition (*see* **Note 18**).
7. From the ATP standard luminescence data, calculate the standard curve and extrapolate the ATP concentration to each experimental sample. Normalize the extrapolated ATP concentration by the number of cells in the initial 100 μL of medium.

3.7 To Measure Lactate Release by iPS Cells

Lactate production is closely related to glycolytic metabolism in the cell, through lactate dehydrogenase (LDH) enzymes that catalyze the interconversion of pyruvate and lactate. In anaerobic glycolysis, production of lactate by the cell can be monitored in the culture environment, as the cell will release lactate in the medium (*see* **Note 19**). L(+)-Lactate is the major lactate stereoisomer formed in human intermediary metabolism, while D(−)-lactate is also present but only at about 1–5 % of the concentration of L(+)-lactate (blood). The Lactate Assay Kit II (BioVision) is based on oxidation of L(+)-lactate by LDH, generating a product that interacts with a colored probe (λmax = 450 nm). The kit detects L(+)-lactate in biological samples such as in serum, cells, culture, and fermentation media, at concentrations from 0.02 to 10 mM lactate (*see* **Note 20**). Here, the measure of lactate concentration in culture medium is described:

1. Culture cells in medium devoid of lactate and/or pyruvate and then collect spent culture medium (*see* **Note 21**). The medium containing FBS should be deproteinized to remove LDH that can degrade lactate. Record cell number and total medium volume at time of medium collection. These will be used to normalize the amount of lactate produced by the cell.
2. Prepare a standard curve by adding 0, 2, 4, 6, 8, and 10 μL of L (+)-lactate standard (1 nmol/μL) into separate wells, and then adjust volume to 50 μL/well with lactate assay buffer to generate 0, 2, 4, 6, 8, and 10 nmol/well of the L(+)-lactate standard.
3. Prepare test sample in 50 μL/well with lactate assay buffer. It is suggested to use two different volumes of test sample, for example, 10 and 40 μL, to ensure the reading is within the standard curve range (*see* **Note 22**).
4. For each well, prepare a total of 50 μL master mix containing 46 μL lactate assay buffer, 2-μL probe, and 2 μL enzyme mix. Add 50 μL of master mix to each well containing the lactate standard or test samples. Mix well by pipetting up and down.
5. Incubate the reaction for 30 min at room temperature, protect from light.
6. Measure OD 570 nm for colorimetric assay in a microplate reader.
7. Correct background by subtracting the value derived from the 0 lactate control from all sample readings. Plot standard curve nmol/well ~OD570nm readings. Apply the sample readings to the standard curve. Calculate the lactate concentrations of the test samples: $C = La/Sv$ (nmol/μL or mM) where La is the lactate acid amount (nmol) of your sample from standard curve and Sv is the sample volume (μL) added into the well. Normalize the lactate concentration to the total volume of medium and the cell number.

3.8 Preparation for Standard Curve Using Quantitative PCR

1. Adjust the concentration of total DNA templates to 100 ng/μL.
2. For amplification using conventional PCR, prepare a reaction mixture with a total volume of 50 μL containing:
   (a) 5 μL 10× NH_4 reaction buffer
   (b) 1.5 μL 50 mM $MgCl_2$ solution
   (c) 0.5 μL 50 mM dNTP mix dilution
   (d) 0.5 μL BIOTAQ™ DNA polymerase
   (e) 1 μL forward primer* (50 pmol/μL)
   (f) 1 μL reverse primer* (50 pmol/μL)
   (g) 2 μL (200 ng) DNA template
   (h) 38.5 μL autoclaved Milli-Q H_2O
3. After amplification of nDNA (β-globin) and mtDNA, run PCR samples on a 2 % agarose gel to confirm the amplified DNA templates.
4. Cut out the targeted band from the gel and purify DNA using Qiagen gel purification kit. Determine the amount of DNA templates purified for nDNA and mtDNA, respectively.
5. Adjust the concentration of purified DNA to 10 ng/μL.
6. Prepare tenfold serial dilutions of the purified DNA from 10^{-1} to 10^{-8} ng/μL for the standard templates.
7. Prepare for quantitative PCR in 20 μL reactions containing:
   (a) 10 μL 2× SensiMix SYBR Hi-ROX master mix
   (b) 1 μL forward primer* (5 pmol/μL)
   (c) 1 μL reverse primer* (5 pmol/μL)
   (d) 2 μL of each standard template (*see* **Note 23**)
8. Reactions are performed in the 72-well Rotor Gene 3000 real-time PCR machine. Reaction conditions are:
   (a) 95 °C for 15 min, followed by:
   - 45 cycles of 95 °C for 15 s
   - Primer-specific annealing temperature for 15 s
   - 72 °C for 15 s

   (b) Extension and data acquisition are performed on FAM/SYBR channel.
   (c) Melt curve analysis from 72 to 99 °C with a 30-s wait for the first step followed by 5 s for each subsequent step of 1 °C increase. Acquire melt data from the FAM/SYBR channel.
9. A standard curve is produced with the efficiency of quantitative PCR determined (*see* **Note 24**). The melt curve determines the specificity of the primers, where only one curve should be observed (*see* **Note 25**).

3.9 Determining mtDNA Copy Number

1. Adjust the concentration of the unknown targeted DNA samples to 10 ng/μL.
2. Prepare reaction mixtures for quantitative PCR, as previously described, with a total volume of 20 μL containing 18 μL of master mix with primers and 2 μL of the unknown samples (20 ng of total DNA). qPCR conditions, as previously described (step 8 of Section 3.8).
3. Concentrations of the mitochondrial and nuclear-specific products will be calculated based on the standard curve computationally.
4. Copy number is then calculated by

   $N = ([\text{ng}/\mu\text{L}] \times 6.023 \times 10^{14})/(N_{\text{bp}} \times 660)$ where:

   (a) N is the number of molecules per reaction.

   (b) $\times 10^{14}$ is the conversion of 1 mol to 1 nmol using Avogadro's constant, which states 1 mol contains 6.023×10^{23}.

   (c) N_{bp} is the product size of the amplicons of interest.

   (d) 660 is the mean molecular weight of a nucleic acid base pairing in Daltons (Da).
5. mtDNA copy number per cell is then calculated by dividing the mtDNA copy number by half of the chromosomal DNA copy number as chromosomal DNA is present in diploidy in cells.

3.10 MeDIP

1. Dilute 10 μg genomic DNA in 100 μL autoclaved Milli-Q H_2O and add 100 μL 2× TE buffer in a Covaris milliTUBE. Use Covaris Adaptive Focused Acoustics (AFA™) S220 system to shear the DNA for 195 s to fragments sized between 200 and 1,000 bp. The setting is 5–10 °C temperature limit range, 105.0 peak power, 5.0 duty factor, and 200 cycles/burst (*see* **Note 26**).
2. Run a 1–1.5 % agarose gel to confirm efficient sonication (*see* **Note 27**).
3. Take at least 3 μg of sonicated DNA and dilute with autoclaved Milli-Q H_2O to a final volume of 450 μL (*see* **Note 28**).
4. Denature DNA by boiling at 95 °C for 10 min and immediately place on ice for 5 min.
5. Add 50 μL of 10× IP buffer and 3 μL (3 μg; 1 μg per 1 μg DNA) of anti-5mC or anti-5hmC antibody.
6. Incubate at 4 °C for 2 h with rotation.
7. Prewash 30 μL of Dynabeads® Protein G (10 μL per 1 μg DNA) two times with 1 mL PBS/BSA buffer for 5 min per wash at 4 °C with rotation.
8. Remove fluid from beads on a magnetic rack and resuspend beads in 30 μL (original volume) of 1× IP buffer.

9. Add beads into each sample and incubate at 4 °C for 16 h with rotation.
10. Remove all fluid on a magnetic rack.
11. Wash the beads with 1 mL of 1× IP buffer for 5 min per wash at 4 °C with rotation. Remove all fluid on a magnetic rack.
12. Resuspend beads in 250 μL proteinase K digestion buffer and add 7 μL of 20 mg/mL proteinase K.
13. Incubate at 50 °C for 3 h on a shaker.
14. Extract DNA with QIAquick PCR purification kit. Elute purified DNA in 50 μL of autoclaved Milli-Q H_2O.

3.11 Analysis of Gene Expression of mtDNA Replication Factors and mtDNA-Encoded Genes

1. cDNA synthesis is performed using SuperScript III kit from Life Technologies.
2. A total of 1 μg of RNA is converted to cDNA (*see* **Note 29**). Take 1 μg of RNA and adjust the volume to 8 μL with RNA-free water and add 1 μL of 50 μM oligo(dT) and 1 μL of 10 mM dNTPs.
3. Incubate the reaction at 65 °C for 5 min and chill on ice for at least 1 min.
4. Add 2 μL of 10× RT buffer, 4 μL of 25 mM magnesium chloride, 2 μL of 0.1 M DTT, 1 μL of RNaseOUT (40 U), and 1 μL of SuperScript III enzyme (200 U) to the reaction (*see* **Note 30**).
5. Incubate the reaction at 50 °C for 50 min followed by 85 °C for 5 min. Chill the reaction on ice.
6. Add 1 μL of RNase H to the reaction and incubate at 37 °C for 20 min. Chill the reaction on ice.
7. Dilute the cDNA with 42 μL of sterile Milli-Q water (1:3 dilutions). cDNA is ready for gene expression or it can be stored at −20 °C.
8. Prepare master mix containing 10 μL of SensiMix, 1 μL of 5 μM forward primer, 1 μL of 5 μM reverse primer, and 6 μL of sterile Milli-Q water.
9. Add 2 μL of cDNA to sterile strip tubes and mix with 18 μL of master mix. Run each reaction in triplicate.
10. A normalization gene (housekeeping gene) should also be run for quantification of gene expression using ΔΔCT method.
11. Run real-time PCR on Rotor Gene 3000 real-time PCR machine (Corbett Research, Cambridge, UK).
12. The reaction conditions for real-time PCR are 95 °C for 10 min followed by 45 cycles of 95 °C for 15 s, followed by gene-specific annealing temperature with an extension for 15 s at 72 °C.

13. The data is acquired in FAM/SYBR channel during extension phase. The second acquisition phase (cycle B) of 78 °C was used to permit quantification of a single specific product only.
14. Gene expression is normalized to a housekeeping gene and data are expressed as relative fold change calculated using ΔΔCT method.

3.12 Next-Generation Sequencing

Long PCR generates the templates for next-generation sequencing.

3.12.1 Long PCR

1. Two long PCR reactions are performed, each in a 0.2-mL PCR tube containing of 1× high-fidelity PCR buffer, 2 mM $MgSO_4$, 10 mM dNTPs, 10 μM of the forward primer, 10 μM of the reverse primer, and 1 U of Platinum *Taq* DNA polymerase high fidelity (*see* **Note 31**).
2. 50 ng of genomic DNA is used as the template (*see* **Note 32**).
3. A negative control contains the components listed in step 1, except for the genomic DNA.
4. Reactions are run as either 20 or 50 μL volumes and are adjusted accordingly with sterile ddH_2O.
5. Reaction conditions are:
   (a) 94 °C for 2 min
   (b) Followed by 35 cycles of:
   - 94 °C for 15 s
   - 63 °C for 30 s
   - 68 °C for 9 min

   (c) The reaction is completed by 1 cycle of:
   (d) 68 °C for 10 min
   (e) 4 °C hold until the reactions

3.12.2 Purification of the PCR Products

1. Add pH indicator I to buffer PB at a ratio of 1:250 to produce "buffer PBI" at a pH of ≤7.5.
2. Transfer the PCR product to a 1.7-mL sterile microcentrifuge tube.
3. Add buffer PBI to the PCR product at a volume ratio of 5:1.
4. If buffer PBI changes color, add 10 μL of 3 M sodium acetate (pH 5.0) to correct the pH. If the color is unaffected, proceed to step 5.
5. Mix by pipetting and then transfer up to 650 μL of the solution to a QIAquick spin column inserted into a 2-mL collection tube.
6. Centrifuge for 1 min at 10,000 × *g*.
7. Discard the flow-through.

8. If there is excess PCR product from step 5, load the remaining solution into the QIAquick spin column and repeat steps 6 and 7. If not, then proceed to step 9.
9. Add 100 % ethanol to buffer PE before use.
10. Add 750 μL of buffer PE to the QIAquick spin column.
11. Repeat steps 6 and 7 twice.
12. Transfer the QIAquick spin column to a 1.7-mL microcentrifuge tube and add 30 μL of sterile ddH_2O.
13. Repeat step 6 and store the purified template for next-generation sequencing at −20 °C until use.

3.12.3 Quantification of Double-Stranded DNA

1. Dilute the Qubit dsDNA HS reagent in Qubit dsDNA HS buffer at a ratio of 1:200 to produce the "Qubit working solution" for the number of samples and standards to be tested (*see* **Note 33**).
2. Standards #1 and #2 are prepared by adding 10 μL of each Qubit standard to 190 μL of Qubit working solution to individual Qubit assay tubes.
3. DNA samples are prepared by adding 1 μL of the DNA to 199 μL of Qubit working solution in individual Qubit assay tubes.
4. Standards and samples are vortexed for 2–3 s and incubated at room temperature for 2 min.
5. Set up the Qubit 2.0 Fluorometer by choosing the assay type "dsDNA High Sensitivity."
6. Choose: run a "new calibration."
7. Choose: insert standard #1 and press "Read."
8. Remove standard #1, then insert standard #2, and choose "Read." Remove standard #2.
9. Insert the DNA sample and choose "read next sample."
10. Repeat step 8 for the quantification of each DNA sample.
11. Calculate the concentration of the sample by using:

$$\text{Concentration(ng/mL)} = \text{QF value} \times (200/x)$$

QF value = the value obtained from the Qubit 2.0 Fluorometer; x = the volume (μL) of sample used for quantification.

*3.12.4 Shearing of Long PCR Amplicons (See **Note 34**)*

1. Pool the two long PCR mtDNA amplicons at equimolar concentrations.
2. Transfer the pooled DNA to a Covaris microTUBE and keep on ice.
3. Fill the Covaris S220 system water tank to level 12 with 4 °C deionized water.

4. Degas the water by choosing the "De-gas" option on the Covaris Sonolab 7 software and run for 30 min.
5. Set the operating conditions to sheer DNA to 200-bp fragments:
   (a) Duty factor = 10 %
   (b) Peak incident power (W) = 175
   (c) Cycles per burst = 200
   (d) Time = 180 s
   (e) Number of cycles = 6
6. Load the Covaris sample tube into the tube holder of the Covaris S220 and apply the holder arm.
7. Lower the arm of the Covaris S220 to partially submerge the Covaris tube into the degassed 4 °C deionized water tank.
8. Select "Start" to initiate DNA fragmentation.
9. When the shearing is complete, remove the Covaris sample tube and place on ice.
10. Repeat steps 6–9 for any other samples.

3.12.5 End Repair of the Sheared DNA

1. Transfer 79 μL of the sheared DNA to a 1.5-mL Eppendorf LoBind tube.
2. Add 20 μL of 1× end repair buffer.
3. Add 1 μL end repair enzyme.
4. Mix by pipetting and incubate for 20 min at room temperature.

3.12.6 Sample Purification

1. Add 180 μL of Agencourt AMPure XP reagent beads to the end-repaired DNA sample.
2. Mix by pipetting five times and centrifuge briefly.
3. Incubate at room temperature for 5 min.
4. Load the sample onto a DynaMag-2 magnetic rack.
5. Leave for 3 min.
6. Discard the supernatant by pipetting.
7. Leaving the tube in the magnetic rack, resuspend the Agencourt AMPure XP reagent beads in 500 μL of 70 % ethanol.
8. Incubate for 30 s.
9. Rotate the tube twice within the DynaMag-2 magnetic rack.
10. Discard the supernatant by pipetting.
11. Repeat steps 7–10.
12. Remove the tube from the magnetic rack and centrifuge briefly and reload the tube into the magnetic rack.
13. Discard residual supernatant by pipetting.

14. Air-dry at room temperature for 5 min.
15. Remove the tube from the magnetic rack and add 25 μL of low TE buffer.
16. Pipette up and down five times to mix.
17. Vortex the tube for 10 s.
18. Centrifuge briefly and place the Eppendorf LoBind tube into the magnetic rack.
19. Incubate at room temperature for 1 min.
20. Transfer the product to a sterile 0.2-mL PCR tube.

3.12.7 Adaptor Ligation

1. Add the following reagents to the 0.2-mL PCR tube containing 25 μL of purified DNA: 1× ligase buffer, 2 μL of Ion P1 adapter, 2 μL of Ion Xpress Barcode 1, 2 μL of dNTP mix, 2 μL of DNA ligase, and 8 μL of nick repair polymerase.
2. Repeat step 1 but replace 2 μL of Ion Xpress Barcode 1 for 2 μL of Ion Xpress Barcode 2 (up to Barcode 16).
3. Add sterile ddH_2O to a total reaction volume of 100 μL.
4. Mix by pipetting.
5. Place the 0.2-mL PCR tubes into a thermal cycler and run the reactions at:
   (a) 25 °C for 15 min
   (b) 72 °C for 5 min
   (c) 4 °C hold
6. Transfer the adaptor-ligated products to a sterile 1.5-mL Eppendorf LoBind tube.
7. Repeat Section 3.12.6 by adding 140 μL of Agencourt AMPure XP reagent beads (step 1) and eluting the purified DNA in 20 μL of low TE buffer (step 16).

3.12.8 Size Selection

1. Load 20 μL of the target DNA into wells of a 2 % E-Gel SizeSelect agarose gel.
2. Dilute the 50-bp reference DNA ladder 1 in 40 using low TE buffer.
3. Load 10 μL of the diluted DNA ladder onto the agarose gel.
4. Load 25 μL of sterile ddH_2O into the top wells of the gel.
5. Load 25 μL of sterile ddH_2O into the bottom wells of the gel, except the collection well used for electrophoresis of the DNA ladder. This is loaded with 10 μL of sterile ddH_2O.
6. Set the "Run Time To Reference Line" to 15 min.
7. End the electrophoresis when the 200-bp reference band reaches the reference line.

8. Refill the collection wells with 25 μL of sterile ddH_2O.
9. Set the "Run Time from Reference Line to Collection Well" to 2 min.
10. Collect the DNA when the 200-bp band has migrated into the collection well and transfer to a 0.2-mL PCR tube.

3.12.9 Amplification of the 200-bp DNA Libraries

1. Add 25 μL of the size-selected DNA to 100 μL of Platinum PCR SuperMix high-fidelity master mix and 5 μL of Library Amplification Primer Mix.
2. Transfer the sample tubes to a thermal cycler and amplify at:
   (a) 95 °C for 5 min
   (b) Followed by 6 cycles of:
      - 95 °C for 15 s
      - 58 °C for 15 s
      - 70 °C for 1 min
   (c) Followed by 4 °C hold until removed from the thermal cycler
3. Repeat Section 3.12.6 by adding 1.5× the sample volume of Agencourt AMPure XP reagent beads, as in step 1, and eluting the purified DNA in 20 μL of low TE buffer, as in step 16.

3.12.10 Quantification of the DNA Libraries

1. Equilibrate the high sensitivity DNA dye concentrate and the high sensitivity DNA gel matrix vials to room temperature for 30 min before use.
2. Mix the high sensitivity DNA dye concentrate by vortexing for 10 s.
3. Add 15 μL of the dye concentrate to a vial of the high sensitivity DNA gel matrix.
4. Vortex the gel-dye mixture for 10 s.
5. Apply the gel-dye mixture to the filter of a spin column and centrifuge at 4,000 × *g* at room temperature for 10 min.
6. Discard the spin column.
7. Place the Agilent high sensitivity DNA chip onto the chip-priming station.
8. Load 9 μL of the gel-dye mix onto the Agilent high sensitivity DNA chip in the well labeled with a circularized "G."
9. Set the syringe plunger at the 1-mL position and close the chip-priming station (*see* **Note 35**).
10. Push the syringe plunger until held in position by the clip.
11. After 60 s, release the syringe plunger so that the plunger reaches the 0.3-mL mark.
12. After 5 s, set the syringe plunger to the 1-mL position.

13. Open the chip-priming station.
14. Load 9 μL of the gel-dye mix onto three wells labeled as "G" on the Agilent high sensitivity DNA chip.
15. Load 5 μL of the high sensitivity DNA marker in the 11 sample wells and in the well labeled with a ladder.
16. Load 1 μL of DNA sample into the 11 sample wells.
17. Place the Agilent high sensitivity DNA chip onto the IKA vortex mixer.
18. Vortex the chip at 2,400 rpm for 1 min.
19. Open the lid of the Agilent 2100 Bioanalyzer and clean the electrodes.
20. Place the Agilent high sensitivity DNA chip into the Agilent 2100 Bioanalyzer.
21. Open the Agilent 2100 Expert Software and select "dsDNA."
22. Select "start."
23. Dispose of the DNA chip at the end of the run.

3.12.11 Preparation of the IKA DT-20 Solution and Ion Sphere Particles

1. Transfer 9 mL of emulsion oil to an IKA DT-20 tube using a 10-mL Eppendorf Combitips Plus pipette tip.
2. Store the IKA DT-20 tube on ice (*see* **Note 36**).
3. Vortex the Ion Sphere Particles for 1 min.
4. Transfer 140 μL of the Ion Sphere Particles to 1.3 mL of wash solution in a 1.5-mL Eppendorf LoBind tube.
5. Pipette five times to mix.
6. Vortex briefly the Ion Sphere Particles.
7. Centrifuge the Ion Sphere Particles at 10,000 × g at room temperature for 5 min.
8. Resuspend the Ion Sphere Particles in 100 μL of supernatant.
9. Adjust the total volume of the Ion Sphere Particles solution to 140 μL with Ion Sphere wash solution.

3.12.12 Preparation of the Aqueous Master Mix for the Emulsion PCR

1. Dilute the DNA library quantified in Section 3.12.10 to 9 pM using sterile ddH_2O.
2. Adjust the total volume of the 9 pM DNA library to 18 μL with sterile ddH_2O.
3. Prepare the aqueous master mix in a 1.5-mL Eppendorf LoBind tube by adding:
   (a) 336 μL of nuclease-free water
   (b) 105 μL 10× amplification buffer
   (c) 105 μL of $MgCl_2$ solution
   (d) 105 μL of dNTPs

(e) 105 μL of clonal amplification primer mix
(f) 5 μL of enrichment primer
(g) 5 μL of thermostable pyrophosphatase (TAP)
(h) 126 μL of polymerase

4. Mix by pipetting five times.
5. Transfer the aqueous master mix to the 140 μL of Ion Sphere Particles prepared in Section 3.12.11.
6. Vortex the Ion Sphere Particles/amplification master mix solution at maximum speed for 5 s.
7. Add the 18 μL of DNA library diluted to 9 pM.

3.12.13 Generation of the Emulsion

1. Place the ice-cold IKA DT-20 tube prepared in Section 3.12.11 on the IKA ULTRA-TURRAX Tube Drive.
2. Remove the adhesive seal from the lid of the IKA DT-20 tube to create an opening in the tube lid.
3. Vortex the Ion Sphere Particles/amplification master mix/DNA solution then centrifuge briefly for 3 s.
4. "Start" the IKA ULTRA-TURRAX Tube Drive.
5. Transfer the master mix solution into the IKA DT-20 tube through the opening while the tube is in motion.
6. Replace the adhesive seal over the opening on the lid of the IKA DT-20 tube.
7. Allow 5 min for mixing and place the IKA DT-20 tube on ice for 5 min.

3.12.14 Emulsion PCR

1. Transfer 100 μL of the emulsion using an Eppendorf Repeater pipettor into each well of an Applied Biosystems MicroAmp 96-Well Reaction Plate.
2. Cover the 96-well PCR plate with Applied Biosystems MicroAmp Optical Cap Strips.
3. Place the PCR plate into a thermal cycler and run at:

   (a) 94 °C for 6 min.

   Followed by 40 cycles of:

   (b) 94 °C for 30 s
   (c) 58 °C for 30 s
   (d) 72 °C for 90 s

   5 cycles of:

   (e) 94 °C for 30 s
   (f) 68 °C for 6 min

   End the reaction with 1 cycle of:

   (g) 10 °C hold, until collected from the machine.

3.12.15 Recovery of the Ion Sphere Particles

1. Mix the emulsion.
2. Transfer ~1 mL of the emulsion from the PVC basin into six 1.5-mL Eppendorf LoBind tubes.
3. Centrifuge at 15,500 × *g* for 2 min.
4. Prepare the "breaking solution" by adding 2.5 mL of recovery solution to 7.5 mL of 1-butanol in a 15-mL tube.
5. Vortex for 1 min.
6. After completing step 3, discard the oil layer without disturbing the pellet.
7. Rinse the Applied Biosystems MicroAmp 96-Well Reaction Plate by transferring 100 μL breaking solution to each well in the first row of the 96-well plate.
8. Mix by pipetting and transfer the breaking solution from step 7 to the next row of the 96-well plate.
9. Repeat rinsing for other rows on the 96-well plate until the bottom row is reached.
10. Transfer the breaking solution to a PVC basin.
11. Repeat steps 7–10.
12. Transfer ~1 mL of breaking solution, collected in the PVC basin, to the six Eppendorf tubes from step 6.
13. Vortex Eppendorf tubes for 1 min.
14. Centrifuge Eppendorf tubes at 15,500 × *g* for 2 min.
15. Discard the top layer without disturbing the pellet.
16. Add 1 mL of recovery solution to the Eppendorf tubes containing the pellets.
17. Vortex the Eppendorf tubes for 1 min.
18. Centrifuge the Eppendorf tubes at 15,500 × *g* for 3 min.
19. Discard the supernatant from each Eppendorf tube but leave ~50 μL of solution behind in each of the six Eppendorf tubes.
20. Transfer the material from the six Eppendorf tubes into one 1.5-mL sterile Eppendorf LoBind tube.
21. Rinse three of the tubes by adding 200 μL of recovery solution to the first tube, mix by pipetting five times, and transfer the recovery solution to the second tube.
22. Pipette the recovery solution five times and transfer to the third tube.
23. Transfer the solution from the third tube to the tube generated in step 20.
24. Repeat steps 21–23 if necessary.
25. Add the recovery solution at a ratio of 1:1 with the combined tube.

26. Vortex for 30 s.
27. Centrifuge at 15,500 × *g* for 3 min.
28. Discard the supernatant but leave ~100 μL of solution in the tube.
29. Pipette up and down five times.
30. Transfer the contents to a sterile 1.5-mL Eppendorf LoBind tube.
31. Rinse the original tube with 100 μL of wash solution and transfer to another Eppendorf tube.
32. Add 1 mL of wash solution to the Eppendorf tube.
33. Repeat steps 26–28.
34. Repeat steps 32 and 33.
35. Place the sample on ice.

3.12.16 Enrichment of the Template-Positive Ion Sphere Particles

1. Prepare melt-off solution (125 mM NaOH and 0.1 % TWEEN 20) by adding 200 μL of 1 M NaOH to 16 μL of 10 % Tween-20 (diluted in sterile ddH_2O) and 1.38 mL of sterile ddH_2O.
2. Vortex Dynabeads MyOne Streptavidin C1 beads.
3. Add 10 μL of MyOne Streptavidin beads to a 1.5-mL Eppendorf LoBind tube.
4. Add 70 μL of wash solution to the MyOne Streptavidin beads.
5. Vortex to mix.
6. Place the Eppendorf tube into the DynaMag-2 magnetic rack for 2 min.
7. Remove the supernatant.
8. Resuspend the MyOne Streptavidin beads by adding 10 μL of wash solution.
9. Transfer the 10 μL of MyOne Streptavidin beads to the Eppendorf tube containing the Ion Sphere Particles (step 35 of Section 3.12.15).
10. Pipette five times to mix.
11. Vortex to mix again.
12. Add 100 μL of annealing buffer to the Ion Sphere Particles.
13. Repeat steps 10 and 11.
14. Place the tube containing the Ion Sphere Particles on a rotating wheel for 10 min at room temperature.
15. Centrifuge the sample for 3 s.
16. Repeat step 6.
17. Transfer the supernatant to a new 1.5-mL Eppendorf LoBind tube marked "unbound."

18. Add 200 μL of wash solution to the Ion Sphere Particles.
19. Pipette 15 times to mix.
20. Repeat steps 6 and 17.
21. Repeat steps 18–20.
22. Add 400 μL of melt-off solution prepared in step 1 to the sample tube.
23. Vortex.
24. Place the tube on a rotating wheel to spin at room temperature for 7 min.
25. Add 400 μL of wash solution to another 1.5-mL Eppendorf LoBind tube labeled "Enriched-1."
26. When step 24 is completed, vortex the tube and centrifuge briefly.
27. Repeat step 6.
28. Transfer the supernatant to the "Enriched-1" tube without disturbing the MyOne Streptavidin beads.
29. Vortex the "Enriched-1" tube.
30. Centrifuge at 15,500 × *g* for 4 min.
31. Remove the supernatant but retain 100 μL containing the enriched Ion Sphere Particles.
32. Add 1 mL of wash solution to the "Enriched-1" tube.
33. Repeat steps 29–31.
34. Pipette to mix.
35. Repeat step 29 and centrifuge for 3 s.
36. Place the "Enriched-1" tube into the DynaMag-2 magnetic rack for 2 min.
37. Transfer the supernatant from "Enriched-1" to a new 1.5-mL Eppendorf LoBind tube labeled "Enriched-2."

3.12.17 Assessment of the Quality of the Enriched Ion Sphere Particles

1. Dilute the oligonucleotides to 100 μM in 1× TE buffer, pH 8.0.
2. Combine the following components to create the hybridization master mix:
   (a) 100 μL of 1× SSPE
   (b) 2 μL of 100 μM B′FAM oligonucleotide
   (c) 2 μL of 100 μL A-CY5 oligonucleotide
3. Add 2 μL of the enriched Ion Sphere Particles to a 0.2 mL PCR tube.
4. For the negative control, add 2 μL of ion wash buffer to a 0.2-mL PCR tube.

5. Add 52 µL of the hybridization master mix from step 2 to both PCR tubes from steps 3 and 4.
6. Pipette five times to mix.
7. Load PCR tubes into a thermal cycler and run at:
   (a) 95 °C for 2 min
   (b) 37 °C for 2 min
8. Transfer the products from the PCR tubes to 1.5-mL Eppendorf LoBind tubes.
9. Add 1 mL 1× TEX buffer to each tube.
10. Vortex.
11. Centrifuge both tubes at 15,600 × *g* for 3 min.
12. Remove 1 mL of the supernatant leaving behind 20 µL containing the Ion Sphere Particles.
13. Repeat steps 9–12 twice.
14. Add 1× TEX to the tubes to a total volume of 200 µL.
15. Transfer samples to separate Qubit assay tubes.
16. Load samples into the Qubit 2.0 Fluorometer.
17. Choose "Ion," followed by "IonFAM," and then "Read."
18. Record the value.
19. Choose "Home," followed by "IonCY5," and then "Read."
20. Repeat step 18.
21. Repeat steps 16–20 with the negative control.
22. Open the Qubit 2.0 Easy Calculator Microsoft Excel Spreadsheet.
23. Enter the values.
24. The templated signal ratio is automatically calculated to determine the quality of the enriched Ion Sphere Particles according to the following guidelines:
   (a) <0.1—insufficient enriched Ion Sphere Particles resulting in suboptimal loading density on the ion chip.
   (b) 0.1–0.3—the optimal to provide the best loading density on the ion chip.
   (c) >0.3—will produce mixed reads due to non-clonal amplification of the template on the Ion Sphere Particles during the emulsion PCR.

3.12.18 Initialization of the Ion Torrent PGM Sequencer

1. Rinse Wash Bottles 1 and 3 twice using ultrapure (18 Ω) water.
2. Add 350 µL of 100 mM NaOH to Wash Bottle 1.
3. Add 50 mL of W3 solution to Wash Bottle 3.
4. Add 1,978 mL ultrapure (18 Ω) water to Wash Bottle 2 and mark the volume level.

5. Empty Wash Bottle 2.
6. Place a magnetic stirrer into the empty Wash Bottle 2.
7. Fill Wash Bottle 2 with argon gas for 5 min using a flowmeter set at 0.5 L per minute.
8. Fill Wash Bottle 2 with ultrapure (18 Ω) water to the 1,978-mL volume mark from step 4. Ensure there are no gaps that would otherwise allow for gas exchange between the bottle and the environment.
9. Place Wash Bottle 2 on a magnetic stirrer, sealing the bottle opening with the argon gas tube while ensuring that the gas nozzle is not in direct contact with the water.
10. Flow of argon gas was set to 0.5 L per minute, and the water was stirred at a speed to create a "whirlpool" effect within Wash Bottle 2.
11. Add 22 mL of W2 solution to Wash Bottle 2.
12. Mix the contents of Wash Bottle 2 for 1 min.
13. Choose "Init PGM" to initialize.
14. Attach a sterile sipper to the bottle cap of Wash Bottle 2 and seal the opening by attaching the bottle to the Ion Torrent PGM to prevent gas exchange.
15. Choose "Next."
16. Repeat step 13 with Wash Bottles 1 and 3.
17. Repeat step 15.
18. The Ion Torrent PGM automatically determines the pH of Wash Bottle 2.
19. Add 20 μL of each dNTP solutions to four separate 50-mL conical tubes.
20. After initialization of the wash solutions is completed, attach sippers to each of the 50-mL conical tubes.
21. Attach all dNTP conical tubes to the Ion Torrent PGM, following the order G, C, A, and T.
22. The Ion PGM automatically fills the conical tubes each with 40 mL of solution.
23. At the end of the initialization, choose "Next" to complete the process and initialize the run.

3.12.19 Preparation of the Positively Enriched Ion Sphere Particles for Sequencing

1. Transfer 50 % of the positively enriched Ion Sphere Particles to a sterile 0.2-mL PCR tube.
2. Vortex the vial containing the control Ion Sphere Particles and centrifuge briefly.
3. Add 5 μL of the control Ion Sphere Particles to the PCR tube from step 1.

4. Add 90 μL of wash solution.
5. Pipette to mix.
6. Centrifuge the PCR tube containing the enriched Ion Sphere Particles at 15,500 × g for 1.5 min.
7. Discard the supernatant, leaving between 20 and 30 μL of supernatant behind.
8. Add 150 μL of annealing buffer to the enriched Ion Sphere Particles.
9. Pipette to mix.
10. Repeat steps 6 and 7, leaving behind only 10 μL of supernatant in the PCR tube.
11. Add 5 μL of sequencing primer and pipette five times to resuspend the Ion Sphere Particles.
12. Load the PCR tube into a thermal cycler and run at:
    (a) 95 °C for 2 min
    (b) 37 °C for 2 min
13. Repeat steps 8 and 9.
14. Repeat step 6 and 7, leaving behind approximately 20 μL of supernatant containing the Ion Sphere Particles.
15. Repeat steps 13 and 14, leaving behind approximately 5–6 μL of supernatant containing the Ion Sphere Particles.
16. Add 1 μL of sequencing polymerase.
17. Pipette to mix.
18. Incubate at room temperature for 5 min. Meanwhile the ion chip check can be performed.

3.12.20 Testing of a New Ion Chip

1. Choose "Experiment" on the main menu of the Ion Torrent PGM screen.
2. Clamp the ion chip on the Ion Torrent PGM.
3. Scan the chip's barcode.
4. Choose "Chip Check."
5. Choose "Next" to continue to the calibration stage.

*3.12.21 Loading the Ion Sphere Particles on the Ion Chip (See **Note 37**)*

1. Add 100 μL of 100 % glycerol to 100 μL of annealing buffer to a 1.5-mL Eppendorf LoBind tube.
2. Vortex.
3. Remove the ion chip from the Ion Torrent PGM and place on the grounding plate.
4. Add 50 μL of annealing buffer into the large opening on the chip.
5. Remove residual buffer.

6. Sonicate the Ion Sphere Particles that have been incubated for 5 min from step 18 of Section 3.12.19, for 10 s in a sonication water bath at 40 kHz.
7. Add 2 μL of the glycerol/annealing buffer mix from step 1 to the sonicated Ion Sphere Particles.
8. Pipette to mix.
9. Add the total mix from step 7 to the large opening of the chip.
10. Repeat step 5.
11. Centrifuge the chip for 10 min.
12. Press "#" to check that the number of cycles is set for 65 on the Ion Torrent PGM.
13. Choose "Autoanalysis" and "Pre-analysis."
14. Choose "Next" to confirm the settings on the Ion Torrent PGM, and then choose "OK."
15. Replace the loaded ion chip.
16. Allow the Ion Torrent PGM to calibrate the chip and to initiate the sequencing run.

*3.12.22 Analysis of Sequence Outputs (See **Note 38**)*

CLC Genomics Workbench 7 (www.clcbio.com)

1. Select "Trim" to trim reads of adaptors.
2. Select "Map Reads" and filter reads to exclude those of a nucleotide length of <15 bp.
3. Select "Variant Mapping."
4. Accept a Phred quality score of ≥15.
5. Set a mismatch cost of 2.
6. Set an insertion/deletion cost of 3.
7. For a single nucleotide polymorphism (SNP) analysis, a minimum mutation threshold of 3 % was applied.

3.13 To Determine Variant Load by High-Resolution Melting Analysis

As the mtDNA is multimeric, it is prone to contain heteroplasmic population of sequence variants. Targeted PCR followed by high-resolution melting of the PCR product can discriminate the presence of a variant nucleotide in the mtDNA. From next-generation sequencing data, HRM analysis can be used to validate the identified variants by designing specific primers around the mtDNA sequences of interest. For quantification of variant frequency, a separate standard curve needs to be run in parallel to the sample. A standard curve is required to isolate reference and variant sequences and to mix them at different ratios (reference/variant), i.e., 100 %:0 %, 90 %:10 % ... 0 %:100 %:

1. Prepare DNA template at 10 ng/μL.
2. Prepare HRM master mix with n = sample number:

(a) $(n + 1) \times 3\ \mu L\ H_2O$
(b) $(n + 1) \times 1\ \mu L$ forward primer
(c) $(n + 1) \times 1\ \mu L$ reverse primer
(d) $(n + 1) \times 4\ \mu L$ HRM enzyme mix

3. Add 20 μL mineral oil to each well of the plate.
4. Add 9 μL HRM master mix to each well of the plate.
5. Add 1 μL DNA template/standard in each well of the plate.
6. Seal the plate with transparent tape then centrifuge plate.
7. Run the plate on PCR thermocycler according to the following cycles:
   (a) 2 min at 95 °C
   (b) 45 cycles of 94 °C for 30 s and annealing primer temperature for 30 s
   (c) 94 °C for 30 s and cooling to 25 °C for heteroduplex formation
8. Insert plate into fluorescence scanner and read fluorescence of melting PCR product at temperature ramping from 70 to 96 °C. Data acquisition increases incrementally by 0.1 °C until the reaction is terminated.
9. In HRM analysis software, normalize fluorescence shift around the temperature of melting (*see* **Note 39**).
10. Along the standard curve, observe the shift of the melting curve corresponding to the sample and determine the level of variant according to the ratios of reference/variant from the standards (*see* **Note 40**).
11. Collect data outputs from data file.

4 Notes

1. These methods are equally applicable to the culture of human embryonic stem cells.
2. Manual passage of iPS colonies allows for the transfer of undifferentiated areas of colonies only.
3. Undifferentiated iPS cells are cultured on mitomycin C-inactivated MEFs.
4. MEFs need to be seeded at 6×10^4 per cm^2 on manual passage plates.
5. MEFs need to be seeded at 2×10^4 per cm^2 for enzymatic passage.
6. Starting material should contain minimal differentiated cells to ensure maintenance of enzymatic cultures in an undifferentiated state.

7. Approximately 25–30 iPS colonies are required for each 25-cm^2 culture flask to be set up.
8. Minimum medium volume added should be equal to original volume of collagenase solution.
9. Recommended dilution range is 1:3–1:6.
10. Set temperature in accordance and magnetic stirrer bar to rotate up to 700 rpm.
11. Minimize the time between cell harvest and injection into the chamber. Maintain resuspended cells at 37 °C in the meantime.
12. If OCR does not increase sufficiently after adding cells to the chamber, more cells can be added (avoid air bubble).
13. The interval of OCR recording should be 2–5 min.
14. A higher concentration of FCCP can be used if OCR does not increase substantially.
15. Traces of rotenone can persist. Efficient cleaning must be achieved.
16. Work in a clean environment and wear gloves to avoid any contamination, as ATP is present in many sources of life.
17. ATP standards should be ready and kept on ice before starting the experiment.
18. Luminescence should be measured 30 min after reagent addition, as light production decreases with time, regardless of cell number.
19. Culture medium can contain phenol red, so a low amount of medium should be used for each sample. Normalize OD with background data from blank wells.
20. Cell concentration during culture will impact on lactate production and lactate concentration in the medium. In the case of glucose starvation (prolonged culture or cell confluence), cells will stop producing lactate from glucose but start consuming lactate.
21. Fresh sample is preferred, but frozen sample (−80 °C) can be used.
22. A high concentration of lactate has an inhibitory effect that leads to lower readings. Lactate overdose will turn from pink to brown.
23. The accuracy of pipetting and preparation of the master mix in advance is critical in order to establish the correct quantitation of nuclear and mitochondrial DNA copy number.
24. The efficiency of amplification of the standards should range between 0.95 and 1, and the R^2 value should be as close to 0.99 as possible to establish accurate quantitation of copy number.

25. The temperature of the extension and data acquisition phases should be set to the temperature where the reduction of fluorescent is first detected.
26. It normally takes 2 h for the sonication system to cool down and degas. Moreover, it is important to plan the timeline well in advance because there are several series of incubation steps in the procedure that includes a 16-h incubation (normally from 5 pm to 9 am next day).
27. The expected result is a smear between 200 and 1,000 bp.
28. Turn on the heat block to preheat to 95 °C for the next step.
29. We typically use a NanoDrop to measure RNA concentration and purity. The absorbance ratio of 260:280 should be ~2.0 indicative of pure RNA, whereas a lower ratio indicates impurities such as phenol or proteins. The absorbance ratio of 260:230 should be in the range of 2.0–2.2 for RNA free of impurities.
30. cDNA synthesis should always be run with an additional reaction with no SuperScript III enzyme (often referred to as no RT control), which is used to determine the presence of contaminating DNA. The presence of contaminating DNA in RNA will affect evaluation of the results.
31. It may be necessary to titrate the $MgSO_4$ concentration (range = 1.5–3 mM).
32. The use of too much template saturates this sensitive reaction.
33. It is essential to avoid quantification by NanoDrop, as it is too inaccurate for this protocol.
34. Another option is to use enzymatic shearing and kits provide the appropriate enzymes to perform this. However, we have achieved better outcomes through shearing.
35. It is essential to ensure that the syringe plunger is held in position.
36. Storage on ice is essential.
37. It is essential to run a previously analyzed sample to ensure reproducibility, as a quality control measure.
38. We extract raw data from the Ion Torrent and solely use the CLC Genomics Workbench for mapping and variant calling.
39. Normalization of the melting curve must be done carefully in order to obtain flat lines on each side of the difference curves.
40. It is suggested to run an internal control on every HRM plate in order to standardize the difference curves from the serial dilutions of reference-variant mtDNA.

Acknowledgments

This work was supported by the Victorian Government's Operational Infrastructure Support Program and NMHRC Project Grants GNT1022222 and GNT 1041471 to J.C.S.J.

References

1. Anderson S, Bankier AT, Barrell BG, de Bruijn MH, Coulson AR, Drouin J, Eperon IC, Nierlich DP, Roe BA, Sanger F, Schreier PH, Smith AJ, Staden R, Young IG (1981) Sequence and organization of the human mitochondrial genome. Nature 290(5806):457–465
2. Kucej M, Butow RA (2007) Evolutionary tinkering with mitochondrial nucleoids. Trends Cell Biol 17(12):586–592. doi:10.1016/j.tcb.2007.08.007, S0962-8924(07)00242-5 [pii]
3. Gill P, Ivanov P, Kimpton C et al (1994) Identification of the remains of the Romanov family by DNA analysis. Nat Genet 6:130–135
4. St John JC, Facucho-Oliveira J, Jiang Y, Kelly R, Salah R (2010) Mitochondrial DNA transmission, replication and inheritance: a journey from the gamete through the embryo and into offspring and embryonic stem cells. Hum Reprod Update 16(5):488–509. doi:10.1093/humupd/dmq002, dmq002 [pii]
5. Spikings EC, Alderson J, St John JC (2007) Regulated mitochondrial DNA replication during oocyte maturation is essential for successful porcine embryonic development. Biol Reprod 76(2):327–335. doi:10.1095/biolreprod.106.054536, biolreprod.106.054536 [pii]
6. Kelly RD, Mahmud A, McKenzie M, Trounce IA, St John JC (2012) Mitochondrial DNA copy number is regulated in a tissue specific manner by DNA methylation of the nuclear-encoded DNA polymerase gamma A. Nucleic Acids Res 40(20):10124–10138. doi:10.1093/nar/gks770
7. Dickinson A, Yeung KY, Donoghue J, Baker MJ, Kelly RD, McKenzie M, Johns TG, St John JC (2013) The regulation of mitochondrial DNA copy number in glioblastoma cells. Cell Death Differ 20(12):1644–1653. doi:10.1038/cdd.2013.115
8. Kelly RD, Sumer H, McKenzie M, Facucho-Oliveira J, Trounce IA, Verma PJ, St John JC (2013) The effects of nuclear reprogramming on mitochondrial DNA replication. Stem Cell Rev 9(1):1–15. doi:10.1007/s12015-011-9318-7
9. Tahiliani M, Koh KP, Shen Y, Pastor WA, Bandukwala H, Brudno Y, Agarwal S, Iyer LM, Liu DR, Aravind L, Rao A (2009) Conversion of 5-methylcytosine to 5-hydroxymethylcytosine in mammalian DNA by MLL partner TET1. Science 324(5929):930–935. doi:10.1126/science.1170116

Methods in Molecular Biology (2016) 1357: 269–283
DOI 10.1007/7651_2014_144

Published online: 20 November 2014

Selection of Phage Display Peptides Targeting Human Pluripotent Stem Cell-Derived Progenitor Cell Lines

Paola A. Bignone, Rachel A. Krupa, Michael D. West, and David Larocca

Abstract

The ability of human pluripotent stem cells (hPS) to both self-renew and differentiate into virtually any cell type makes them a promising source of cells for cell-based regenerative therapies. However, stem cell identity, purity, and scalability remain formidable challenges that need to be overcome for translation of pluripotent stem cell research into clinical applications. Directed differentiation from hPS cells is inefficient and residual contamination with pluripotent cells that have the potential to form tumors remains problematic. The derivation of scalable (self-renewing) embryonic progenitor stem cell lines offers a solution because they are well defined and clonally pure. Clonally pure progenitor stem cell lines also provide a means for identifying cell surface targeting reagents that are useful for identification, tracking, and repeated derivation of the corresponding progenitor stem cell types from additional hPS cell sources. Such stem cell targeting reagents can then be applied to the manufacture of genetically diverse banks of human embryonic progenitor cell lines for drug screening, disease modeling, and cell therapy. Here we present methods to identify human embryonic progenitor stem cell targeting peptides by selection of phage display libraries on clonal embryonic progenitor cell lines and demonstrate their use for targeting quantum dots (Qdots) for stem cell labeling.

Keywords: Progenitor cells, Stem cells, Targeting peptides, Affinity selection, Phage display

1 Introduction

Human pluripotent stem (hPS) cells have a great potential as a source of specialized cells and engineered tissues for treating degenerative diseases or repairing injury (1). Establishment of stem cell identity and purity is essential for developing cells for transplantation that must meet stringent FDA requirements for safety and efficacy. The potential for contamination by pluripotent stem cells with tumor forming capacity or inappropriate stem cell types capable of ectopic tissue growth presents a formidable challenge to the translation of hPS derived stem cells to the clinic (2). For these reasons it is essential to have the means of identifying and isolating pure hPS-derived cell populations. One approach is the isolation of single cell clones from differentiating hPS cells as a means of assuring cell purity; however, reproducibly isolating the same cell type in this manner may be difficult and labor intensive. We reasoned that clonal progenitor cell lines can provide an excellent resource for identifying cell targeting reagents that could be used as tools for repeated isolation of the same cell type from any hPS cell source. Phage display technology is a powerful method for selecting cell targeting peptides from large combinatorial libraries of peptides (3). We, therefore, have used phage library selection against clonally pure embryonic progenitor cell lines to identify embryonic progenitor cell targeting peptides (4). The resulting peptides may be useful for identification, isolation, and expansion of the desired cell type from a variety of normal and diseased donor-derived hPS lines.

Phage display is a reiterative affinity-based selection process that allows the identification of cell-binding peptides present in large combinatorial libraries displayed on phage particles. Originally described by Smith in 1985, simple bacteriophage can be engineered to display a guest peptide by genetically fusing the encoding DNA to a phage coat protein (5). Since then, a variety of libraries have been engineered that display random peptides, cDNAs, and antibodies on filamentous phage (commonly M13) as well as other phage types such as T7 (6, 7). Selection against a variety of simple or complex targets including mammalian cells has been described (8). Selection consists of repeated binding of a large complex library containing up to 10^9 displayed peptides against a target protein or cell culture, removal of unbound phage particles, and amplification of the bound phage population by infection of host bacteria. This results in enrichment of the library for target-binding peptides with each successive round until the library complexity collapses to relatively few peptides. Because the peptide encoding DNA is encapsulated inside the phage particle, there is a direct linkage between the peptide (phenotype) and its sequence (genotype) thus allowing identification of binding peptides by sequencing the DNA from the bound phage particles. Peptide display libraries consisting of random linear or cyclic peptides (7- and 12-mers) displayed by fusion to the pIII minor coat protein of the M13 phage are commercially available, and they offer a convenient way to search for peptides that bind to a specific target (9). Alternatively, natural occurring binding peptides or antibodies can also be identified by selection from cDNA display libraries (6, 7, 10). When phage display techniques are applied to complex targets such as receptors on a cell surface, it is best to limit the number of selection rounds to 2 or 3 to minimize the chance that highly replicative peptide phage species will dominate the selected library and thus maximize the diversity of the selected peptides (11).

We used phage library selection against clonally derived hPS derivatives as a strategy for identifying progenitor cell-type specific markers. Scalable and clonal progenitors can sometimes be identified by their morphology in long-term differentiating hPS cultures (12), but identification of hES-derived self-renewing progenitor cell types by morphology alone is rare. However, a large library of clonal progenitor lines has been derived from hPS cells using a two-step combinatorial cell cloning approach (13). The hES cells were first differentiated under a variety of culture conditions to produce heterogeneous candidate cultures. The candidate cultures were then grown under an array of different growth media conditions to select progenitor cell types that were both clonally pure and scalable. The project yielded over 140 clonally pure and scalable human embryonic progenitor cell lines. Gene expression array analysis showed each line to be distinct, thus indicating the high degree of diversity of clonally pure self-renewing progenitor cell types that can be obtained starting from hPS cells. These lines are an excellent source of human progenitor cell types that can be used to develop progenitor targeting peptides.

The library of previously isolated progenitor cell lines includes lines that express ectodermal, mesodermal, endodermal, and neural crest markers (13). Cell lines with osteochondral potential have been characterized and shown to be distinct from mesenchymal stromal/stem cells (MSCs) (14). Additional lines have been described that have the potential to express choroid plexus or leptomeningeal markers upon differentiation (15). Full characterization including preclinical studies of one of the initial clonal lines, the osteochondral embryonic progenitor cell line, 4D20.8, has been reported (16). These cells are capable of regenerating cartilage and bone in a rat femoral condyle defect model (16). Unlike bone marrow-derived MSCs which undergo very limited passages in culture, the 4D20.8 cells are chondrogenic to at least passage 33 and show only minor CNVs through passage 38. Also, the data support differentiation of 4D20.8 to definitive articular cartilage unlike bone marrow-derived MSCs which tend to differentiate to hypertrophic chondrocytes (17). Thus, these progenitor cell lines could potentially be used for patient-specific cell therapy or personalized drug development upon reisolation from donor iPS cells which could be used directly or banked as source HLA-matched progenitor cells. A means of tagging candidate cells followed by isolation of tagged cells from differentiated iPS cell cultures would streamline the derivation process by reducing the time and expense associated with single cell cloning and screening.

We present here methods for selection of peptides against an osteogenic progenitor cell line, SM30, using a commercially available 12-mer peptide phage display library (Fig. 1a). We also describe methods for validating the selected peptides for cell binding and peptide specificity (Fig. 1b–d) and the use of SM30 binding peptides for fluorescent labeling of cells with peptide-quantum dot complexes (Fig. 1e). The selected peptides can be used to monitor and validate embryonic cell differentiation under the conditions used to differentiate SM30 cells (4). Our long-term goal is to use the identified peptides to re-isolate SM30-like cells from any hPS cell line including patient-matched iPS cell lines for use in regenerative medicine-related therapies.

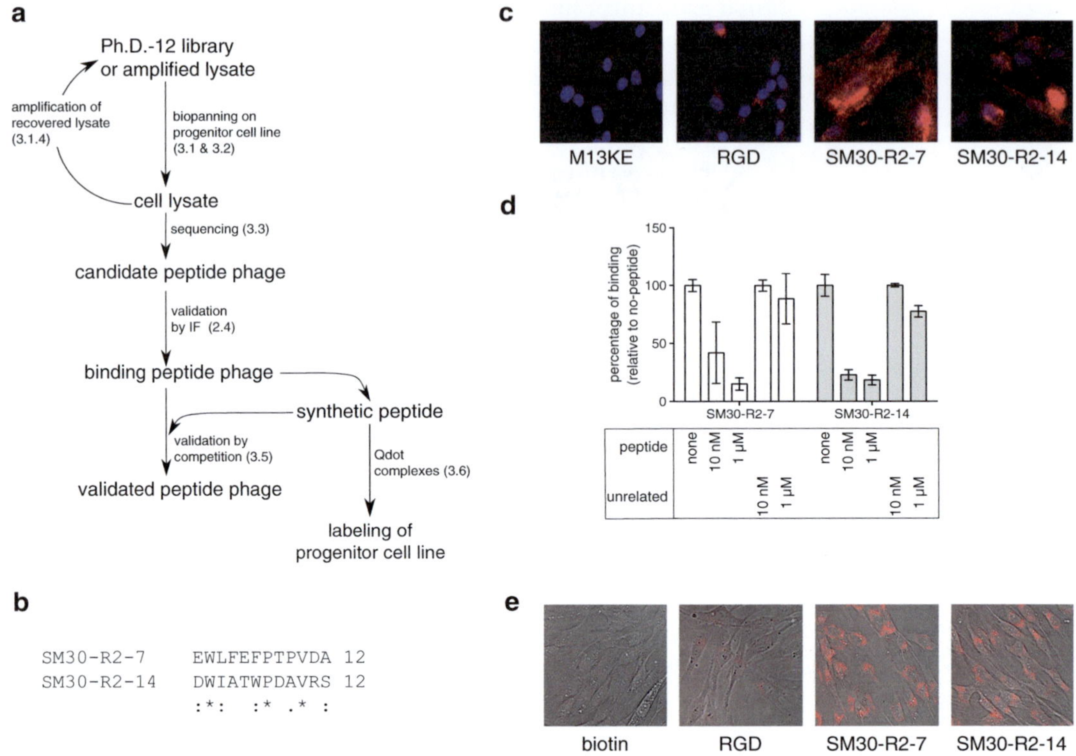

Fig. 1 Selection and validation of embryonic progenitor cell-binding peptides from a phage display library. (**a**) Schematic illustration of methods used to identify peptide phages that bind to the embryonic progenitor cell line, SM30, by enrichment in two to three rounds of biopanning; method section is in parenthesis. (**b**) Sequence alignment (ClustalW) of two representative candidate peptides (SM30-R2-7 and SM30-R2-14) identified in round 2 of the panning. (**c**) Cell binding is verified by immunofluorescence detection of SM30-R2-7 and SM30-R2-14 peptide phages following incubation with SM30 cells. Empty phage (M13KE) and RGD peptide phage are included as negative and positive controls. (**d**) Peptide phage cell-binding specificity is indicated by competition with the corresponding free peptide and lack of competition by an unrelated peptide (negative control). Phage binding is quantitated by titration of cell-bound phage recovered from cell lysates. (**e**) SM30 cells are labeled by peptide-Qdot complexes using SM30-R2-7 peptide and SM30-R2-14 peptide and weakly labeled by RGD peptide-Qdots compared to free biotin-Qdot complexes (negative control)

2 Materials

2.1 Selection of Cell-Binding Peptides from a Peptide Phage Display Library

1. Peptide phage display library (Ph.D.-12 Phage Display Library, NE Biolabs, cat# E8110S).
2. Monolayer culture of PureStem® SM30 progenitor cells for selection (ESI.BIO, cat# ES-256), plated on 0.1 % gelatin-coated 10 cm dish, approx. 1×10^6 cells/dish, grown for 48 h.
3. DMEM/F12 medium.
4. Library buffer: Embryonic progenitor cell growth medium, 2 % BSA.

5. Washing buffer: 1 % BSA in PBS, 0.9 mM $CaCl_2$, 0.73 mM $MgCl_2$.
6. Dissociation buffer: PBS, 1 mM EDTA.
7. Lysis buffer: 30 mM Tris-HCL pH 7.5, 2 mM EDTA, protease inhibitors.

2.2 Amplification of Phage Particles from Lysates or Phage Plaques and Quantitation by Titration

1. Overnight culture of *E. coli* ER2738 in LB medium with 10 μg/ml tetracycline.
2. Plating bacteria culture (50 ml) prepared fresh in an Erlenmeyer flask by diluting an overnight culture of *E. coli* ER2738 1:1,000 in LB medium with 10 μg/ml tetracycline and growing at 37 °C with shaking to OD = 0.5 (approx. 3–4 h) and then stored on ice.
3. Disposable Erlenmeyer flask, sterile, 250 ml.
4. LB media: 5 g Bacto Tryptone, 2.5 g yeast extract, 2.5 g NaCl, 0.5 g $MgCl_2$-$6H_2O$, pH 7.0.
5. LB media plates: LB media, 2 % agar.
6. Top agar: LB media, 0.8 % agar.
7. Tetracycline hydrochloride, stock solution at 10 mg/ml in ethanol.
8. Isopropyl β-D-1-thiogalactopyranoside (IPTG), stock solution at 50 mg/ml in water.
9. 5-Bromo-4-chloro-3-indolyl β-D-galactopyranoside (X-gal), stock solution at 40 mg/ml.
10. PEG solution: 1.5 M NaCl, 30 % polyethylene glycol (PEG) 8000.
11. Sterile, disposable labware: 15 ml and 50 ml conical centrifuge tubes, pipettes, snap-cap 15 ml round bottom tubes (*see* **Note 1**).

2.3 Sequencing of Recovered Phage

1. Primers that anneal to phage DNA outside of the insert: M13KE-Ext01: 3′-TTGTCATTGTCGGCGCAACT-5′ and M13KE-Ext02: 3′-GCATTCCACAGACAGCCCTCA-5′.
2. Reagents for PCR.
3. Sequencing primer: -96gIII 3′-CCCTCATAGTTAGCG-TAACG-5′.

2.4 Validation of Phage Binding by Immunofluorescence

1. Monolayer culture of PureStem® SM30 progenitor cells for validation, plated at 100,000 cells/well in 0.1 % gelatin-coated wells of 24-well plates.
2. Embryonic progenitor cell growth medium, 2 % BSA.
3. Phage particle preparation at concentration of 10^{12} pfu (from small or large-scale amplification), store at 4 °C.
4. Blocking buffer: 5 % normal goat serum in PBS.

5. Washing buffer: 1 % BSA in PBS.
6. Primary antibody: rabbit anti-Fd bacteriophage antibody (Sigma, cat# B7786), use at 1:700 dilution.
7. Secondary antibody: goat anti-rabbit AlexaFluor568 conjugated antibody, use at 1:1,000 dilution.
8. 4′,6-Diamidino-2-phenylindole dihydrochloride (DAPI), stock solution at 5 mg/ml in water.

2.5 Validation of Peptide Specific Phage Binding by Peptide Competition

1. Monolayer culture of PureStem® SM30 progenitor cells for validation, plated at 100,000 cells/well in 0.1 % gelatin-coated wells of 24-well plates.
2. Embryonic progenitor cell growth medium, 2 % BSA.
3. Phage particle preparations at concentration of 10^{12} pfu, store at 4 °C.
4. Synthetic peptide solutions at 20 mM in DMSO, store at −20 °C.

2.6 Cell Labeling with Peptide Targeted Qdot Complexes

1. Monolayer culture of PureStem® SM30 progenitor cells, plated at 100,000 cells/well in 0.1 % gelatin-coated wells of 24-well plate.
2. Synthetic peptide solutions at 20 mM in DMSO, store at −20 °C.
3. Qdot streptavidin conjugate (Qdot605-ITK-SA) and binding buffer (Life Technologies, cat# Q10001MP).
4. Streptavidin magnetic beads.
5. Magnetic separation rack for microcentrifuge tubes.

3 Methods

3.1 Selection of Cell-Binding Peptides from a Peptide Phage Display Library: Round 1

Cell targeting peptides are selected from a 12-mer peptide phage display library by three rounds of affinity selection (Fig. 1a) consisting of a pre-adsorption of the library against adult dermal fibroblasts cells to subtract nonspecific binding phages from the library followed by binding of the pre-adsorbed library on undifferentiated SM30 cells. Binding (and internalizing) peptide phages are recovered from the SM30 cell lysate. The enriched binding phage fraction is amplified from the lysate by bacterial infection and subjected to subsequent rounds of fibroblast subtraction followed by selection and recovery of SM30 binding phages.

3.1.1 Subtraction of Nonspecific Binders by Pre-adsorption Against hDFa Cells

1. Place hDFa cells on ice for 5 min.
2. Wash cells with 5 ml DMEM/F12 medium.
3. Add 2 × 10^{11} pfu of peptide phage display library in 5 ml in library buffer at 4 °C.

4. Incubate on ice for 45 min with gentle rocking or occasional mixing.
5. Remove unbound library from hDFa cells.

3.1.2 Selection of Library

1. Wash SM30 cells with 5 ml DMEM/F12 medium.
2. Add the unbound library from hDFa cells to SM30 cells.
3. Incubate 2 h at 37 °C with gentle rocking or occasional mixing.
4. Remove unbound library and transfer to 15 ml conical tube.

3.1.3 Harvesting Phage Particles from Target Cells

1. Wash cells 10× with 20 ml washing buffer, incubating cells in wash buffer for 3 min each with gentle rocking or occasional mixing.
2. Remove washes by aspiration.
3. Keep an aliquot of last wash for testing phage concentration.
4. Scrape cells in 2 ml dissociation buffer and transfer to a 15 ml conical tube.
5. Wash plate with 3 × 2 ml dissociation buffer to collect remaining cells.
6. Pellet cells (600 × *g*, 5 min, RT).
7. Wash cell pellet with 10 ml PBS (without Ca^{2+} and Mg^{2+}).
8. Resuspend pellet in 300 μl of lysis buffer.
9. Incubate on ice for 30 min; using 1 ml syringe, pass lysate through 27G needle 10× to reduce viscosity.
10. Incubate on ice for 30 min.
11. Pellet insoluble material in microfuge 5 min 4 °C at 13,000 × *g* and transfer lysate to fresh tube.
12. Store lysate on ice until the following day (*see* **Note 2**).

3.1.4 Quantitation of Phage Particles in Lysate by Titration

1. Prewarm LB plates in 37 °C incubator
2. Melt top agar and keep in 50 °C water bath. Add IPTG to give final concentration of 0.4 mM and X-gal to 160 μg/ml.
3. Prepare dilutions of lysate to be titer as follows (*see* **Note 3**):

   Neat = 10 μl of lysate

   Dil 1 = 2.5 μl of lysate + 22.5 μl PBS;

   Dil 2, dil 3 and dil 4 = 1/10 serial dilution of dil 1

   Wash (10 μl of last wash, *see* Section 3.1.3, step 3)
4. Aliquot 200 μl of plating bacteria for each phage dilution into microcentrifuge tubes.
5. Add 10 μl phage dilution to bacteria, mix gently with pipet, and keep at room temp (10 min) while pipetting top agarose.

6. Pipette 3 ml of warm top agarose into 15 ml snap-cap, round bottom tubes; one for each phage dilution; keep tubes in 50 °C water bath.
7. Retrieve plates from the 37 °C incubator and pull out a few at a time so the others stay warm.

   Steps 8 and 9 should be done quickly, one tube at the time:

8. Add 210 μl of bacteria with phage into 3 ml top agarose. Mix immediately by gently vortexing to minimize bubbles but well enough to thoroughly mix the bacteria and agarose.
9. Pour the agarose/bacteria/phage mixture onto a prewarmed plate and quickly swirl and shake to distribute evenly over the plate. Let agarose set for 10 min.
10. Invert plates and incubate overnight at 37 °C.
11. Count blue plaques the next day and calculate titer (*see* **Note 4**).

3.1.5 Amplification of Recovered Phage Particles

1. Dilute overnight culture 1:100 in 50 ml LB with 5 mM $MgCl_2$ and 10 μg/ml tetracycline.
2. Add 150 μl of lysate with recovered phage.
3. Incubate at 37 °C for 5 h with vigorous shaking (200 rpm).
4. Pellet cells in a 50 ml conical tube at 4,750 × *g* for 30 min at 4 °C.
5. Carefully remove 35 ml of supernatant without disturbing the pellet and place in fresh 50 ml conical tube on ice.
6. Add 7 ml of PEG solution.
7. Store on ice O/N.
8. Pellet at 10,000 × *g* for 1 h at 4 °C.
9. Pour off supernatant into waste beaker and remove all liquid.
10. Resuspend the pellet in 6 ml of PBS with protease inhibitors.
11. Add 1.2 ml of PEG solution and mix thoroughly.
12. Store on ice for 2 h or overnight at 4 °C.
13. Pellet at 10,000 × *g* as before.
14. Pour off supernatant into waste beaker and remove all liquid.
15. Resuspend in 1 ml PBS with protease inhibitors.
16. Incubate 10 min at 37 °C. Chill on ice.
17. Pellet insoluble material at full speed in microfuge for 5 min 4 °C.
18. Transfer supernatant to new tube and titer.
19. Store at 4 °C.
20. Titer phage preparation (*see* Section 3.1.4), preparing seven to ten serial 1:10 dilutions of amplified phage particles (*see* **Note 3**).

3.2 Selection of Cell-Binding Peptides from a Peptide Phage Display Library: Round 2 and 3

Perform selection of cell-binding peptides as before with 2×10^{10} pfu of amplified phage from the lysate of the previous round, in 5 ml in library buffer. Enrichment of binding peptides and elucidation of related peptides are usually seen after the second round of panning; phage plaques from titration plates corresponding to rounds 2 and 3 are amplified and sequenced (*see* Section 3.3).

3.3 Sequencing of Phage DNA from Phage Plaques

Individual phage plaques isolated from the titration plates of the cell lysate (from Section 3.1.3) are grown as individual phage cultures, and the variable insert region is amplified by PCR and sequenced using standard methods. A small fraction of the phage plaques are sequenced (24–50 clones) from round 2 and 3 of the panning. The encoded peptide sequences are deduced from the DNA sequences and analyzed for relative abundance and peptide similarities.

3.3.1 Plaque Amplification

1. Using a sterile pipette tip, transfer a blue plaque to 100 μl LB in a microcentrifuge tubes.
2. Dilute O/N bacterial culture 1:100 in LB with 5 mM $MgCl_2$ and 10 μg/ml tetracycline.
3. Grow for 1 h at 37 °C with shaking (200 rpm).
4. Distribute 1 ml of growing bacteria into the microcentrifuge tube with the phage plaque.
5. Incubate tubes at 37 °C for 6 h to 10 h with shaking, placing the tubes on their side.
6. Centrifuge to collect the bacteria (5 min, 13,000 × *g*) and transfer supernatant (2 × 0.45 ml) to fresh microcentrifuge tubes.
7. To one tube, add 0.45 ml 50 % glycerol, mix, and store at −20 °C for long-term storage.

3.3.2 Rapid Amplification of Templates by PCR and Sequencing

1. Use amplified plaque as template for PCR with primers that anneal to phage DNA outside of the insert (M13KE-Ext01 and M13KE-Ext02) using standard methods.
2. Run PCR products in 1 % agarose to determine empty phage clones, if any. PCR products from phage particles with inserts are approx. 600 bp in size, while those amplified from empty phage clones are 300 bp.
3. PCR products can be sequenced directly using 200 ng of single stranded DNA and 5 pmol of sequencing primer 96gIII primer in each sequencing reaction.

3.3.3 Peptide Sequence Analysis

Sequences from each phage plaque are extracted and analyzed for similarity by multiple sequence alignment (e.g., ClustalW) to identify related peptides (Fig. 1b). Repetition of several unique sequences and/or families of related sequences indicates that the library complexity has collapsed. At this point, validation of the peptide phages is recommended; more rounds of selection/amplification will not identify new binding peptides due to the reduction in diversity in each amplification step (11).

3.3.4 Small-Scale Amplification of Phage Particles

Phage amplification is carried out for those phages displaying sequences of interest. Initially, a small-scale amplification can be used to generate enough purified phage particles to validate the binding to SM30 cells.

1. Dilute overnight culture 1:100 in 15 ml LB with 5 mM $MgCl_2$ and 10 μg/ml tetracycline in a 50 ml conical tube.
2. Add 5 μl of phage (amplified plaque, from Section 3.3.1) to diluted culture.
3. Incubate at 37 °C for 5 h with vigorous shaking (200 rpm); the lid of the tube should be loosely screwed to allow sufficient aeration of the bacterial culture.
4. Pellet cells at 4,750 × *g* for 30 min at 4 °C.
5. Carefully remove 10.5 ml of supernatant without disturbing the pellet and place in fresh 15 ml conical tube on ice.
6. Add 2.1 ml of PEG solution.
7. Store at least 2 h on ice.
8. Pellet at 10,000 × *g* for 1 h at 4 °C.
9. Pour off supernatant into waste beaker and remove all liquid.
10. Resuspend the pellet in 0.5 ml of PBS with protease inhibitors and transfer to 1.5 ml microcentrifuge tube.
11. Add 100 μl of PEG solution and mix thoroughly.
12. Store on ice for 2 h or overnight at 4 °C.
13. Pellet at 10,000 × *g* for 30 min.
14. Pour off supernatant into waste beaker and remove all liquid.
15. Resuspend in 0.3 ml PBS with protease inhibitors.
16. Incubate 10 min at 37 °C. Chill on ice.
17. Pellet insoluble material at full speed in microcentrifuge tube for 5 min 4 °C.
18. Transfer supernatant to new tube and titer.
19. Store at 4 °C.

3.3.5 Large-Scale Amplification of Phage Particles

Large-scale phage amplifications are recommended for all experiments once the phage binding to SM30 cells has been validated.

For large-scale amplification of a single phage particle, the steps listed for the amplification of recovered phage particles in lysate is followed by inoculating the bacterial culture with 20 μl of amplified phage (*see* Section 3.3.4) or 50 μl of amplified plaque phage (*see* Section 3.3.1). This preparation of phage particles provides a phage preparation of high titer suitable for all subsequent experiments.

3.4 Validation of Phage Binding by Immunofluorescence

Binding of individual candidate phage particles to SM30 cells can be assessed and validated by immunofluorescence, in conditions similar to the ones used during the panning (Fig. 1c). A standard immunofluorescence protocol on fixed and permeabilized cells with an antibody that detects the bacteriophage filament allows the detection of phage particles that had internalized or are bound to the surface of SM30 cells regardless of the individual peptide sequence expressed. Positive and negative control phages are used to assure that specific phage binding is detectable and that there is low background from non-specific phage binding (*see* **Note 6**).

1. Remove medium from cells plated in 24-well plate.
2. Add phage dilution to test to cells (2×10^{10} pfu/well, 0.5 ml/well) in growth medium with 2 % BSA.
3. Incubate phage on cells for 2 h at 37 °C.
4. Wash wells with 6 × 2 ml/well washing buffer, leaving some wash buffer in the wells to prevent drying out.
5. Fix cells with 4 % p-formaldehyde in PBS (0.5 ml/well) for 20 min RT, in the dark.
6. Wash 1 × 2 ml/well PBS.
7. Place plates on ice and permeabilize cells with cold methanol (0.5 ml/well) for 10 min.
8. Wash 2 × 2 ml/well PBS.
9. Block with blocking solution (1 ml/well) for 1 h, RT or 4 °C O/N.
10. Add dilution of primary antibody in blocking buffer (0.5 ml/well) for 1 h, RT (*see* **Note 5**)
11. Wash 2 × 2 ml/well with PBS.
12. Add dilution of secondary antibody in 0.5× blocking buffer (0.5 ml/well) for 1 h, RT in the dark.
13. Wash 6 × 2 ml/well PBS.
14. Stain nuclei with DAPI dilution (1:5,000 dilution DAPI stock in PBS, 0.5 ml/well) for 5 min, RT or 4 °C, O/N.
15. Store plate, wrapped in aluminum foil until ready to visualize under fluorescent microscope.

3.5 Validation of Peptide Specific Phage Binding by Peptide Competition

The specificity of the phage binding to the cell surface can be validated using competition experiments with synthetic peptides (Fig. 1d). Competition experiments measure the ability of the free peptide to bind to the surface of SM30 cells; they are performed under conditions that do not permit the internalization of the bound peptide, blocking the binding of the peptide displayed at the phage particle if there is a free peptide bound to the receptor, hence, reducing the amount of phage detected. In this way, phage binding is determined to be peptide specific. Negative control peptides that are not expected to compete with the selected phage displayed peptide should also be included to control for non-specific peptide effects on phage binding (*see* **Note 7**).

1. Remove cells plated in 24-well plate from incubator and place on ice for 5 min.
2. Add synthetic peptide to cells at different concentrations (e.g., 5 nM, 5 μM) in cold growth medium with added BSA (0.5 ml/well) and incubate at 30 min at 4 °C.
3. Add phage to the peptide dilution (carefully mixing the phage aliquot with peptide dilution outside the well so as not to disturb the attached cells, 2×10^{10} pfu/well, 0.5 ml/well.
4. Incubate phage on cells for 1 h at 4 °C.
5. Wash wells with 6 × 2 ml/well washing buffer, leaving some wash buffer in the wells to prevent drying out.
6. At this point, phage bound to the cells can be determined qualitatively by immunofluorescence of fixed cells (as in Section 3.4) or quantitatively by tittering the recovered phage from the cell lysates (as in Section 3.1.3, using dilutions of the lysate for titration) (*see* **Note 8**).

3.6 Cell Labeling with Peptide Targeted Qdot Complexes

Monomeric C-terminal fluorescent-labeled peptides are often suitable for binding to and labeling the desired target cells, but some peptides may require multivalent peptide display that approximates or exceeds the multivalent display of the peptide on the phage particle (3–5 peptides per particle) for efficient binding. Peptide-targeted Qdots allows higher valency and sensitivity than monomeric peptides. Typically, Qdots particles contain five to ten streptavidin molecules covalently linked to an organic coating. As each streptavidin binds up to four biotinylated peptides, the resulting Qdot particle can present between 20 and 40 peptides for binding to the cell surface. We have obtained highly efficient labeling of SM30 cells with SM30 selected peptides that are readily imaged (Fig. 1e). Because the peptide/Qdot complexes are internalized, live labeled cells can also be removed from the plate and analyzed by FACS ((4), data not shown). The ability to label live cells and the stability of the Qdots and intense fluorescent signal make them ideal for labeling and tracking progenitor stem cell populations.

Peptides are synthesized with a free N-terminus and a biotinylated lysine at the C-terminus separated from the 12-mer peptide by a small Gly_4 linker ([H_2N-X_{12}-GGGG-K-biotin]) to resemble the displayed peptide in the phage particle (*see* **Note 9**).

1. Prepare peptide complexes before use by incubating (for each well of a 24-well plate) 20 μl binding buffer with 1 μl synthetic peptide diluted 1:10 in PBS; mix by vortexing and add 1 μl Qdot605-ITK-SA.
2. Incubate for 30 min at 4 °C or on ice.
3. Wash 5 μl SA beads with 2 × 500 μl PBS, and resuspend in 5 μl.
4. Transfer SA beads to peptide incubation reaction and incubate for 30 min at 4 °C.
5. To remove excess peptide using magnetic beads, place tubes in magnet, wait 5 min for magnetic particles to concentrate in the side of the tube, and transfer Qdot-complex mixture to fresh tube.
6. Wash beads once with 5 μl PBS; place on magnet and as before, remove wash and pool with complex mixture.
7. Remove medium from cells plated in 24-well plate.
8. Add peptide complex (volume is approx. 30 μl per well) diluted in 370 μl of growth medium and transfer to cells.
9. Incubate for 16 h at 37 °C
10. Wash cells with 2 × 1 ml/well PBS to remove unbound peptide complexes and replenish growth medium.
11. Visualize targeted cell population under fluorescent microscope or quantitate by flow cytometry.

4 Notes

1. Use sterile and disposable plasticware for all phage work to prevent contamination of the library. Contamination with wild-type phage is easily detected by the presence of clear plaques in titration plates. Contamination with empty phage or phage with a specific sequence can be determined by sequencing of the phage ssDNA.
2. Store lysate on ice at 4 °C to reduce degradation of phage particles, replenishing ice as it melts. It is recommended to proceed to the amplification of phage particle from the lysate as soon as possible (ideally, the following day) to ensure the amplified phage from the lysate resembles as closely as possible the composition of the phage in the lysate. An inherent bias in the amplified phage preparation from the lysate is normally observed toward the faster propagating clones having certain peptide sequences that favor phage replication. Empty phage with no-peptide insert replicates the fastest and can quickly dominate the library. For these reasons it is important to minimize amplification and not to re-amplify the library once it is prepared from the lysate. We include a subtraction step (or

negative panning) to remove the nonspecific binders from the library in an attempt to reduce the number of selection and amplification cycles needed to obtain peptides that preferentially bind the target cells (18).

3. Vortex after preparing each dilution and use a fresh filtered pipette tip for each dilution. For serial dilutions, use 10 μl of phage lysate or dilution and 90 μl PBS.
4. For example, titer (in pfu/ml) = number of plaques × dilution factor for dilution/volume (in ml)
5. Anti-M13 monoclonal antibody reacts with the bacteriophage major coat protein VIII; hence, all the phage particles are detected equally, regardless of the presence of the particular peptide sequence displayed on pIII.
6. Control wells with other phage particles of known sequences are included to assess the binding conditions and staining process. Negative controls include empty M13KE phage (no displayed peptide) or a 12-mer polyglycine (Gly12) peptide display phage. The positive control should be a peptide phage displaying a peptide that is expected to bind the target cell line. For example, we use the integrin binding peptide, DGARYCRGDCFDG, which binds to most mammalian cells in adherent culture (19).
7. Control competition reactions with synthetic peptides are included to assess the nonspecific binding for the labeling conditions. Negative controls that should not compete for phage binding include scrambled versions of the peptide being tested. An unrelated peptide can also be included as a negative control.
8. The percentage of recovered phage for the competition assay is normalized by the recovered phage in the no-peptide control.

   For example, recovered phage (%) = phage titer (5 nM)/ phage titer (no - peptide) × 100
9. Negative control cell-binding reactions using untargeted streptavidin-Qdots (by themselves or complexed to free biotin) are included to assess nonspecific cell binding for the labeling conditions. An integrin-binding RGD peptide (as in **Note 6**) complexed to Qdots is used as a positive control.

Acknowledgments

This work was supported, in part, by a grant from the California Institute for Regenerative Medicine (CIRM Grant Number TR1-1276).

References

1. Shtrichman R, Germanguz I, Itskovitz-Eldor J (2013) Induced pluripotent stem cells (iPSCs) derived from different cell sources and their potential for regenerative and personalized medicine. Curr Mol Med 13:792–805
2. Knoepfler PS (2009) Deconstructing stem cell tumorigenicity: a roadmap to safe regenerative medicine. Stem Cells 27:1050–1056
3. Smith GP, Petrenko VA (1997) Phage display. Chem Rev 97:391–410
4. Bignone PA, Krupa RA, Sternberg H, Funk WD, Snyder EY, West MD, Larocca D (2013) Identification of human embryonic progenitor cell targeting peptides using phage display. PLoS One 8:e58200
5. Smith GP (1985) Filamentous fusion phage: novel expression vectors that display cloned antigens on the virion surface. Science 228:1315–1317
6. Hoogenboom HR, de Bruine AP, Hufton SE, Hoet RM, Arends JW, Roovers RC (1998) Antibody phage display technology and its applications. Immunotechnology 4:1–20
7. Crameri R, Achatz G, Weichel M, Rhyner C (2002) Direct selection of cDNAs by phage display. Methods Mol Biol 185:461–469
8. McGuire MJ, Li S, Brown KC (2009) Biopanning of phage displayed peptide libraries for the isolation of cell-specific ligands. Methods Mol Biol 504:291–321
9. New England Biolabs (2012) Ph.D. phage display libraries. Instruction manual—version 1.1
10. Faix PH, Burg MA, Gonzales M, Ravey EP, Baird A, Larocca D (2004) Phage display of cDNA libraries: enrichment of cDNA expression using open reading frame selection. Biotechniques 36:1018–1022, 1024, 1026–1019
11. Derda R, Tang SK, Li SC, Ng S, Matochko W, Jafari MR (2011) Diversity of phage-displayed libraries of peptides during panning and amplification. Molecules 16:1776–1803
12. Cheng X, Ying L, Lu L, Galvao AM, Mills JA, Lin HC, Kotton DN, Shen SS, Nostro MC, Choi JK, Weiss MJ, French DL, Gadue P (2012) Self-renewing endodermal progenitor lines generated from human pluripotent stem cells. Cell Stem Cell 10:371–384
13. West MD, Sargent RG, Long J, Brown C, Chu JS, Kessler S, Derugin N, Sampathkumar J, Burrows C, Vaziri H, Williams R, Chapman KB, Larocca D, Loring JF, Murai J (2008) The ACTCellerate initiative: large-scale combinatorial cloning of novel human embryonic stem cell derivatives. Regen Med 3:287–308
14. Sternberg H, Kidd J, Murai JT, Jiang J, Rinon A, Erickson IE, Funk WD, Wang Q, Chapman KB, Vangsness CT Jr, West MD (2013) Seven diverse human embryonic stem cell-derived chondrogenic clonal embryonic progenitor cell lines display site-specific cell fates. Regen Med 8:125–144
15. Sternberg H, Jiang J, Sim P, Kidd J, Janus J, Rinon A, Edgar R, Shitrit A, Larocca D, Chapman KB, Binette F, West MD (2014) Human embryonic stem cell-derived neural crest cells capable of expressing markers of osteochondral or meningeal-choroid plexus differentiation. Regen Med 9:53–66
16. Sternberg H, Murai JT, Erickson IE, Funk WD, Das S, Wang Q, Snyder E, Chapman KB, Vangsness CT Jr, West MD (2012) A human embryonic stem cell-derived clonal progenitor cell line with chondrogenic potential and markers of craniofacial mesenchyme. Regen Med 7:481–501
17. Pelttari K, Winter A, Steck E, Goetzke K, Hennig T, Ochs BG, Aigner T, Richter W (2006) Premature induction of hypertrophy during in vitro chondrogenesis of human mesenchymal stem cells correlates with calcification and vascular invasion after ectopic transplantation in SCID mice. Arthritis Rheum 54:3254–3266
18. Lindner T, Kolmar H, Haberkorn U, Mier W (2011) DNA libraries for the construction of phage libraries: statistical and structural requirements and synthetic methods. Molecules 16:1625–1641
19. Holig P, Bach M, Volkel T, Nahde T, Hoffmann S, Muller R, Kontermann RE (2004) Novel RGD lipopeptides for the targeting of liposomes to integrin-expressing endothelial and melanoma cells. Protein Eng Des Sel 17:433–441

Methods in Molecular Biology (2016) 1357: 285–293
DOI 10.1007/7651_2014_147

Published online: 20 November 2014

piggyBac Transposon Mediated Reprogramming and Flow Cytometry Analysis of CD44 and ICAM1 Cell-Surface Marker Changes

Sara Brightwell and Keisuke Kaji

Abstract

Generation of iPSCs is inefficient and the molecular mechanisms underlying reprogramming are not well understood. While several studies have demonstrated that reprogramming is not entirely a random process and contains predictable stepwise changes, varying degrees of cellular heterogeneity that arise in different reprogramming systems can obscure the process. Among several reprogramming systems available, delivery of polycistronic reprogramming factor expression cassettes with piggyBac transposon into mouse embryonic fibroblasts (MEFs) is one of the simplest and most robust reprogramming approaches that provide a low background of partially reprogrammed cells. Using two novel cell surface markers, ICAM1 and CD44, clear cell population changes undergoing reprogramming can be observed over a time course upon induction of the reprogramming factors. Consequently, this technique allows for easy identification of factors that enhance or delay reprogramming, and can be a useful strategy in elucidating key mechanisms for efficient generation of iPSCs.

Keywords: Induced pluripotent stem cells (iPSCs), Reprogramming, Mouse embryonic fibroblasts (MEF), piggyBac transposon, Flow cytometry, ICAM1 (CD54), CD44, Nanog-GFP

1 Introduction

piggyBac (PB) transposon is the most active and widely used DNA transposon for efficient gene delivery. It can be handled as conventional plasmids without safety cabinets and shows high integration efficiency when co-transfected together with PB transferase which catalyzes the integration of PB transposon (1). Among several strategies to introduce Yamanaka reprogramming factors, piggyBac (PB) transposon is one of the easiest and safest tools to generate iPS cells (iPSCs) (2–4). Particularly, in combination with polycistronic cassettes of reprogramming factors taking advantage of self-cleaving 2A-peptides, generation of iPSCs is robust and highly reproducible from mouse embryonic fibroblasts, which are most commonly used to study molecular mechanisms of reprogramming. Recently we reported a PB reprogramming system using cells expressing all four reprogramming factors, Oct4, Sox2, Klf4, and c-Myc, as visualized by ires-mOrange. We demonstrated that almost all colonies expressing mOrange can activate a pluripotency marker, Nanog-GFP (5), indicating minimal background of partially reprogrammed cells (6). Using this system we have also demonstrated stepwise progression of reprogramming towards iPSCs with flow cytometry using ICAM1, CD44, and Nanog-GFP reporter as markers (6). This approach allows detailed analysis of reprogramming kinetics, or the effect of factors of interest, which can be added or removed from the standard reprogramming conditions, providing a strong tool to investigate molecular mechanisms of reprogramming. In this protocol, we describe how to reprogram MEFs with PB transposon and analyze the marker expression changes during reprogramming with flow cytometry.

2 Materials

2.1 Reprogramming Materials

1. *Mouse embryonic fibroblasts (MEFs)*. We use MEFs from E12.5 embryos with Nanog-GFP reporter to identify iPSC colonies live (5). While Oct4-GFP reporter is commonly used to identify iPSC colonies, we found upregulation of endogenous Oct4 occurs much earlier than many other pluripotency genes in the following protocol as also shown in other systems (6–8). Reporter system and/or markers for iPSC need to be chosen with great care. Reprogramming efficiency with the following method is largely affected by passage number/proliferation rate of MEFs. We usually use MEFs less than passage 3 for reprogramming experiments. We recommend culturing MEFs in the presence of Fgf2 and heparin (MEF medium as below) for propagation to delay senescence.
2. *Basic medium* consists of 500 mL of GMEM (Sigma, G5154) supplemented with 51 mL fetal calf serum (FCS, Invitrogen, 10270, Batch 40F0240K) (*see* **Note 1**), 5.5 mL MEM non-essential amino acids (100×, Invitrogen, 11140-036), 1,140 μL 50 mM 2-mercaptoethanol (Life Technologies, 31350010), 550 μL LIF (100,000 units/mL, homemade), 2.5 mL penicillin/streptomycin (10,000 U/mL Penicillin, 10,000 μg/mL Streptomycin, Invitrogen, 15140-122), 5.5 mL 100 mM sodium pyruvate (Invitrogen 11360-039), and 5.5 mL 200 mM L-glutamine (Invitrogen, 25030-024).
3. *MEF medium* is supplemented with 10 ng/mL Fgf2 (Peprotec 100-18-B) and 1 μg/mL heparin (Sigma, H3149) to the basic medium before use. Fgf2 10,000× stock (100 μg/mL), heparin 1,000× stock (1 mg/mL) are stored at −80 °C for long term, and at 4 °C for less than 1 month. Do not repeat freeze and thaw.
4. *Transfection medium* is MEF medium without penicillin/streptomycin and heparin.

5. *Reprogramming medium* is supplemented with 500 nM Alk4/5/7 inhibitor (Alki, A83-01, TOCRIS Bioscience, 2939), 10 μg/mL vitamin C (VitC, Sigma, 1000731348), and 1 μg/mL Doxycycline (Dox, Sigma, D9891-1G) (*see* **Notes 2** and **3**). Alki 10,000× stock (5 mM), VitC 5,000× stock (50 mg/mL), Dox 1,000× stock (1 mg/mL) are stored at −80 °C for long term, and at −20 °C for less than 1 month.
6. *Dulbecco's phosphate buffered saline* (PBS, Sigma, D8537).
7. *Trypsin-EDTA* (Invitrogen, 15090-046).
8. *Fugene HD* (Promega, E2311).
9. *Plasmids*

   PB-TAP IRI attP 2LMKOSimO (piggyBac (PB) transposon for Dox-inducible expression of 2A peptide linked Myc, Klf4, Oct4, Sox2 reprogramming factors followed by ires-mOrange) Available upon request to *keisuke.kaji@ed.ac.uk*. PB-CAG-rtTA (PB transposon for constitutive rtTA expression vector), pCMV-hyPBase (constitutive piggyBac transposase (PBase) expression vector). PB-CAG-rtTA and pCMV-hyPBase are available from Wellcome Trust Sanger Institute (http://www.sanger.ac.uk/form/-WcLcvb-BStKQEt0xeg5MjA).
10. *Opti-MEM* (Life Technologies, 31985062).

2.2 FACS Analysis Materials

1. *Anti-mouse CD54 (ICAM1) biotin*, eBioscience, 13-0541-82.
2. *Anti-Human/Mouse CD44 APC*, eBioscience, 17-0441-82.
3. *Streptavidin PE-Cy7*, eBioscience, 25-431-82.
4. *FACS buffer* (2 % FCS in PBS).
5. *Polystyrene round bottom FACS tubes* with or without strainer lid (BD Falcon, 352235 or 352054).

3 Methods

3.1 Reprogramming of MEF with PB Transposon

1. Day-2; Seed 1.5×10^5 MEF per well of a 6-well plate in 2.5 mL transfection medium. Penicillin/streptomycin and heparin can decrease cell viability when Fugene/DNA mix is added, therefore we use transfection medium from this point.
2. Day-1; Prepare the following DNA mix in an Eppendorf tube per well; PB-TAP IRI attP 2LMKOSimO 0.5 μg, PB-CAG-rtTA 0.5 μg, pCMV-hyPBase 0.5 μg (total 1.5 μg DNA), and add 100 μL of Opti-MEM (a). In another tube, prepare 100 μL of Opti-MEM and add 6 μL of Fugene HD (b). Mix a and b, and add to MEFs immediately. It is not necessary to incubate the DNA Fugene mix at room temperature, or change medium.

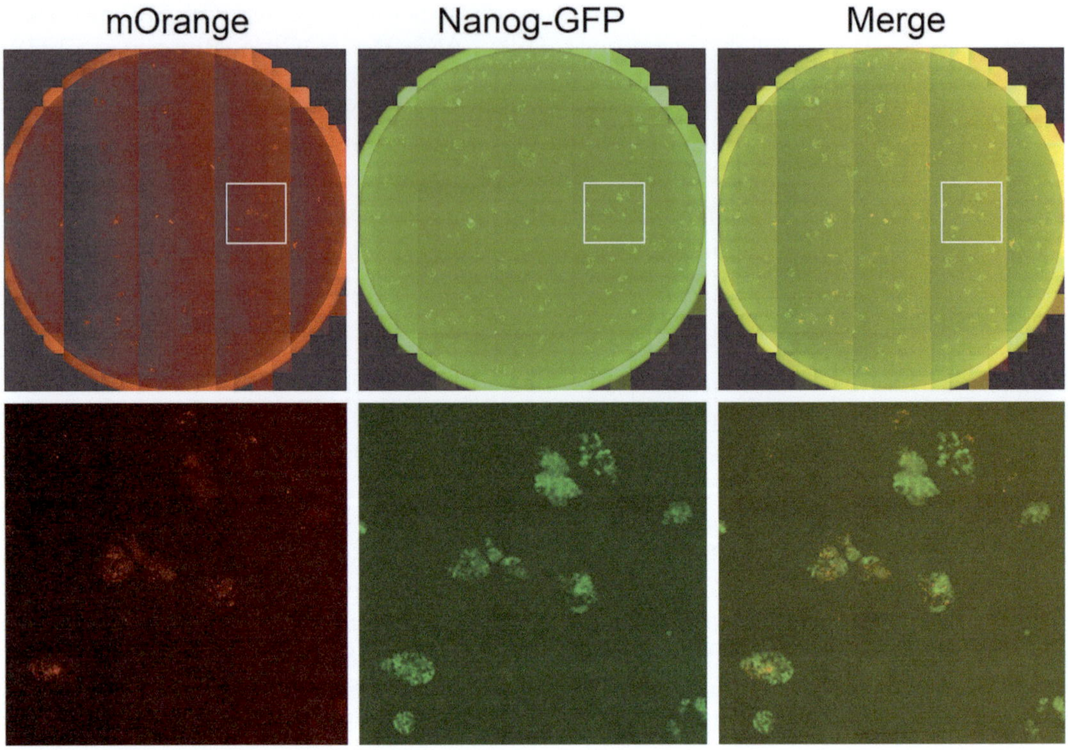

Fig. 1 Typical wells at day 14 of reprogramming with PB-TAP IRI attP 2LMKOSimO vector. *Top*: whole well images from a 6-well plate. *Bottom*: high magnification images of the *white square* of the *top* images. Images were taken with Celigo S Imaging Cytometer (Nexcelom). In this system, mOrange expression (*red*) is downregulated when Nanog-GFP (*green*) is upregulated without removing dox from the culture medium

3. Day 0; Change medium to reprogramming medium. Toxicity of transfection with Fugene HD is minimal and the wells should be 70–80 % confluent by this time.
4. mOrange expression should be visible 24 h after changing medium with a fluorescence microscope. Transfection efficiency at day 2 is usually about 10 % by flow cytometry. Change medium every 2 days (*see* **Note 4**). Clusters of mOrange$^+$ ESC-like cells should appear by day 5, and we start observing Nanog-GFP reporter$^+$ colonies at around day 8. Most colonies have Nanog-GFP$^+$ cells by day 14 (Fig. 1). If MEFs with pluripotency gene reporter are not available, immunofluorescence against Nanog (eBiosciences, 14-5761-80) or Dppa4 Cosmo Bio, CAC-TMD-PB-DP4) around day 12–14 is recommended to evaluate number of fully reprogrammed iPSC colonies.

3.2 Harvesting Samples for FACS Time Course Analysis

The above Fugene transfection protocol gives 30–100 iPSC colonies depending on condition of MEFs. mOrange$^+$ cell number is low at the early time points of reprogramming, and gradually increase since cells undergoing reprogramming proliferate faster.

To analyze the cell surface marker changes during reprogramming taking a time course, we recommend preparation of at least the following well numbers for each time point; day 2 ×10, day 4, ×8, day 6 ×5, day 8 ×3, day 10 and onwards ×1 each (*see* **Note 5**).

1. Remove media and wash cells in PBS, then aspirate.
2. Lift the cells with 500 μL of trypsin-EDTA per 6 well. Incubate for 1–3 min at 37 °C then pipette to dissociate cells to single cells (*see* **Note 6**).
3. Quench with 2.5–5 mL of media into a universal tube and count cells (*see* **Note 7**).
4. Centrifuge at 250 *g* for 3 min.
5. Aspirate supernatant and resuspend cell pellet in 100 μL FACS buffer per staining required, then transfer to a V-bottom 96-well plate.
6. Store plate on ice; cells are now ready for cell surface marker antibody staining.

3.3 Antibody Staining of Samples for FACS Time Course Analysis

Samples are stained in 100 μL aliquots so prepare enough staining solution for 100 μL/sample + 1, e.g., if you need to stain 4 samples prepare 500 μL of *each* staining solution required. All centrifugation steps are carried out at 1,300 rpm for 3 min.

1. Prepare ICAM1/CD44 antibody staining solution as specified in Table 1 and foil cover/keep out of light on ice until use.
2. Centrifuge 96-well plate and remove FACS buffer with an aspirator.
3. Resuspend cells in 100 μL of ICAM1/CD44 staining solution and incubate plate on ice in the dark for 15 min (*see* **Note 8**).
4. Prepare streptavidin PE-Cy7 secondary stain and foil cover/keep out of light on ice until use.
5. Centrifuge plate and wash cells with 100 μL PBS. Repeat centrifugation and remove PBS.
6. Resuspend cells in 100 μL of streptavidin PE-Cy7 secondary staining solution and incubate plate on ice in the dark for 5 min.

Table 1
Antibody staining dilutions for ICAM1/CD44 time course analysis

Antibody	Conjugate	Dilution in FACS buffer
CD44	APC	1/300
ICAM1	Biotin	1/100
Streptavidin	PE-Cy7	1/1,500

7. Centrifuge and wash cells in PBS twice as above.
8. Resuspend cells in 100 μL of FACS buffer, transfer to 5 mL FACS tubes and store on ice until analysis.

3.4 Control Samples Required for FACS Time Course Analysis

Appropriate controls should always be used (*see* **Note 9**).

1. *Unstained E14 ES cells*; this control should be negative for all markers.
2. *ICAM1 PE-Cy7 stained ES cells*; this is a positive control for ICAM1. All cells should be ICAM1 positive. This control also allows for compensation of leakiness into the Red 670/30 (CD44) and Yellow-Green 582/15 (mOrange) filters.
3. *CD44 stained MEFs*; this is a positive control for CD44. All cells should be CD44 positive. This control also allows for compensation of leakiness into the Yellow-Green 780/60 (ICAM1) filter.
4. *Unstained reprogramming (mOrange$^+$) sample*; this control allows for compensation of leakiness of the Yellow-Green 582/15 (mOrange) signal into the Blue 530/30 (Nanog-GFP) and Yellow-Green 780/60 (ICAM1) filters.

3.5 FACS Time Course Analysis of Reprogramming Samples

This protocol is based on the use of a BD LSRFortessa cell analyzer. The voltages suggested are for use with this machine and so optimization of parameters will be required for use of other flow cytometry analysers.

1. Set the cytometer up with the parameters in Table 2 (*see* **Note 10**).
2. Run all control samples first to set up appropriate compensation and base line acquisitions for your reprogramming samples.
3. Run your reprogramming sample(s) through the analyzer and gate your cells firstly to isolate the intact/live cells (this will be your P1 gate) and then gate mOrange$^+$ transfected (reprogramming) cells (this will be your P2 gate) (Fig. 2, *see* **Note 11**).
4. From the mOrange$^+$ population you can observe Nanog-GFP$^+$ cells when they arise (Fig. 2).

Table 2
Guide cytometer settings for BD LSRFortessa

BD LSRFortessa™		Excitation line			
		488 nm	561 nm	640 nm	Voltage
Band pass (BP) filter	530 ± 30	eGFP			300
	582 ± 15		mOrange		455
	780 ± 60		PECy7		470
	670 ± 30			APC	400

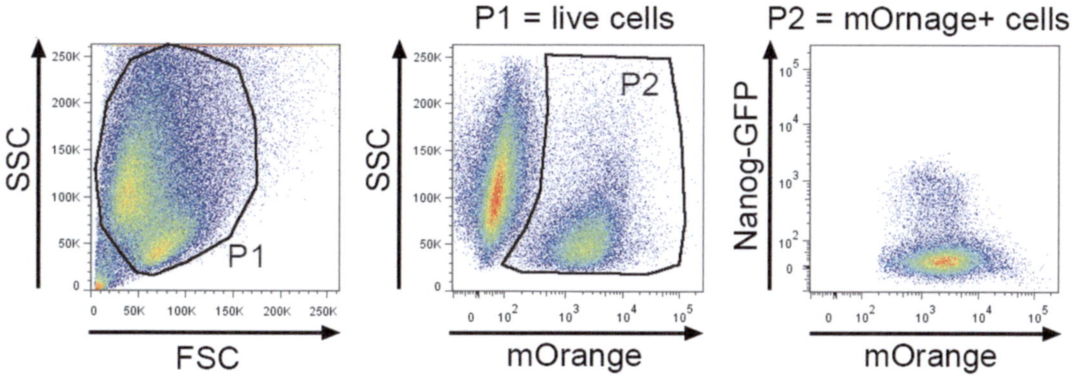

Fig. 2 Typical gating strategy for cells undergoing reprogramming. Gate firstly for intact/live cells in the SSC/FSC channels (P1) and then gate the mOrange$^+$ transfected/reprogramming cells by plotting mOrange against SSC (P2). This plot was from day 10 samples and Nanog-GFP$^+$ cells will typically emerge around 7–10 days of reprogramming. Note Nanog-GFP$^+$ cells have lower mOrange expression

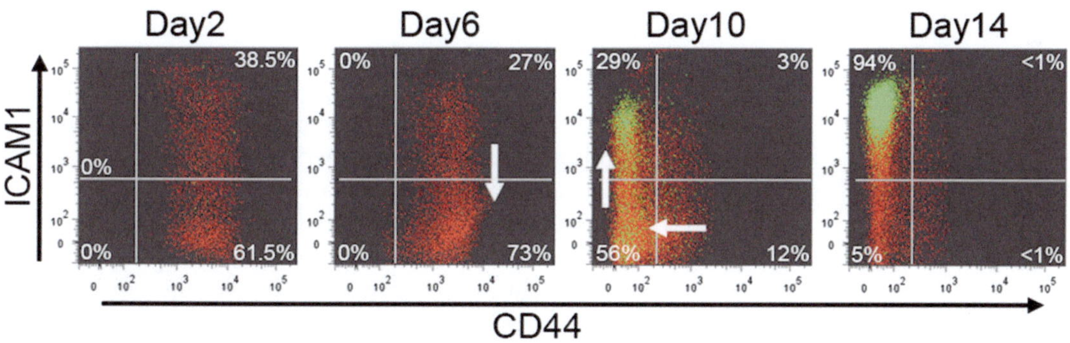

Fig. 3 ICAM1 and CD44 mark cells undergoing reprogramming. Cells initially exhibit heterogeneous expression of ICAM1 and high CD44. Decreasing ICAM1 expression is then followed by downregulation of CD44 around day 8. Finally, upregulation of ICAM1 marks cells entering the final "iPSC" between day 8 and 14, which is concurrent with expression of Nanog-GFP (*green*). Figures are from ref. 6 with modification. See [6] for more detail

5. A typical FACS data set for ICAM1/CD44 in our hands is shown in Fig. 3 (*see* **Note 12**).

4 Notes

1. FCS lot affects reprogramming efficiency. We have experienced that all mOrange$^+$ cells died off before expressing Nanog-GFP even with an FCS lot that supported ES cell self-renewal. If the above mentioned protocol does not make any iPSC-like colonies, we recommend testing other FCS.
2. This condition gives the highest reprogramming in most of MEFs in our hands so if you are interested in observing the

effects of a particular factor on reprogramming then having a lower efficiency by omitting Alki and/or vitC might be more appropriate.

3. We use doxycycline in a range from 300 ng/mL up to 1 μg/mL depending on the condition of our starting MEF and experimental requirements. In general, we find that 300 ng/mL is sufficient and preferred when starting MEF are in good condition and reprogram well. Higher concentrations of doxycycline are used when reprogramming conditions are less optimal, in order to maintain good induction of the four factors.
4. It is important to keep the reprogramming cultures in a good condition. At later time points, if color of media gets yellow quickly, change the media every day.
5. CD44 downregulation usually starts to be observed between day 6 and day 8. ICAM1 upregulation can be observed before day 10.
6. This time of trypsin-EDTA treatment does not affect staining with the antibodies described here.
7. Counting cells is not strictly required, although it is useful; until you are familiar with the technique, it gives you a good idea of how many cells you can harvest from "X" number of wells at "X" time point during the preliminary experiments which will help you to plan and optimize future experiments. Also, if you are comparing the effect of additional factors on reprogramming, counting the cells at each time point will give some indication if your factor of interest is having any effect on proliferation of cells undergoing reprogramming (mOrange^{+}) or MEFs (mOrange^{-}). You can calculate absolute cell numbers of each gate based on the total number and % of cells in each gate.
8. If using the antibodies specified in this protocol, you can prepare the ICAM1 and CD44 antibodies in the same staining solution—there is no need to stain them separately.
9. This is important for two reasons: (1) over the duration of a time course experiment the power of the lasers in the FACS instrument may be reduced on any given day, even if the voltage is the same. This can produce variation in your time course data but can be identified by use of appropriate controls. (2) Sometimes some of the fluorophores we use emit light that is detected by filters other than the one we intend. This "leakiness" can be detected by appropriate controls and compensated for.
10. These settings are only a guide of the voltages we use in our lab. Take time to adjust the voltages accordingly to your own instrument, even if it is also a BD LSRFortessa.
11. Ensure your settings allow 10,000 events from your P2 (mOrange^{+}) gate to be saved. In this example we plotted the

mOrange channel against the SSC channel to set P2. The mOrange population can be plotted against any other channel. It is advisable to use the channel with the clearest separation between the positive and negative cells.

12. Variations between experiments due to MEF conditions, transfection efficiency etc. can cause the reprogramming kinetics to change and therefore the FACS data may vary by a day or two. In case you use this system to evaluate impact of factors of interest, always control experiments need to be carried out at the same time.

Acknowledgements

We thank E. Chantzoura and L. Tosti for providing images, A. Nagy, K. Woltjen, and K. Yusa for advice on the use of PB transposon for reprogramming. SB is funded by an MRC PhD Studentship. KK is supported by ERC grants ROADTOIPS (no. 261075) and the Anne Rowling Regenerative Neurology Clinic.

References

1. Wu SC, Meir YJ, Coates CJ, Handler AM, Pelczar P, Moisyadi S, Kaminski JM (2006) piggyBac is a flexible and highly active transposon as compared to sleeping beauty, Tol2, and Mos1 in mammalian cells. Proc Natl Acad Sci U S A 103(41):15008–15013. doi:10.1073/pnas.0606979103
2. Kaji K, Norrby K, Paca A, Mileikovsky M, Mohseni P, Woltjen K (2009) Virus-free induction of pluripotency and subsequent excision of reprogramming factors. Nature 458(7239):771–775. doi:10.1038/nature07864
3. Woltjen K, Michael IP, Mohseni P, Desai R, Mileikovsky M, Hamalainen R, Cowling R, Wang W, Liu P, Gertsenstein M, Kaji K, Sung HK, Nagy A (2009) piggyBac transposition reprograms fibroblasts to induced pluripotent stem cells. Nature 458(7239):766–770. doi:10.1038/nature07863
4. Yusa K, Rad R, Takeda J, Bradley A (2009) Generation of transgene-free induced pluripotent mouse stem cells by the piggyBac transposon. Nat Methods 6(5):363–369. doi:10.1038/nmeth.1323
5. Chambers I, Silva J, Colby D, Nichols J, Nijmeijer B, Robertson M, Vrana J, Jones K, Grotewold L, Smith A (2007) Nanog safeguards pluripotency and mediates germline development. Nature 450(7173):1230–1234. doi:10.1038/nature06403
6. O'Malley J, Skylaki S, Iwabuchi KA, Chantzoura E, Ruetz T, Johnsson A, Tomlinson SR, Linnarsson S, Kaji K (2013) High-resolution analysis with novel cell-surface markers identifies routes to iPS cells. Nature 499(7456):88–91. doi:10.1038/nature12243
7. Buganim Y, Faddah DA, Cheng AW, Itskovich E, Markoulaki S, Ganz K, Klemm SL, van Oudenaarden A, Jaenisch R (2012) Single-cell expression analyses during cellular reprogramming reveal an early stochastic and a late hierarchic phase. Cell 150(6):1209–1222. doi:10.1016/j.cell.2012.08.023
8. Di Stefano B, Sardina JL, van Oevelen C, Collombet S, Kallin EM, Vicent GP, Lu J, Thieffry D, Beato M, Graf T (2014) C/EBPalpha poises B cells for rapid reprogramming into induced pluripotent stem cells. Nature 506(7487):235–239. doi:10.1038/nature12885

Methods in Molecular Biology (2016) 1357: 295–310
DOI 10.1007/7651_2015_199

Published online: 20 February 2015

Generation and In Vitro Expansion of Hepatic Progenitor Cells from Human iPS Cells

Ayaka Yanagida, Hiromitsu Nakauchi, and Akihide Kamiya

Abstract

Stem cells have the unique properties of self-renewal and multipotency (producing progeny belonging to two or more lineages). Induced pluripotent stem (iPS) cells can be generated from somatic cells by simultaneous expression of pluripotent factors (Oct3/4, Klf4, Sox2, and c-Myc). They share the same properties as embryonic stem (ES) cells and can differentiate into several tissue cells, i.e., neurons, hematopoietic cells, and liver cells. Therefore, iPS cells are suitable candidate cells for regenerative medicine and analyses of disease mechanisms.

The liver is the major organ that regulates a multitude of metabolic functions. Hepatocytes are the major cell type populating the liver parenchyma and express several metabolic enzymes that are necessary for liver functions. Although hepatocytes are essential for maintaining homeostasis, it is difficult to alter artificial and transplanted cells because of their multifunctionality, donor shortage, and immunorejection risk. During liver development, hepatic progenitor cells in the fetal liver differentiate into both mature hepatocytes and cholangiocytes. As hepatic progenitor cells have bipotency and high proliferation ability, they could present a potential source for generating transplantable cells or as a liver study model. Here we describe the induction and purification of hepatic progenitor cells derived from human iPS cells. These cells can proliferate for a long term under suitable culture conditions.

Keywords: Hepatic progenitor cells, Human iPS cells, Induction of hepatic differentiation, Long-term proliferation, Purification of progenitor cell

1 Introduction

The liver is the largest internal organ and performs multiple functions, including metabolism, production of bile acids, storage of lipids, glycogen, and protein, and detoxification of drugs, ammonia, and alcohol. The liver also has high proliferation ability. Even after 70 % partial hepatectomy, the remaining parts of the liver can increase and regenerate within a week in rats. However, in the case of intractable diseases, the liver cannot regenerate efficiently. Hepatocyte and organ transplantations are the most effective treatments because it is difficult to compensate multiple liver functions using an artificial organ. However, current donor shortages and immunorejections remain barriers to transplantation therapy.

Human iPS cells could be generated from human somatic cells such as dermal fibroblasts by simultaneous introduction of three to four genes (i.e., Oct3/4, Klf4, and Sox2, with or without c-Myc) (1, 2). These cells can give rise to tissues originating from three germ layers: the ectoderm, mesoderm, and endoderm. Using this pluripotency, it has been reported that hepatocytes or hepatic cells can be generated from iPS cells (3). However, hepatocytes have low proliferative potential and lose function in vitro. It remains difficult to expand hepatic cells derived from iPS cells while maintaining their function in vitro. We approached these challenges by the generation of hepatic progenitor cells from human iPS cells. Hepatic progenitor cells in the fetal liver possess high proliferation ability and capacity of differentiating into both hepatocytes and cholangiocytes in the middle to late fetal stage (4, 5). Therefore, even a small number of hepatic progenitor cells are likely to be sufficient for cell transplantation following in vitro expansion and maturation.

Differentiation processes from human iPS cells into hepatocytic cells are generally modified in vivo developmental processes. During in vivo liver development, the foregut endoderm cells receive inductive signals such as fibroblast growth factors (FGFs) from the adjacent cardiac mesoderm and bone morphogenetic proteins (BMPs) from the septum transversum mesenchyme (6, 7). These cells then migrate into the septum transversum mesenchyme and form the liver bud. During the middle liver development, hepatic progenitor cells in the liver bud expand by the stimulation of hepatocyte growth factor (HGF). Several studies have described methods to differentiate human iPS cells into hepatocytic cells using these cytokines and growth factors (3).

Surface markers such as CD13, CD133, and Dlk specific to mouse hepatic progenitor cells have recently been reported (8–11). We also reported mouse hepatic progenitor cells can proliferate in vitro under supplementation with ROCK inhibitor and mouse embryonic fibroblast (MEF) feeder cells (11). Using these methods, we recently established a method to purify human iPS cell-derived hepatic progenitor cells through CD13 and CD133 expression by using fluorescence-activated cell sorting (FACS). We also found that these purified hepatic progenitor cells could proliferate over long-term culture while maintaining functionality in the presence of ROCK and ALK inhibitors and MEF feeder cells. The expansion and bipotency into hepatocytes and cholangiocytes of human iPS cell-derived hepatic progenitor cells have been confirmed (12). Here we discuss the production and analysis of human iPS cell-derived hepatic progenitor cells.

2 Materials

2.1 Preparation of Mitomycin C-Treated MEF Feeder Cells

1. Mitomycin C: Wako Pure Chem. (Tokyo, Japan). A 200× stock solution (2 mg/ml) is prepared with ultrapure water.
2. ICR mice: Nihon SLC (Shizuoka, Japan). Embryonic day (E) 12.5 embryos are used. The approval of a suitable committee is usually required for animal experiments. Our experiments were performed with the approval of the Institutional Animal Care and Use Committee of the Institute of Medical Science, the University of Tokyo (permit number: PA09-21) and the Institutional Animal Care and Use Committee of Tokai University (permit number: 122047).
3. Surgical instruments: Dissecting scissors and forceps. For embryo dissection, INOX 5 forceps and a microscope are used.
4. 0.05 % trypsin-EDTA: Sigma (St. Louis, MO, USA).
5. Phosphate-buffered saline (PBS): 10× PBS (Sigma) is diluted with ultrapure water, autoclaved, and stored at 4 °C.
6. Fetal bovine serum (FBS; Life Technologies, Carlsbad, CA): FBS is inactivated by incubation at 55 °C for 30 min.
7. MEF culture medium: Dulbecco's modified Eagle's medium (DMEM; Sigma) supplemented with 10 % FBS and 1× penicillin-streptomycin-glutamine (Sigma).
8. Cryotubes: Corning Incorporated. (Corning, NY) #430289.
9. Stock solution: DMEM supplemented with 10 % dimethyl sulfoxide (DMSO; Wako Pure Chem.) and 20 % FBS (100-mm dish/vial).

2.2 Maintenance and Differentiation of Human iPS Cells into Hepatic Lineage Cells

1. Human iPS cells: TkDA3-4 is established from human dermal fibroblasts by introduction of Oct3/4, Klf4, and Sox2 using a retroviral system (13). These cells are cryopreserved in liquid nitrogen using DAP213 cryopreserve solution (14).
2. 0.1 % gelatin: Gelatin derived from porcine skin (Sigma) is dissociated with PBS and autoclaved.
3. Human iPS cell medium: DMEM/F-12 medium (Sigma) supplemented with the following:
   (a) 0.1 mM nonessential amino acids (Life Technologies).
   (b) 1× penicillin-streptomycin-glutamine.
   (c) 20 % knockout serum replacement (Life Technologies).
   (d) 0.1 mM 2-mercaptoethanol (Life Technologies).
   (e) 5 ng/ml recombinant human basic FGF (Wako Pure Chem.).

4. Y-27632: ROCK inhibitor (Wako Pure Chem.). A 1,000× stock solution (10 mM) is prepared with ultrapure water.
5. Differentiation medium: RPMI 1640 (Sigma) containing 2 % B27 supplement (Life Technologies).
6. Recombinant human activin A: PeproTech (Rocky, NJ). A 1,000× stock solution (100 mg/ml) is prepared.
7. Recombinant human basic FGF: PeproTech. A 1,000× stock solution (10 mg/ml) is prepared.
8. Recombinant human BMP-4: PeproTech. A 1,000× stock solution (20 mg/ml) is prepared.
9. Recombinant human HGF: PeproTech. A 1,000× stock solution (40 mg/ml) is prepared.

2.3 Isolation of Hepatic Progenitor Cells from Human iPS Cell-Derived Hepatic Lineage Cells

1. Staining medium: PBS containing 3 % FBS.
2. 0.05 % trypsin-EDTA: Sigma.
3. Stock propidium iodide solution (×1,000): Propidium iodide (Sigma) is dissolved with PBS (1 mg/ml). The stock solution is diluted with staining medium.
4. 5-ml FACS tube: Falcon Round-Bottom Polystyrene Tubes (Thermo Fisher Science, Waltham, MA).
5. 48-μm nylon filter: N-No.305T (Sanshyo, Tokyo, Japan) is sterilized by autoclaving.
6. Antibodies: *See* Table 1.

Table 1
List of antibodies used for immunostaining and flow cytometry experiments

Primary antibodies for flow cytometry	Clone	Source	Catalog number
CD13-PE	WM15	BD Pharmingen	555394
CD133/1-APC	AC133	Miltenyi Biotec	130-090-826
Primary antibodies for immunostaining	**Dilution**	**Source**	**Catalog number**
α-fetoprotein (AFP) (C3)	1/600	Sigma	A8452
HNF4α (C-19)	1/600	Santa Cruz	sc-6556
Secondary antibodies	**Dilution**	**Source**	**Catalog number**
Anti-mouse/Alexa Fluor 488	1/1,000	Invitrogen	A21202
Anti-goat/Alexa Fluor 546	1/1,000	Invitrogen	A11056

2.4 In Vitro Long-Term Culture of Hepatic Progenitor Cells

1. Standard culture medium: 1:1 mixture of DMEM/F-12 and DMEM. The medium is supplemented with the following:
   (a) 10 % FBS (Nichirei Biosciences, Tokyo, Japan).
   (b) 0.5× Insulin-Transferrin-Selenium X (Life Technologies).
   (c) 10 mM nicotinamide (Sigma).
   (d) 10^{-7} M dexamethasone (Sigma).
   (e) 2.5 mM HEPES buffer solution (Life Technologies).
   (f) 1× penicillin-streptomycin-glutamine.
   (g) 0.5× nonessential amino acids.
   (h) 0.25 μM A-83-01 (Tocris Bioscience, Bristol, UK).
   (i) 10 μM Y-27632.
   (j) 40 ng/ml recombinant human HGF.
   (k) 20 ng/ml recombinant human epidermal growth factor (EGF; PeproTech).

2.5 Characterization of Human iPS Cell-Derived Hepatic Progenitor Cells

2.5.1 Analyses of Hepatocytic Differentiation Steps Derived from Human iPS Cells Using Real-Time (RT) PCR

1. TRIzol: Life Technologies.
2. HepG2: Several cell banks supply this cell line. We obtained it from RIKEN Cell Bank (Ibaragi, Japan).
3. High Capacity cDNA Reverse Transcription Kit: Life Technologies.
4. TaqMan® Gene Expression Master Mix: Life Technologies.
5. Universal Probe Library: Roche Diagnostics (Basel, Switzerland). The probe numbers and primer sequences for each gene are provided in Table 2.

2.5.2 Analyses of Hepatocytic Marker Gene Expression Using Immunofluorescence Staining

1. 4 % Paraformaldehyde: # 16320145, Wako Pure Chem.
2. Triton X-100: #35501-15, Wako Pure Chem.
3. Blocking medium: 5 % donkey serum solution in PBS. Donkey serum, #S30-100 ml (Millipore, Bedford, MA). After heat inactivation by incubation at 56 °C for 30 min, donkey serum is diluted with PBS.
4. 4′, 6-Diamidino-2-phenylindole dihydrochloride (DAPI, Sigma): A 1,000× solution (2 mg/ml) is prepared with methanol.

2.5.3 In Vitro Differentiation of Human iPS Cell-Derived Hepatic Progenitor Cells into Mature Hepatocytes

1. Hepatocyte maturation medium: DMEM supplemented with following:
   (a) 10 % FBS.
   (b) 1× nonessential amino acids.
   (c) 1× penicillin-streptomycin-glutamine.
   (d) 10^{-7} M dexamethasone.

Table 2
PCR primers for detection of human gene expression

Human genes	Forward primer (5′–3′)	Reverse primer (5′–3′)	Probe number
AFP	tgtactgcagagataagtttagctgac	tccttgtaagtggcttcttgaac	61
COMT	tgcacacactaccaatcgttc	gcctgggcccttgtagat	10
CXCR4	ggtggtctatgttggcgtct	actgacgttggcaaagatga	18
CYP3A4	gatggctctcatcccagactt	agtccatgtgaatgggttcc	2
CYP3A7	caaaagactctgagacccacaa	agccagcaaaaataaagataattga	50
CYP7A1	gcaggcacctgtagtcttagc	cggagacgggatctcactaa	64
EPHX1	gatgacccagaagcatgagc	gcgtgtgcaatagctcaaaa	8
FMO5	attagccaaacagccaagca	acacgattcaggatccaagc	73
GSC	cctccgcgaggagaaagt	cgttctccgactcctctgat	29
hHex	cggacggtgaacgactaca	agaaggggctccagagtagag	61
HNF3β	cccaatcttgacacggtga	aaataaagcacgcagaaacca	85
HNF4α	gagatccatggtgttcaagga	gtgccgagggacaatgtagt	68
HPRT1	tgaccttgatttattttgcatacc	cgagcaagacgttcagtcct	73
MAOA	attaagtgcatgatgtattacaaggag	tggagcatcttcatcttcaatg	5
MAOB	ctggcagtcagaaccagagtc	gagggcaaatgtctctccaa	9
MIXL1	ctgaggagccatgactgaca	gcatggaagtcagaaaggaca	14
ONECUT1	cctggagcaaactcaaatcc	ttctttccttttgcatgctg	88
Sox17	acgccgagttgagcaaga	tctgcctcctccacgaag	61
SULT1A1	aagtgtcctacggatcctggt	tctcccttttcgggttctc	24
SULT1A2	gacctgggaaagcttcctg	tggtaccaggacccatagga	19
T	gctgtgacaggtacccaacc	ggagaattgttccgatgagc	23

Afp α-feto protein, *COMT* catechol-O-methyltransferase, *CXCR4* chemokine (C-X-C motif) receptor 4, *CYP* cytochrome P450, *EPHX1* epoxide hydrolase 1, microsomal (xenobiotic), *FMO5* Flavin-containing monooxygenase 5, *GSC* goosecoid homeobox, *hHex* hematopoietically expressed homeobox, *HNF* hepatocyte nuclear factor, *HPRT1* hypoxanthine phosphoribosyltransferase 1, *MAO* monoamine oxidase, *MIXL1* Mix paired-like homeobox 1, *ONECUT1* one cut homeobox 1, *Sox17* SRY-box 17, *SULT1A1* sulfotransferase family, cytosolic, 1A, phenol-preferring, member1

2. Recombinant human oncostatin M (OSM): R&D Systems (Minneapolis, MN, USA). A 1,000× stock solution (20 mg/ml) is prepared.
3. Syringes: 18-G and 25-G needles and 1-ml syringes (Terumo Corporation, Tokyo, Japan).
4. The probe numbers and primer sequences for each gene are shown in Table 2.

3 Methods

3.1 Preparation of Mitomycin C-Treated MEF Feeder Cells

1. E12.5 ICR mouse embryos are dissected and washed with PBS. Following this, the head and internal organs are surgically removed.
2. The torso is minced and dissociated in 0.05 % trypsin-EDTA for 30 min at room temperature.
3. After centrifugation, the cells are washed with DMEM/10 % FBS. The cells are inoculated into MEF culture medium. The cells are seeded onto 100-mm dishes (one embryo/dish) and culture at 37 °C in a 10 % CO_2 incubator.
4. On the next day, contaminating hematopoietic and other cells are removed by medium change. Cells are cultured with new MEF culture medium for 24 h.
5. On the next day, the cells are washed with PBS and incubated in 1 ml 0.05 % trypsin-EDTA for 5 min at 37 °C. The cells are collected in 15-ml tubes with MEF culture medium and centrifuged at 280 × *g* for 5 min.
6. Precipitated cells are dissociated with stock solution and preserved (one embryo cell/cryotube). These tubes are used as primary MEF stocks.
7. A primary MEF stock tube is thawed with MEF culture medium, and the cells are inoculated onto a 150-mm dish.
8. After the cells have expanded confluently (almost 2 days of culture), the cells are passaged to three 150-mm dishes.
9. Step 8 is repeated. The cells are passaged up to three times in 1:3 ratios. Up to 27 150-mm dishes.
10. After the cells have expanded confluently, the cells are incubated with 0.01 mg/ml mitomycin C in MEF culture medium at 37 °C for 2 h.
11. The cells are washed three times with PBS and incubated in 2 ml 0.05 % trypsin-EDTA for 5 min at 37 °C. The cells are collected in 50-ml tubes with MEF culture medium and centrifuged at 280 × *g* for 5 min.
12. Precipitated cells are dissociated with stock solution and preserved (3–4 × 10^6 cells/cryotube).

3.2 Culture and Differentiation of Human iPS Cells into Hepatic Cell Lineages

3.2.1 Culture of Human iPS Cells

1. Human iPS cells are cultured on mitomycin C-treated MEF feeder cells. Therefore, feeder cells are prepared before the day of iPS cell passage. Cryopreserved mitomycin C-treated MEF cells (from Section 3.1) are thawed and inoculated onto 60-mm 0.1 % gelatin-coated dishes (3 × 10^5 cells/dish).
2. After ≥12 h, cryopreserved human iPS cells are washed and inoculated with human iPS cell medium. The addition of

10 μM Y-27632 increases human iPS cells survival after thawing. The cells are cultured at 37 °C in a 5 % CO_2 incubator.

3. Human iPS cell medium is changed daily.
4. Human iPS cells are maintained on MEF feeder cells. Semi-confluent state human iPS cells are passaged onto the new MEF feeder cells or used for hepatocytic differentiation.

3.2.2 Differentiation of Human iPS Cell Culture into Hepatocytic Cells

1. Semi-confluent human iPS cells are washed three times with RPMI 1640. Following this, RPMI 1640 containing 2 % B27 (**Note 1**) supplemented with 100 ng/ml recombinant human activin A is added on days 0–2 at 37 °C in a 5 % CO_2 incubator (**Note 2**).
2. On day 2, the medium is gently removed and changed to new RPMI 1640 containing 2 % B27 supplemented with 100 ng/ml recombinant human activin A for 2 days (**Note 3**).
3. On day 4, human iPS cell-derived cells are washed three times with RPMI 1640. Finally, human iPS cells are cultured with RPMI 1640 containing 2 % B27 supplemented with 10 ng/ml basic FGF and 20 ng/ml recombinant human BMP-4 for 2 days.
4. On day 6, the medium is gently removed and changed to new RPMI 1640 containing 2 % B27 supplemented with 10 ng/ml basic FGF and 20 ng/ml recombinant human BMP-4 for 2 days.
5. On day 8, human iPS cell-derived cells are washed three times with RPMI 1640. Finally, human iPS cells are cultured with RPMI 1640 containing 2 % B27 supplemented with 40 ng/ml recombinant human HGF for 2 days.
6. On day 10, the medium is gently removed and changed to new RPMI 1640 containing 2 % B27 supplemented with 40 ng/ml recombinant human HGF for 2 days. These cells are used for the isolation of hepatic progenitor cells.

3.3 Isolation of Hepatic Progenitor Cells from Human iPS Cell-Derived Hepatic Lineage Cells

1. After 12 days of culture (from Section 3.2.2), differentiated cells are washed with PBS and 0.05 % trypsin-EDTA is added (1 ml/60-mm dish). Dishes are incubated at 37 °C for 5 min.
2. Following this, 2 ml staining medium is added to the trypsinized cells, and gently pipetting is performed. The dissociated cells are collected into a 15-ml tube and centrifuged at 280×g for 5 min.
3. The supernatant is aspirated, and washing is repeated with staining medium.
4. The cell number is counted, followed by centrifugation at 280×g for 5 min.
5. The supernatant is aspirated, and fluorescent-conjugated antibodies are added to the cell pellets (Antibody aliquots:

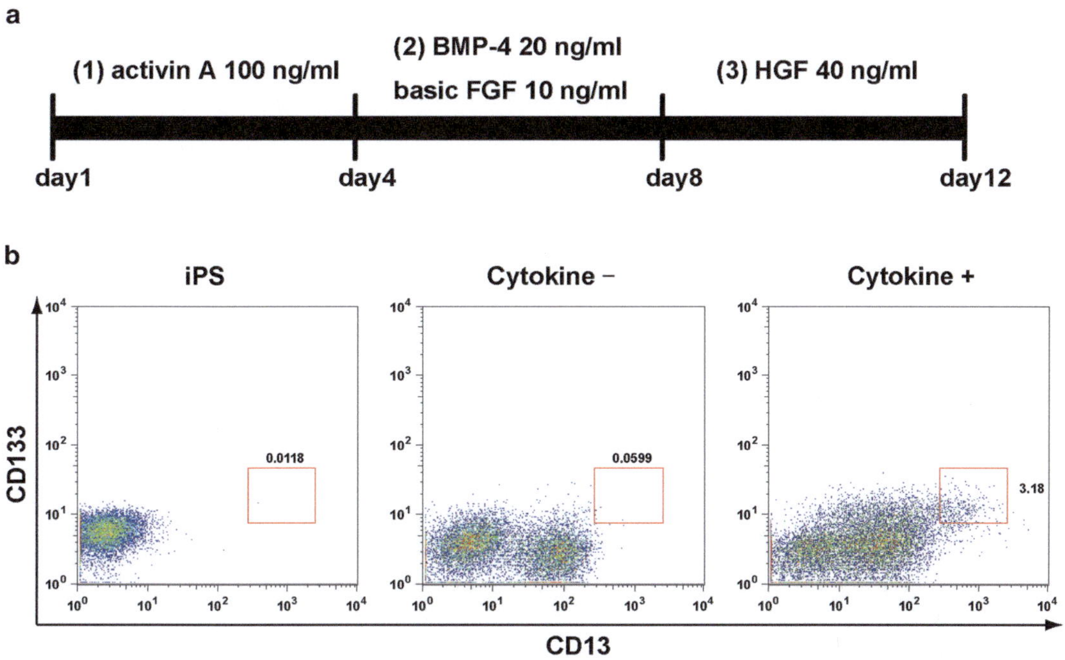

Fig. 1 Differentiation and isolation of hepatic progenitor cells from human iPS cells. (**a**) Schematic of the experimental procedure. Human iPS cells were sequentially stimulated with various cytokines: (1) activin A, (2) basic FGF and BMP-4, and (3) HGF. (**b**) Expression of hepatic progenitor markers in undifferentiated human iPS cells and differentiated cells. After 12 days of culture with or without cytokines, cells were stained with antibodies against CD13 and CD133 and then analyzed by flow cytometry. Ratios of $CD13^{high}CD133^{+}$ cells are shown

APC-conjugated anti-CD133/1 and PE-conjugated anti-CD13 in Table 1).

6. The cells are incubated at 4 °C for 1 h, washed with staining medium, and centrifuged at 280 × *g* for 5 min.
7. The supernatant is aspirated, and staining medium with propidium iodide is added to the pellets.
8. Suspended cells are filtrated with 48-μm nylon filter and collected in 5-ml FACS tubes.
9. Doublets are excluded by FSC, SSC, and pulse width. Dead cells are excluded by propidium iodide staining. $CD13^{high}CD133^{+}$ cells are sorted as a hepatic progenitor cell fraction using FACS (i.e., a MoFlo™ fluorescence-activated cell sorter; DAKO, Glostrup, Denmark). Gates are shown in Fig. 1.

3.4 In Vitro Long-Term Culture of Hepatic Progenitor Cells

1. Mitomycin C-treated MEF cells (2 × 10^5 cells/well) are plated onto 0.1 % gelatin-coated 12-well plates the day before sorting. MEF feeder cells are incubated for at least 12 h at 37 °C in a 10 % CO_2 incubator.
2. In total, 750 μl of standard culture medium is added to MEF-cultured 12-well plates.

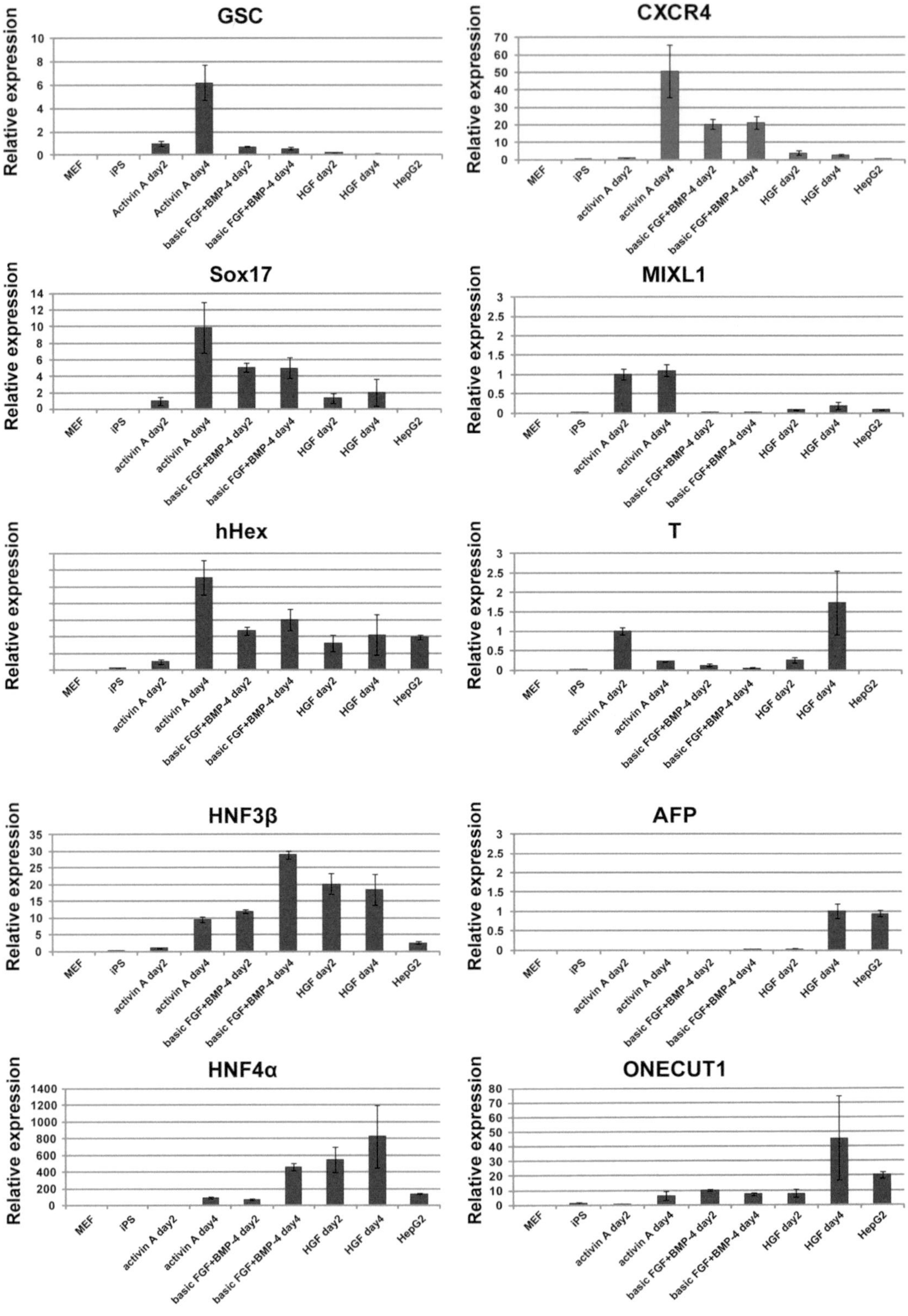

Fig. 2 Expression levels of undifferentiated and differentiated cell markers in human iPS cell-derived cells. The expression levels of marker genes for mesodermal and endodermal cells (GSC, CXCR4, Sox17, MIXL1, T,

3. CD13^{high}CD133^{+} hepatic progenitor cells (from Section 3.3) are sorted and inoculated onto MEF-cultured 12-well plates.
4. Medium is replaced every 3 days.
5. After 7–10 days of culture, several hepatic progenitor colonies are formed from individual CD13^{high}CD133^{+} cells.
6. For passage, mitomycin C-treated MEF cells (2×10^5 cells/well) are plated onto 0.1 % gelatin-coated 12-well plates the day before passage.
7. The cells are washed with PBS, and 500 μl 0.05 % trypsin-EDTA is added to individual wells of 12-well plates. Plates are incubated at 37 °C for 5 min.
8. In total, 1 ml MEF culture medium is added to individual wells, and the cells are dissociated by pipetting.
9. The dissociated cells are collected in 15-ml tubes and centrifuged at 280×g for 5 min.
10. The supernatant is aspirated, and the pellet is suspended in standard culture medium.
11. The suspended cells are inoculated in individual wells of 12-well plates in 750 μl standard culture medium.
12. After 7–10 days of culture, the passaged cells form new hepatocytic colonies. These colonies are used for hepatic progenitor cell characterization.

3.5 Characterization of Human iPS Cell-Derived Hepatic Progenitor Cells

3.5.1 Analyses of Hepatocytic Differentiation Steps Derived from iPS Cells Using RT PCR

1. During differentiation steps (from Section 3.2.2), iPS cells-derived cells stimulated with activin A, basic FGF + BMP-4, and HGF are washed with PBS, and total RNAs are extracted with TRIzol (2 ml/60-mm dish). For the control, total RNAs derived from MEF and HepG2 cells are also extracted.
2. Total RNA (1 μg) is used for single-strand cDNA synthesis using the High Capacity cDNA Reverse Transcription Kit.
3. The expression levels of marker genes for mesodermal and endodermal cells (GSC, CXCR4, Sox17, MIXL1, T, HNF3β, and hHex) and hepatic cells (AFP, HNF4α, and ONECUT1) are examined in normal human iPS cells and differentiated iPS cell-derived cells (Fig. 2). The expression of hypoxanthine

Fig. 2 (continued) HNF3β, and hHex) and hepatic cells (AFP, HNF4α, and ONECUT1) were examined in normal human iPS cells and differentiated iPS cells. For quantitative PCR analyses, mRNA was purified from MEFs, HepG2 cells, normal iPS cells, and iPS cells stimulated with activin A on days 2 and 4, basic FGF and BMP-4 on days 2 and 4, and HGF on days 2 and 4. The results are represented as the mean expression ± SD (triplicate samples)

phosphoribosyltransferase 1 (HPRT1) is used as an internal control. RT PCR is performed using Universal Probe Library and TaqMan® Gene Expression Master Mix according to the manufacturer's protocol.

3.5.2 Analyses of Hepatocytic Marker Gene Expression Using Immunofluorescence Staining

1. Formed colonies (from Section 3.4) are washed three times with PBS.
2. In total, 500 μl 4 % paraformaldehyde is added to individual wells and incubated at room temperature for 10 min.
3. Individual wells are washed three times with PBS.
4. In total, 500 μl 0.25 % Triton X-100/PBS is added to individual wells and incubated at room temperature for 10 min.
5. Individual wells are washed three times with PBS.
6. The blocking medium is added to individual wells and incubated at room temperature for 1 h.
7. The blocking medium is then aspirated and incubated with primary antibodies (Table 1) diluted with the blocking medium overnight at 4 °C.
8. Individual wells are washed three times with PBS.
9. The cells are then incubated with secondary antibodies (Table 1) diluted with the blocking medium for 40 min at room temperature.
10. Individual wells are washed three times with PBS.
11. The cells are incubated with 1× DAPI in PBS for nuclear staining.
12. Colonies are imaged under Carl Zeiss Axio Observer Z1 using AxioVision version 4.8 software (Carl Zeiss, Jena, Germany). Results are shown in Fig. 3.

3.5.3 In Vitro Differentiation of Human iPS Cell-Derived Hepatic Progenitor Cells into Mature Hepatocytes

1. The colonies are washed with PBS, and 500 μl 0.05 % trypsin-EDTA is added to individual wells of 12-well plates. Plates are incubated at 37 °C for 5 min.
2. In total, 1 ml MEF culture medium is added to individual wells, and the cells are dissociated by pipetting.

   The dissociated cells are collected in 15-ml tubes, the cell number is counted, and the cells are centrifuged at 280×g for 5 min.
3. The supernatant is aspirated, and the pellet suspended in hepatocyte maturation medium.
4. Individual drops (40 μl) containing 1×10^4 cells are plated on the inside of lids of 100-mm dishes containing PBS (to avoid desiccation) with or without 20 ng/ml recombinant human OSM (**Note 4**).

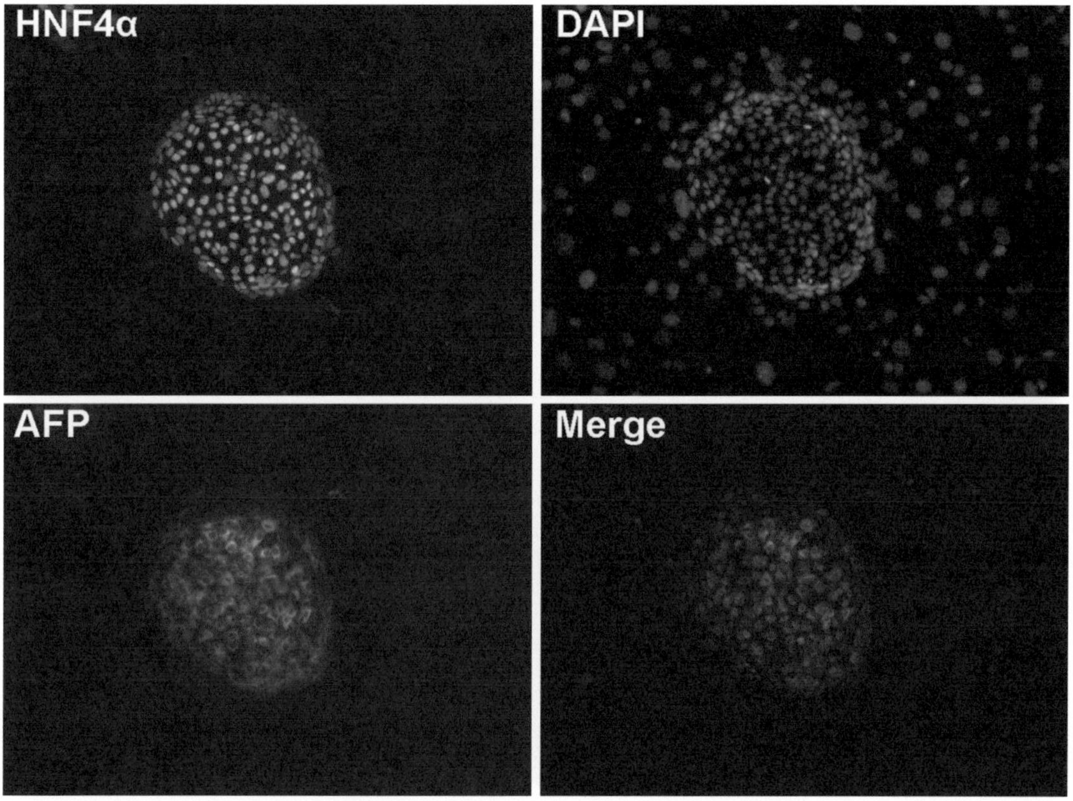

Fig. 3 Representative images of a colony derived from a single $CD13^{high}CD133^{+}$ cell Colonies were stained with antibodies against AFP and HNF4α. Nuclei were counterstained with DAPI

5. After 3 days of culture, spheroids derived from hepatic progenitor cells are formed.
6. Induced spheroids are collected in 1.5-ml tubes by gravity drop, and 1 ml TRIzol is added to individual tubes. The cells are dissociated using 18-G and 25-G syringes.
7. Total RNA is extracted using Trizol according to the manufacturer's protocol.
8. First-strand cDNAs are synthesized using the High Capacity cDNA Reverse Transcription Kit. Mature hepatocyte marker genes (Cyp3A4, Cyp3A7, Cyp7A1, EPHX1, MAOA, MAOB, SULT1A1, SULT1A2, COMT, and FMO5) and an internal control gene (HPRT1) are detected using Universal Probe Library. Results are shown in Fig. 4.

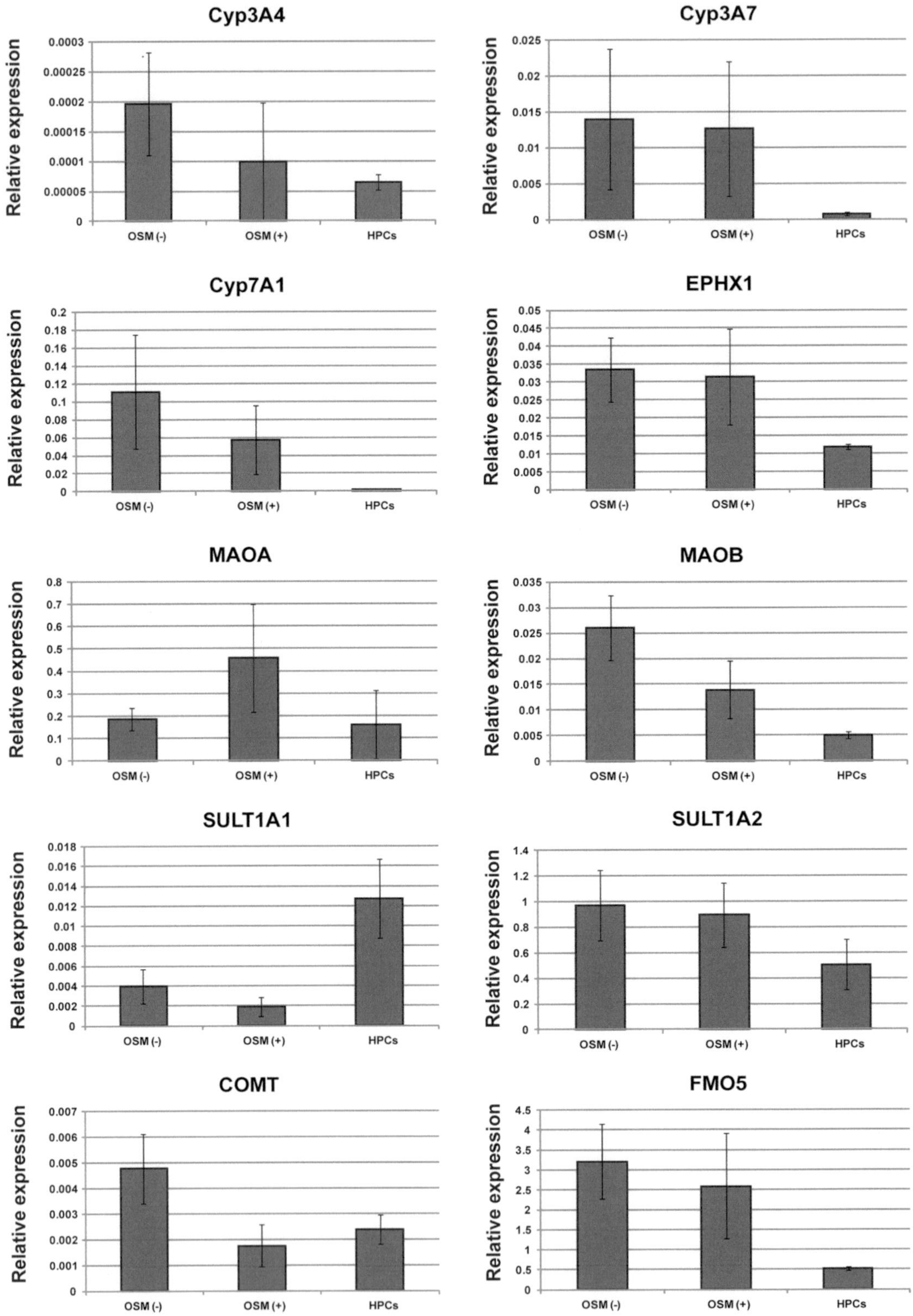

Fig. 4 Expression levels of hepatic functional genes in differentiated hepatic progenitor cells (HPCs). Levels of mRNA encoding phase 1 and 2 enzymes in human iPS cell-derived hepatic progenitor cells and spheroids

4 Notes

1. Before the experiments, B27 lots should be checked. Compare the morphology of differentiated hepatic lineage cells and the percentage of $CD13^{high}CD133^{+}$ cells among several lots.
2. In a previous report, differentiating human iPS cells were cultured under hypoxic conditions after day 4 (15). However, hepatic progenitor cells can be generated under normoxic conditions during the entire differentiation procedure.
3. On the next day, after the addition of activin A to human iPS cells, a large number of cells will be found floating in the wells. The remaining parts of the cells will differentiate into hepatic lineage cells. When induced human iPS colonies are overgrown, the center of the colony can be easily removed.
4. In mice, OSM was reported to induce maturation of hepatic progenitor cells (16). OSM is also used to induce differentiation of human iPS cells into mature hepatocytes (15). However, we found no dramatic effect of OSM on the maturation of human iPS cell-derived hepatic progenitor cells.

Acknowledgments

This work was supported in part by the Education and Research Support Center of Tokai University. Some figures have been reproduced from (12). This work was supported by a Grant-in-Aid for JSPS Fellows and a Grant-in-Aid from the Ministry of Education, Culture, Sports, Science and Technology, Japan (13J05787 for AY, 26293178 and 25670373 for AK).

Fig. 4 (continued) derived from human iPS cell-derived hepatic progenitor cells are shown as fold values relative to the levels in uncultured human hepatocytes. Spheroid formation was induced by hanging drop culture in the presence or absence of OSM. The results are represented as the mean expression $\pm$ SD (spheroid culture, $n = 6$; hepatic progenitor cells, $n = 3$; uncultured human hepatocytes, $n = 2$)

References

1. Takahashi K, Tanabe K, Ohnuki M et al (2007) Induction of pluripotent stem cells from adult human fibroblasts by defined factors. Cell 131:861–872
2. Nakagawa M, Koyanagi M, Tanabe K et al (2007) Generation of induced pluripotent stem cells without Myc from mouse and human fibroblasts. Nat Biotechnol 26:101–106
3. Gerbal-Chaloin S, Funakoshi N, Caillaud A et al (2014) Human induced pluripotent stem cells in hepatology. Am J Pathol 184:332–347
4. Lemaigre FP (2009) Mechanisms of liver development: concepts for understanding liver disorders and design of novel therapies. Gastroenterology 137:62–79
5. Si-Tayeb K, Lemaigre FP, Duncan SA (2010) Organogenesis and development of the liver. Dev Cell 18:175–189
6. Rossi JM (2001) Distinct mesodermal signals, including BMPs from the septum transversum mesenchyme, are required in combination for hepatogenesis from the endoderm. Gene Dev 15:1998–2009
7. Jung J (1999) Initiation of mammalian liver development from endoderm by fibroblast growth factors. Science 284:1998–2003
8. Tanimizu N (2003) Isolation of hepatoblasts based on the expression of Dlk/Pref-1. J Cell Sci 116:1775–1786
9. Watanabe T, Nakagawa K, Ohata S et al (2002) SEK1/MKK4-mediated SAPK/JNK signaling participates in embryonic hepatoblast proliferation via a pathway different from NF-κB-induced anti-apoptosis. Dev Biol 250:332–347
10. Kakinuma S, Ohta H, Kamiya A et al (2009) Analyses of cell surface molecules on hepatic stem/progenitor cells in mouse fetal liver. J Hepatol 51:127–138
11. Okada K, Kamiya A, Ito K et al (2012) Prospective isolation and characterization of bipotent progenitor cells in early mouse liver development. Stem Cell Dev 21:1124–1133
12. Yanagida A, Ito K, Chikada H et al (2013) An in vitro expansion system for generation of human iPS cell-derived hepatic progenitor-like cells exhibiting a bipotent differentiation potential. PloS One 8:e67541
13. Takayama N, Nishimura S, Nakamura S et al (2010) Transient activation of c-MYC expression is critical for efficient platelet generation from human induced pluripotent stem cells. J Exp Med 207:2817–2830
14. Fujioka T, Yasuchika K, Nakamura Y et al (2004) A simple and efficient cryopreservation method for primate embryonic stem cells. Int J Dev Biol 48(10):1149–1154
15. Si-Tayeb K, Noto FK, Nagaoka M et al (2009) Highly efficient generation of human hepatocyte-like cells from induced pluripotent stem cells. Hepatology 51:297–305
16. Kamiya A, Kinoshita T, Ito Y et al (1999) Fetal liver development requires a paracrine action of oncostatin M through the gp130 signal transducer. EMBO J 18:2127–2136

Methods in Molecular Biology (2016) 1357: 311–327
DOI 10.1007/7651_2015_203

Published online: 17 February 2015

Induced Pluripotent Stem (iPS) Cell Culture Methods and Induction of Differentiation into Endothelial Cells

Ishita Chatterjee, Fei Li, Erin E. Kohler, Jalees Rehman, Asrar B. Malik, and Kishore K. Wary

Abstract

The study of stem cell behavior and differentiation in a developmental context is complex, time-consuming, and expensive, and for this reason, cell culture remains a method of choice for developmental and regenerative biology and mechanistic studies. Similar to ES cells, iPS cells have the ability to differentiate into endothelial cells (ECs), and the route for differentiation appears to mimic the developmental process that occurs during the formation of an embryo. Traditional EC induction methods from embryonic stem (ES) cells rely mostly on the formation of embryoid body (EB), which employs feeder or feeder-free conditions in the presence or absence of supporting cells. Similar to ES cells, iPS cells can be cultured in feeder layer or feeder-free conditions. Here, we describe the iPS cell culture methods and induction differentiation of these cells into ECs. We use anti-mouse Flk1 and anti-mouse VE-cadherin to isolate and characterize mouse ECs, because these antibodies are commercially available and their use has been described in the literature, including by our group. The ECs produced by this method have been used by our laboratory, and we have demonstrated their in vivo potential. We also discuss how iPS cells differ in their ability to differentiate into endothelial cells in culture.

Keywords: Angiogenesis, Endothelial cells, iPS cells, Nanog, Oct4, Sox2, Klf4, Flk1, CD31, VE-cadherin

1 Introduction

Totipotent stem cells (ESCs) can give rise to differentiated and specialized cells with restricted developmental potential (1–7). At some point, the specialized cells no longer differentiate or dedifferentiate, and this state has been referred to as terminal differentiation. The process of terminal differentiation has been thought to be an irreversible process. In contrast to this long-held view, retroviral-mediated, forced expression of key transcription factors such as *Klf4*, *c-Myc*, *Nanog*, *Oct4*, and *Sox2* into somatic cells, such as fibroblasts, can convert (reprogram) these cells into induced pluripotent stem (iPS) cells (8–12). The additive activities of these transcription factors were thought to be necessary and sufficient to reprogram human or mouse somatic cells to iPS cells. In addition to these classical transcription factors described by the Yamanaka and Thomson groups (8, 9, 11), additional transcription factors and miRNAs and small molecules have been added to the list (12–15). Accordingly, a combination of two or three transcription factors (often called Yamanaka factors) may be sufficient to reprogram fibroblast cells into human or mouse iPS cells. For example, in some cell types, Oct4 and Sox2 might be sufficient to establish an iPS cell line (16), while in others Sox2 is dispensable (17, 18). It is apparently clear that Oct4 occupies the most upstream position in terms of its ability to reprogram somatic cells, while other Yamanaka factors are required for developmental differentiation events downstream of Oct4 (19, 20). More recently, forced expression of the transcription factors *Sall4*, *Nanog*, *Esrrb*, and *Lin28* in mouse fibroblast cells has been shown to generate high-quality iPS cells (21). The mechanisms of differentiation in iPS and ES cells could differ from those of various iPS cells derived from different somatic cells, but their similarities and differences have not been precisely delineated. Currently, the underlying mechanisms of iPS generation remain an area of great interest.

Upon orthotopic implantation into nude mice, similar to embryonic stem cells (ESCs), iPS cells form teratomas (8–11). Immunohistochemical analyses of the teratoma sections using markers for the three germ layers, e.g., ectoderm, mesoderm, and endoderm, provide a good indication of iPS cell stemness. In addition, functional tests, including tetraploid complementation assays and the production of chimeric and germline mice, establish that iPS cells can acquire an ESC-like state (8–11, 21, 22). Therefore, it is not surprising that genuine interest for the application of iPS technology has emerged in many areas of regenerative, reparative, and transplantation medicine. Nevertheless, inefficiency remains the main bottleneck for converting somatic cells into iPS cells, e.g., of 1,000–10,000 somatic cells, only a single iPS cell can be fully reprogrammed using the most efficient method. For this reason, the production of patient-derived stem cells not only is an expensive task but also remains an uphill battle. Although a retroviral method is considered the most efficient way to produce iPS cells, chimeric mice and mice derived through the use of these iPS cells often produce tumors (8–11). One of the caveats of this approach is that the retroviruses, for instance, long terminal repeats, are known to integrate randomly into the genome, which could activate oncogenes or inactivate tumor suppressor genes to initiate neoplastic transformation. Thus, these observations have provided the impetus to the development of non-integrating vectors such as piggyback, episomal non-integrating and non-integrating Sendai Virus as well as mini-genes, and small-molecule compounds (23–27). Thus, the development of a highly efficient iPS reprogramming technique that also evades these undesirable genetic alterations should be a rewarding research endeavor.

The observations that iPS cells have the capacity to self-renew and undergo differentiation in response to specific growth factors in culture dishes make these cells an ideal source of progenitor cells for cell-based therapy, drug screening, and disease modeling; thus they have vast therapeutic potential. Therefore, in our laboratory, we have used iPS cells as a source for VE-cadherin+ and Flk1+ endothelial cells (ECs) and showed their ability to incorporate into CD31+ neovessels in Matrigel plugs (28) and into newly formed blood vessels in a mouse model of hind limb ischemia (28). Thus, based purely upon our recent publication (28), here, we outline methods for iPS culture, including the conditions used to differentiate iPS cells into ECs, as well as for the isolation, purification, and characterization of VE-cadherin+ and Flk1+ ECs.

2 Materials

A clean cell culture laboratory with laminar flow and vacuum connected to a liquid waste container (a glass flask) through a HEPA filter, an electrical outlet for a pipette aid, and a gas connection for a Bunsen burner. Some media and growth factors are filter-sterilized using a 0.2-μm filter. Please ensure that the iPS cell lines you are using do not secrete live virus or are contaminated with mycoplasma. Please follow aseptic techniques throughout the procedure.

2.1 Equipment

(a) For all cell culture work, we recommend a Zeiss Primo Vert (Carl Zeiss MicroImaging Inc., Thornwood, NY), Nikon TS100, or Olympus CKX41 inverted phase contrast cell culture microscope, with long working distance (LWD) lenses and 5×, 10×, and 20× magnifications. Preferably with a digital camera for recording images.

(b) Refrigerated cell culture centrifuge.

(c) Water bath (20 l), 37 °C.

(d) Humidified CO_2 (5 %) cell culture incubator set at 37 °C.

(e) Liquid N_2 storage tank.

(f) Frosty freezing container (Nalgene).

(g) Nalgene System cryovials (1.8 ml).

(h) Flow cytometer. We use Beckman Coulter MoFlo (Becton Dickinson, Franklin Lakes, NJ).

(i) 70 % Ethanol in a spray bottle.

2.2 iPS Cell-Freezing Medium

Prepare cell-freezing medium, which contains 30 % fetal bovine serum (FBS), 10–20 % Dimethyl Sulfoxide (DMSO), 60 % DMEM (alternatively, use ES-DMEM).

1. Store at +4 °C for 3 months.

2.3 Medium and Growth Factors

(a) Cell culture dishes: 6-, 12-, and 24-well cell culture-certified dishes.

(b) Sterile serological pipettes (1, 5, and 10 ml; all disposable plastics must be sterile).

(c) iPS cells can be generated in the laboratory using commercially available Yamanaka factors. The mouse induced Pluripotent Stem (iPS) cells (iMZ-9 and iMZ-21 lines) used in this example were obtained through Material Transfer Agreement (MTA) from Dr. Kristin Baldwin (The Scripps Research Institute, La Jolla, CA).

(d) Embryonic Stem (ES) cell-qualified Fetal Calf Serum (FCS) (Invitrogen).

(e) Attachment factor (AF) from Life Technologies.

(f) Human placenta-derived type IV collagen (Millipore/Chemicon).

(g) Dulbecco's Phosphate-buffered saline (DPBS, pH 7.4, with no Ca++ or Mg++).

(h) TrypLE™ Select Cell Dissociation Reagent (Life Technologies). For passaging with a feeder layer, use type IV collagenase (Invitrogen) diluted with DMEM-F12.

(i) Falcon cell strainer (70-μm nylon mesh).

(j) 2 mM Ethylenediaminetetraacetic Acid (EDTA) in PBS, pH 7.4.

(k) Recombinant Leukemia Inhibitory Factor (LIF; ESGRO®; Millipore, Billerica, MA).

(l) High-glucose Dulbecco's Modified Eagle Medium (DMEM, Invitrogen, Carlsbad, CA). Alternatively, use KnockOut DMEM (Life Technologies).

(m) iPS cell Complete Medium: prepare high-glucose Dulbecco's modified Eagle's Medium (DMEM) (alternatively, use Knockout-DMEM, Invitrogen, Carlsbad, CA) containing 15 % ES-qualified FCS, 1.0 mM Sodium Pyruvate, 2.5 mM Glutamax (Invitrogen), 0.1 mM nonessential amino acids, penicillin (10.0 μg/ml), streptomycin (5.0 μg/ml), 0.1 mM β-mercaptoethanol (Invitrogen), and 1,000 U/ml recombinant leukemia inhibitory factor (LIF). Filter-sterilize the solution using a 0.22-μm filter, aliquot the solution into 100 ml volumes, and store them at +4 °C for 1 week.

(n) Mitomycin-treated mouse embryonic fibroblast cells (MEFs) can be purchased or prepared in-house. Alternatively, use irradiated MEFs. These MEFs can be used for up to 3–4 passages. Store aliquots of Mitomycin-treated or irradiated MEFs in liquid nitrogen.

(o) A refrigerated cell culture centrifuge should be within close proximity of the cell culture room.

(p) All cell culture and stem cell media should be stored inside a +4 °C refrigerator. Do not store the cell culture media with *E. coli* bacterial plates or leftover *E. coli* cultures.

(q) Some universities may have preferred vendors for cell culture dishes, stem cell culture media, and growth factors.

3 Methods

3.1 Aseptic Techniques and Biohazard Procedures

We emphasize here that ES and iPS cell culture is an expensive undertaking. Therefore, it is important to minimize waste in stem cell and biological research. With regard to compliance and safety issues, please follow your institutional guidelines and rules. The following general points may be applied to all stem cell and iPS culture laboratories.

1. Make sure there is no mycoplasma contamination in iPS cells.
2. We highly recommend a dedicated stem cell culture area with low traffic that is secured behind a closed door. Strict cleanliness is required at all times, including a regular schedule for the removal of used pipettes, dishes, and any liquid or solid wastes from the cell culture room. All cell culture waste should be removed on a daily basis. This cleanliness will minimize possible contamination from mycoplasma, yeast, bacteria, or fungi. Cell culture areas often experience too much human traffic; therefore, effort should be made to minimize unnecessary movement. Before and after work, spray 70 % ethanol and clean the inside of the cell culture hood. Leave the UV lamp on overnight. The best time to work is in the morning, as the hood had been exposed to UV lamp overnight.
3. Cell culture hoods are usually equipped with HEPA filters with a 4–5-year shelf life. The cell culture HEPA filter should be cleaned and recertified on a regular basis. All personnel must wear a clean laboratory coat and latex or nitrile gloves before using the cell culture hood. The 37 °C water bath must be cleaned on a regular basis, and the shelves of the CO_2 incubator should be autoclaved every month.
4. Every institute has an oversight committee for blood-borne pathogens, laboratory safety, recombinant DNA, and animal and stem cell research. All personnel working in a wet-bench laboratory must have completed the required training. Some universities require yearly recertification, and the laboratory safety and recombinant DNA oversight committees may also conduct online training and/or classroom training and instructions.

5. If you are new to the cell culture laboratory or if you are training a new member of the laboratory, please provide clear instructions about cleanliness and aseptic techniques. There should be no mouth pipetting or discarding of waste in the regular trash or dumping into a laboratory sink.
6. All liquid wastes must be neutralized and disinfected with Clorox (bleach, final concentration 15 %). All solid wastes, including serological pipettes, dishes, and paper towels, should be autoclaved in a biohazard bag and disposed according to the institutional rules.
7. Finally, absence of warning does not mean that everything is safe and sterile.

3.2 Culture Conditions for iPS Cells

We grow iPS cells in feeder-free conditions (28). High-quality iPS cells grow well if under optimal conditions. Like ES cells, the cell doubling time of iPS cells is short, 18–20 h. However, if the density of the iPS cells is too high, they tend to grow as 3D cellular aggregates. Cellular aggregates of iPS cells more often give rise to cells with heterogeneous morphologies and may be indistinguishable from parental iPS cells. It is important that iPS cells be subcultured every 4–5 days to maintain their growth in exponential phase. If iPS cells are not growing in exponential phase, it is desirable to periodically clone them. We recommend freezing iPS cells in several aliquots on a regular basis.

3.2.1 iPS Cell Culture

This protocol is optimized for $>10^5$ iPS cells per 6-well cell culture dish. We recommend recording phase contrast images of ES and iPS cell growth every other day during the course of culture (*see* **Notes 1** and **2**).

1. Prepare iPS cell complete culture media as described above. If using a commercially available kit, add all components to the 500-ml basal media container. If the kit has no Penicillin, Streptomycin, and GlutaMax solutions, add these reagents to the media prior to use.
2. Warm the media in a 37 °C water bath for 30–60 min. Do not keep the media in the water bath for more than 1 h at 37 °C, as prolonged exposure to 37 °C will reduce the effectiveness of the growth factors and GlutaMax.
3. iPS cells can be maintained in an undifferentiated state on mitomycin-C-arrested or irradiated MEFs. Mitotically arrested MEFs are plated 3–4 days prior to passaging of iPS cells.
4. Add 100 μl of attachment factor (AF) solution to the well; after 5 min, remove by aspirating. If plating multiple cell lines, AF can be kept on the dish for as long as 1 h. Add 5 ml of sterile 1× PBS to the dish, swirl side-to-side, and thereafter remove the PBS by aspirating. Excess AF should be removed by

washing with PBS; however, some laboratories do not recommend washing with PBS. AF-coated wells are good for use for up to 24 h.

5. Label all dishes or wells with the cell line name and number, date, and passage #.
6. Add 2 ml of warm, complete media to the well (6-well plate). If using MEFs as a feeder layer, the MEF media must be removed before the iPS cells can be plated. The iPS cells can be cultured in MEF dishes conditioned with iPS medium. We recommend conditioning new feeder dishes prior to passaging iPS cells into them.
7. Thaw a frozen cryovial containing iPS cells (if stored in liquid N_2) by immersing the vial in a 37 °C water bath for 5 min or to the point when the ice crystals disappear. Spray the cryovial with 70 % ethanol, remove the excess ethanol using a clean Kim-Wipe or paper towel, and bring the cryovial to the sterile cell culture hood.
8. Empty the contents of the cryovial into 15-ml sterile Falcon centrifuge tube containing warm complete media.
9. Centrifuge the cell suspension at ~1,000 rpm ($400 \times g$) in a refrigerated centrifuge for 5 min. After centrifugation, discard the supernatant and save the cell pellet.
10. Add 6 ml of iPS cell media to the tube and completely resuspend the cell pellet.
11. Remove the iPS cell suspension using a 5-ml pipette; thereafter, dispense 2 ml into each of 3 wells of a 6-well plate.
12. Return the dish to the humidified 37 °C CO_2 incubator.
13. The next day, aspirate the media to remove the dead cells and gently add prewarmed media. Examine the cells under the microscope.
14. The iPS cell cycle is approximately 18–20 h, and iPS cell colonies (three-dimensional structures) should be visible within 3–4 days. While growing, do not let two or more colonies fuse or mix because this event triggers differentiation.
15. Cells should be passaged after 3–4 days. If iPS cells dominate the culture as 3D structures, the iPS cells must be passaged by enzymatic dissociation. However, if the cells are too dense or growing quickly, some of the cells will differentiate spontaneously at the periphery of the 3D colony. In that case, the 3D iPS colonies must be separated manually from the differentiated cells (these often have an adherent phenotype) using the "car-wash method" with a cell scraper. If 3D iPS structures (cellular aggregates) dominate the culture, we recommend expanding the iPS cells into 35-mm dishes as described below.

3.2.2 Passaging iPS Cells

To maintain high-quality iPS cells, we recommend passaging the iPS cells on a regular basis. Maintenance of high-quality iPS cells in culture will be crucial for endothelial differentiation experiments (*see* **Notes 1** and **2**). Under carefully monitored iPS cell culture conditions, iPS cells retain pluripotency and self-renewing capacity.

1. Prepare dishes as described in Section 3.2.1 (step 4).
2. Prewarm the complete media as needed.
3. Label the wells with the cell line name, number, date, passage #, and split ratio.
4. Using a microscope, examine the morphology of the iPS cell colonies. Remove any differentiated cells from the iPS cell colonies to be passaged. Both ES and iPS cells can be dissociated enzymatically or manually by scraping the cell colonies. If your wells have more than 25 colonies/well in a 6-well plate, we recommend manually scraping the iPS cell colonies.
5. Aspirate the media from the wells using a Pasteur pipette and rinse the well once with DPBS without Ca++ and Mg++. If working with multiple cell lines or wells, change the pipette for each dish or each well.
6. Add 200 μl of prewarmed TrypLE™ (without Phenol red, Life Technologies) cell dissociation solution to the dish containing the iPS cell colonies. The volume of the cell detachment solution can be adjusted depending on the size of the well or dish.
7. Incubate the cells for 4–5 min in a CO_2 incubator at 37 °C. If the cells are detaching from the dish, the cell dissociation solution is likely inactive. If that is the case, please use a new batch of TrypLE cell dissociation solution.
8. After detachment, collect all cells into a 15-ml sterile centrifuge tube. Resuspend the cells in complete media and centrifuge at 1,000 rpm (400 × *g*) for 5 min at +4 °C in a refrigerated centrifuge.
9. Discard the supernatant by aspirating and resuspend the cell pellet by gently adding the appropriate volume of media. If the cells form clumps, pass the cell suspension through a Falcon Cell strainer (70-μm nylon mesh).
10. Replate the cells at the appropriate density (1:2 ratio) and return the dishes to a CO_2 incubator at 37 °C.
11. Change the media every 24 h.
12. If large-scale amounts of iPS cells are needed, repeat steps 1–11.

3.2.3 Freezing iPS Cells

Freezing aliquots of high-quality iPS cells in liquid nitrogen should be performed on a regular basis to maintain reliable stocks of iPS cells. We recommend changing the media 3–4 h before freezing iPS cells.

1. Detach the iPS cells using TrypLE cell dissociation solution, as described in Section 3.2.2 (steps 5–10)
2. Resuspend the cell pellet (2×10^6 cells/ml) in cell-freezing reagent (30 % FBS, 10 % DMSO, 60 % DMEM). If preparing fresh cell-freezing solution, cool the solution on ice for 20 min prior to use. Remember DMSO is a toxic solvent. Open the DMSO container inside the laminar flow hood.
3. Use a permanent marker to label the cryovials with the cell line name, passage number, date, and your initials.
4. Transfer the vials to a Frosty freezing cryocontainer filled with isopropanol.
5. Leave the Frosty freezing container in a −86 °C freezer overnight.
6. On the next day, transfer the cryovials to a liquid nitrogen tank.

3.3 Induction of Differentiation of iPS Cells into Endothelial Cells (ECs)

There are several methods in the literature that describe the induction of differentiation of ES and iPS cells into ECs (29–37). Depending upon the main objectives of the downstream experiments, the methods for inducing ES and iPS cells could be different (29–37). To minimize contamination of unknown cells, we use monoclonal antibodies that recognize the extracellular domains of Flk-1 and VE-cadherin. It is important to note that, in vitro, iPS-derived ECs express the Flk-1 and VE-cadherin endothelial markers, which are commonly used to distinguish adult ECs from both mature and immature cell sources. It would be fair to state here that, currently, there is no definitive marker for endothelial progenitor cells (EPCs) that can be used for revascularization therapy or therapeutic angiogenesis. In the following section, we describe a method for isolating Flk-1+ and VE-cadherin+ cells from mouse iPS cells (*see* **Note 3**). In addition to VE-cadherin and Flk1, we use functional assays, including tube formation and Matrigel plug assays (*see* **Note 4**). We have also tested the efficacy of these cells using a mouse model of hind limb ischemia.

3.3.1 Materials

1. Cell dissociation solution: 1.0 mM EDTA in PBS, pH 7.4 or TrypLE.
2. Type IV collagen-coated dishes (BD Bioscience, San Jose, CA).
3. Serum-free basal medium: 75 % Iscove's Modified Dulbecco's Media (IMDM) and 25 % Ham's F12 medium (Life Technologies).
4. B-27 Supplement (no Vitamin A).
5. Bovine Serum Albumin (BSA) (Life Technologies).
6. α-Monothioglycerol (MTG) (Sigma, St Louis, MO).

7. Ascorbic Acid (Sigma, St Louis, MO) (42).
8. Human BMP-4 (R&D Systems).
9. Human VEGF^{165} (Miltenyi Biotec).
10. Human Basic FGF (Millipore).
11. Rat anti-mouse Flk1 antibody (clone avas12a, BD Bioscience) and goat-anti-mouse VE-cadherin (R&D Systems, Minneapolis, MN) antibody.
12. Secondary antibodies conjugated to Alexa-Flour, APC, FITC, or PE.

3.3.2 Differentiation and Sorting of ECs

Given that ECs are anchorage-dependent cells, the induction of ES and iPS cell differentiation entails the use of a defined supporting matrix or purified extracellular matrix (ECM) protein. Thus, to initiate differentiation in ES and iPS cells, we used type IV collagen as a supporting ECM protein, which is known to induce differentiation in ES cells to mesodermal lineages. We recommend recording the differentiating cells in culture every 24 h using an inverted cell culture phase contrast microscope. Prepare in advance to do FACS sorting, e.g., sign up and coordinate with the core facility, because the cell sorter must be aseptically cleaned. Alternative to FACS, antibody-coated, e.g., anti-CD31, magnetic beads may be used. To isolate ECs, all antibodies must be FACS validated and free of sodium azide or any preservatives. If the antibody contains sodium azide, it must be dialyzed and filter-sterilized. As commercial antibodies are expensive, we recommend using antibodies produced by a bioreactor or cell culture supernatant of hybridoma cells that secrete monoclonal antibodies for routine isolation of large-scale ECs. The Flk1 hybridoma can be purchased from ATCC or may be obtained through the Iowa Developmental Hybridoma Bank.

1. Serum-free medium: prepare with 75 % IMDM, 25 % Ham's F12 medium, B-27 Supplement, 0.05 % BSA, 4.5×10^4 M MTG and 0.5 mM ascorbic acid, human BMP-4 (2 ng/ml), human VEGF^{165} (50 ng/ml), and human basic FGF (10 ng/ml).
2. Coat a 35-mm dish with type IV collagen (5.0 μm/ml) for 1 h, wash with sterile 1× DPBS, pH 7.4. Alternatively, use premade type IV collagen-coated dishes from BD Biosciences.
3. Plate actively growing, high-quality iPS cells (*see* Section 3.2.2, step 12) at a density of 3.5×10^4/35-mm well onto the type IV collagen-coated dishes.
4. Adherent cells will emerge as soon as 2–3 days. Change the differentiation media every 24 h and take pictures every 24 h.
5. The expression of Flk-1 begins to appear as early as 2.0–3.0 days, while VE-cadherin expression begins at

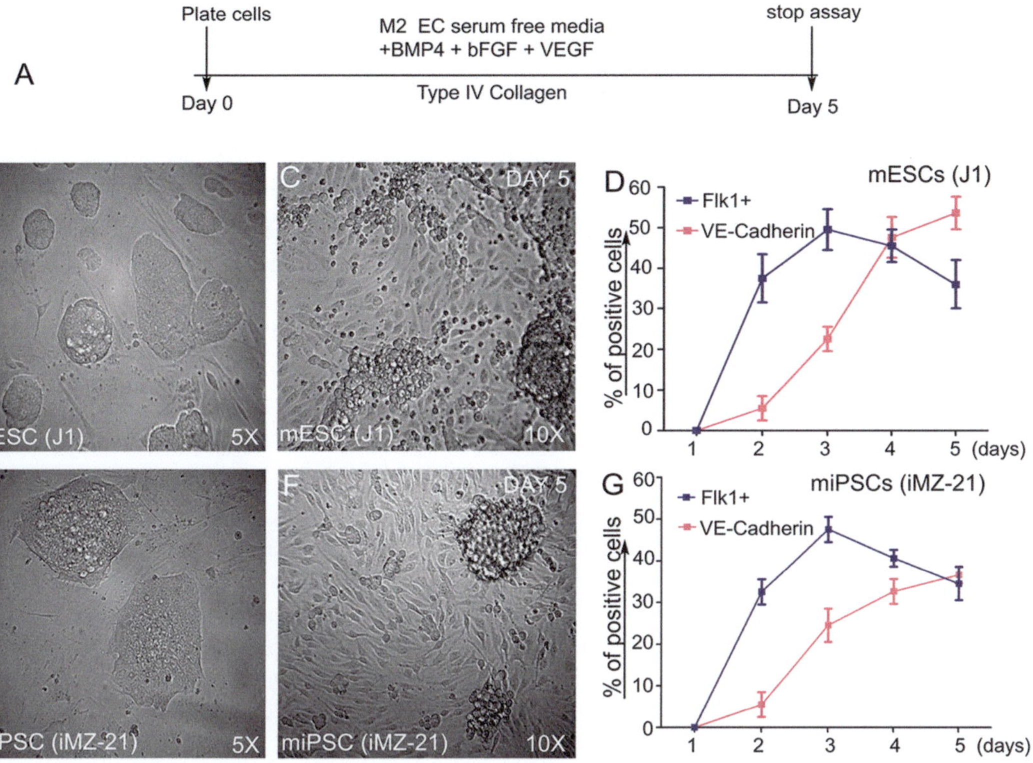

Fig. 1 Induction of Flk1$^+$VE-cadherin$^+$ vascular EC progenies from iPS and ES cells. Timeline of emergence of Flk1 + VE-cadherin + vascular ECs. (**a**) Undifferentiated mES (J1 line) or miPS (iMZ-21) cells were cultured for 5 days in IV Col-coated dishes in media containing BMP4, bFGF, and VEGF^{165} to induce generation of vascular EC progenies. (**b, c**). Representative phase contrast microscopy of mES cell-derived adherent vascular progenies after day 5 in culture at indicated magnifications (**b, c**). Representative phase contrast microscopy of miPS cell-derived vascular progenies after day 5 in culture at the indicated magnifications (**e, f**). FACS profile of the emergence of Flk1$^+$VE-cadherin$^+$ vascular progenies (**d, g**). All experiments were repeated $>$5 times. Data indicate the mean $\pm$ S.E.M. $n = 5$. (Reprinted from Kohler EE et al., PloS One 2013 Dec 30;8 (12):e85549)

2.5–3.5 days (Fig. 1). For optimal expression of Flk-1 and VE-cadherin proteins, we recommend culturing these cells up to day 4–5.

6. Physically remove 3D cellular aggregates, as these are a mixture of cells.
7. Add cell dissociation solution (1 mM EDTA, pH 7.4) to the dish and incubate for 20 min in a CO_2 incubator at 37 °C.
8. To stop the action of EDTA or cell dissociation enzyme, add complete serum-free media (differentiation media); thereafter, centrifuge at 400 × *g* for 5 min in a refrigerated centrifuge.

9. After centrifugation, wash the cells in cold phosphate-buffered saline (PBS). If clumps appear, pass the cells through a Falcon cell strainer (70-μm pore size).
10. To isolate high-quality ECs, the detached cells are made into a single-cell suspension as described in step 9 and are thereafter subjected to FACS or magnetic separation.
11. The cell suspension is adjusted to 2×10^6 cells/ml. Use an appropriate concentration of primary antibody (1–2 μg/ml). Incubate on ice for 1 h (in the dark if the antibody is directly conjugated to Alexa fluor); thereafter, wash once with cold PBS and save the cell pellet.
12. Next, resuspend the cell pellet in secondary antibody solution, incubate on ice for 30–60 min (in dark), wash once with PBS, and resuspend the cells in FACS buffer.
13. For FACS, please have negative and positive control samples and determine the gating parameters for cell sorting. Gating should be fixed at the level of fluorescence above which almost all cells are negative. These parameters can be adjusted depending on the FACS machine and objective of the downstream experiments as well as on the experience of the investigator. We use the Moflo FACS high-speed sorter.
14. As ECs are larger than leukocytes, adjust the autofluorescence to the first quadrant of the fluorescence to be monitored while gating the ECs and ensure that the cells of interest are included in the gate. If required, check the fluorescence of the positive control sample and compensate. Initiate the acquisition of Flk1 + and VE-cadherin+ ECs; 10^5/sample is optimal. Save the data and analyze the dot blot of the gated ECs.
15. If ECs are not 100 % pure, subject these cells through one more cycle of FACS with the anti-Flk1 and anti-VE-cadherin antibodies (Fig. 2).

To determine the phenotypic characteristics and colony formation of the sorted cells, plate the cells onto type IV collagen-coated coverslips in the presence of serum-free differentiation media. Use rat anti-mouse CD41 to assess hematopoietic cells and anti-VE-cadherin and anti-CD31 antibodies to determine the identity of the ECs. FACS-sorted cells can be passaged 1–2 times. We recommend further characterization of these cells using anti-VE-cadherin, anti-CD31, anti-Flk1, and anti-vWF antibodies. In addition, gene expression profiling, proteomics, and epigenetic studies can be carried out. In all of these characterization experiments, positive and negative controls must be included.

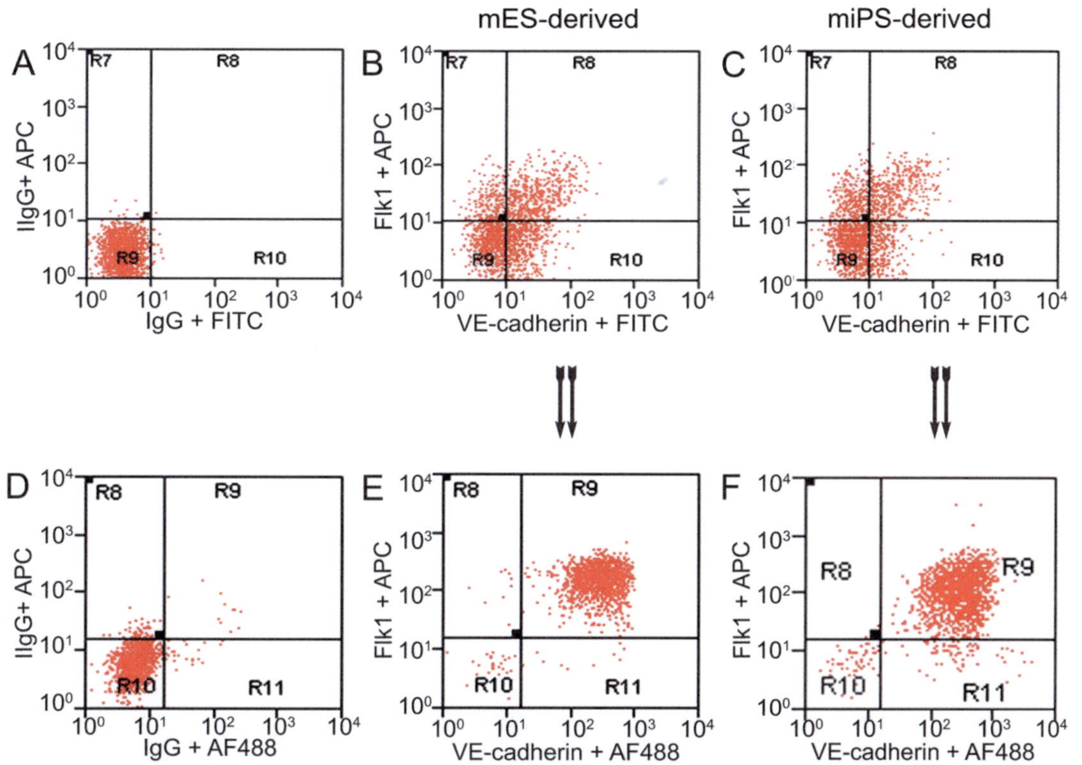

Fig. 2 FACS analysis of emerging vascular EC progenies from mES and iPS cells. Adherent cells (2×10^5) were detached and subjected to two-step FACS-aided purification. Control FACS profile on day 5 of cells derived from mES (J1) cells (**a**). Representative FACS profiles of day 5, with vascular progenies assessed using anti-Flk1 and anti-VE-cadherin antibodies obtained from mES (J1) cells (**b**) and derived from miPS (iMZ-21) cells (**c**); Control FACS profile on day 5 of cells derived from miPS (iMZ-21) cells (**d**). Representative FACS after the second step of purification derived from mES (**e**) and iPS cells (**f**). The yield of Flk1$^+$VE-cadherin$^+$ after the second round of FACS was 100 % for both mES and miPS-derived vascular progenies. Morphology of mES- and miPS-derived vascular ECs (**g–j**). Flk1$^+$VE-cadherin$^+$ vascular progenies derived from mES and miPS cells were cultured overnight in IV Col-coated dishes, immunostained with anti-VE-cadherin (*green*) and anti-CD31 (*red*) of cells derived from mES cells (**g**, **h**) and miPS cells (**i**, **j**). DAPI, nucleus (*blue*). Magnifications are as indicated; the scale bar is 200 μm. Experiments were repeated three times. (Reprinted from Kohler EE et al., PloS One 2013 Dec 30;8(12):e85549)

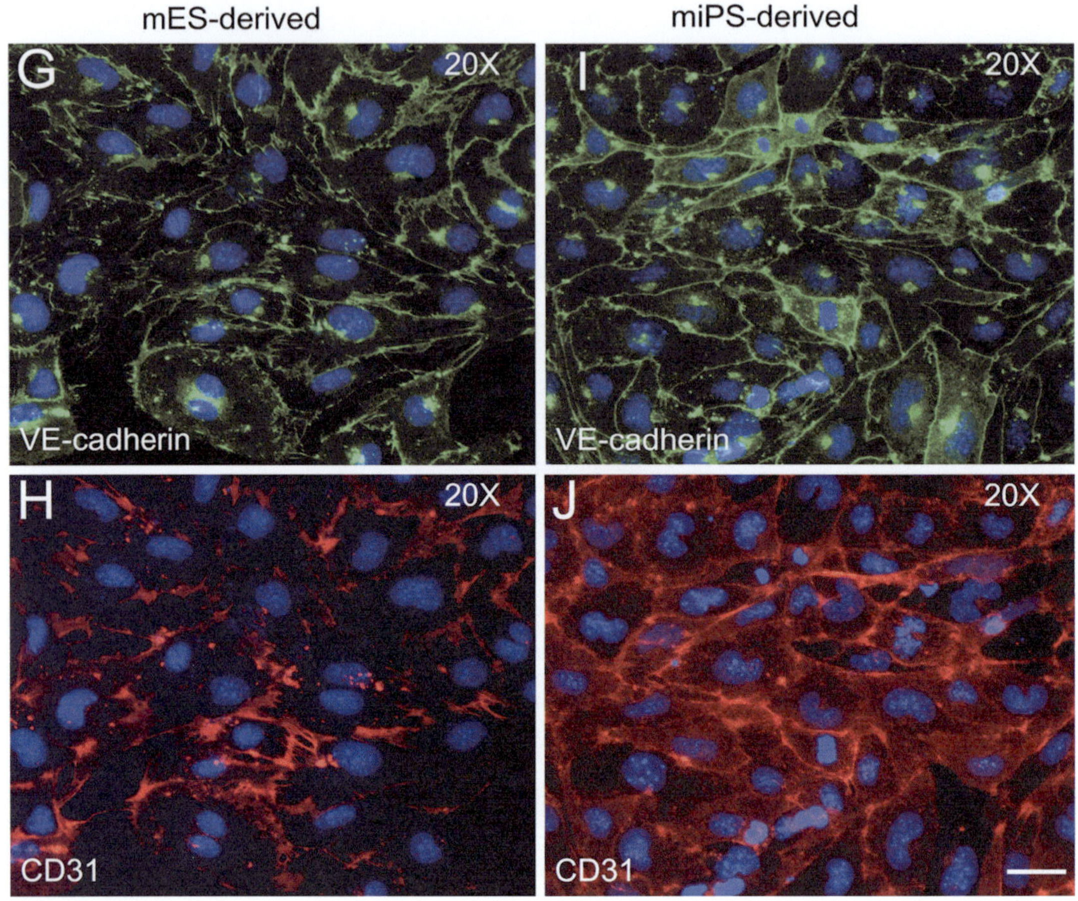

Fig. 2 (continued)

4 Notes

1. Based purely upon our recent publication (28), here, we have described iPS cell culture and the induction of EC differentiation in vitro as well as the isolation of Flk1- and VE-cadherin-positive ECs. Unlike tumor cell culture, one has to be patient with iPS cell culture.
2. It is necessary to have live iPS cells and optimal cell culture conditions. The quality of the media and growth factors could alter the behavior of the iPS cells and their differentiation potential. However, if the conditions are optimal, the iPS cells will grow so that important experiments can be carried out in a timely manner.
3. There are a multiplicity of protocols for the differentiation and characterization of ECs. However, the quality of iPS cells is critical for the emergence of ECs. It is important to realize that

not all iPS cell lines will give rise to the same number of ECs, and we have found that some iPS cell lines perform better than others. For this reason, we recommend starting with a high-quality iPS cell culture, optimal growth and maintenance conditions with no contamination. If the FACS-sorted cells are not clean, these cells will quickly die in culture or become contaminated. If contamination occurs, we recommend thawing a new aliquot of cells and begin everything anew. We do not recommend sorted ECs be passaged more than 1–2 times. Currently, we have not established iPS-derived ECs as cell lines.

4. To assess the functionality of ECs, several different assays can be carried out, including in vitro Matrigel tube formation assays and in vivo Matrigel plug angiogenesis assays (28). For in vivo studies, these cells can be transfected by a retrovirus encoding red fluorescent protein (RFP) (28). For all functional studies, we recommend the use of freshly isolated cells.

Acknowledgments

The research in the authors' laboratory is supported by grants from the National Institutes of Health (NIH) and the American Heart Association (AHA) to KKW.

[According to the PloS One Open Access Journal policy no permission is required to reproduce text and the Figures (Kohler EE et al., PLoS One 8(12):e85549.)]

References

1. Evans MJ, Kaufman MH (1981) Establishment in culture of pluripotential cells from mouse embryos. Nature 292:154–156
2. Bradley A, Robertson E (1986) Embryo-derived stem cells: a tool for elucidating the developmental genetics of the mouse. Curr Top Dev Biol 20:357–371
3. Nagy A, Gócza E, Diaz EM, Prideaux VR, Iványi E, Markkula M, Rossant J (1990) Embryonic stem cells alone are able to support fetal development in the mouse. Development 110(3):815–821
4. Nagy A, Rossant J, Nagy R, Abramow-Newerly W, Roder JC (1993) Derivation of completely cell culture-derived mice from early-passage embryonic stem cells. Proc Natl Acad Sci U S A 90(18):8424–8428
5. Wilmut I, Schnieke AE, McWhir J, Kind AJ, Campbell KH (1997) Viable offspring derived from fetal and adult mammalian cells. Nature 385:810–813
6. Thomson JA, Itskovitz-Eldor J, Shapiro SS, Waknitz MA, Swiergiel JJ, Marshall VS, Jones JM (1998) Embryonic stem cell lines derived from human blastocysts. Science 282 (5391):1145–1147
7. Amit M, Carpenter MK, Inokuma MS, Chiu CP, Harris CP, Waknitz MA, Itskovitz-Eldor J, Thomson JA (2000) Clonally derived human embryonic stem cell lines maintain pluripotency and proliferative potential for prolonged periods of culture. Dev Biol 227 (2):271–278
8. Waddington CH (1957) The strategy of the genes. Geo Allen and Unwin, London
9. Takahashi K, Yamanaka S (2006) Induction of pluripotent stem cells from mouse embryonic and adult fibroblast cultures by defined factors. Cell 126(4):663–676
10. Okita K, Ichisaka T, Yamanaka S (2007) Generation of germline-competent induced pluripotent stem cells. Nature 448(7151):313–317

11. Yu J, Vodyanik MA, Smuga-Otto K, Antosiewicz-Bourget J, Frane JL, Tian S, Nie J, Jonsdottir GA, Ruotti V, Stewart R, Slukvin II, Thomson JA (2007) Induced pluripotent stem cell lines derived from human somatic cells. Science 318(5858):1917–1920
12. Wernig M, Meissner A, Foreman R, Brambrink T, Ku M, Hochedlinger K, Bernstein BE, Jaenisch R (2007) In vitro reprogramming of fibroblasts into a pluripotent ES-cell-like state. Nature 448(7151):318–324
13. Wang Y, Baskerville S, Shenoy A, Babiarz JE, Baehner L, Blelloch R (2008) Embryonic stem cell-specific microRNAs regulate the G1-S transition and promote rapid proliferation. Nat Genet 40:1478–1483
14. Judson RL, Babiarz JE, Venere M, Blelloch R (2009) Embryonic stem cell-specific microRNAs promote induced pluripotency. Nat Biotechnol 27:459–461
15. Subramanyam D, Lamouille S, Judson RL, Liu JY, Bucay N, Derynck R, Blelloch R (2011) Multiple targets of miR-302 and miR-372 promote reprogramming of human fibroblasts to induced pluripotent stem cells. Nat Biotechnol 29(5):443–448
16. Gonzalez F, Boue S, Belmonte JC (2011) Methods for making induced pluripotent stem cells: reprogramming a la carte. Nat Rev Genet 12:231–242
17. Giorgetti A et al (2009) Generation of induced pluripotent stem cells from human cord blood using OCT4 and SOX2. Cell Stem Cell 5:353–357
18. Utikal J, Maherali N, Kulalert W, Hochedlinger K (2009) Sox2 is dispensable for the reprogramming of melanocytes and melanoma cells into induced pluripotent stem cells. J Cell Sci 122:3502–3510
19. Eminli S, Utikal J, Arnold K, Jaenisch R, Hochedlinger K (2008) Reprogramming of neural progenitor cells into induced pluripotent stem cells in the absence of exogenous Sox2 expression. Stem Cells 26 (10):2467–2474
20. Kim JB, Greber B, Araúzo-Bravo MJ, Meyer J, Park KI, Zaehres H, Schöler HR (2009) Direct reprogramming of human neural stem cells by OCT4. Nature 461:649–653
21. Li Y, Zhang Q, Yin X, Yang W, Du Y, Hou P, Ge J, Liu C, Zhang W, Zhang X, Wu Y, Li H, Liu K, Wu C, Song Z, Zhao Y, Shi Y, Deng H (2011) Generation of iPSCs from mouse fibroblasts with a single gene, Oct4, and small molecules. Cell Res 21(1):196–204
22. Theunissen TW, Jaenisch R (2014) Molecular control of induced pluripotency. Cell Stem Cell 14(6):720–734
23. Boland MJ, Hazen JL, Nazor KL, Rodriguez AR, Gifford W, Martin G, Kupriyanov S, Baldwin KK (2009) Adult mice generated from induced pluripotent stem cells. Nature 461:91–94
24. Woltjen K, Michael IP, Mohseni P, Desai R, Mileikovsky M, Hämäläinen R, Cowling R, Wang W, Liu P, Gertsenstein M, Kaji K, Sung HK, Nagy A (2009) piggyback transposition reprograms fibroblasts to induced pluripotent stem cells. Nature 458(7239):766–770
25. Kaji K, Norrby K, Paca A, Mileikovsky M, Mohseni P, Woltjen K (2009) Virus-free induction of pluripotency and subsequent excision of reprogramming factors. Nature 458 (7239):771–775
26. Woltjen K, Hämäläinen R, Kibschull M, Mileikovsky M, Nagy A (2011) Transgene-free production of pluripotent stem cells using piggyBac transposons. Methods Mol Biol 767:87–103
27. Papapetrou EP, Sadelain M (2011) Generation of transgene-free human induced pluripotent stem cells with an excisable single polycistronic vector. Nat Protoc 6:1251–1273
28. Kohler EE, Wary KK, Li F, Chatterjee I, Urao N, Toth PT, Ushio-Fukai M, Rehman J, Park C, Malik AB (2013) $Flk1^+$ and $VE\text{-}cadherin^+$ endothelial cells derived from iPSCs recapitulates vascular development during differentiation and display similar angiogenic potential as ESC-derived cells. PLoS One 8(12):e85549
29. Yamashita J, Itoh H, Hirashima M, Ogawa M, Nishikawa S, Yurugi T, Naito M, Nakao K, Nishikawa S (2000) Flk1-positive cells derived from embryonic stem cells serve as vascular progenitors. Nature 408:92–96
30. Levenberg S, Golub JS, Amit M, Itskovitz-Eldor J, Langer R (2002) Endothelial cells derived from human embryonic stem cells. Proc Natl Acad Sci U S A 99:4391–4396
31. Choi KD, Yu J, Smuga-Otto K, Salvagiotto G, Rehrauer W, Vodyanik M, Thomson J, Slukvin I (2009) Hematopoietic and endothelial differentiation of human induced pluripotent stem cells. Stem Cells 27(3):559–567
32. Yin L, Ohanyan V, Pung YF, Delucia A, Bailey E, Enrick M, Stevanov K, Kolz CL, Guarini G, Chilian WM (2012) Induction of vascular progenitor cells from endothelial cells stimulates coronary collateral growth. Circ Res 110 (2):241–252

33. Yang HM, Moon SH, Choi YS, Park SJ, Lee YS, Lee HJ, Kim SJ, Chung HM (2013) Therapeutic efficacy of human embryonic stem cell-derived endothelial cells in humanized mouse models harboring a human immune system. Arterioscler Thromb Vasc Biol 33:2839–2849
34. Adams WJ, Zhang Y, Cloutier J, Kuchimanchi P, Newton G, Sehrawat S, Aird WC, Mayadas TN, Luscinskas FW, García-Cardeña G (2013) Functional vascular endothelium derived from human induced pluripotent stem cells. Stem Cell Reports 1(2):105–113
35. Prasain N, Lee MR, Vemula S, Meador JL, Yoshimoto M, Ferkowicz MJ, Fett A, Gupta M, Rapp BM, Saadatzadeh MR, Ginsberg M, Elemento O, Lee Y, Voytik-Harbin SL, Chung HM, Hong KS, Reid E, O'Neill CL, Medina RJ, Stitt AW, Murphy MP, Rafii S, Broxmeyer HE, Yoder MC (2014) Differentiation of human pluripotent stem cells to cells similar to cord-blood endothelial colony-forming cells. Nat Biotechnol 32(11):1151–1157
36. Belair DG, Whisler JA, Valdez J, Velazquez J, Molenda JA, Vickerman V, Lewis R, Daigh C, Hansen TD, Mann DA, Thomson JA, Griffith LG, Kamm RD, Schwartz MP, Murphy WL (2014) Human vascular tissue models formed from human induced pluripotent stem cell derived endothelial cells. Stem Cell Rev (in press)
37. Kusuma S, Facklam A, Gerecht S (2014) Characterizing human pluripotent-stem-cell-derived vascular cells for tissue engineering applications. Stem Cells Dev (in press)

Methods in Molecular Biology (2016) 1357: 329–338
DOI 10.1007/7651_2015_227

Published online: 12 April 2015

Derivation of Neural Stem Cells from Mouse Induced Pluripotent Stem Cells

Işıl Karanfil and Tugba Bagci-Onder

Abstract

Neural stem cells (NSCs) derived from induced pluripotent stem cells offer therapeutic tools for neurodegenerative diseases. This review focuses on embryoid body (EB)-mediated stem cell culture techniques used to derive NSCs from mouse induced pluripotent stem cells (iPSCs). Generation of healthy and stable NSCs from iPSCs heavily depends on standardized in vitro cell culture systems that mimic the embryonic environments utilized during neural development. Specifically, the neural induction and expansion methods after EB formation are described in this review.

Keywords: Induced pluripotent stem cells, Neural stem cells, Embryoid body, Stem cell protocols, Neural induction

1 Introduction

Stem cell therapy is considered ideal for the treatment of neurodegenerative diseases due to the lack of neuron renewal in the body. To successfully deliver neural stem cells/neurons in the central nervous system (CNS), stem cells must be isolated, differentiated, and enriched for populations with high rates of homogeneity. The derivation of specific neuronal cell types invariably requires the production of neural stem cells (NSCs). NSCs are capable of generating neurons, astrocytes, and oligodendrocytes.

In the last decade of stem cell research, NSCs were mainly generated from Embryonic Stem (ES) cells through directed differentiation. Such derived NSCs were shown to be capable of differentiating into synaptically integrated neurons (1). Similarly, adherent and self-renewing NSCs with long-term symmetrical cell division capacities were derived from ES cells, using neural induction medium that contains growth factors such as basic FGF (FGF-2) and EGF (2, 3).

Future of stem cell therapies for neural disorders requires generation of patient- and disease-specific NSCs, and ongoing studies in this field have been specifically focused on this issue. To reach this distant but achievable goal, generation of induced pluripotent stem cells (iPSCs) is a revolutionary step towards personalized medicine (4). Through the transduction of a cocktail of transcriptional factors named as the four Yamanaka factors (Oct4, Sox2, Klf4, and c-Myc) into somatic cells via retroviruses followed by exposure to ES medium, Takahashi et al. efficiently reprogrammed mouse and human somatic cells to iPSCs (5). Because of their similarities in proliferation, gene expression, epigenetic status of pluripotent cell-specific genes, and telomerase activity, iPSCs provide an excellent alternative to ES cells (6). Applications of iPSCs have already progressed rapidly in development of cell replacement therapies, disease modeling, and drug screening (5). The production of iPSCs for basic or clinical applications is an exciting field, which is extensively reviewed by many other researchers (7).

iPSC-derived NSCs have been shown to be effective in the animal models of several neurodegenerative diseases including Parkinson's Disease (8), Huntington's Disease (9), Multiple Sclerosis (10, 11), and ALS (12). In this review, we aim to highlight the specific processes during the generation of NSCs from mouse iPSCs. Large number of neural cells can be obtained after somatic cells are programmed into iPSCs, which are then induced to form NSCs and finally specific neural cells (13). Given recent methods for the production of iPSC-derived NSCs available (14–17), this review aims to describe the directed NSC differentiation methods.

This method involves four basic steps: iPSC maintenance in undifferentiated state, Embryoid Body (EB) formation, neural induction, NSC enrichment and expansion (Fig. 1) as described below.

2 Materials

2.1 Feeder Fibroblast, iPSC, and EB Culture Components

1. Feeder Mouse Embryonic Fibroblasts (MEFs) (Millipore: EmbryoMax®, Strain CF1, Mitomycin C treated).
2. Dulbecco's Phosphate-Buffered Saline (DPBS) without calcium and magnesium (Gibco).
3. 0.25 % Trypsin–EDTA (Gibco).

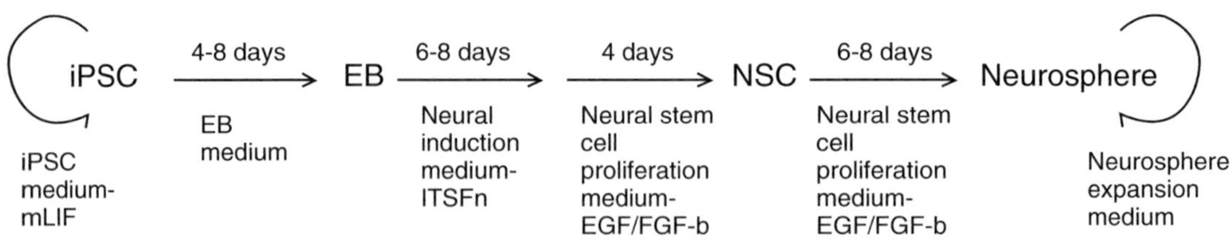

Fig. 1 Outline of the four steps of NSC derivation from iPSCs. *iPSC* induced pluripotent stem cells, *EB* embryoid body, *NSC* neural stem cell, *mLIF* mouse leukemia inhibitory factor, *ITSFn* insulin–transferrin–selenium–fibronectin, *EGF* epidermal growth factor, *FGF-b* fibroblast growth factor-basic

4. 0.1 % gelatin: Dissolve 0.1 % gelatin (Sigma) in distilled water, mix and autoclave. Store at room temperature.
5. DMEM-High Glucose (Stem Cell Technologies).
6. ES-Cult™ Fetal Bovine Serum (FBS) (Stem Cell Technologies).
7. Sodium pyruvate (Gibco).
8. Penicillin–Streptomycin (Gibco).
9. L-Glutamine (Gibco).
10. MEM-Non-Essential Amino Acids (Gibco).
11. mLIF (Leukemia Inhibiting Factor) (Stem Cell Technologies) (**Note 1**).
12. 2-mercaptoethanol (2-ME).

2.2 Neural Induction Medium and Neurosphere Culture Components

1. ITS (Insulin–Transferrin–Selenium) Supplement-A (Stem Cell Technologies).
2. ES-Cult™ Basal Medium (Stem Cell Technologies).
3. Fibronectin (Stem Cell Technologies) (**Note 2**).
4. Poly-L-ornithine (Sigma) (15 μg/mL freshly prepared in PBS).
5. Laminin (Sigma) (20 μg/mL freshly prepared in PBS) (**Note 2**).
6. NeuroCult™ NSC Proliferation Medium (Stem Cell Technologies).
7. NeuroCult™ Proliferation Supplements (Stem Cell Technologies).
8. Heparin (Stem Cell Technologies).
9. EGF (R&D Systems) (**Note 3**).
10. FGF-b (R&D Systems) (**Note 3**).

2.3 Medium Preparation

Prepare all media under a laminar flow hood.

1. *Feeder Fibroblast Culture Medium* (*D10*): Combine all reagents at indicated ratios: 90 % DMEM (with L-Glutamine and Pyruvate), 10 % FBS, 1 % penicillin–streptomycin solution. Mix the components and filter through with 0.22 μm filter. Store at +4 °C up to 4 weeks.
2. *Induced Pluripotent Stem Cell Medium* (*iPSC Medium*): Combine all components at indicated final concentrations: 15 % ES-Cult™ Fetal Bovine Serum, 0.1 mM MEM-Non-Essential Amino Acids, 2 mM L-Glutamine, 10 ng/mL (10^3 U/mL) mLIF, 0.1 M 2-Mercaptoethanol (2-ME), DMEM high glucose. Mix the components and filter through with 0.22 μm filter. Store at +4 °C up to 2 weeks (**Note 4**).
3. *Embryoid Body Medium* (*EB medium*): Same as iPSC medium without the addition of mLIF.

4. *Neural Induction Medium* (*ITSFn Medium*): Add 5 mL of ITS Supplement-A to 500 mL of ES-Cult Basal Medium. On the day of use, add 50 μL of fibronectin (1 mg/mL) per 10 mL of ITS medium to achieve final concentration of 5 μg/mL. Store the ITS medium at +4 °C and use within 2 weeks (**Note 5**).
5. *Neurosphere Expansion Medium* (*NSC Medium*): Prepare NeuroCult™ NSC Proliferation medium by adding 1 mL of NeuroCult™ Proliferation Supplements to every 9 mL of NeuroCult™ Basal Medium. Add cytokines at indicated final concentrations: 20 ng/mL EGF, 20 ng/mL FGF-b. Also, add Heparin at a 1:1,000 dilution. Add 0.5 % Penicillin–Streptomycin. Mix the components and filter through a 0.22 μm filter. Store at +4 °C up to 2 weeks (**Note 6**).

3 Methods

3.1 Maintenance of iPSCs

3.1.1 Preparation of Plates with Inactivated MEFs

1. Coat the cell culture dishes (a 6-well plate) with 0.1 % gelatin and incubate at room temperature for 30 min.
2. Thaw a vial of inactivated MEFs (mitotically arrested by mitomycin C treatment) by holding the cryogenic vial in a 37 °C water bath until only a small amount of ice is left. Immediately but slowly add 1 mL of warm D10 media by drop, mix, and transfer to Falcon tube with 10 mL D10 medium. Spin at 1,200 rpm (300 × *g*) for 5 min.
3. During centrifugation, aspirate gelatin and add 1 mL D10 medium per 10 cm^2 (1 well of a 6-well plate). Resuspend MEF pellet in an appropriate volume for the resulting cell concentration of 1.5–2 × 10^5 cells/mL. Plate 1 mL per 10 cm^2. Transfer to incubator and incubate overnight (**Note 7**).

3.1.2 Maintenance of iPSCs in the Undifferentiated State

1. Thaw a vial of mouse iPSCs rapidly in a water bath at 37 °C and add cells drop wise to a tube containing iPSC medium. Spin down at 1,200 rpm (300 × *g*) to pellet the cells. Resuspend the pellet in appropriate volume of fresh iPSC medium containing mLIF.
2. Remove MEF feeder cells from the incubator and aspirate the D10 medium from the wells of the 6-well plates. Add 2 mL iPSC medium to each well.
3. Distribute 0.2 mL of the iPSC suspension to each well of the 6-well plate. After plating iPSCs, gently swirl the plate back-and-forth and side-to-side. Incubate the cells at 37 °C, 5 % CO_2 in a humidified incubator.
4. Change iPSC media every day until the iPSCs reach 80 % confluency (**Note 8**).

5. For passaging of iPSCs, aspirate the medium and wash the cells twice with PBS.
6. Remove PBS completely and add 0.5 mL of 0.25 % Trypsin–EDTA solution to each well of the 6-well plate, and incubate at 37 °C for 2 min. During trypsinization, remove a new 6-well plate of feeder cells prepared as described in Section 3.1.1. Aspirate the medium and add 2 mL of mouse iPSC medium to each well. Collect iPSCs with iPSC medium and suspend them by pipetting up and down. Distribute iPSC suspension as described above. Subculture mouse iPSCs ~1:10 every 2–3 days. Track the passage number (**Note 9**).

3.2 Differentiation of iPSCs

3.2.1 Embryoid Body Formation

1. Harvest the iPSCs grown on feeder cells using Trypsin–EDTA as described above. Check under the microscope to ensure that a single cell suspension has been achieved to optimize the efficiency of EB formation.
2. Perform a cell count using Trypan Blue to measure viability. Cell viability should be greater than 90 % for optimal differentiation afterwards.
3. Use of the Ultra-Low Adherent 6-well plates (Stem Cell Technologies) or regular petri dishes for optimal EB formation. Prehydrate the plates with EB medium 30 min prior to plating (**Note 10**).
4. Plate the freshly harvested iPSCs at a density of 5×10^5 cells per well of the pre-hydrated 6-well plates. For uncoated petri dishes, adjust the cell density accordingly (i.e., use 2×10^6 cells/100 mm petri dish) (**Note 11**).
5. Return the cells to the incubator and keep for 4 days. Every 2 days, EBs can be fed by gently removing the cells and collecting them into a 15 mL Falcon tube (**Note 12**). Allow the EBs to settle to the bottom of the tube for approximately 5 min and carefully remove the medium. Replace with fresh EB medium and replate the cells into each well or dish.
6. After an additional 4-day culture, harvest the EBs from the wells of the plates or petri dishes and place into a sterile polypropylene tube. Allow the EBs to settle as described above and carefully aspirate the medium. Proceed with Neural Induction Protocol as described below.

3.2.2 Enrichment for Neural Precursors with Neural Induction

1. Prepare 24-well tissue culture plates for Neural Induction. First, coat each well of the plate poly-L-ornithine and laminin. Add 500 μL of 15 μg/mL poly-L-ornithine solution per well and incubate for 1 h at 37 °C. Then, remove the poly-L-ornithine solution and rinse with PBS three times. Second, add 500 μL of 20 μg/mL laminin solution to each well and

incubate for 2 h at 37 °C. Remove the laminin solution and rinse once with PBS prior to use (**Note 13**).

2. After the EBs have been harvested as described above, add 3 mL of ITSFn medium (with freshly added fibronectin) to the settled EBs.
3. Count a minimal volume sample (10–20 μL) to determine the density of the EB suspension.
4. Plate 20–50 EBs per well of the coated 24-well plates in 1 mL ITSFn medium (**Note 14**).
5. Culture for 6–8 days. Feed the cultures every second day by gently removing the medium and replacing it with fresh ITSFn medium containing fibronectin (**Note 15**).

3.2.3 Expansion of NSCs

1. Prepare a new 24-well tissue culture plate by coating each well with poly-L-ornithine and laminin as described above.
2. After the EBs have been cultured on 24-well plates in the ITSFn medium for 6–8 days, aspirate off the medium and wash the adherent cells gently with PBS twice.
3. Add sufficient volume of Trypsin–EDTA to just cover the cells in the tissue culture treated plates and incubate at 37 °C for 2 min (**Note 16**).
4. Collect the trypsinized cells with NSC medium into a 15 mL conical tube and disperse the clumps by pipetting up and down (Some clumps may remain).
5. Allow the clumps to settle to the bottom of the tube for 3 min.
6. Transfer the supernatant (**Note 17**) containing cells (but not the clumps) to a new 15 mL sterile conical tube. Pellet cells by centrifugation at 1,200 rpm (300 × *g*) for 5 min.
7. Aspirate the supernatant and resuspend the cells in 1–2 mL of NSC medium.
8. Perform a cell count using Trypan Blue to measure cell viability.
9. Plate 5×10^5 viable cells per well (that are freshly coated) in NSC medium containing EGF, FGF-b, and Heparin. Culture cells at 37 °C, 5 % CO_2 in a humidified incubator.
10. Feed the cells by aspirating the media from the wells and adding freshly prepared NSC medium every day for 4 days.

3.2.4 Neurosphere Culture

1. After the cells grow as attached on coated wells for 4 days, scrape the cells into suspension in NSC medium. First, aspirate the media from the wells and add fresh NSC medium to each well of the 24-well plate (1 mL each). With a scraper, gently lift the cells from the wells and transfer into a sterile 15 mL conical tube. Gently pipet up and down and dissociate the cells from each other slowly (**Note 18**).

2. Add 1 mL more NSC medium for each 1 mL of the existing suspension.
3. Perform a cell count using Trypan Blue to measure cell viability.
4. Transfer 1×10^6 cells to a T-25 cell culture flask in 8 mL fresh NSC medium. Culture cells at 37 °C, 5 % CO_2 in a humidified incubator.
5. Culture for 6–8 days. Feed the cultures every second day by gently removing 1/3 of the medium and replacing it with fresh NSC medium (**Note 19**).
6. To subculture neuropsheres, transfer neurospheres into 15 mL conical tubes and allow the neurospheres to settle for 5 min.
7. Aspirate the medium and add 50–100 μL of fresh NSC medium to the neurosphere pellet. Gently pipette up and down with a 200 μL pipettor for ~30 times to allow for gentle and complete dissociation of NSC neurospheres into single cells.
8. Transfer 1:5 of the cell suspension into a new T-25 flask and repeat **steps 5–8** (**Note 20**).

3.2.5 Freezing iPSC-Derived NSCs

1. Grow NSCs cells to the exponential phase in T-25 flasks and subculture as described.
2. 1 day after subculturing, collect the cells in a 15 mL conical tube, count the cell number, and centrifuge at 1,200 rpm for 5 min to make NSC pellet.
3. Discard the supernatant, add freshly prepared freezing medium (10 % DMSO, 40 % NSC medium, 50 % FBS) to the cell pellet to the concentration of 1×10^6 cells per mL.
4. Aliquot cell suspension as 1 mL per freezing vial.
5. Put the vials in a cell-freezing container and store them at −80 °C overnight.
6. Transfer the vials to liquid nitrogen for long-term storage (**Note 21**).

4 Notes

1. These medium components should be aliquoted immediately upon arrival and kept at +4 °C. Make sure that one tube is used only once for medium preparation.
2. These medium components should be aliquoted immediately upon arrival and kept at −80 °C. These components should be thawed on ice slowly prior to use. Make sure that one tube is used only once for medium preparation.

3. These medium components arrive as lyophilized and should be resuspended according to instructions, aliquoted immediately and kept at −80 °C. These components should be thawed on ice slowly prior to use. Make sure that one tube is used only once for medium preparation. In all cell culture media that contains these components (EGF and FGF-b), they may lose activity if incubated for long periods of time at 37 °C.
4. iPSC medium can be prepared without mLIF, and mLIF can be added at the time of use. This way, the medium can be kept for longer than 2 weeks at +4 °C.
5. Fibronectin should be added fresh, otherwise its effect cannot be observed.
6. NSC medium can be made, filtered, and aliquoted. Alternatively, 10–50 mL aliquots can be stored at −80 °C, and used after slowly thawing on ice.
7. While aspirating gelatin solution, make sure the wells are not air-dried. Add medium quickly after gelatin aspiration.
8. iPSCs should not be overgrown, to avoid nonspecific and spontaneous differentiation.
9. Typically, for iPSC subculturing, one well of a 6-well plate is split into two fresh 6-well plates. Residual feeder MEFs will be transferred to the next plate; however, they will not grow and populate the wells since they are mitotically inactive.
10. Petri dishes are good alternatives to the 6-well low attachment plates. If needed, EBs can be formed on an orbital shaker placed into the incubator and gently shaking the EB containing petri dishes in a shorter time.
11. This cell concentration has been determined for optimal EB formation. At higher densities, the EBs may form large aggregates.
12. Alternatively, medium can be changed by tilting the cell culture dish and letting it stand at a 45° angle in the hood for several minutes for EBs to collect at the bottom of the dish.
13. Ideally, the plates should be coated freshly before use. The wells should not be air-dried during coating process. Alternatively, plates can be coated, PBS can be added to each well and then plates can be stored at −80 °C. Before use, the plates can be thawed slowly at room temperature, PBS can be aspirated, wells can be rinsed, and cells can be seeded immediately.
14. Exceeding the optimal EB number/well will cause overconsumption of media, and not enough space for differentiation in the wells, due to confluence. This might also cause nonspecific differentiation.

15. Individual outgrowth of cells from attached EBs will be observed during this time. These cells will spread and fill the empty spaces in the wells, and also start adopting neural stem cell-like morphology as attached monolayers.
16. Trypsin should not be kept longer than indicated, as this might cause harm to incipient neural precursors.
17. Residual EB aggregates or clumps sink to the bottom of the tubes, whereas the supernatant contains the NSCs. Therefore, it is important to collect the supernatant here.
18. Scraping the attached cells into suspension should be performed very gently, to avoid the harming of NSC processes.
19. Primary neurospheres will form over this time. The cells will start to proliferate and form spheroids. Occasional cells will attach to the bottom of the flask but they will lift back to suspension in 1–2 days. If they remain attached for longer, this might be a sign of nonspecific differentiation and these cells should be eliminated from the culture. If this is the case, only the neurospheres growing in suspension should be transferred to a new flask.
20. The differentiation ability of neurospheres formed here into neuronal and glial lineages can be characterized and validated as described [14, 16].
21. The frozen iPSC derived NSCs can be restarted into culture and grown for 5–8 passages.

Acknowledgements

This work is supported by FP7 Marie Curie CIG and Koç University School of Medicine.

References

1. Okabe S, Forsberg-Nilsson K, Spiro AC, Segal M, McKay RD (1996) Development of neuronal precursor cells and functional postmitotic neurons from embryonic stem cells in vitro. Mech Dev 59:89–102
2. Conti L, Pollard SM, Gorba T, Reitano E, Toselli M, Biella G, Sun Y, Sanzone S, Ying QL, Cattaneo E, Smith A (2005) Niche-independent symmetrical self-renewal of a mammalian tissue stem cell. PLoS Biol 3:e283
3. Conti L, Reitano E, Cattaneo E (2006) Neural stem cell systems: diversities and properties after transplantation in animal models of diseases. Brain Pathol 16:143–154
4. Ferreira LM, Mostajo-Radji MA (2013) How induced pluripotent stem cells are redefining personalized medicine. Gene 520:1–6
5. Takahashi K, Yamanaka S (2013) Induced pluripotent stem cells in medicine and biology. Development 140:2457–2461
6. Takahashi K, Tanabe K, Ohnuki M, Narita M, Ichisaka T, Tomoda K, Yamanaka S (2007) Induction of pluripotent stem cells from adult human fibroblasts by defined factors. Cell 131:861–872
7. Gonzalez F, Boue S, Izpisua Belmonte JC (2011) Methods for making induced pluripotent stem cells: reprogramming a la carte. Nat Rev Genet 12:231–242

8. Wernig M, Zhao JP, Pruszak J, Hedlund E, Fu D, Soldner F, Broccoli V, Constantine-Paton M, Isacson O, Jaenisch R (2008) Neurons derived from reprogrammed fibroblasts functionally integrate into the fetal brain and improve symptoms of rats with Parkinson's disease. Proc Natl Acad Sci U S A 105:5856–5861
9. An MC, Zhang N, Scott G, Montoro D, Wittkop T, Mooney S, Melov S, Ellerby LM (2012) Genetic correction of Huntington's disease phenotypes in induced pluripotent stem cells. Cell Stem Cell 11:253–263
10. Liu Q, Spusta SC, Mi R, Lassiter RN, Stark MR, Hoke A, Rao MS, Zeng X (2012) Human neural crest stem cells derived from human ESCs and induced pluripotent stem cells: induction, maintenance, and differentiation into functional Schwann cells. Stem Cells Transl Med 1:266–278
11. Wang S, Bates J, Li X, Schanz S, Chandler-Militello D, Levine C, Maherali N, Studer L, Hochedlinger K, Windrem M, Goldman SA (2013) Human iPSC-derived oligodendrocyte progenitor cells can myelinate and rescue a mouse model of congenital hypomyelination. Cell Stem Cell 12:252–264
12. Dimos JT, Rodolfa KT, Niakan KK, Weisenthal LM, Mitsumoto H, Chung W, Croft GF, Saphier G, Leibel R, Goland R, Wichterle H, Henderson CE, Eggan K (2008) Induced pluripotent stem cells generated from patients with ALS can be differentiated into motor neurons. Science 321:1218–1221
13. Matsui T, Akamatsu W, Nakamura M, Okano H (2014) Regeneration of the damaged central nervous system through reprogramming technology: basic concepts and potential application for cell replacement therapy. Exp Neurol 260:12–18
14. Tsuji O, Miura K, Okada Y, Fujiyoshi K, Mukaino M, Nagoshi N, Kitamura K, Kumagai G, Nishino M, Tomisato S, Higashi H, Nagai T, Katoh H, Kohda K, Matsuzaki Y, Yuzaki M, Ikeda E, Toyama Y, Nakamura M, Yamanaka S, Okano H (2010) Therapeutic potential of appropriately evaluated safe-induced pluripotent stem cells for spinal cord injury. Proc Natl Acad Sci U S A 107:12704–12709
15. Miura K, Okada Y, Aoi T, Okada A, Takahashi K, Okita K, Nakagawa M, Koyanagi M, Tanabe K, Ohnuki M, Ogawa D, Ikeda E, Okano H, Yamanaka S (2009) Variation in the safety of induced pluripotent stem cell lines. Nat Biotechnol 27:743–745
16. Onorati M, Camnasio S, Binetti M, Jung CB, Moretti A, Cattaneo E (2010) Neuropotent self-renewing neural stem (NS) cells derived from mouse induced pluripotent stem (iPS) cells. Mol Cell Neurosci 43:287–295
17. Conti L, Cattaneo E (2010) Neural stem cell systems: physiological players or in vitro entities? Nat Rev Neurosci 11:176–187

Methods in Molecular Biology (2016) 1357: 339–351
DOI 10.1007/7651_2014_143

Published online: 23 November 2014

Production of Retinal Cells from Confluent Human iPS Cells

Sacha Reichman and Olivier Goureau

Abstract

Human induced pluripotent stem (hiPS) cells could be used as an unlimited source of retinal cells for the treatment of retinal degenerative diseases. Although much progress has been made in the differentiation of pluripotent stem cells towards different retinal lineages, the production of retinal cells from hiPS cells for therapeutic approaches require the development of easy and standardized protocols. In this chapter, we describe a simple and effective protocol for retinal differentiation of hiPS cells bypassing embryoid body formation and the use of exogenous molecules and substrates. In 2 weeks, confluent hiPS cells cultured in pro-neural medium can generate both retinal pigmented epithelial cells and self-forming neural retina-like structures containing retinal progenitor cells. These progenitors can be differentiated into all retinal cell types, including retinal ganglion cells and precursors of photoreceptors, which could find important applications in regenerative medicine. This differentiation system and the resulting hiPS-derived retinal cells will also offer opportunity to study the molecular and cellular mechanisms underlying human retinal development, and the establishment of in vitro models of human retinal degenerative diseases.

Keywords: Three-dimensional retina, Retinal progenitor, Retinal ganglion cell, Photoreceptor, Human iPS cells, Eye field, Retinal pigmented epithelium, Glaucoma, Retinopathies

1 Introduction

Irreversible blindness caused by retinal diseases is mainly due to the impairment or loss of function of specific retinal cells. The destruction of retinal ganglion cells (RGCs) in glaucoma causes loss of vision in more than five million people worldwide. Retinal diseases cause by the cell death of photoreceptors and/or supporting retinal pigmented epithelium (RPE), such as diabetic retinopathies and age-related macular degeneration (AMD), are increasingly significant causes of incurable sight loss. In addition, inherited retinopathies in which photoreceptors and RPE cells died are other causes of inevitable blindness worldwide. Rescuing the degenerated retina by specific cell replacement is one of the most promising approaches (1, 2), particularly in disease states where there is a significant cell damage. Pluripotent stem cells, like human embryonic stem (hES) cells or human induced pluripotent stem (hiPS) cells, have the ability to be expanded indefinitely in culture and could be used as an unlimited source of retinal cells for the treatment of retinal degenerative diseases. Furthermore, hiPS cells derived from somatic cells of adult individuals constitute a powerful tool to study retinal diseases by the generation of mutated hiPS cells that should recapitulate defined pathogenic steps at the molecular level.

Recent data have indicated that hES and hiPS can be differentiated into RPE cells by different methods (3). Concerning neuroretinal cells, a growing body of convergent data has demonstrated the ability of hES or hiPS cells to be committed after various manipulations of the culture environment into the neural retinal lineage and further differentiated into cells expressing photoreceptor markers (3–5). A majority of these protocols are based on the promotion of eye field territory by inhibition of BMP and Wnt pathways, in combination with activation of IGF signaling from embryoid bodies (EBs) derived from human pluripotent stem cells (6, 7). Innovative 3D protocols have been recently developed with human pluripotent stem cells leading to a successful generation of self-organizing retinal tissue in a dish. These structures obtained from EBs or cell aggregates in the presence of specific exogenous molecules and different substrates, such as Matrigel, resemble optic cup structures (8) or share characteristics of developing optic vesicles (9).

These protocols require multiple steps like EBs formation and selection, addition of inductive molecules and/or Matrigel, as well as EBs coating on adherent substrates and trained handling, which are not always compatible with the manufacturing process for therapeutic approach or drug screening that need a large scale production of cells of interest. We recently developed a protocol, which eliminates all these complicated steps required for retinal differentiation, with the simultaneous generation of hiPS-derived RPE (hiRPE) cells and neural retina (NR)-like structures containing retinal progenitor cells (RPCs) (10). These RPCs can be differentiated into all retinal cell types, including RGCs and photoreceptors (precursors and matures), needed for future regenerative medicine respectively for glaucoma and retinopathies.

The methods described in this chapter outline the basic protocol for the use of hiPS cells to generate NR-like structures, containing RPCs and for further differentiating them into retinal cells. We have also described simple methods to generate and amplify hiRPE cells. Additional information for analyzing commitment and differentiation of hiPS-derived RPCs using RT-PCR and immunohistochemistry are also presented.

2 Materials

2.1 Culture of hiPS Cells

1. hiPS cells were produced by episomal reprogramation (11) of adult human dermal fibroblast primary cell line (10) and cultivated on feeders.

2. Feeders are mitomycin-inactivated mouse embryonic fibroblasts (MEF) derived from CF-1 strain mouse embryos at E13.5 (Zenith Biotech).
3. Fibroblast medium: Dulbecco's Modified Eagle's Medium (DMEM), high glucose, Glutamax II/10 % fetal bovine serum/1 mM sodium pyruvate/1 % MEM Non-Essential Amino Acids (MEM NEAA), 100 U/ml penicillin and 100 μg/ml streptomycin (all from Life Technologies).
4. 0.1 % gelatin solution: Gelatin (Sigma) is dissolved in phosphate buffer saline (PBS) at 0.1 %.
5. The iPS medium is composed by ReproStem medium (ReproCELL) supplemented by 10 ng/ml of human recombinant fibroblast growth factor-2 (FGF-2, Preprotech).
6. 3-cm culture dishes (Corning).

2.2 Generation of NR-Like Structures from hiPS Cells

1. Proneural N2-containing (ProN2) medium is a serum-free medium composed by DMEM/Nutrient Mixture F-12 (DMEM/F-12, 1:1 with L-glutamine), 1 % MEM NEAA, 1 % N2 supplement, 100 U/ml penicillin and 100 μg/ml streptomycin (all from Life Technologies).
2. ReproStem medium.
3. FGF-2.
4. Shaker 3D Nutator (VWR).
5. 24-well plates and 3-cm culture dishes (Corning).

2.3 Generation of hiPS-Derived RPE Cells

1. TrypLE (Life Technologies).
2. ProN2 medium.
3. 24-well plates (Corning).
4. 0.1 % gelatin solution.

2.4 RNA Extraction and TaqMan Assay

1. NucleoSpin RNA XS Kit (Macherey-Nagel).
2. QuantiTech reverse transcription kit (Qiagen).
3. TaqMan Gene expression Master Mix (Life Technologies).
4. Primers and MGB probes labeled with FAM (Life Technologies, Table 1).
5. PCR 96-well reaction plates for TaqMan Gene expression Assay (Life Technologies) (*see* **Note 1**).
6. DNA/RNAse-free water (Life Technologies).

2.5 Inclusion

1. 4 % Paraformaldehyde solution (*see* **Note 2**).
2. Sucrose solution: Sucrose at 30 % (w/v) in PBS (*see* **Note 3**).
3. Plastic molds.

Table 1
List of TaqMan® gene expression ID assays used for qRT-PCR

Gene symbols	Assays IDs (LifeTechnologies)
18S	18S-Hs99999901_s1
BEST1	BEST1-Hs00188249_m1
BLUE OPSIN	OPN1SW-Hs00181790_m1
BRN3A	POU4F1-Hs00366711_m1
BRN3B	POU4F2-Hs00231820_m1
CALRETNIN	CALB2-Hs00242372_m1
CONE ARRESTIN	ARR3-Hs00182888_m1
CRX	CRX-Hs00230899_m1
GAD2	GAD2-Hs00609534_m1
GLAST1	SLC1A3-Hs00188193_m1
LHX2	LHX2-Hs00180351_m1
LIM1	LHX1-Hs00232144_m1
MERTK	MERTK-Hs01031973_m1
MITF	MITF-Hs01117294_m1
NEUROD1	NEUROD1-Hs00159598_m1
NRL	NRL-Hs00172997_m1
PAX6	PAX6-Hs00240871_m1
PDEF	SERPINF1-Hs01106934_m1
PKCα	PRKCA-Hs00925195_m1
R/G OPSIN	OPN1MW-Hs00241039_m1
RAX	RAX-Hs00429459_m1
RECOVERIN	RCVRN-Hs00610056_m1
RHO	RHO-Hs00892431_m1
RPE65	RPE65-Hs01071462_m1
SIX3	SIX3-Hs00193667_m1
VSX2	VSX2-Hs00766959_s1

4. Freezing solution: Sucrose at 30 % (w/v) and gelatin at 7.5 % (w/v) in PBS (*see* **Note 4**).
5. Isopentane.
6. Dry ice.

Table 2
List of primary antibodies used for immunohistochemistry analysis

Antigen	Species	Dilution	Source
AP2	Mouse monoclonal	1:100	DSHB
BRN3A	Mouse monoclonal	1:250	Millipore
CALRETININ	Mouse monoclonal	1:500	Abcys
CONE ARRESTIN	Rabbit polyclonal	1:2,000	Millipore
CRX	Mouse monoclonal	1:5,000	Abnova
GLUTAMINE SYNTHETASE	Mouse monoclonal	1:500	Millipore
KI67	Mouse monoclonal	1:200	BD Pharmagen
LIM1 (LHX1)	Mouse monoclonal	1:20	DSHB
LHX2	Goat polyclonal	1:100	Santa Cruz
MITF	Mouse monoclonal	1:200	DAKO
OPSIN G/R	Rabbit polyclonal	1:5,000	Millipore
OTX2	Rabbit polyclonal	1:5,000	Millipore
PAX6	Rabbit polyclonal	1:1,000	Millipore
PKCα	Rabbit polyclonal	1:5,000	Santa Cruz
RAX/RX	Rabbit polyclonal	1:10 000	Abcam
RHODOPSIN	Mouse monoclonal	1:250	From Dr R Molday
RECOVERIN	Rabbit polyclonal	1:2,000	Millipore
SOX9	Rabbit polyclonal	1:1,000	Millipore
VSX2 (CHX10)	Goat polyclonal	1:2,000	Santa Cruz
ZO1	Rabbit polyclonal	1:250	Life Technologies

2.6 Immunostaining

1. PBS.
2. Blocking solution: PBS/0.2 % gelatin (w/v)/0.25 % Triton ×100.
3. Washing solution: PBS/0.1 % Tween-20.
4. Primary antibodies (Table 2).
5. Alexa Fluorophore conjugated secondary antibodies (Molecular Probes/Life Technologies).
6. 4′,6-diamidino-2-phenylindole (DAPI): 100 μg/ml in H_2O.
7. Slide mounting: Gel mount medium and glass coverslips.

3 Methods

3.1 Maintenance and Expansion of hiPS Cells

Human iPS cells are cultured on feeders in the iPS medium in a 37 °C incubator in 5 % CO_2.

1. Coat 3-cm culture dishes with 1 ml of 0.1 % gelatin solution and leave for 15 min at room temperature.
2. Remove the gelatin solution and plate feeders at 2×10^4 cells/cm^2 the day before hiPS cell thawing.
3. Thaw a cryogenic sample of hiPS cells by quickly transferring from liquid nitrogen tank to water bath at 37 °C during 30 s.
4. Transfer hiPS cells from the cryotube to 15 ml tube containing 3 ml of prewarmed and pre-equilibrated iPS medium.
5. Centrifuge tube 3 min at $110 \times g$.
6. Remove the supernatant.
7. Resuspend the cells with 500 μl of conditioned iPS medium from a 3-cm culture dish containing feeders and retransfer gently the hiPS cells to the culture dish. Put in incubator at 37 °C and 5 % CO_2 (*see* **Note 5**).
8. Change medium every day and pass the hiPS cells every week.

3.2 Generation of NR-Like Structures from hiPS Cells

1. Expand hiPS cells to confluence in iPS medium in standard culture condition as described above. At the confluence, the time status is noted as Day 0 (D0, Fig. 1a, b) (*see* **Note 6**).
2. At D0, hiPS cells were placed in iPS medium without FGF-2, allowing spontaneous differentiation.
3. At D2, remove medium and add fresh ProN2 medium previously warm and equilibrate in the incubator at 37 °C and 5 % CO_2.
4. Change the medium every 2 other days.
5. On D14, emergent self-forming NR-like structures are distinguishable by neuroepithelium buds surrounded by pigmented cells (Fig. 1c).

3.3 Maturation of NR-Like Structures by Floating Cultures

At D14, NR-like structures essentially contain RPCs, which coexpress PAX6 and RX (10). These RPCs can be differentiated and maturated into the seven neuroretinal cell types in serum-free ProN2 medium by floating cultures.

1. At D14, emergent NR-like structures with surrounded pigmented cells are isolated manually with a needle.
2. Each NR-like structure is then transferred in a single well of 24-well plates and cultured as a floating structure in 1 ml of ProN2 medium supplemented with 10 ng/ml of FGF-2 to favor neuroretinal development.

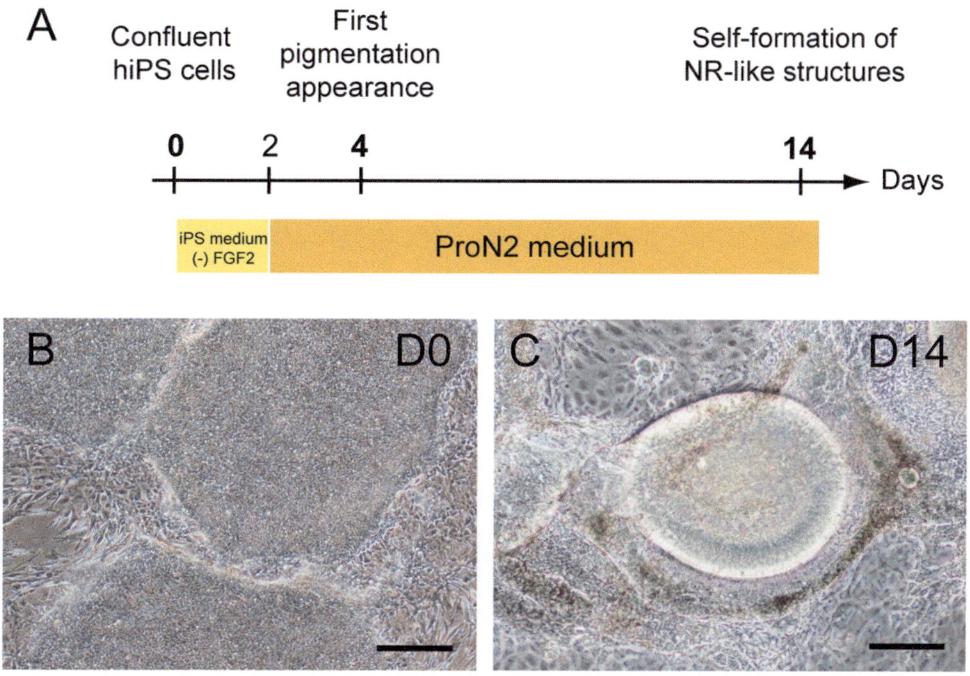

Fig. 1 Generation of self-forming NR-like structures and hiRPE from hiPS cells. (**a**) Diagram of the culture conditions used for generating NR-like structures from confluent hiPS cells. (**b**) Confluent hiPS cells at D0. (**c**) Emergent NR-like structure at D14 with a typical ring of pigmented cells. Scale bars: 100 μm

3. Put the 24-well plates on a 3D Nutator shaker in a 37 °C incubator in 5 % CO_2 during the 2 first days (from D14 to D16) to avoid the adhesion of NR-like structures to the bottom of the well. Change half of the medium every 2 other days.
4. At D21, remove half of the medium and add fresh ProN2 medium without FGF-2.
5. Change half of the medium every 2 other days.
6. Keep the NR-like structures in floating culture the time required to obtain retinal cell types needed (Fig. 2).

3.4 Generation of hiPS-Derived RPE

Simultaneously to the generation of NR-like structures, hiPS-derived retinal pigmented epithelium (hiRPE) patches first appear as soon as D4. But, it is only between D14 to D21 that hiRPE patches can be easily isolated and amplified.

1. Coat each well from 24-well plates with 500 μl of 0.1 % gelatin solution and leave for 15 min at room temperature (*see* **Note 7**).
2. Remove the gelatin solution and add 1 ml of ProN2 medium by well and put plates in the incubator at 37 °C/5 % CO_2 to warm and equilibrate the medium.
3. Between D14 and D21, in the 3-cm culture dishes containing differentiated hiPS cells cut, with a needle, hiRPE patches.

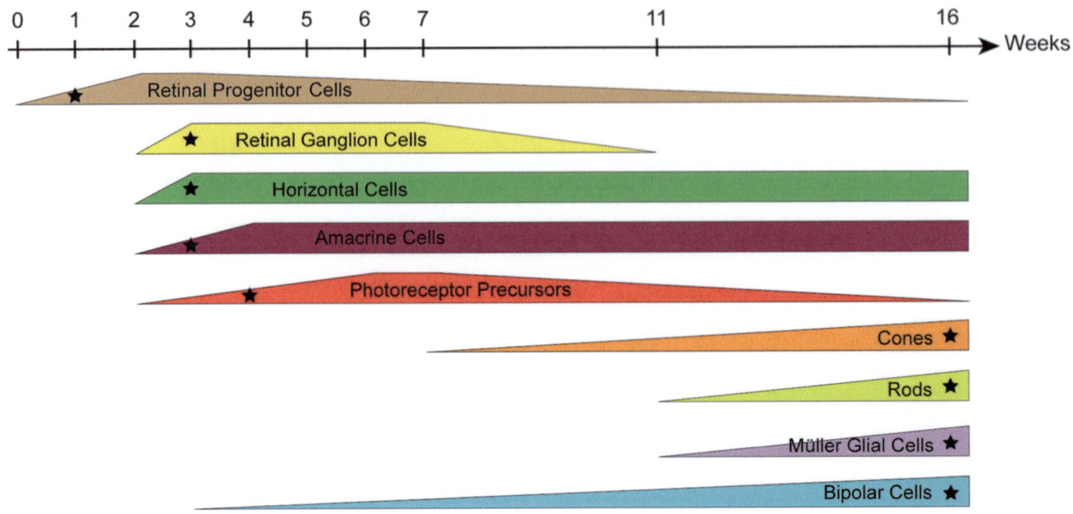

Fig. 2 Temporal expression of neuroretinal markers in differentiating hiPS-derived NR-like structures. *Colored blocks* represent evolution of each specific retinal cell type evaluated by RT-qPCR and immunohistochemistry. *Black stars* indicate the time when each specific retinal cell was undoubtedly detected by immunofluorescence

4. Isolate and transfer by pipetting patches in one well of 24-well plate previously coated with gelatin. Put back the plates in the incubator at 37 °C/5 % CO_2 (*see* **Note 8**).
5. Change ProN2 medium every 2 other days.
6. At confluence, remove the medium and wash the plate once with PBS.
7. Added 200 μl of TrypLE to each well and incubate a minimum of 5 min at 37 °C/5 % CO_2.
8. Dissociated mechanically the sheet of hiRPE cells by up- and downpipetting.
9. Add 800 μl of ProN2 medium to stop TrypLE activity by dilution as recommended by supplier.
10. Place the cell suspension in a 15 ml tube and centrifuge 5 min at 110 × *g*.
11. Remove the supernatant and gently resuspend the cells in 3–4 ml of prewarmed and pre-equilibrated ProN2 medium.
12. Distribute 1 ml in another gelatin-coated 24-well plate to split the cells at the ratio of 1:3–1:4 (*see* **Note 8**).
13. Change media every 2 other days until hiRPE cells reach confluence and perform the next passage (*see* **Notes 9** and **10**).

3.5 RNA Extraction and Analysis of Retinal Differentiation Using TaqMan Assay

All the RNAs are extracted using Kit according to the manufacturer's protocol. Use the human housekeeping gene *18S* to normalize results (Table 1). Quantification of gene expression is based on the DeltaCt Method in three minimum independent biological experiments adapted to Reichman et al. (10).

1. For NR-like structures, collect a minimum of ten structures in 200 μl of lysis buffer supplemented by 4 μl of reducing agent TCEP supplied whit kit. Homogenize the lysate by up- and downpipetting. For hiPS-derived RPE cells, remove the medium, wash once with PBS, and add directly on cells the 200 μl of lysis buffer previously supplemented with 4 μl of TCEP per well. Homogenize the lysate by up- and downpipetting.
2. Extract RNA according to the manufacturer's protocol (*see* **Note 11**).
3. Check RNA yields and quality by spectrophotometer and agarose gel.
4. Take 500 ng of fresh RNA sample.
5. Synthesize cDNA using the QuantiTech reverse transcription kit following the manufacturer's recommendations.
6. Dilute synthesized cDNA at 1:20 in DNA/RNAse-free water.
7. Make the reaction mix for qPCR TaqMan® Assay as follows:

   Using TaqMan® Gene Expression Assays (*see* **Note 1**), for one well mix.

   9 μl of diluted cDNA (1:20).

   1 μl of TaqMan® Gene Expression Assay (20×).

   10 μl of TaqMan® Gene expression Master Mix (2×).
8. Run samples following standard condition of the manufacturer's recommendations [1 × 2 min at 50 °C; 1 × 10 min at 95 °C; 40×(1 × 15 s at 95 °C; 1 × 1 min at 60 °C)].
9. Primers of specific genes (Table 1) were used to characterize each differentiating retinal cell types (10): (a) eye-field identity (RPCs): *SIX3, LHX2, MITF, RAX, PAX6,* and *VSX2* (also known as *CHX10*); (b) retinal ganglion cells: *BRN3A, BRN3B,* and *CALRETININ*; (c) amacrine cells: *GAD2* and *CALRETININ*; (d) horizontal cells: *LIM1* and *CALRETININ*; (e) cone photoreceptor cells: *NEUROD1 RECOVERIN, CRX, BLUE OPSIN, RED/GREEN OPSIN,* and *CONE ARRESTIN;* (f) rod photoreceptor cells, *NRL, NEUROD1, RECOVERIN, CRX,* and *RHODOPSIN*; (g) bipolar cells, *PKCα*, (h) Müller glial cells: *GLAST1* and (i) RPE cells: *MERTK, RPE65, BEST1,* and *PEDF.*
10. Figure 2 illustrates the time of appearance of specific cell types in hiPS-derived NR-like structures cells.

3.6 Immunohistochemistry Analysis

To analyze retinal differentiation by immunohistochemistry, NR-like structures at different time of maturation are fixed, included in gelatin block, freezed and sliced with cryostat (12). Each slice is dropped off on microscope slides in series.

1. Fix NR-like structures in 4 % paraformaldehyde solution for 15 min at 4 °C.
2. Rinse structures twice with PBS.
3. Put one retinal structure in 1.5 ml tube containing 1 ml of sucrose solution during at least 2 h.
4. Embed structures in plastic molds using freezing solution prewarmed at 37 °C. Cool molds at room temperature and place them at 4 °C in humid chamber at least 2 h.
5. Remove embedded structures from plastic molds and freeze them by immersion in an isopentane bath at −50 °C during 1 min (*see* **Note 12**).
6. Collect 10-μm-thick cryosections on microscope slides and stock at −20 °C.
7. Perform immunofluorescence staining by first incubating slides in PBS at 37 °C for 5–10 min to remove gelatin.
8. Incubate with blocking solution for 1 h at room temperature.
9. Incubate with primary antibody solution overnight at 4 °C. Antibodies are diluted to the suggested working concentration (Table 2) in the blocking solution.
10. Wash slides three times 5 min each in washing solution.
11. Incubate with secondary antibody solution for 1 h at room temperature. Antibodies are diluted at 1:600 in blocking solution.
12. Wash slides two times 5 min each in washing solution.
13. Incubate slides for 5 min at room temperature, with DAPI diluted in washing solution at 1:10,000.
14. Wash slides three times 5 min each in PBS.
15. Mounted in Fluoromount-G (Southern Biotech).
16. Antibodies directed against selected proteins (Table 2) were used to characterize each specific differentiating retinal cell type (10): (a) RPCs: PAX6, RAX, KI67, LHX2; (b) retinal ganglion cells: BRN3A; (c) amacrine cells: AP2 and CALRETININ; (d) horizontal cells: LIM1 and CALRETININ; (e) cone photoreceptor cells: RECOVERIN, CRX, RED/GREEN OPSIN, and CONE ARRESTIN; (f) rod photoreceptor cells: RECOVERIN, CRX, and RHODOPSIN; (g) bipolar cells,: PKCα; (h) Müller glial cells: GLUTAMINE SYNTHASE and SOX9; and (i) RPE cells: MITF and ZO1.
17. Examples of immunofluorescence staining are shown in Fig. 3.

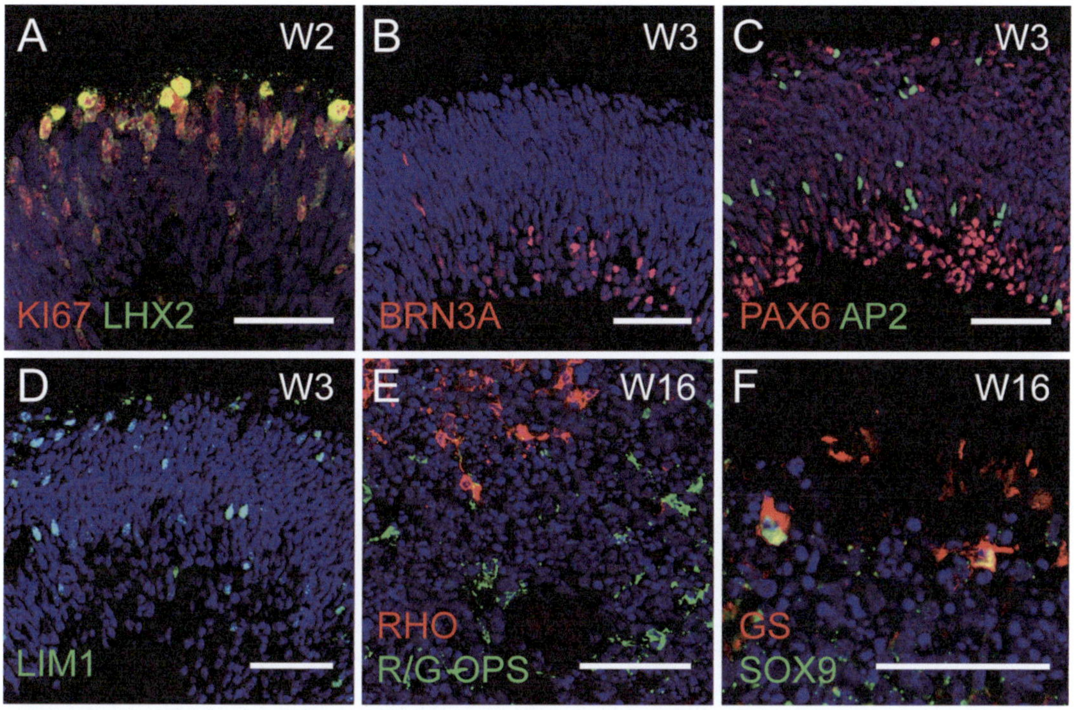

Fig. 3 Immunohistochemical analysis of cryosectioned NR-like structures after different weeks of differentiation. (**a**) RPCs triggered by Ki67 and LHX2 immunolabeling. (**b**) BRN3A immunostaining identifying retinal ganglion cells. (**c**) Laminar organization of retinal cells in NR-like structures showed by gradient of PAX6-positive cells and amacrine cells emergence by AP2 staining. (**d**) LIM1-positive cells corresponding to differentiating horizontal cells. (**e**) RHODOPSIN (RHO) and Red/Green (R/G) OPSIN staining reflecting the maturation of rod and cone photoreceptors respectively. (**f**) Müller glial cells observation by staining with glutamine synthetase (GS) and SOX9 markers. Scale bars: 50 μm week (W)

4 Notes

1. PCR can be performed also with TaqMan® Array fast assay technology using 96-well fast plates containing preloaded primers. In this case, for each well, the PCR mix is composed by 5 μl of diluted cDNA (1:20) and 5 μl of TaqMan® Gene expression Master Mix (2×, Life Technologies). Use the standard condition of the manufacturer's recommendations: 1 × 2 min at 50 °C; 1 × 10 min at 95 °C; 40 × (1 × 15 s at 95 °C; 1 × 1 min at 60 °C).
2. Aliquots of 4 % paraformaldehyde solution can be stored at −20 °C for up to 6 months.
3. Aliquots of sucrose solution can be stored at −20 °C for up to 6 months.
4. Aliquots of freezing solution can be stored at −20 °C for up to 6 months.

5. During thawing hiPS cells, a rock inhibitor, such as Y-27632 at 10 μM, can be added to the pre-equilibrated iPS medium in 3-cm culture dishes to reduce cell apoptosis.
6. Depending on the growth rate of each hiPS cell line, the state of confluence is generally obtained around 2 weeks without morphological sign of differentiation (*see* Fig. 1b).
7. Gelatin, PolyD-Lysin plus Laminin, or Matrigel can be used to coat 24-well plates for hiRPE cell cultures (3).
8. For hiRPE cell culture, wait a minimum of 24 h before moving the plate after passages to allow adhesion of pigmented patches to the bottom of the well.
9. Keep plates between 3 and 5 weeks in culture to reach confluence.
10. The hiRPE cells can be amplified until passage 3–4 without epithelial–mesenchymal transition (EMT) in our conditions. Nevertheless, recent publication showed that addition of rock inhibitors allowed to extend the number of passage of hiPS-derived RPE cells preventing EMT (13).
11. To obtain cDNA with QuantiTech reverse transcription kit (Qiagen), 500 ng of total RNA, in a maximum of 12 μl, is required.
12. The −50 °C freezing temperature is obtained by using an adapted container containing isopentane surrounded by dry ice. Do not exceed −60 °C during freezing and store blocks at −80 °C before sectioning.

Acknowledgements

The authors would like to thank the members of the Goureau lab for their input during the setting-up of the methods described in this chapter and J-A Sahel for his continuous support. This work was supported by the ANR [GPiPS: ANR-2010-RFCS005], by French state funds managed by the ANR within the Investissements d'Avenir programme [ANR-11-IDEX-0004-02] in the frame of the LABEX LIFESENSES [ANR-10-LABX-65], and by Regional Council of Ile-de-France.

References

1. Ramsden CM, Powner MB, Carr A-JF et al (2013) Stem cells in retinal regeneration: past, present and future. Development 140:2576–2585
2. Pearson RA (2014) Advances in repairing the degenerate retina by rod photoreceptor transplantation. Biotechnol Adv 32:485–491
3. Rowland TJ, Buchholz DE, Clegg DO (2012) Pluripotent human stem cells for the treatment of retinal disease. J Cell Physiol 227:457–466
4. Borooah S, Phillips MJ, Bilican B et al (2013) Using human induced pluripotent stem cells to treat retinal disease. Prog Retin Eye Res 37:163–181

5. Jin Z-B, Takahashi M (2012) Generation of retinal cells from pluripotent stem cells. Prog Brain Res 201:171–181
6. Reh TA, Lamba D, Gust J (2010) Directing human embryonic stem cells to a retinal fate. Methods Mol Biol 636:139–153
7. Meyer JS, Shearer RL, Capowski EE et al (2009) Modeling early retinal development with human embryonic and induced pluripotent stem cells. Proc Natl Acad Sci U S A 106:16698–16703
8. Nakano T, Ando S, Takata N et al (2012) Self-formation of optic cups and storable stratified neural retina from human ESCs. Cell Stem Cell 10:771–785
9. Meyer JS, Howden SE, Wallace KA et al (2011) Optic vesicle-like structures derived from human pluripotent stem cells facilitate a customized approach to retinal disease treatment. Stem Cells 29:1206–1218
10. Reichman S, Terray A, Slembrouck A et al (2014) From confluent human iPS cells to self-forming neural retina and retinal pigmented epithelium. Proc Natl Acad Sci U S A 111:8518–8523
11. Yu J, Hu K, Smuga-Otto K et al (2009) Human induced pluripotent stem cells free of vector and transgene sequences. Science 324:797–801
12. Orieux G, Picault L, Slembrouck A et al (2014) Involvement of Bcl-2-associated transcription factor 1 in the differentiation of early-born retinal cells. J Neurosci 34:1530–1541
13. Croze RH, Buchholz DE, Radeke MJ et al (2014) ROCK inhibition extends passage of pluripotent stem cell-derived retinal pigmented epithelium. Stem Cells Transl Med 3:1066–1078

Methods in Molecular Biology (2016) 1357: 353–374
DOI 10.1007/7651_2014_142

Published online: 03 December 2014

Differentiation of iPSC to Mesenchymal Stem-Like Cells and Their Characterization

Kim Hynes, Danijela Menicanin, Stan Gronthos, and Mark P. Bartold

Abstract

Mesenchymal stem cells (MSC) are a unique population of adult stem cells that have the capacity to differentiate into numerous cell types as well as the ability to modulate the immune system. As such, MSC represent a promising stem cell population for use in the clinical treatment of a range of disorders involving tissue regeneration as well as the immune system. The lack of accessibility to MSC is currently limiting the use of MSC in mainstream clinical treatment strategies. It is therefore imperative for the future success of stem cell-based treatment approaches that are more reliable, and accessible sources of MSC are identified. The present chapter describes a method for generating MSC-like cells from induced pluripotent stem cells (iPSC), with equivalent growth and functional properties to parental MSC populations.

Keywords: Mesenchymal stem cells, Induced pluripotent stem cells, Differentiation, Characterization, Serial passaging

1 Introduction

Mesenchymal stem cells (MSC) have generated considerable interest in recent years due to their potential for use in clinical applications to treat a wide range of diseases (1). MSC possess unique biological properties which make them highly amenable for use in immunotherapy and regenerative medicine. Specifically, MSC have the capacity to differentiate into various tissues of mesodermal origin (2), making them well suited to use in regenerative therapies. Furthermore, MSC are capable of modulating humoral and cellular immune responses (3) through the secretion of anti-inflammatory molecules, meaning that they are ideally suited for use in the treatment of a range of immune system-based disorders. Finally, MSC do not express class II antigens or their co-stimulatory molecules (4) which allow the cells to evade the host's immune system allowing them to be used in allogeneic transplantation settings.

A search of the clinical trials database (https://clinicaltrials.gov) using the search term "mesenchymal stem cells" reveals 416 current registered clinical trials involving MSC. The clinical uses of MSC being investigated encompass a wide range of indications, including but not limited to musculoskeletal disease, liver disease, autoimmune diseases, cardiovascular disease, rheumatoid arthritis, type I diabetes, systemic lupus erythematosus, and a range of neurodegenerative diseases such as multiple sclerosis, Alzheimer disease, and Parkinson's disease (5).

As the interest in MSC grows, regarding their use in clinical treatment strategies, identification of reliable and accessible sources of these cells is becoming a critical issue. MSC were first identified within bone marrow aspirates, where MSC-like populations have been identified in a range of additional tissues including adipose tissue (6), umbilical cord blood (7, 8), placenta (9), umbilical cord Wharton's jelly (10, 11), a range of dental tissues (12–18), prostate (19), lung (20), intestinal tissue (21), eye (22), nasal mucosa (23), and knee joint (24–27). While MSC-like cells have now been isolated from a wide range of tissues, it is still a major challenge to obtain sufficient quantities of cells for widespread clinical use. Ex vivo expanded MSC-like populations from different tissues have a varied but limited (50–80 population doublings) capacity to proliferate in vitro, thereby restricting the number of cells which can be obtained. Identifying new and more readily expandable sources of MSC-like cells is critical to the future success of MSC-based treatment approaches.

The discovery that induced pluripotent stem cells (iPSC) can be generated from adult somatic cells represents a promising alternative for obtaining larger populations of adult stem cells for use in regenerative medicine and stem cell-based therapies. iPSC were first generated in 2006 through groundbreaking work which established that adult cells could be reprogrammed back to a pluripotent state through the forced expression of a cocktail of four transcription factors: *Oct3/4*, *Sox2*, *c-Myc*, and *Klf4* (28). While the discovery of iPSC was first established in mouse cells, it was quickly replicated using human cells (29, 30).

IPSC possess distinct advantages over MSC as they can be generated from any tissue type in the body, and more importantly, they have an unlimited proliferation capacity so could potentially serve as an inexhaustible source of stem cells for use in the clinic. While iPSC have a distinct advantage with their high proliferative capacity compared to MSC-like cells, they also have distinct disadvantages as their traits of self-renewal and pluripotency also result in genetic instability and tumorigenicity, dramatically reducing their clinical utility.

Recently, a number of research groups have been able to successfully differentiate iPSC into MSC-like cells through a range of methods (31–40). The resulting MSC-like cells derived from iPSC (iPSC-MSC) are emerging as a promising new stem cell population for use in clinical stem cell treatment approaches as they possess traits which are advantageous over traditional MSCs and iPSC. Specifically, iPSC-MSC can be generated from readily accessible tissue sources. Additionally, iPSC-MSC have a greater proliferative capacity than traditional BM-MSC; iPSC-MSC are capable of proliferating for approximately 40 passages (120 population doublings) without the onset or replicative senescence or loss of plasticity (37). Finally, iPSC-MSC appear more genetically stable than the parental iPSC and do not exhibit the potential to form tumors (39–42), making iPSC-MSC a potentially safer alternative for use in a clinical setting.

The aim of this methods paper is to describe a simplified and reproducible method for generating MSC-like cells from iPSC. The method described in this manuscript utilizes a serial passaging protocol to establish iPSC-MSC and was modified from a protocol previously used to generate MCS-like cells from embryonic stem cells (41). We have previously used the method described below to successfully generate iPSC-MSC from iPSC lines derived from four different somatic tissues (35, 36).

2 Materials

2.1 Induction of iPSC to MSC-Like Cells

1. Gelatin—0.1 % working stock made up in sterile water (gelatin from bovine skin type A, Sigma-Aldrich Cat# G1890) (*see* **Note 1**).
2. 0.22 μm Millex-GV syringe filter (Millipore Cat# SLGV025NB).
3. Dulbecco's phosphate buffered saline (1× PBS—Sigma-Aldrich Cat# D8537).
4. Collagenase I (4 mg/mL stock solution in DMEM—Dulbecco's Modified Eagle's Medium, Sigma-Aldrich Cat# D6546) (*see* **Note 2**).
5. MSC wash media: consisting of αMEM (Minimum Essential Medium Eagle—alpha modification, Sigma-Aldrich Cat# M4526) base medium supplemented with 5 % (v/v) fetal bovine serum (FBS—JRH Biosciences Cat# 12003) and 50 U/mL penicillin and 50 μg/mL streptomycin (penicillin-streptomycin, Sigma-Aldrich Cat# P4333).
6. 14 mL polypropylene round-bottom tubes (BD Falcon Cat# 352059).
7. MSC complete medium: consisting of αMEM base medium supplemented with 10 % (v/v) FBS, 2 mM L-glutamine (L-glutamine, Sigma-Aldrich Cat# G7513), 100 μM L-ascorbic acid phosphate (Wako Cat# 013-12061), 1 mM sodium pyruvate (Sigma-Aldrich Cat# S8636), 50 U/mL penicillin and 50 μg/mL streptomycin, MEM nonessential amino acids (NEAA—Gibco Cat# 1140-050), and HEPES solution (Sigma-Aldrich Cat# H0887) (*see* **Note 3**).
8. TrypLE™ Select 1× (Gibco Cat# 12563-029).

2.2 Initial Flow Analysis

1. Blocking buffer: consisting of Hanks' balanced salt solution (HBSS—Sigma-Aldrich Cat# H9394) supplemented with 5 % (v/v) FBS, 1 % bovine serum albumin (BSA), 50 U/mL penicillin, 50 μg/mL streptomycin, and 5 % normal human serum (SA Pathology, SA) (*see* **Note 4**).
2. 5 mL polypropylene round-bottom flow tubes (Interpath Services Cat# P7512UU).
3. HHF wash medium: consisting of HBSS supplemented with 5 % (v/v) FBS, 50 U/mL penicillin, and 50 μg/mL streptomycin (*see* **Note 5**).
4. FACS fix solution: consisting of PBS supplemented with 1 % (v/v) formalin (ACE Chemicals Company), 0.1 M D-glucose (Calbiochem Cat# 346351), and 0.02 % sodium azide (Scharlau Cat# SO00910100).

2.3 Osteogenic Induction

1. 24-well tissue culture plates (Costar Cat# 3524).
2. Osteogenic induction medium: αMEM supplemented with 5 % FBS, 100 μM L-ascorbate-2-phosphate, 1 mM sodium pyruvate, 50 U/mL penicillin, 50 μg/mL streptomycin, 2 mM L-glutamine, dexamethasone 10^{-7} M (Hospira), and 1.8 mM inorganic phosphate (*see* **Note 6**).
3. Osteogenic control medium: αMEM supplemented with 5 % FBS, 100 μM L-ascorbate-2-phosphate, 1 mM sodium pyruvate, 50 U/mL penicillin, 50 μg/mL streptomycin, and 2 mM L-glutamine (*see* **Note 6**).
4. 1× PBS.
5. 10 % neutral buffered formalin (10 % formaldehyde, ACE Chemicals Company).
6. Alizarin red stain: consisting of 2 g of Alizarin red S (Sigma-Aldrich Cat# A5533) resuspended in 100 mL of RO water (*see* **Note 7**).
7. 96-well flat bottom tissue culture plates (Costar Cat# 3596).
8. 0.6 M HCL (Chem Supply Cat# HT020).
9. 96-well flat bottom, white polystyrene assay plate (Costar Cat# 3912).
10. Calcium Arsenazo III Reagent Set (Pointe Scientific Cat# C7529).
11. Calcium standards (made up from calcium chloride Scharlau Chemie Cat# CA01940500).
12. Invitrogen Quant-iT Pico Green dsDNA Assay Kit (Molecular Probes, Life Technologies Cat# P11496).
13. Proteinase K (Qiagen Cat# 19131).
14. Black polystyrene microtiter plate (Corning Cat# 3915).

15. 6-well tissue culture plates (Costar Cat# 3516).
16. TRIzol Reagent (Ambion/RNA, Life Technologies Cat# 15596018).

2.4 Adipogenic Induction

1. 24-well tissue culture plates.
2. Adipogenic induction medium: αMEM supplemented with 10 % FBS, 100 μM L-ascorbate-2-phosphate, 1 mM sodium pyruvate, 50 U/mL penicillin, 50 μg/mL streptomycin, 2 mM L-glutamine, dexamethasone 0.1 μM, and 60 μM indomethacin (Sigma-Aldrich Cat# I8280) (*see* **Note 8**).
3. Adipogenic control medium: αMEM supplemented with 10 % FBS, 100 μM L-ascorbate-2-phosphate, 1 mM sodium pyruvate, 50 U/mL penicillin, 50 μg/mL streptomycin, and 2 mM L-glutamine (*see* **Note 8**).
4. 1× PBS.
5. 10 % neutral buffered formalin (10 % Formaldehyde, ACE Chemicals Company).
6. Oil-Red-O stain: contains 0.5 g of Oil-Red-O stain (INC Biochemicals Cat# 155984) in 100 mL of isopropanol (propan-2-OL, Analytical Reagent) and 67 mL RO water.
7. Nile Red staining solution: contains 50 μg/mL Nile Red (Sigma-Aldrich Cat# N3013) and dissolved in DMSO (dimethyl sulfoxide, Chem Supply Cat# DA013-P). Immediately prior to staining, dilute the stock Nile Red solution to a working concentration of 25 ng/mL in 1× PBS.
8. 96-well flat bottom tissue culture plates.
9. 1× PBS.

2.5 Chondrogenic Induction

1. 10 mL conical polypropylene tubes.
2. Chondrogenic medium: DMEM, high glucose medium supplemented with 1× ITS + Premix, 100 μM L-ascorbate-2-phosphate, 50 U/mL penicillin, 50 μg/mL streptomycin, 2 mM L-glutamine, dexamethasone 10^{-5} M, and 0.125 % BSA.
3. TGFβ3 (recombinant human TGF-β3, R&D Systems Cat# 243-B3-002) (*see* **Note 9**).
4. 1× PBS.
5. Collagenase/dispase: 6 mg/mL collagenase type I and 8 mg/mL dispase II (Gibco Cat# 17105-041) made up in HANKs base medium.
6. TRIzol Reagent (Ambion/RNA, Life Technologies Cat# 15596018).
7. MSC wash medium.
8. Xylene (xylene—sulfur-free, Chem Supply).

9. Ethanol (ethanol—denatured, Chem Supply).
10. 0.5 % H_2O_2 (Chem Supply).
11. Anti-collagen II monoclonal antibody type II (Chemicon International MAB1330).
12. Goat anti-mouse IgG biotin (Southern Biotech).
13. Streptavidin HRP.
14. DAB substrate-chromogen (Dako Glostrup, Denmark).
15. Hematoxylin (ProSciTech Cat# C1071).
16. Leica CV MounT—mounting medium (Leica Cat# 14046430011).
17. 96-well flat bottom tissue culture plates.
18. Papain digest solution (final concentration of 20 U/mL): dissolve 62.1 mg of papain (papain from Carica papaya, 3.1 U/mg) (Sigma-Aldrich Cat# 76220) in 333 μL of 3 M sodium acetate (Chem Supply), 48 μL of 0.5 M EDTA (ethylenediaminetetraacetic acid—Chem Supply), 1 mL of 200 mM of *N*-acetyl-L-cysteine (Sigma-Aldrich Cat# A7250), and 8.619 mL of RO water.
19. Invitrogen Quant-iT Pico Green dsDNA Assay Kit (Molecular Probes, Life Technologies Cat# P11496).
20. $^{35}SO_4$ (sulfur-35 radionuclide—PerkinElmer Life Sciences NEX042005MC).
21. CPC solution: 1 g cetylpyridinium chloride (Sigma-Aldrich Cat# C0732), 16 mL ethanol, and 24 mL water.
22. CSA stock solution (5 mg/mL solution): 50 mg chondroitin sulfate A (Sigma-Aldrich Cat# C4382) reconstituted in 10 mL RO water.
23. Glass fiber filter (PerkinElmer Cat# 6005422).
24. Microscint™ (PerkinElmer Life and Analytical Sciences Cat# 6013621).

3 Methods

3.1 Induction of iPSC to MSC-Like Cells

The induction of iPSC to MSC-like cells involves two major steps; the first is a 2-week differentiation stage and the second step involves multiple rounds of serial passaging in order to select for the MSC-like cells and select against undifferentiated pluripotent cells. Figure 1 demonstrates the typical cell morphology present at various stages through the differentiation process.

3.1.1 Differentiation of iPSC

1. The differentiation procedure requires the use of gelatin-coated T-25 culture flasks, which need to be preprepared. To do this,

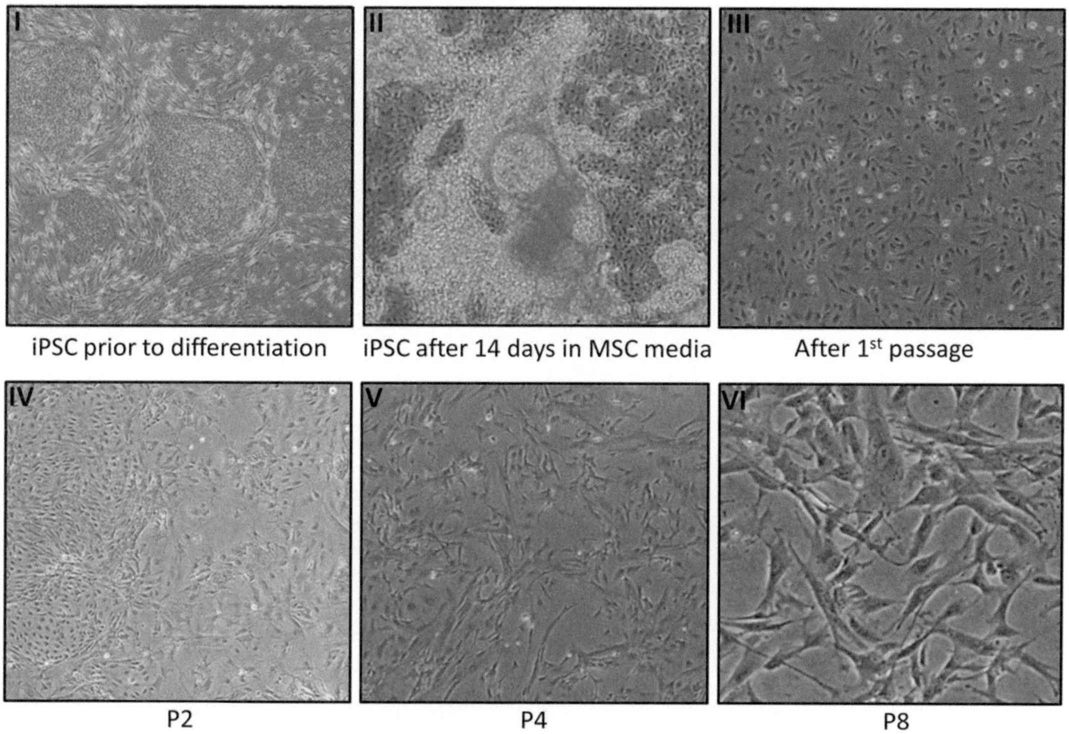

Fig. 1 Cell morphology changes that occur during the induction of iPSC to MSC-like cells. Representative light microscopy images demonstrating the cell morphology changes that occur during the differentiation of iPSC into MSC-like cells. (I) Representative cell morphology of iPSC colonies prior to induction. (II) A representative image of the cell morphology of iPSC after they have been cultured in MSC complete medium for 14 days. (III) Typical appearance of the differentiated iPSC cells once they have been passaged onto gelatin-coated flasks after they have been growing in MSC complete medium for 14 days. (IV–VI) Cell morphology across different stages of induction, namely, passage 2 (IV), passage 4 (V), and passage 8 (VI)

add 5 mL of 0.1 % sterile gelatin to a T-25 culture flask and place in a 37 °C incubator for the gelatin to adhere for at least 30 min. Remove the excess gelatin and wash the flask with PBS prior to use (*see* **Note 10**).

2. Wash a confluent T-25 culture flask of non-induced human iPSC with sterile PBS.
3. If the iPSC are being cultured in the presence of mouse embryonic fibroblast (MEF) cells, these and any differentiated iPSC need to be removed prior to induction of differentiation. To achieve this, add 1 mL of collagenase I to the flask and return it to a 37 °C incubator for 5–10 min to dissociate the MEF cells and any differentiated iPSC while leaving the undifferentiated iPSC colonies attached to the flask.

4. Once the majority of the MEF cells and the differentiated cells have dissociated from the flask, remove and discard the suspended cells by washing the flask twice with 5 mL of PBS.
5. Add 5 mL of MSC wash buffer to the flask and use a cell scrapper to dissociate the iPSC colonies from the flask.
6. Transfer the iPSC colonies in the MSC wash buffer to a 14 mL polypropylene round-bottom tube.
7. Add an additional 5 mL of MSC wash buffer to the flask and scrape the flask again to remove any remaining iPSC colonies, transfer medium to the 14 mL tube, and centrifuge at 400 × *g* for 5 min at 4 °C.
8. Remove supernatant.
9. Resuspend the iPSC colonies in 1 mL of MSC culture medium, use a 1 mL pipette tip, and aspirate the cells and partially dissociate the iPSC colonies through mechanical agitation (*see* **Note 11**).
10. Plate all of the partially dissociated iPSC colonies out on to the gelatin-coated T-25 flask which was prepared in **step 1**. Add an additional 4 mL of MSC complete medium to the T-25 flask.
11. Leave these cells to differentiate for 2 weeks, changing the medium every 3–4 days as required.

3.1.2 Serial Passaging to Select MSC-Like Cells

After the 2-week differentiation period, the iPSC colonies will have spontaneously differentiated and generated outgrowths consisting of a mixture of heterogeneous cell types. Serial passaging will then be used to select for MSC-like cells.

1. After the 2-week differentiation, the heterogeneous cell types should be passaged using TrypLE to obtain a single-cell suspension.
2. Once cells have lifted off the bottom of the flask, as determined by light microscopy, the enzymatic digestion is terminated neutralized through the addition of FBS containing MSC wash medium.
3. Pellet cells by centrifugation at 400 × *g* for 5 min at 4 °C, resuspend cells in 500 μL MSC complete medium, and keep on ice (*see* **Note 12**).
4. Cells should be plated back out at a 1:3 ratio with all of the cells from a T-25 flask being passaged into a T-75 flask.
5. Prior to passaging the differentiated iPSC, a T-75 flask needs to be pre-coated in gelatin, as described above. For a T-75 flask, add 12 mL 0.1 % gelatin.
6. Passaging should continue to be performed as soon as the cells become 70–80 % confluent.

7. Note—Gelatin coating is only required after the first passage; from then on, tissue culture flasks should be used uncoated.
8. The day after passaging the MSC complete medium often requires changing as the non-MSC-like cells should die off during passaging resulting in numerous floating cells being present.
9. Through the serial passaging process, the cells should begin to look more homogeneous and fibroblast-like (*see* Fig. 1 for representative images of the desired cell morphologies).

3.2 Characterization of Cells Generated

The cells generated then need to be shown to be MSC-like cells. Unfortunately, there is not a single definitive assay which can be performed to demonstrate that particular cells are MSC-like. Instead, the characterization of the possible MSC-like cells involves three main aspects which assess the cells culture capacity, their expression of certain markers, and finally their capacity to undergo differentiation into certain lineages. These experiments have been established in order to satisfy the International Society of Cellular Therapies minimal criteria for defining multipotent MSC (43, 44). In order for cells to be deemed as MSC, they firstly have to be plastic adherent when maintained in standard culture conditions. Secondly, more than 95 % of the MSC cells have to express the markers CD73, CD90, and CD105; additionally, less than 2 % of the cells should express CD14, CD34, and CD45, as determined by flow cytometry. Finally, the MSC must have the ability to differentiate into osteoblasts, adipocytes, and chondrocytes when cultured under standard in vitro differentiation protocols (43, 44).

The first criteria for defining MSC which is based on their ability to adhere to and grow on plastic under standard culture conditions will have been assessed as part of the differentiation process. Only those cells that are plastic adherent will have been able to survive the serial passaging stage.

3.2.1 Initial Flow Cytometric Analysis During Differentiation

During the serial passaging, it is good to periodically perform flow cytometric analysis to assess changes in expression of MSC and pluripotent markers. The aim of this initial flow cytometric analysis is to identify if the number of cells expressing MSC markers is increasing, while the expression of pluripotency markers is decreasing. Figure 2 shows representative images of the flow cytometric analysis results obtained from both the starting undifferentiated iPSC (iPSC) and from MSC-like cells derived from iPSC (MSC-like cells).

1. To perform flow cytometric analysis, single-cell suspensions of the cells are generated through enzymatic digestion with TrypLE, as described above.
2. Pellet cells through centrifugation at 400 × *g* for 5 min at 4 °C, resuspend cells in 1 mL blocking buffer per T175 cm^2

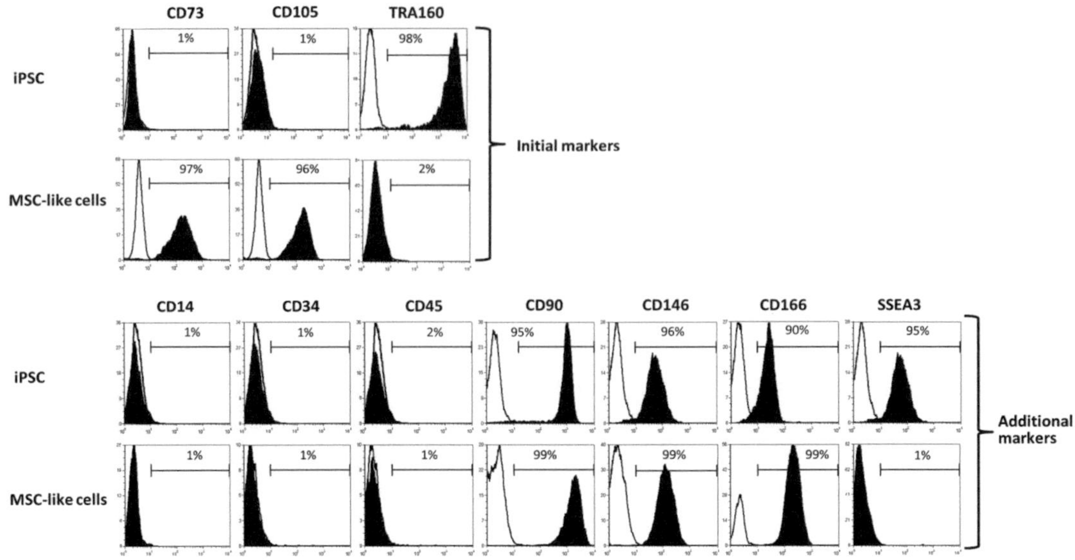

Fig. 2 Representative flow cytometric analysis results before and after differentiation. Representative flow cytometric profiles obtained from starting undifferentiated iPSC (iPSC) and from the MSC-like cells obtained after successful differentiation (MSC-like cells). The *open histograms* represent the isotype controls, while the *solid histograms* represent the individual markers being assessed

flask, and incubate the resuspended cells on ice in blocking buffer for 30 min to reduce the possibility of Fc receptor-mediated binding of antibodies.

3. While the incubation is occurring, remove a 10 μL aliquot and perform a 1:10 dilution in 0.4 % trypan blue/PBS. Determine the number of cells present by using a hemocytometer.
4. A minimum of 1×10^5 cells are required per antibody stain being used. If you have sufficient cells for flow analysis, 2×10^5 cells should be used. For the initial assessment, two MSC markers (CD73 and CD105) and one pluripotency marker (TRA160) should be analyzed along with their respective isotype controls (IgG_1 for CD73 and CD105 and IgM for TRA160).
5. After calculating the number of cells present, aliquot 1×10^5–2×10^5 cells into a 5 mL polypropylene round-bottom tube for each antibody being used.
6. Once the 30 min incubation in block buffer has finished, add the appropriate amount of primary monoclonal antibodies to achieve a concentration of 20 μg/mL, and incubate in ice for 1 h. The list of these antibodies is outlined in Table 1.
7. Wash the cells twice with 1 mL of ice-cold HHF, and then incubate with the appropriate secondary detection reagent, goat anti-mouse IgG- or goat anti-mouse IgM-PE (phycoerythrin) (*see* **Note 13**) conjugated antibody (1:50 dilution, Southern Biotechnology, Birmingham, AL, USA) for 45 min on ice.

Table 1
List of initial screening antibodies

Name	Isotype	Source
IgG_1 isotype control	IgG_1	BD Pharmigen Cat# 562652
CD73	IgG_1	BD Pharmigen Cat# 550256
CD105	IgG_1	BD Pharmigen Cat# 555690
IgM isotype control	IgM	BD Pharmigen Cat# 550340
TRA-160	IgM	Millipore Cat# MAB4360

Use all primary antibodies and isotype-matched controls at a concentration of 20 μg/mL. Use secondary detection antibodies at 10 μg/mL

8. Then wash the cells in 1 mL of HHF and then fix the cells in 500 μL of FACS fix solution.
9. Perform the analysis on a fluorescence-activated cell sorter fitted with a 250 MW argon laser.

3.2.2 Comprehensive Flow Cytometric Analysis During Differentiation

Once the initial flow cytometric analysis for CD73, CD105, and TRA160 suggests that the cells have differentiated into MSC-like cells (with more than 95 % of the cells expressing CD73 and CD105 and the expression of TRA160 being limited to less than 2 % of the cells), you should proceed to assess the expression of more MSC and hematopoietic markers.

1. Repeat the flow staining as described above for the initial flow cytometric analysis but include additional antibodies, as described in Table 2.

3.2.3 Tri-lineage Differentiation

1. Osteogenic differentiation
   Three different sets of results are used to demonstrate that the MSC-like cells have the capacity to undergo osteogenic differentiation. Firstly, images are used to demonstrate the presence of Alizarin red stained calcium deposits; secondly, quantitative assessment of the levels of calcium produced are performed; and finally, expression analysis of osteogenic associated genes is performed.

   Alizarin red staining of calcium deposits generated by MSC-like cells

   1. For this assay, seed MSC-like cells at 8 × 10^3 cells per cm^2, in 24-well plates, in MSC complete medium. For this assay, 6 wells of cells are required per MSC-like cell line being assessed. Three of these wells of cells will be used as negative controls as osteogenesis will not be induced; the remaining three wells will be used for osteogenic induction.

Table 2
List of more comprehensive antibody panel

Name	Isotype	Source
IgG_{2a} isotype control	IgG_{2a}	BD Pharmigen Cat# 563464
CD14	IgG_{2A}	Beckman Coulter Cat# IM0650U
CD34	IgG_1	Beckman Coulter Cat# IM1871U
CD45	IgG_1	Beckman Coulter Cat# IM2078U
CD90	IgG_1	BD Pharmigen Cat# 555594
CD146	IgG_{2a}	(45, 46)
CD166	IgG_1	BD Pharmigen Cat# 559260
Goat anti-mouse IgG PE	IgG	Southern Biotech Cat# 1030-09
Goat anti-mouse IgM PE	IgM	Southern Biotech Cat# 1020-09
SSEA3	IgM	Millipore Cat# MAB4303

Use all primary antibodies and isotype-matched controls at a concentration of 20 μg/mL. Use secondary detection antibodies at 10 μg/mL

2. Once the cells have reached ~80 % confluence, the MSC complete medium is removed, and 500 μL of osteogenic induction medium is added to three wells and 500 μL of osteogenic control medium is added to the remaining three wells.
3. Culture cells in their respective medium for 28 days with twice weekly medium changes.
4. After 28 days, wash the cells three times with 1× PBS and fix them with 10 % neutral buffered formalin for 1 h.
5. Rinse the cells three times in RO water.
6. Stain cells with filtered Alizarin red stain either overnight at 4 °C or for 1 h at room temperature on a vibrating plate vortex.
7. Rinse the cells five times in RO water or until the excess Alizarin red stain is removed. Upturn the 24-well plate to allow it to dry before photographing.

Quantitative assessment of calcium production
This assay involves two components, the first assess the level of acid solubilized calcium produced by the cells, while the second component determines the level of DNA present in each well. By determining the amount of DNA present, the level of calcium produced can be standardized to the amount of cells present in each well. This is particularly beneficial when comparing the osteogenic potential between different cell lines which have different proliferation rates.

1. For this assay, seed MSC-like cells at 8×10^3 cells per cm^2, in 96-well flat bottom plates, in MSC complete medium. This assay requires 6 wells of cells per MSC-like cell line being investigated. Once again, three of these wells of cells will be used as negative controls, and the remaining three wells will be used for osteogenic induction.
2. Once the cells have reached 80 % confluence, the MSC complete medium is removed, and 200 μL of osteogenic induction medium is added to three wells and 200 μL of osteogenic control medium is added to the remaining three wells.
3. Culture cells in their respective medium for 28 days with twice weekly medium changes.
4. Wash the cells three times with 1× PBS, being careful not to dislodge the cells from the plate.
5. Dissolve the mineralized matrix in 100 μL/well of 0.6 M HCL for 2 h at room temperature or overnight at 4 °C.
6 Transfer the acid solubilized solution to a new 96-well plate and add 200 μL 1× PBS to the original plate and store at 4 °C; this will be used to perform the fluorometric DNA quantitation described below.
7. Transfer 4 μL of the acid solubilized solution to a 96-well microtiter plate and quantitate calcium levels using the Arsenazo 111 assay.
8. Add 4 μL in triplicate of calcium standard at concentrations of 0.25, 0.5, 0.75, 1.0, 1.5, 2.0, and 2.5 mM to establish a standard curve.
9. Add 200 μL of calcium Arsenazo III reagent per well to the test samples and the calcium/phosphorous standards.
10. Incubate the plates at room temperature for 2 min and read the absorbance at 650 nm on a microplate reader (*see* **Note 14**).
11. Go back to the original 96-well plate that the cells were cultured in and perform the fluorometric DNA quantitation assay using Pico Green, as described below.
12. Remove the 200 μL of PBS from each well and wash twice with 1× PBS.
13. Add 100 μL of proteinase K (100 μg/mL) per well and incubate for 1 h at 37 °C.
14. After 1 h, vigorously pipette each sample to help dissociate and digest the cells, and incubate for a further 1 h at 37 °C.
15. Vigorously pipette each sample again to ensure a homogenous cell mixture, and check using light microscopy that a single-cell suspension has been achieved.

16. Generate DNA standards of 10, 5, 2.5, 1.25, 0.625, and 0.312 μg/mL, using high molecular weight DNA.
17. Add 50 μL of each DNA standard in triplicate to a black polystyrene microtiter plate, and also add 3 wells of 50 μL of the proteinase K used to digest the cells (this acts as a base line control).
18. Add 50 μL of digested sample to the appropriate wells.
19. Gently agitate the plate, and protect it from light while it incubates for 2–5 min at room temperature.
20. Analyze using a spectrofluorometer or fluorescence microplate reader using standard fluorescein wavelengths (excitation ~480 nm, emission ~520 nm) (*see* **Note 15**).
21. Use the data obtained from the Arsenazo 111—mineral quantitation and the Pico Green DNA quantitation to determine the amount of calcium produced relative to the DNA quantity for each sample.

2. Adipogenic differentiation

Three different sets of results are used to demonstrate that the MSC-like cells have the capacity to undergo adipogenic differentiation. Firstly, images are used to demonstrate the presence of Oil-Red-O stained lipid-laden vacuoles within adipocytes; secondly, quantitative assessment is used to demonstrate the levels of Nile Red stained lipid; and finally, expression analysis of adipogenic-associated genes is performed.

Oil-Red-O stained lipid-laden vacuoles within adipocytes generated by the MSC-like cells

1. For this assay, seed MSC-like cells at 8×10^3 cells per cm^2, in 24-well plates, in MSC complete medium. For this assay, you need 6 wells of cells per MSC-like cell line. Three of these wells of cells will be cultured in the adipogenic induction medium, while the remaining three wells will be cultured in adipogenic control medium.
2. Once the cells have reached 80 % confluence, the MSC complete medium should be removed, and 500 μL of adipogenic induction and adipogenic control medium are added to their respective wells.
3. Culture cells in their respective medium for 28 days with twice weekly medium changes.
4. After 28 days, wash the cells once with PBS, and fix them with 10 % neutral buffered formalin for 15 min at room temperature.
5. Stain the cells with filtered Oil-Red-O for at least 2 h at room temperature.

6. Rinse the cells three times with RO water or until the excess Oil-Red-O stain is removed.
7. Store in water at 4 °C till images can be captured.

Quantitative assessment of adipocyte production

1. For this assay, seed MSC-like cells at 8×10^3 cells per cm^2, in a 96-well flat bottom plate, in MSC complete medium. This assay requires 6 wells of cells per MSC-like cell line being assessed, 3 wells for adipogenic induction and 3 wells as controls.
2. Once the cells have reached 80 % confluence, the MSC complete medium is removed, and 200 μL of adipogenic induction medium or 200 μL of adipogenic control medium is added to their respective wells.
3. Culture cells in their respective medium for 28 days with twice weekly medium changes.
4. Aspirate medium and gently wash the cells 3× with PBS.
5. Fix cells in 10 % neutral buffered formalin for 1 h at room temperature.
6. Stain the cells with Nile Red for 15 min in the dark.
7. Visualize stained lipid using an inverted fluorescence microscope (excitation at 485 nM and emission 525 nM). Counterstain the cells with DAPI.
8. Use image capture software to quantitate the number of Nile Red positive adipocytes relative to the number of cells present, as determined by the DAPI staining.

3. Chondrogenic differentiation

Two different experiments are used to demonstrate that the MSC-like cells have the capacity to undergo chondrogenic differentiation. Firstly, MSC-like cells are cultured as cell pellets in order to form chondrocyte pellets. The chondrocyte pellets formed are assessed for the expression of chondrocyte-associated genes by quantitative real-time PCR and through histological assessment. The second set of experiments involves quantitative assessment of the level of glycosaminoglycan (GAG) production achieved by the MSC-like cells.

Chondrocyte pellet formation

1. To induce chondrogenesis, resuspend 5×10^6 MSC in DMEM medium in a 10 mL polypropylene conical tube, and centrifuge at 600 × *g*, at 4 °C for 5 min to form cell pellets. For each MSC-like cell line being assessed, 4 chondrogenic pellets need to be set up, two of which will be used for gene expression analysis (one cultured in chondrogenic induction conditions and one cultured in control conditions) and two which will be used for histological analysis.

2. Carefully aspirate the supernatant; avoid disturbing the cell pellet.
3. Gently add 200 μL of chondrocyte medium to each tube.
4. Add 10 ng/mL of TGFβ3 to the tubes in which chondrogenic induction is being performed; the TGFβ3 needs to be added fresh with every medium change.
5. Culture cells in their respective medium for 28 days with twice weekly medium changes.
6. Wash pellet cultures designated for RT-PCR, three times in 1× PBS.
7. Digest the pellet in 200 μL collagenase/dispase at 37 °C for 1 h.
8. Wash the cells in 5 mL of MSC wash medium, and centrifuge samples at 400 × *g* for 5 min.
9. Aspirate the medium and repeat the washing of the cells.
10. Add 100 μL of TRIzol to the dissociated pellet, and transfer to a suitable tube for storage at −80 °C.
11. Perform RNA extraction and cDNA synthesis on the dissociated pellets.
12. Employ RT-PCR analysis to assess the levels of expression of collagen markers aggrecan, collagen II, collagen X, and SOX9.
13. Pellet cultures designated for histological analysis should be washed and then fixed in 10 % neutral buffered formalin overnight at 4 °C; then histologically process and paraffin embed the pellets.
14. 5 μm sections should be obtained using a microtome.
15. Deparaffinize the sections through two 2 min incubations in xylene followed by three 1 min incubations in fresh absolute ethanol, and finished off with a final wash in RO water.
16. To neutralize endogenous peroxidase activity, incubate the sections in 0.5 % H_2O_2 (v/v) in methanol for 30 min.
17. Block nonspecific binding by incubating sections with 5 % (v/v) normal goat serum in 1× PBS for 1 h at RT.
18. Incubate the sections overnight with either an isotype-matched, nonbinding control monoclonal antibody (1B5, IgG1), or the anti-collagen type II monoclonal antibody at a 1 in 200 dilution in 5 % (v/v) normal goat serum in 1× PBS.
19. Wash the slides three times in 1× PBS and incubate in secondary antibody, goat anti-mouse IgG biotin diluted 1 in 200 in 5 % (v/v) normal goat serum in 1× PBS, for 2 h at room temperature.

20. Wash the slides three times in 1× PBS and incubate in tertiary antibody, Streptavidin HRP, 1 in 500 in 5 % (v/v) normal goat serum.
21. Wash the slides three times in 1× PBS and incubate with the DAB substrate solution for 5–10 min, as per the manufacturer's instructions.
22. Counterstain briefly with hematoxylin and mount in DePex mounting medium.

Quantitative assessment of glycosaminoglycan (GAG) production

1. Seed MSC-like cells at 5×10^4 cells per well of a 96-well plate in MSC complete medium; 6 wells are required for each MSC-like cell line being assessed. Three wells will be used as controls, and three wells will be induced to undergo chondrogenic differentiation. Set up two identical plates, one will be used to quantify the amount of glycosaminoglycans (GAG) produced the second will be used to quantify the amount of DNA present.
2. Following overnight adhesion, replace the MSC complete medium with chondrogenic medium, and add 10 ng/mL TGFβ3 to the wells which are to undergo chondrogenic induction.
3. For the DNA plate, culture the cells in their respective medium for 72 h and then digest the cells in 100 μL papain digest solution (100 U/mL) at 65 °C for 6 h.
4. Transfer 50 μL of the cell lysate from the DNA plate to a white microtiter plate and measure the DNA content using the Invitrogen Quant-iT Pico Green dsDNA Assay Kit as described above.
5. For the GAG plate after 48 h, replace the medium in GAG plate with 100 μL of medium containing 1 μCi $^{35}SO_4$ (sulfur-35 radionuclide), and incubate the plates at 37 °C overnight. Note—ensure that radiation safety procedures are followed.
6. Add 100 μL of papain solution to the medium in each well.
7. Incubate at 65 °C for 2–3 h.
8. Add 40 μL of CPC to each well, and mix well on a shaker for 5 min.
9. Thaw the stock solution of CSA and dilute it 1:5 to make the working solution. Add 10 μL CSA per well and mix for 5 min on a shaker.
10. Harvest the contents of the plate onto a glass fiber filter using a cell harvester.

11. Store the glass fiber filter appropriately and allow it to dry for a minimum of 2 h.
12. Once dry, the glass fiber filter can be read using a scintillation counter; to do this, the filter is placed within a holder and 25 μL Microscint™ is added per well.
13. The amount of $^{35}SO_4$-labeled GAG per well will be measured using a TopCount NXT Microplate Scinitillation and Luminescence Counter (PerkinElmer Life and Analytical Sciences).
14. Finally, normalize the glycosaminoglycan levels to DNA content to determine relative GAG production per cell.

Expression analysis of genes associated with osteogenic, adipogenic and chondrogenic differentiation

1. For this assay, seed MSC-like cells at 8×10^3 cells per cm^2, in 6-well plates, in MSC complete medium. For this assay, you need two wells of cells per MSC-like cell line for each of the three types of differentiation (osteogenic, adipogenic, and chondrogenic). One of the wells will be used as a negative control, and the second well will be used for induction.
2. Once the cells have reached 80 % confluence, the MSC complete medium should be removed, and 1 mL of the relevant induction medium is added to one well and 1 mL of control medium is added to the remaining well.
3. Culture cells in their respective medium for 28 days with twice weekly medium changes.
4. Wash the cells twice with PBS and then use TRIzol to isolate total RNA from the cells.
5. Purify the RNA using standard protocols and then use a reverse transcription protocol to generate cDNA.
6. Perform quantitative real-time PCR to assess the expression of genes associated with the three different differentiation methods. Osteogenic genes include bone morphogenetic protein (BMP2), bone sialoprotein (BSP2), osteopontin (OPN), runt-related transcription factor 2 (RUNX2), osteocalcin (OCN), and osterix (OSX). Adipogenic-associated genes include CCAAT/enhancer-binding protein-alpha (C/EBPα), adiponectin (ADIPOQ), adipsin, and peroxisome proliferator-activated receptor-gamma 2 (PPARγ2). Chondrogenic-associated genes include collagen II (COLII), collagen X (COLX), SRY-BOX-9 (SOX9), aggrecan (ACAN).

4 Notes

1. To prepare the 0.1 % gelatin working stock solution, firstly prepare a 1 % (10×) solution of gelatin in distilled H_2O, i.e. 5 g of gelatin in 500 mL of distilled H_2O. The gelatin will not completely dissolve at this stage. Autoclave the 1 % stock and then store it at 4 °C until required. To prepare the 0.1 % working solution, warm up the 1 % stock in a 37 °C water bath to help dissolve the gelatin. Once warmed, dilute the 1 % stock in distilled H_2O and pass it through a 0.22 μm filter before use.
2. Collagenase I should be passed through a 0.22 μm filter before use.
3. All medium should be used within 2 weeks of being made up. If your medium is more than 2 weeks old, add fresh 2 mM L-glutamine.
4. Human serum requires heat inactivation at 56 °C for 30 min prior to use. After heat inactivation, the serum should be centrifuged at 1,000 × *g* for 10 min and the supernatant collected and used.
5. HHF should be kept on ice when being use for flow cytometry staining.
6. Both the osteogenic induction medium and the osteogenic control medium need to be made up fresh every 2 weeks.
7. The Alizarin red stain should be passed through a 0.22 μm filter before use to remove any particulate matter.
8. Both the adipogenic induction medium and the adipogenic control medium need to be made up fresh every 2 weeks.
9. TGFβ3 is added directly to the pellets undergoing chondrogenic induction; it is not added to the stock medium.
10. If the gelatin-coated flasks are not being used immediately, they can be stored for up to a week in the fridge; add 5 mL of PBS to each flask to ensure the gelatin does not dry out.
11. Do not attempt to dissociate the colonies to a single-cell suspension; the dissociation step is only to partially dissociate the iPSC colonies. iPSC do not like to be cultured as single cells; dissociating iPSC into a single-cell suspension will result in considerable cell death.
12. Because the cells have been allowed to grow without passaging for 2 weeks, they will have grown to a high confluence, and you

may end up with a clumpy solution of cells when you try to resuspend the cells in 500 μL of MSC complete medium. If this occurs, pass the cell suspension through a 70 μM filter prior to platting them back out.

13. Fluorescein isothiocyanate (FITC)-conjugated secondary antibodies can be used instead of PE-conjugated if required.
14. Check that the absorbance values obtained for all of your samples being analyzed are within the values obtained from the various standard concentrations assessed. If any of the sample values are about the absorbance obtained for the 2.5 nM concentration, you will have to dilute all of your samples and reanalyze them.
15. Once again, you need to check that the values obtained for your samples fall within the standard curve you have set up. If any of your samples have a higher absorbance than that of the 10 μg/mL standard, then you will have to dilute your test samples and repeat the analysis.

References

1. Sutton MT, Bonfield TL (2014) Stem cells: innovations in clinical applications. Stem Cell Int 2014:516278. doi:10.1155/2014/516278
2. Pittenger MF, Mackay AM, Beck SC, Jaiswal RK, Douglas R, Mosca JD, Moorman MA, Simonetti DW, Craig S, Marshak DR (1999) Multilineage potential of adult human mesenchymal stem cells. Science 284(5411):143–147
3. Wada N, Gronthos S, Bartold PM (2013) Immunomodulatory effects of stem cells. Periodontol 2000 63(1):198–216. doi:10.1111/prd.12024
4. Tse WT, Pendleton JD, Beyer WM, Egalka MC, Guinan EC (2003) Suppression of allogeneic T-cell proliferation by human marrow stromal cells: implications in transplantation. Transplantation 75(3):389–397. doi:10.1097/01.TP.0000045055.63901.A9
5. Farini A, Sitzia C, Erratico S, Meregalli M, Torrente Y (2014) Clinical applications of mesenchymal stem cells in chronic diseases. Stem Cell Int 2014:306573. doi:10.1155/2014/306573
6. Katz AJ, Tholpady A, Tholpady SS, Shang H, Ogle RC (2005) Cell surface and transcriptional characterization of human adipose-derived adherent stromal (hADAS) cells. Stem Cells 23(3):412–423. doi:10.1634/stemcells.2004-0021, 23/3/412 [pii]
7. Bieback K, Kern S, Kluter H, Eichler H (2004) Critical parameters for the isolation of mesenchymal stem cells from umbilical cord blood. Stem Cells 22(4):625–634. doi:10.1634/stemcells.22-4-625, 22/4/625 [pii]
8. Romanov YA, Svintsitskaya VA, Smirnov VN (2003) Searching for alternative sources of postnatal human mesenchymal stem cells: candidate MSC-like cells from umbilical cord. Stem Cells 21(1):105–110. doi:10.1634/stemcells.21-1-105
9. Miao Z, Jin J, Chen L, Zhu J, Huang W, Zhao J, Qian H, Zhang X (2006) Isolation of mesenchymal stem cells from human placenta: comparison with human bone marrow mesenchymal stem cells. Cell Biol Int 30(9):681–687. doi:10.1016/j.cellbi.2006.03.009
10. Hsieh JY, Fu YS, Chang SJ, Tsuang YH, Wang HW (2010) Functional module analysis reveals differential osteogenic and stemness potentials in human mesenchymal stem cells from bone marrow and Wharton's jelly of umbilical cord. Stem Cells Dev 19(12):1895–1910. doi:10.1089/scd.2009.0485
11. Weiss ML, Anderson C, Medicetty S, Seshareddy KB, Weiss RJ, VanderWerff I, Troyer D, McIntosh KR (2008) Immune properties of human umbilical cord Wharton's jelly-derived cells. Stem Cells 26(11):2865–2874. doi:10.1634/stemcells.2007-1028

12. Gronthos S, Mankani M, Brahim J, Robey PG, Shi S (2000) Postnatal human dental pulp stem cells (DPSCs) in vitro and in vivo. Proc Natl Acad Sci U S A 97(25):13625–13630. doi:10.1073/pnas.240309797, 240309797 [pii]
13. Miura M, Gronthos S, Zhao M, Lu B, Fisher LW, Robey PG, Shi S (2003) SHED: stem cells from human exfoliated deciduous teeth. Proc Natl Acad Sci U S A 100(10):5807–5812. doi:10.1073/pnas.0937635100, 0937635100 [pii]
14. Seo BM, Miura M, Gronthos S, Bartold PM, Batouli S, Brahim J, Young M, Robey PG, Wang CY, Shi S (2004) Investigation of multipotent postnatal stem cells from human periodontal ligament. Lancet 364 (9429):149–155. doi:10.1016/S0140-6736(04)16627-0, S0140673604166270 [pii]
15. Morsczeck C, Gotz W, Schierholz J, Zeilhofer F, Kuhn U, Mohl C, Sippel C, Hoffmann KH (2005) Isolation of precursor cells (PCs) from human dental follicle of wisdom teeth. Matrix Biol 24(2):155–165. doi:10.1016/j.matbio.2004.12.004, S0945-053X(04)00159-3 [pii]
16. Sonoyama W, Liu Y, Fang D, Yamaza T, Seo BM, Zhang C, Liu H, Gronthos S, Wang CY, Wang S, Shi S (2006) Mesenchymal stem cell-mediated functional tooth regeneration in swine. PLoS One 1:e79. doi:10.1371/journal.pone.0000079
17. Xiong J, Mrozik K, Gronthos S, Bartold PM (2012) Epithelial cell rests of malassez contain unique stem cell populations capable of undergoing epithelial-mesenchymal transition. Stem Cells Dev 21(11):2012–2025. doi:10.1089/scd.2011.0471
18. Zhang Q, Shi S, Liu Y, Uyanne J, Shi Y, Le AD (2009) Mesenchymal stem cells derived from human gingiva are capable of immunomodulatory functions and ameliorate inflammation-related tissue destruction in experimental colitis. J Immunol 183(12):7787–7798. doi:10.4049/jimmunol.0902318, jimmunol.0902318 [pii]
19. Santamaria-Martinez A, Barquinero J, Barbosa-Desongles A, Hurtado A, Pinos T, Seoane J, Poupon MF, Morote J, Reventos J, Munell F (2009) Identification of multipotent mesenchymal stromal cells in the reactive stroma of a prostate cancer xenograft by side population analysis. Exp Cell Res 315(17):3004–3013. doi:10.1016/j.yexcr.2009.05.007
20. Popova AP, Bozyk PD, Goldsmith AM, Linn MJ, Lei J, Bentley JK, Hershenson MB (2010) Autocrine production of TGF-beta1 promotes myofibroblastic differentiation of neonatal lung mesenchymal stem cells. Am J Physiol Lung Cell Mol Physiol 298(6):L735–L743. doi:10.1152/ajplung.00347.2009
21. Lanzoni G, Alviano F, Marchionni C, Bonsi L, Costa R, Foroni L, Roda G, Belluzzi A, Caponi A, Ricci F, Luigi Tazzari P, Pagliaro P, Rizzo R, Lanza F, Roberto Baricordi O, Pasquinelli G, Roda E, Paolo Bagnara G (2009) Isolation of stem cell populations with trophic and immunoregulatory functions from human intestinal tissues: potential for cell therapy in inflammatory bowel disease. Cytotherapy 11(8):1020–1031. doi:10.3109/14653240903253840
22. Polisetty N, Fatima A, Madhira SL, Sangwan VS, Vemuganti GK (2008) Mesenchymal cells from limbal stroma of human eye. Mol Vis 14:431–442
23. Jakob M, Hemeda H, Janeschik S, Bootz F, Rotter N, Lang S, Brandau S (2010) Human nasal mucosa contains tissue-resident immunologically responsive mesenchymal stromal cells. Stem Cells Dev 19(5):635–644. doi:10.1089/scd.2009.0245
24. Morito T, Muneta T, Hara K, Ju YJ, Mochizuki T, Makino H, Umezawa A, Sekiya I (2008) Synovial fluid-derived mesenchymal stem cells increase after intra-articular ligament injury in humans. Rheumatology (Oxford) 47 (8):1137–1143. doi:10.1093/rheumatology/ken114
25. Nimura A, Muneta T, Otabe K, Koga H, Ju YJ, Mochizuki T, Suzuki K, Sekiya I (2010) Analysis of human synovial and bone marrow mesenchymal stem cells in relation to heat-inactivation of autologous and fetal bovine serums. BMC Musculoskelet Disord 11:208. doi:10.1186/1471-2474-11-208
26. Segawa Y, Muneta T, Makino H, Nimura A, Mochizuki T, Ju YJ, Ezura Y, Umezawa A, Sekiya I (2009) Mesenchymal stem cells derived from synovium, meniscus, anterior cruciate ligament, and articular chondrocytes share similar gene expression profiles. J Orthop Res 27(4):435–441. doi:10.1002/jor.20786
27. Ju YJ, Muneta T, Yoshimura H, Koga H, Sekiya I (2008) Synovial mesenchymal stem cells accelerate early remodeling of tendon-bone healing. Cell Tissue Res 332(3):469–478. doi:10.1007/s00441-008-0610-z
28. Takahashi K, Yamanaka S (2006) Induction of pluripotent stem cells from mouse embryonic and adult fibroblast cultures by defined factors. Cell 126(4):663–676. doi:10.1016/j.cell.2006.07.024, S0092-8674(06)00976-7 [pii]
29. Takahashi K, Tanabe K, Ohnuki M, Narita M, Ichisaka T, Tomoda K, Yamanaka S (2007) Induction of pluripotent stem cells from adult human fibroblasts by defined factors. Cell 131

(5):861–872. doi:10.1016/j.cell.2007.11.019, S0092-8674(07)01471-7 [pii]

30. Yu J, Vodyanik MA, Smuga-Otto K, Antosiewicz-Bourget J, Frane JL, Tian S, Nie J, Jonsdottir GA, Ruotti V, Stewart R, Slukvin II, Thomson JA (2007) Induced pluripotent stem cell lines derived from human somatic cells. Science 318(5858):1917–1920. doi:10.1126/science.1151526, 1151526 [pii]
31. Chen YS, Pelekanos RA, Ellis RL, Horne R, Wolvetang EJ, Fisk NM (2012) Small molecule mesengenic induction of human induced pluripotent stem cells to generate mesenchymal stem/stromal cells. Stem Cells Transl Med 1 (2):83–95. doi:10.5966/sctm.2011-0022, sctm.2011-0022 [pii]
32. Firth AL, Yuan JX (2012) "Ether-a-go-go" proliferation of iPSC-derived mesenchymal stem cells. Focus on "Regulation of cell proliferation of human induced pluripotent stem cell-derived mesenchymal stem cells via ether-a-go-go 1 (hEAG1) potassium channel". Am J Physiol Cell Physiol 303(2):C113–C114. doi:10.1152/ajpcell.00160.2012, ajpcell.00160.2012 [pii]
33. Giuliani M, Oudrhiri N, Noman ZM, Vernochet A, Chouaib S, Azzarone B, Durrbach A, Bennaceur-Griscelli A (2011) Human mesenchymal stem cells derived from induced pluripotent stem cells down-regulate NK-cell cytolytic machinery. Blood 118 (12):3254–3262. doi:10.1182/blood-2010-12-325324, blood-2010-12-325324 [pii]
34. Guzzo RM, Gibson J, Xu RH, Lee FY, Drissi H (2013) Efficient differentiation of human iPSC-derived mesenchymal stem cells to chondroprogenitor cells. J Cell Biochem 114:480. doi:10.1002/jcb.24388
35. Hynes K, Menicanin D, Han J, Marino V, Mrozik K, Gronthos S, Bartold PM (2013) Mesenchymal stem cells from iPS cells facilitate periodontal regeneration. J Dent Res 92(9):833–839. doi:10.1177/0022034513498258
36. Hynes K, Menicanin D, Mrozik K, Gronthos S, Bartold PM (2014) Generation of functional mesenchymal stem cells from different induced pluripotent stem cell lines. Stem Cells Dev 23:1084. doi:10.1089/scd.2013.0111
37. Lian Q, Zhang Y, Zhang J, Zhang HK, Wu X, Lam FF, Kang S, Xia JC, Lai WH, Au KW, Chow YY, Siu CW, Lee CN, Tse HF (2010) Functional mesenchymal stem cells derived from human induced pluripotent stem cells attenuate limb ischemia in mice. Circulation 121(9):1113–1123. doi:10.1161/CIRCULATIONAHA.109.898312, CIRCULATIONAHA.109.898312 [pii]
38. Liu Y, Goldberg AJ, Dennis JE, Gronowicz GA, Kuhn LT (2012) One-step derivation of mesenchymal stem cell (MSC)-like cells from human pluripotent stem cells on a fibrillar collagen coating. PLoS One 7(3):e33225. doi:10.1371/journal.pone, 0033225 PONE-D-11-14316 [pii]
39. Villa-Diaz LG, Brown SE, Liu Y, Ross AM, Lahann J, Parent JM, Krebsbach PH (2012) Derivation of mesenchymal stem cells from human induced pluripotent stem cells cultured on synthetic substrates. Stem Cells 30 (6):1174–1181. doi:10.1002/stem.1084
40. Wei H, Tan G, Manasi QS, Kong G, Yong P, Koh C, Ooi TH, Lim SY, Wong P, Gan SU, Shim W (2012) One-step derivation of cardiomyocytes and mesenchymal stem cells from human pluripotent stem cells. Stem Cell Res 9 (2):87–100. doi:10.1016/j.scr.2012.04.003, S1873-5061(12)00042-6 [pii]
41. Karlsson C, Emanuelsson K, Wessberg F, Kajic K, Axell MZ, Eriksson PS, Lindahl A, Hyllner J, Strehl R (2009) Human embryonic stem cell-derived mesenchymal progenitors–potential in regenerative medicine. Stem Cell Res 3 (1):39–50. doi:10.1016/j.scr.2009.05.002
42. Kuo ML (2011) Efficient derivation & concise gene expression profiling of human embryonic stem cell-derived mesenchymal progenitors (EMPs). Cell Transplant 20:1529. doi:10.3727/096368910X564067, ct0107yenetalpdf [pii]
43. Dominici M, Le Blanc K, Mueller I, Slaper-Cortenbach I, Marini F, Krause D, Deans R, Keating A, Prockop D, Horwitz E (2006) Minimal criteria for defining multipotent mesenchymal stromal cells. The International Society for Cellular Therapy position statement. Cytotherapy 8(4):315–317. doi:10.1080/14653240600855905, Q2183N8UT042W62H [pii]
44. Horwitz EM, Le Blanc K, Dominici M, Mueller I, Slaper-Cortenbach I, Marini FC, Deans RJ, Krause DS, Keating A (2005) Clarification of the nomenclature for MSC: the International Society for Cellular Therapy position statement. Cytotherapy 7(5):393–395. doi:10.1080/14653240500319234, H049267664665226 [pii]
45. Filshie RJ, Zannettino AC, Makrynikola V, Gronthos S, Henniker AJ, Bendall LJ, Gottlieb DJ, Simmons PJ, Bradstock KF (1998) MUC18, a member of the immunoglobulin superfamily, is expressed on bone marrow fibroblasts and a subset of hematological malignancies. Leukemia 12(3):414–421
46. Shi S, Gronthos S (2003) Perivascular niche of postnatal mesenchymal stem cells in human bone marrow and dental pulp. J Bone Miner Res 18(4):696–704. doi:10.1359/jbmr.2003.18.4.696

Methods in Molecular Biology (2016) 1357: 375–381
DOI 10.1007/7651_2014_146

Published online: 23 November 2014

Hepatic Differentiation from Human Ips Cells Using M15 Cells

Kahoko Umeda, Nobuaki Shiraki, and Shoen Kume

Abstract

Here, we describe a procedure of human iPS cells differentiation into the definitive endoderm, further into albumin-expressing and albumin-secreting hepatocyte, using M15, a mesonephros- derived cell line. Approximately 90 % of human iPS cells differentiated into SOX17-positive definitive endoderm then approximately 50 % of cells became albumin-positive cells, and secreted ALB protein. This M15 feeder system for endoderm and hepatic differentiation is a simple and efficient method, and useful for elucidating molecular mechanisms for hepatic fate decision, and could represent an attractive approach for a surrogate cell source for pharmaceutical studies.

Keywords: Hepatic differentiation, Endoderm differentiation, Feeder cells, M15 cells

1 Introduction

Human iPS cells are potential sources of hepatocytes for applications in regenerative medicine and drug development (1). We previously reported a procedure in which ES cells are sequentially induced into the regional specific gut endoderm lineages, such as the pancreas, liver, and intestine, by use of M15, a mesoderm derived cell line (2–4).

M15 is used as a source for signals for in vitro ES differentiation. M15 directs human ES cells to differentiate into the definitive endodermal lineages with the addition of activin and LY294002, a potent PI3 kinase inhibitor, further into the hepatic lineages with the addition of dexamethasone (Dex) and Hepatocyte growth factor (HGF) (2). Approximately 80 % of the human ES cells differentiated into alpha feto protein (AFP)-positive hepatic precursor cells on day 20. On day 40, approximately 9 % of the total cells became Albumin (ALB)-positive hepatocytes and secreted a substantial level of ALB protein (2). Here, we describe an optimized protocol which is more efficient and results in generating a higher portion (85.9 %) of SOX17-positive definitive endoderm by altering the endoderm differentiation medium (higher concentration of activin, B27 supplement contained RPMI medium) and yielding higher ALB transcription levels (5).

2 Materials

1. M15 cells (ECACC cell no. 95102517).
2. Culture Dish (90-mm, Nunc, 150350) (150-mm, Nunc, 168381).

   (24-well dish, Corning, 3526).
3. PBS (*see* **Note 1**).
4. 0.05 % trypsin/0.53 mM EDTA (Invitrogen, 25300-062).
5. EF medium.

DMEM (Invitrogen, 11995-075)	500 mL
FBS (Hyclone)	58 mL
Penicillin and streptomycin (PS: Nacalai Tesque, 26252-94) (*see* **Note 2**)	5.8 mL
L-Glutamine (Nacalai Tesque, 16948-04) (*see* **Note 2**)	5.8 mL

6. 2× Freeze solution.

EF medium	28 mL
DMSO (Sigma, D2650)	10 mL
FBS (Hyclone)	2 mL

7. Mitomycin C solution.

   Dissolve mitomycin C (2 mg, Sigma, M4287) in 2 mL PBS.
8. Mitomycin C containing medium.

EF medium	200 mL
Mitomycin C solution	2 mL

   The final concentration of mitomycin C will be 10 μg per mL.
9. CTK solution (*see* **Note 3**).

2.5 % Trypsin (Invitrogen, 15090-046)	10 mL
10 mg/mL Collagenase IV (Invitrogen, 17104-019) (*see* **Note 4**)	0.5 mL
Knockout Serum Replacement (KSR,Invitrogen, 10828-028)	20 mL
100 mM $CaCl_2$ (filtrated) (*see* **Note 5**)	1 mL
PBS	59 mL

10. Human iPS medium.

Knockout DMEM/F12 (Sigma-Aldrich)	500 mL
KSR (Invitrogen, 10828-028)	125 mL
PS (Nacalai Tesque, 26252-94) (*see* **Note 2**)	6.25 mL
L-glutamine (Nacalai Tesque, 16948-04) (*see* **Note 2**)	6.25 mL
Nonessential amino acids (NEAA; Invitrogen, 11140-050) (*see* **Note 2**)	6.25 mL
0.1 M β-mercaptoethanol (ME) (*see* **Note 6**)	625 μL

11. Supplements for human iPS medium.

    bFGF (Peprotech, 100-18B-2).

    Stock solution at 5 μg/mL in 0.1 % (w/v) BSA/PBS. Aliquot into 100 μL and store at −80 °C. Once thawed, keep at 4 °C. Add to human iPS medium at a final concentration of 5 ng/mL.

12. Endoderm Differentiation basal Medium (store at 4 °C).

RPMI 1640 medium (Invitrogen, 11875-093)	500 mL
PS (Nacalai Tesque, 26252-94 (*see* **Note 2**)	5 mL
L-Glutamine (Nacalai Tesque, 16948-04) (*see* **Note 2**)	5 mL
NEAA (Invitrogen, 11140-050) (*see* **Note 2**)	5 mL
0.1 M ME (*see* **Note 6**)	500 μL

13. Supplements for endoderm differentiation Medium (store at 4 °C).

    Activin (R&D, 338-AC).

    Stock solution at 100 μg/mL in 0.1 % (w/v) BSA/PBS. Aliquot into 100 μL and store at −80 °C. Once thawed, keep at 4 °C. Add to endoderm differentiation medium at a final concentration of 100 ng/mL.

    B27 supplement (Invitrogen, 17504-044).

    Stock solution at 100 % (50×). Aliquot into 500 μL and store at −20 °C. Once thawed, keep at 4 °C. Add to endoderm differentiation medium at a final concentration of 2 % (v/v, 1×).

14. Hepatic Differentiation basal Medium (store at 4 °C).

DMEM (Invitrogen, 11885-092, low glucose)	500 mL
KSR (Invitrogen, 10828-028)	58 mL
PS (Nacalai Tesque, 26252-94) (*see* **Note 2**)	5.8 mL

(continued)

L-Glutamine (Nacalai Tesque, 16948-04) (*see* **Note 2**)	5.8 mL
NEAA (Invitrogen, 11140-050) (*see* **Note 2**)	5.8 mL
0.1 M ME (*see* **Note 6**)	580 μL
100 mg/mL Glucose (*see* **Note 7**)	5.8 mL

15. Supplements for hepatic differentiation medium.

    Dexamethasone (Dex, Sigma, #D8893).

    Stock solution at 1 mM in EtOH. Aliquot into 100 μL and store at −80 °C. Once thawed, keep at 4 °C. Add to differentiation medium at a final concentration of 1 μM.

    HGF (Peprotech, #100-39).

    Stock solution at 10 μg/mL in 0.1 % (w/v) BSA/PBS. Aliquot into 100 μL and store at −80 °C. Once thawed, keep at 4 °C. Add to differentiation medium at a final concentration of 10 ng/mL.

3 Methods

3.1 Preparation of Mitomycin C Treated M15 Cells (MMC-M15 Cells)

(a) Thawing M15 cells.

1. Prepare 4 mL of EF medium in a 15 mL tube.
2. Remove a vial of frozen M15 stock and put the vial in a 37 °C water bath until most (but not all) cells are thawed.
3. Wipe the vial with ethanol, open the cap, and transfer the cell suspension to a tube prepared in step 1.
4. Centrifuge at 180 × *g* for 5 min and then discard the supernatant.
5. Resuspend the cells with 10 mL of EF medium, and transfer a 90-mm dish, and incubate the cells in a 37 °C 5 % CO_2 incubator.

(b) Passage of M15 cells.

1. When cells are confluent, aspirate the culture medium, wash the cells with PBS, add 0.05 % trypsin/0.53 mM EDTA (1 mL per 90-mm dish, 3 mL per 150-mm dish), and incubate for 5 min at 37 °C, 5 % CO_2.
2. After incubation, add EF medium into the M15 cells dish (4 mL per 90-mm dish, 6 mL per 150-mm dish), suspend the cells by gently pipetting, and transfer the cell suspension to a 15 mL tube or 50-mL tube.
3. Centrifuge the cells at 180 × *g* for 5 min.
4. Discard the supernatant, break the pellet by finger tapping, and resuspend the cells in an appropriate amount of EF medium.

5. Seed cells at 1.5×10^6 cells per 150-mm dish, and incubate at 37 °C, 5 % CO_2 incubator until they are confluent.

(c) Mitomycin C-inactivation of M15 cells.

1. Discard the medium and add mitomycin C containing medium, and incubate for 2 h at 37 °C, 5 % CO_2.
2. After incubation, aspirate all of mitomycin C containing medium off the cells, and wash the cells twice with PBS.
3. Aspirate off PBS, add 3 mL 0.05 % trypsin/0.53 mM EDTA, and incubate for 5 min at 37 °C, 5 % CO_2.
4. Neutralize the trypsin by adding 3 ml EF medium, and break up the cells to a single cell suspension by pipetting up and down. Pool the cells suspension into 50-mL tubes and count the number of cells.
5. Centrifuge at 180 × *g* for 5 min and then discard the supernatant.
6. Resuspend the cells with EF medium to the concentration at 2×10^7 cells per mL.
7. Add equal volume 2× freeze solution, and mix gently.
8. Transfer 1 mL of the cell suspension into cryovial.
9. Put cryovials into a Nalgene controlled-rate freezer box and then put the box into a −80 °C freezer. The next day, transfer the vials of frozen MMC-M15 cells into the −150 °C freezer for long-term storage. When use frozen cells, thaw 1 vial to two 24-well plates.

3.2 Preparation of Gelatin-Coat Plates

1. Transfer enough 0.1 % gelatin solution to cover the bottom of the plates (i.e., 0.5 mL/well for 24-well dish. Let sit at 37 °C for 2 h (*see* **Note 8**).
2. Remove excess gelatin solution, and add 0.25 mL fresh M15 medium into 24-well gelatin-coated plates (*see* **Note 8**).

3.3 Preparation of MMC Treated M15 Feeder Plates

1. Remove a vial of MMC-M15 cells from −150 °C freezer and plunge into 37 °C water bath, agitating the vials until the frozen suspension becomes slurry.
2. Transfer MMC-M15 cells into a 15 mL tube pre-added with 4 mL EF medium.
3. Collect cells by centrifugation at 180 × *g* for 5 min.
4. Resuspend the pellet with EF medium, cell count, and adjust to a final cell density of 4.0×10^5 cells/mL
5. Plate 0.5 mL MMC-M15 cell suspension into 24-well gelatin-coated plates added with EF medium (Section 3.2).
6. Incubate at 37 °C under 5 % CO_2.

7. On the next day, MMC-M15 cells reach confluence and are ready to be used as feeders for human iPS differentiation (*see* **Note 9**).

3.4 Plating of Human iPS Cells (See Note 10)

1. Remove medium from the human iPS cells.
2. Wash with PBS.
3. Add 1 mL CTK solution to the culture dish. Let stand for 6 min at 37 °C and confirm under microscope for detachment of cells.
4. Remove CTK solution from the iPS cells.
5. Add 2 mL human iPS medium and disaggregate iPS clumps into smaller pieces (5–20 cells) by a cell scraper and pipetting by a P1000 pipet.
6. Add 2 mL human iPS medium and collect the cells by centrifugation, at 180 × *g* for 5 min.
7. Resuspend the pellet with 10 mL human iPS medium.
8. Remove M15 medium from the MMC-M15 cells plates (Section 3.3) and add 0.25 mL fresh human iPS medium into MMC-M15 24-well plate (*see* **Note 11**).
9. Add 0.5 mL human iPS cells suspension into MMC-M15 24-well plate.
10. Incubate at 37 °C under 5 % CO_2.
11. Remove medium from the human iPS cells on the next day.
12. Wash with PBS.
13. Change medium with fresh endoderm differentiation medium supplemented with both Activin and B27 on day 1, 3, 5, 7, 9.
14. Change medium with fresh hepatic differentiation medium supplemented with both Dex and HGF from day 10 to 30, every 2 days.

4 Notes

1. Dissolve three tablets PBS (Sigma, P4417-100TAB) in 600 mL ultrapure water, autoclave, and store at room temperature.
2. Aliquot into 5.8 mL and store at −20 °C. Avoid freeze and thaw.
3. Aliquot into 1 mL and store at −20 °C. Avoid freeze and thaw.
4. Dissolve 10 mg of collagenase IV in 1 mL of distilled water, and through with a 0.22 μm pore filter. Aliquot and store at −20 °C.
5. Dissolve 0.11 g of $CaCl_2$ (Nacalai Tesque, 06729-55) in 10 mL of distilled water, and through with a 0.22 μm pore filter.

6. Dilute 2-mercaptoethanol (Sigma, M7522) to 0.1 M with PBS. (i.e., 2-mercaptoethanol (Sigma, M7522) 100 μL/PBS 14.1 mL. Store at 4 °C and use within 1 month.)
7. Dissolve 10 g D-(+)-Glucose (Sigma, G5146-1KG) in 100 mL PBS and filtrate, and store at 4 °C (100 mg/mL).
8. Dissolve 0.2 g gelatin (Sigma, G9391) in 200 mL ultrapure water. Let stand at room temperature for 1 h, and autoclave, store at room temperature.

   We routinely add gelatin solution on the day before plating of M15 cells, incubate until plating. And just before plating, remove gelatin solution and substitute with fresh differentiation medium.
9. If you are in a rush, you can use M15 feeder dishes 2 h after plating. But we routinely plate MMC treated M15 feeders on the previous day.
10. We plate approximately 70 % confluent human iPS cells in 90-mm dish into one 24-well plate.
11. This is necessary to prevent nonuniform platings.

References

1. Davila JC, Cezar GG, Thiede M, Strom S, Miki T, Trosko J (2004) Use and application of stem cells in toxicology. Toxicol Sci 79:214–223
2. Shiraki N, Umeda K, Sakashita N, Takeya M, Kume K, Kume S (2008) Differentiation of mouse and human embryonic stem cells into hepatic lineages. Genes Cells 13:731–746
3. Shiraki N, Yoshida T, Araki K, Umezawa A, Higuchi Y, Goto H, Kume K, Kume S (2008) Guided differentiation of embryonic stem cells into Pdx1-expressing regional-specific definitive endoderm. Stem Cells 26:874–885
4. Ogaki S, Shiraki N, Kume K, Kume S (2013) Wnt and notch signals guide embryonic stem cell differentiation into the intestinal lineages. Stem Cells 31:1086–1096
5. Umeda K, Suzuki K, Yamazoe T, Shiraki N, Higuchi Y, Tokieda K, Kume K, Mitani K, Kume S (2013) Albumin gene targeting in human embryonic stem cells and induced pluripotent stem cells with helper-dependent adenoviral vector to monitor hepatic differentiation. Stem Cell Res 10:179–194

Methods in Molecular Biology (2016) 1357: 383–394
DOI 10.1007/7651_2014_145

Published online: 20 November 2014

Determination of Functional Activity of Human iPSC-Derived Hepatocytes by Measurement of CYP Metabolism

Pooja Chaudhari, Neha Prasad, Lipeng Tian, and Yoon-Young Jang

Abstract

The advent of induced pluripotent stem cell (iPSC) technology has enabled the modeling of an array of specific human disease phenotypes, aiding in the increasingly important and indispensable understanding of disease progression and pathogenesis. Pluripotent stem cell-derived hepatocytes present a new avenue for drug screening and personalized drug testing toward precision medicine. CYP450 microsomal enzymes play a critical role in drug metabolism. Hence, CYP activity measurement of iPSC-derived hepatocytes is a vital prerequisite, to ensure metabolic functionality before proceeding to drug testing. Herein, we describe the protocol for measurement of different CYP450 enzyme activities in human iPSC-derived hepatocytes.

Keywords: iPSCs, Hepatic differentiation, CYP450, Liver function, Drug metabolism

1 Introduction

In vertebrates, definitive endoderm in the primitive gut tube gives rise to hepatoblasts (1). Hepatoblasts themselves are bipotent/multipotent, giving rise to both hepatocytes and cholangiocytes (2). During this hepatic development, differential stage-specific and liver-specific markers characterize the different stages of maturation. Alpha-fetoprotein (AFP), an early hepatic progenitor marker, is expressed in the liver prior to birth, and the expression rapidly reduces after birth. Albumin (ALB), on the contrary, has been considered as a mature hepatocyte marker, with its expression levels increasing as hepatocytes mature; the expression levels of ALB in fetal hepatocytes are drastically low as compared to those of mature hepatocytes (3).

The human genome contains 57 cytochrome P450 (CYP450) genes and 58 pseudogenes, classified into 18 families (4). Most CYP450 isoforms are not detected in fetal liver or hepatocytes owing to their extremely reduced expression levels; however, their expression becomes more pronounced after birth (5), indicating that increased CYP450 expression is a better marker of functional hepatocytes than ALB expression. These CYP450 enzymes govern oxidative metabolism of intrinsic substrates and extrinsic xenobiotics.

CYP450 enzymes deactivate most drugs directly or by aiding in their excretion from the body. Conversely, some drugs may be biologically activated by CYP450 enzymes, to form either an active or a toxic variant of the compound (6). CYP3A enzyme family accounts for the largest contributor to the CYP450 enzymes and is responsible for metabolizing approximately half of the drugs metabolized in adults (7, 8). The CYP450 enzyme isoforms also have varied expression levels dependent upon hepatocyte maturation state. CYP3A7, for example, is a major enzyme in fetal and newborn liver (9, 10), while CYP3A4 is expressed throughout development and accounts for approximately 10–50 % of total adult liver CYPs (11–14).

A vast majority of drugs are metabolized and degraded by these CYP enzymes in a "first-pass" effect, and approximately 30–40 % of drugs in clinical trials have been withdrawn from further clinical development due to drug-induced epatotoxicity or other undesirable pharmacokinetic properties (15, 16). Thus, a varied number of liver-derived model systems such as liver slices, immortalized cell lines, and primary human hepatocytes have been utilized to study drug metabolism. However, due to disadvantages such as the short-term viability, diffusional barriers of liver slices, and the lack of liver-specific functions in the immortalized cell lines, primary hepatocytes have been the de facto standard for in vitro pharmacokinetic and toxicity testing of drugs for the last decade (17–20).

However, the use of human primary hepatocytes for drug testing also presents some obstacles, such as the rapid loss of metabolic or functional activity ex vivo, batch-to-batch difference associated with their limited resource and individual variations, and their noncompliance with expansion in culture (21). Human pluripotent stem cell-derived hepatocytes thus present a hitherto unexplored avenue for consistent drug testing, due to the potential of pluripotent stem cells being a seemingly endless source of genetically defined hepatocytes. The extensive interindividual variations observed in human drug metabolism are one of the major obstacles for drug development, which can lead to adverse effects, drug failure, or specific population-related organ toxicity issues (5, 22). These metabolic variations can arise from a multitude of factors such as physiological states, disease status, genetic abnormalities and polymorphisms, environmental factors, or concurrent drug therapies (5). Human iPSC-derived hepatocytes can be used to create a representative library of varied populations and subpopulations with these inherent differences in drug metabolism, which can then be used for predictive drug testing.

Here we describe the protocol for CYP450 enzyme activity testing in human iPSC-derived hepatocytes via our preestablished hepatic differentiation protocol (23–26).

2 Materials

This study was performed in accordance with the Johns Hopkins Intuitional Stem Cell Research Oversight regulations and followed approved protocols by the Johns Hopkins Institutional Review Board.

2.1 Human Induced Pluripotent Stem Cell (iPSC) Lines

- The human iPSC lines used in this study were previously generated from diverse healthy donor tissues (23–26) and cultured in a feeder-free condition (mTeSR1 medium- and Matrigel-coated plates).

2.2 Matrigel

- BD Matrigel, hESC-qualified Matrix (BD Biosciences, Cat #354277).
- DME/F-12 1:1 (1×) (HyClone, Cat #SH30023.01).
- BioLite 12 Well Multidish (Thermo Fisher Scientific, Cat #130185).

2.3 iPSC Culture Medium

- Prepare mTeSR medium by adding 1× mTeSR Supplement (Stemcell Technologies, Cat# 05850, component #05852/100 ml) and 1 % Pen Strep (Life Technologies, Cat #15140-122) to mTeSR1 Basal Medium (Stemcell Technologies, Cat #05850, component #05851/400 ml).
- Until required, store at 4 °C. Bring the medium to room temperature before use.
- Collagenase Type IV (Sigma, Cat.No. C5138-5 g), store at 4 °C. Prepare 1 mg/ml collagenase IV solution with DMEM/F12 (Cellgro, Cat #10-092-CV) and filter for sterilization. Store the collagenase solution at 4 °C.
- Accutase solution (Sigma, Cat.No. A6964-100 ml), store at −20 °C.

2.4 iPSC Differentiation Medium

- The differentiation medium used to induce hepatocytes from iPSCs is varied. RPMI (Gibco-Life Technologies, Cat #61870-036) is used as a basal medium.
- Activin A (R&D Systems, Cat# 338AC/CF), CHIR-99021 (Tocris, Cat #4423), B27 supplement (Gibco-Life Technologies, Cat# 17504.044), hepatocyte growth factor (HGF) (R&D Systems, Cat# 294-HG/CF), and fibroblast growth factor 4 (FGF4) (R&D Systems, Cat# 235-F4/CF) are used in concentrations listed in Section 3.3. 1 % FBS may be used in conjunction with this medium to improve cell viability.

2.5 Hepatocyte Culture Medium (HCM)

- Prepare HCM by adding 1× GlutaMAX (Gibco, Cat #35050), 15 mM 4-(2-hydroxyethyl)-1-piperazineethanesulfonic acid

(Hepes) (Sigma-Aldrich, Cat #H0887-100 ml), 1× ITS (Mediatech, Cat #25-800-CR), 10^{-7} M dexamethasone in DMSO (Sigma-Aldrich), and 0.1 % gentamicin (Sigma-Aldrich, Cat #G1397) to Williams' E medium (Invitrogen, Cat #12551-032).

- Filter the medium and store at 4 °C. Bring the medium to room temperature before use.
- Supplement the HCM with 10 ng/ml hepatocyte growth factor (HGF), 10 ng/ml fibroblast growth factor 4 (FGF4) and 10 ng/ml Oncostatin M (OSM) (R&D Systems, Cat #295-OM), prior to adding the medium to the culture wells.

2.6 Promega CYP Assay Kits

- P450-Glo™ CYP3A4 Assay (Luc-PFBE), Cell-Based/Biochemical Assay (#V8901/8902), P450-Glo™ CYP1A2 Assay (# V8772), P450-Glo™ CYP2C9 Assay (# V8791), P450-Glo™ CYP2D6 Assay (# V8892) P450-Glo™ CYP2C19 Assay (# V8881).
- Store at −20 °C until required.
- Thaw the substrates and the detection reagents provided with the assays on the day of use by transferring to 4 °C. After use, store at −20 °C.

2.7 Assay Plate

- A Costar 96-well white solid flat well plate (Corning, #3917) is used for CYP measurement.

2.8 Luminometer

- Turner Biosystems Instrument (model# 9101-002).

2.9 Quantitative Real-Time PCR

- TRIzol Reagent (Ambion, Cat # 15596018).
- High Capacity cDNA Reverse Transcription Kit (Applied Biosystems, Cat # 4368814). Store at -20 °C, and thaw on ice before use.
- StepOnePlus Real-Time PCR System (Applied Biosystems, Cat # 4371435).
- TaqMan Fast Universal PCR Master Mix (2×) (Applied Biosystems, Cat # 4367846). Store at 4 °C, thaw to room temperature as required.
- TaqMan gene expression assay reagent for 18S rRNA (Hs03003631_gl), albumin (Hs00609411_ml), CYP3A4 (Hs00430021_ml), CYP2D6 (Hs03043789_gl), CYP1A2 (Hs01070374_ml), CYP7A1 (Hs00167982_ml), and CYP2E1 (Hs00559367_ml). Store at −20 °C and thaw on ice when required.

3 Methods

3.1 Matrigel

3.1.1 Preparation of the MG Aliquots

1. Thaw the Matrigel by keeping the bottle at 4 °C overnight.
2. Prepare 1/4 dilution of Matrigel in DME/F-12 by adding 15 ml DME/F12 to the bottle containing 5 ml Matrigel and mix well.
3. Aliquot 5 ml of 1/4 MG into 50 ml tubes and store at −20 °C.

3.1.2 Preparation of Matrigel Plates

1. Thaw the Matrigel aliquots when required and dilute with DME/F-12, to make a final concentration of 1/400. Ensure that the Matrigel is placed in an ice box at all times (*see* **Note 1**).
2. Label BioLite 12-well plates and add 1 ml of the diluted Matrigel per well.
3. Mix the Matrigel well after every 2–3 plates.
4. Seal the plates with aluminum foil, incubate them at room temperature for 1 h, and store the plates at 4 °C.

3.2 Human iPSC Maintenance

1. Culture the iPSCs in 12-well plates pre-coated with Matrigel, using mTeSR1 medium. Replace the medium every day until the cells reach the desired confluency for passaging or differentiation.
2. To passage the cells, add 0.5 ml Accutase to each well and incubate at 37 °C for 3–4 min.
3. Transfer the cells to a 5 ml centrifuge tube and spin them at 1,000 rpm for 5 min at room temperature.
4. Bring the Matrigel-coated plates to room temperature. Remove the Matrigel from the wells and add 0.5 ml mTeSR medium to each well.
5. After centrifugation, discard the supernatant and resuspend the pellet in mTeSR medium.
6. Add 1–10 μM Y-27632 dihydrochloride (TOCRIS, Cat #1254), mix gently with a pipette, and transfer 0.5 ml of the cell suspension per well, in as many plates as desired.
7. Gently move the plates back and forth to ensure uniform distribution and incubate the plates at 37 °C.
8. Observe the cells the next day, and depending on the confluency, replace mTeSR medium or proceed with hepatic directed differentiation.

3.3 Human iPSC Differentiation

1. Once cells reach a 40–60 % confluence status, start the directed hepatic differentiation process (*see* **Note 2**).
2. Day 0: Prepare RPMI medium supplemented with 100 ng/ml Activin A and 1–2 μM CHIR-99021. Mix well by vortexing.

Add 1 ml to iPSC cells per well of a 12-well plate. Change medium everyday.

3. Day 1–4/5: Prepare RPMI medium supplemented with 100 ng/ml Activin A and B27 supplement ranging from 0.05 % to 2 %. Mix well by vortexing. Add 1.5–2.5 ml/well of this medium to the cells depending on cell confluency and viability. This medium may be further supplemented with 1 % defined FBS for a more protective effect (*see* **Note 3**). Change medium everyday
4. Day 4/5–9/10: Prepare RPMI medium supplemented with HGF and FGF-4 10 ng/ml each and B27 supplement ranging from 0.1 % to 1 %. Mix well by vortexing. Add 1.5–2.5 ml/well of this medium to the cells depending on cell number and viability. Change medium everyday.
5. Day 9/10–36: Prepare HCM medium supplemented with the cytokines HGF, FGF4, and OSM, each at a final concentration of 10 ng/ml. Add 1–3 % FBS to this medium for a protective effect and to improve viability. Mix well by vortexing. Add 1–2 ml medium per well to the cells gently. Change medium everyday (*see* **Note 4**).

3.4 CYP Measurement

CYP activity can be measured from day 0 to day 36 for iPSC-derived multistage hepatic cells (Figs. 1 and 2). Proceed as per the Promega kit protocol.

3.4.1 Detection Reagent Reconstitution

1. Bring the lyophilized Luciferin detection reagents and reconstitution buffer (3A4, 1A2, and 2C9) or reconstitution buffer with esterase (2D6 and 2C19) to room temperature.

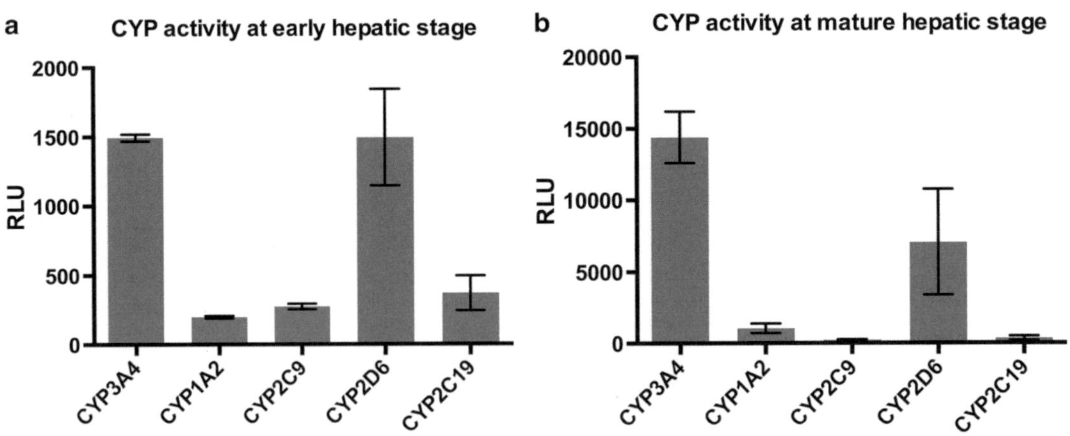

Fig. 1 Multiple CYP450 activities at immature- and mature-hepatocyte-like cells derived from human iPSCs. (**a**) Metabolic activities of CYP3A4, CYP1A2, CYP2C9, CYP2D6, and CYP2C19 measured at early hepatocyte stage (day 13–14). (**b**) Activities of the same five CYP450 enzymes were measured at more mature hepatocyte stage (day 20–22). CYP3A4 and CYP2D6 are highly expressed at the early hepatic stage (**a**) and significantly increase as the iPSC-derived hepatocytes mature (**b**). Relative luminescence units (RLU), means ± SEM

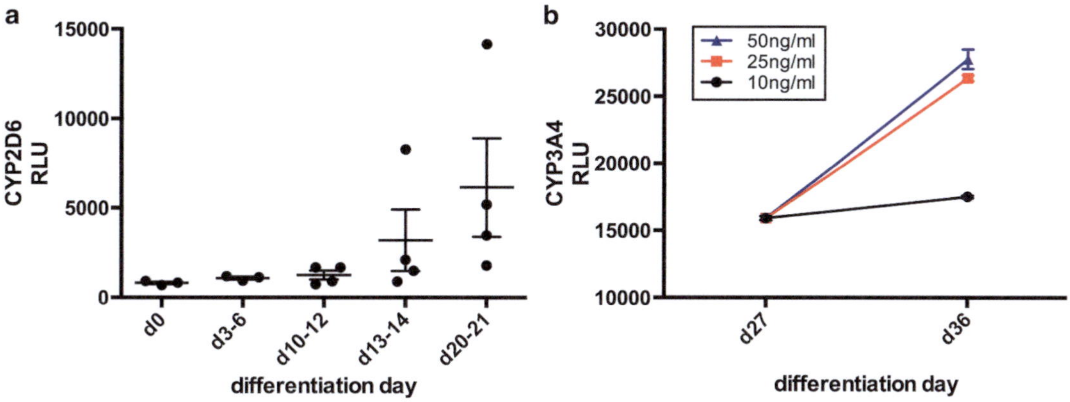

Fig. 2 Kinetics and variability of CYP2D6 activity at different stages of hepatic differentiation process from human iPSCs and potential for increasing CYP3A4 activity by growth factor-based stimulation. (**a**) CYP2D6 activities measured at various hepatic differentiation stages from human iPSCs. Both the CYP2D6 activity and the individual variability increase with maturation of iPSC-derived hepatocytes. (**b**) Day 27 iPSC-hepatic cells were treated with different doses of HGF (10, 25, and 50 ng/ml) for 9 days and the day 36 cells were analyzed for CYP3A4 activity. The CYP3A4 activity further increases in response to higher HGF concentrations in the hepatic differentiation medium. Relative luminescence units (RLU), means $\pm$ SEM

2. Add the contents of one bottle of reconstitution buffer to the contents of bottle containing the lyophilized Luciferin detection reagent. Label the reconstituted detection reagents with the name of the corresponding CYP detected and do not mix the two different detection reagents.
3. Mix by inverting several times to obtain a homogeneous solution. For long-term storage, store at –20 °C for up to 3 months. Move to 4 °C on the day of using 3 h before use.
4. Mix the thawed Luciferin detection reagent well before use by pipetting (*see* **Note 5**).

3.4.2 Preparation of P450-Glo Luminogenic Substrates

1. Luciferin-ME (1A2), Luciferin-H (2C9), and Luciferin-PFBE (3A4) are offered in the kit as aqueous solutions. Thaw the substrate solutions as needed, and keep on ice in dark. Use as is, according to dilution ratio.
2. Luciferin-H EGE (2C19) and Luciferin-ME EGE (2D6) are offered as dried pellets in the kit. Dissolve these pellets in acetonitrile to a final concentration of 10 mM. Store unused substrate at –20 °C, protected from light.
3. Thaw on the day needed.
4. Vortex well and spin down before using.

3.4.3 Detection of CYP in Adherent Hepatocytes

1. Vortex and spin down all CYP substrates before using. Proceed in dark.
2. Aliquot 500 μl of HCM basal medium supplemented with ITS (no cytokines, no FBS) in each Eppendorf for each CYP to be measured.

3. Measure each CYP in distinct wells and include one well to serve as control. Label as such.
4. Add each CYP substrate to the labeled Eppendorfs (*see* **Note 6**) according to the dilution ratios indicated in the Promega kit: 3A4 (Luciferin-PFBE, 1:40), 1A2 (Luciferin-ME, 1:50), 2C9 (Luciferin-H, 1:50), 2D6 (Luciferin-ME EGE, 1:333), and 2C19 (Luciferin-H EGE, 1:1,000).
5. Mix each CYP substrate mixture well by pipetting with 1 ml tips three to four times.
6. To the control Eppendorf, add only 500 μl of HCM supplemented with ITS (no FBS, no cytokines/growth factors).
7. From the cells of which CYP activity is to be measured, remove the old medium from the wells, and wash each well with 1 ml PBS. Remove the PBS.
8. Add 488 μl of each of the previously prepared CYP substrate mixture to the labeled wells individually. For the control well, add 488 μl of HCM supplemented with ITS from the appropriate Eppendorf (*see* **Note 7**).
9. Incubate the plate for 3 h in the dark in an incubator at 37 °C, 5 % CO_2.
10. Before the incubation time ends, prepare 6 new labeled Eppendorf tubes to collect the cell-treated CYP substrate mixture and a new white 96-well assay plate.
11. Also make sure to mix the thawed reconstituted detection reagents by pipetting individually.

3.4.4 Measurement of CYP Activity in Adherent Hepatocytes

1. Proceed in dark. Collect the cell-treated CYP substrate mixture from the wells to the appropriate individually labeled Eppendorfs. For the cells, add HCM supplemented with ITS, HGF, FGF4, and OSM (with or without FBS) and continue to culture for further experiments.
2. Add 50 μl of each individual cell-treated CYP substrate mixture to the wells of appropriately labeled 96-well plate. Make sure to have at least three repeats for each CYP to be detected. There should be two sets of 3 wells each for the control cell-treated CYP substrate mixture, to get the baseline measurements of each of the two different detection reagents.
3. Keep both the detection reagents ready to use, and set the timer to 0. Start the timer when starting to add detection reagent in the first well. The plate needs to be incubated for 20 min with the detection reagents before measuring the CYP activity (*see* **Note 5**).
4. Make sure to add the appropriate detection reagents to each cell-treated CYP substrate mixture. Add 50 μl detection reagent to wells containing substrates for control, 3A4, 1A2, and 2C9.

No color change will be detected. Add 50 μl detection reagent (with esterase) to wells containing substrates for control, 2D6, and 2C19. On addition of detection reagent with esterase, the color will change from pink to yellow (*see* **Notes 8** and **9**).

5. Add the detection reagent in the order in which the luminometer will read the wells, and finish adding both the detection reagents within the time required for the luminometer to finish reading the entire plate.
6. Keep the plate covered in foil, as the reaction is light sensitive. After exactly 20 min of incubation, measure the CYP activity using the luminometer. *See* **Note 9**. Data can be normalized per albumin-expressing hepatocytes using immunostaining or flow cytometry as described in our previous papers (23–26).

3.4.5 Quantification of CYP mRNA Expression in Adherent Hepatocytes

1. Collect the adherent iPSC-derived hepatocytes using trypsin 0.05 %. Spin down the cells to obtain a pellet.
2. Resuspend the pellet in the appropriate amount of TRIzol and proceed for RNA extraction using the manufacturer's instructions.
3. Measure the RNA concentration and proceed for cDNA synthesis using the High Capacity cDNA Reverse Transcription Kit, using the manufacturer's instructions.
4. Proceed to Q-PCR setup for a 20 μl reaction for each well in a 96-well plate as below: Target Taqman gene expression assay reagent 0.5 μl, Taqman Fast Universal PCR Master Mix (2×) 5 μl, cDNA 2ug, and distilled water up to 14.2 μl. Put the plate in the StepOne Plus Real-Time PCR System, set up, and run the experiment as required.
5. Analyze the data to find fold increase of the target gene over control gene (18S). The data can be compared or normalized with albumin expression (Fig. 3).

4 Notes

1. When preparing Matrigel plates, ensure that Matrigel is kept at 4 °C or below by keeping the Matrigel in an ice box, to prevent Matrigel from clumping.
2. When starting directed hepatic differentiation of iPSCs, make sure all the wells are similar confluency states. 40–60 % confluency depending on the proliferative nature of the cell line is a good starting point for beginning endoderm differentiation.
3. During hepatic differentiation, the differentiation medium may be supplemented with 1–2 % FBS to improve cell viability.

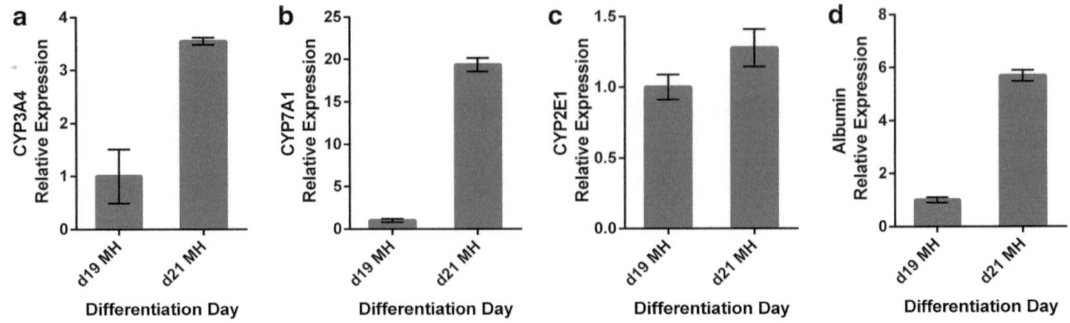

Fig. 3 Multiple CYP450 expression of human iPSC-derived mature hepatocyte-like cells measured using quantitative PCR. CYP450 mRNA expression including CYP3A4 (**a**), 7A1 (**b**), and 2E1 (**c**) were measured in day 19 and day 21 mature hepatic cells derived from human iPSCs. Albumin (**d**) expression was measured as a control and 18S was used for normalization. Human iPSC-derived mature hepatocyte-like cells (MH)

4. Furthermore, if cell number is too low at the hepatic progenitor or mature hepatocyte stage, pool the cells together using 0.05 % trypsin. Spin down these collected cells at 50–100 × *g* for 10 min and replate them onto collagen I-coated plates. Supplementing the differentiation medium with 1–2 % FBS at this stage may improve cell adherence and viability.
5. Keep the detection reagents protected from light at all times, even when thawing. Also, keep the CYP measurement plate protected from light on addition of the detection reagent during the 20-min incubation period.
6. CYP 2D6 and 2C19 substrates are less viscous and one needs extra caution while pipetting these substrates.
7. Measure all CYP450 activities in wells with similar confluency and differentiation status.
8. No color change will be detected on addition of detection reagent without esterase to the corresponding CYP3A4, CYP1A2, or CYP2C9 cell-treated substrates.
9. On addition of detection reagent with esterase to the corresponding CYP2D6 or CYP2C19 cell-treated substrates, the color changes to yellow.

Acknowledgments

This work was supported in part by grants from Maryland Stem Cell Research Funds (2010-MSCRFII-0101 and 2013-MSCRFII-0170 and 2014-MSCRFF-0655) and by NIH (R43 ES023514, R21AA020020).

References

1. Lavon N, Benvenisty N (2005) Study of hepatocyte differentiation using embryonic stem cells. J Cell Biochem 96(6):1193–1202. doi:10.1002/jcb.20590
2. McLin VA, Zorn AM (2006) Molecular control of liver development. Clinics Liver Dis 10 (1):1–25. doi:10.1016/j.cld.2005.10.002, v
3. Inamura M, Kawabata K, Takayama K, Tashiro K, Sakurai F, Katayama K, Toyoda M, Akutsu H, Miyagawa Y, Okita H, Kiyokawa N, Umezawa A, Hayakawa T, Furue MK, Mizuguchi H (2011) Efficient generation of hepatoblasts from human ES cells and iPS cells by transient overexpression of homeobox gene HEX. Mol Ther 19(2):400–407. doi:10.1038/mt.2010.241
4. Nelson DR (2009) The cytochrome p450 homepage. Human Genomics 4(1):59–65
5. Ingelman-Sundberg M, Oscarson M, McLellan RA (1999) Polymorphic human cytochrome P450 enzymes: an opportunity for individualized drug treatment. Trends Pharmacol Sci 20 (8):342–349
6. Elens L, van Gelder T, Hesselink DA, Haufroid V, van Schaik RH (2013) CYP3A4*22: promising newly identified CYP3A4 variant allele for personalizing pharmacotherapy. Pharmacogenomics 14(1):47–62. doi:10.2217/pgs.12.187
7. Domanski TL, Finta C, Halpert JR, Zaphiropoulos PG (2001) cDNA cloning and initial characterization of CYP3A43, a novel human cytochrome P450. Mol Pharmacol 59 (2):386–392
8. Finta C, Zaphiropoulos PG (2000) The human cytochrome P450 3A locus. Gene evolution by capture of downstream exons. Gene 260 (1–2):13–23
9. Stevens JC, Hines RN, Gu C, Koukouritaki SB, Manro JR, Tandler PJ, Zaya MJ (2003) Developmental expression of the major human hepatic CYP3A enzymes. J Pharmacol Exp Ther 307(2):573–582. doi:10.1124/jpet.103.054841
10. Lacroix D, Sonnier M, Moncion A, Cheron G, Cresteil T (1997) Expression of CYP3A in the human liver–evidence that the shift between CYP3A7 and CYP3A4 occurs immediately after birth. Eur J Biochem 247(2):625–634
11. Shimada T, Yamazaki H, Mimura M, Inui Y, Guengerich FP (1994) Interindividual variations in human liver cytochrome P-450 enzymes involved in the oxidation of drugs, carcinogens and toxic chemicals: studies with liver microsomes of 30 Japanese and 30 Caucasians. J Pharmacol Exp Ther 270 (1):414–423
12. Paine MF, Khalighi M, Fisher JM, Shen DD, Kunze KL, Marsh CL, Perkins JD, Thummel KE (1997) Characterization of interintestinal and intraintestinal variations in human CYP3A-dependent metabolism. J Pharmacol Exp Ther 283(3):1552–1562
13. Shimada T, Yamazaki H, Mimura M, Wakamiya N, Ueng YF, Guengerich FP, Inui Y (1996) Characterization of microsomal cytochrome P450 enzymes involved in the oxidation of xenobiotic chemicals in human fetal liver and adult lungs. Drug Metab Dispos 24(5):515–522
14. Ince I, Knibbe CA, Danhof M, de Wildt SN (2013) Developmental changes in the expression and function of cytochrome P450 3A isoforms: evidence from in vitro and in vivo investigations. Clin Pharmacokinet 52(5):333–345. doi:10.1007/s40262-013-0041-1
15. Kondo Y, Iwao T, Nakamura K, Sasaki T, Takahashi S, Kamada N, Matsubara T, Gonzalez FJ, Akutsu H, Miyagawa Y, Okita H, Kiyokawa N, Toyoda M, Umezawa A, Nagata K, Matsunaga T, Ohmori S (2014) An efficient method for differentiation of human induced pluripotent stem cells into hepatocyte-like cells retaining drug metabolizing activity. Drug Metab Pharmacokinet 29(3):237–243
16. Wilke RA, Lin DW, Roden DM, Watkins PB, Flockhart D, Zineh I, Giacomini KM, Krauss RM (2007) Identifying genetic risk factors for serious adverse drug reactions: current progress and challenges. Nat Rev Drug Discov 6 (11):904–916. doi:10.1038/nrd2423
17. Ferrini JB, Pichard L, Domergue J, Maurel P (1997) Long-term primary cultures of adult human hepatocytes. Chem Biol Interact 107 (1–2):31–45
18. Guillouzo A (1998) Liver cell models in in vitro toxicology. Environ Health Perspect 106(Suppl 2):511–532
19. Kern A, Bader A, Pichlmayr R, Sewing KF (1997) Drug metabolism in hepatocyte sandwich cultures of rats and humans. Biochem Pharmacol 54(7):761–772
20. LeCluyse EL (2001) Human hepatocyte culture systems for the in vitro evaluation of cytochrome P450 expression and regulation. Eur J Pharm Sci 13(4):343–368
21. Sengupta S, Johnson BP, Swanson SA, Stewart R, Bradfield CA, Thomson JA (2014) Aggregate culture of human embryonic stem cell-derived hepatocytes in suspension are an improved in vitro model for drug metabolism

and toxicity testing. Toxicol Sci 140 (1):236–245. doi:10.1093/toxsci/kfu069

22. Lazarou J, Pomeranz BH, Corey PN (1998) Incidence of adverse drug reactions in hospitalized patients: a meta-analysis of prospective studies. JAMA 279(15):1200–1205
23. Choi SM, Liu H, Chaudhari P, Kim Y, Cheng L, Feng J, Sharkis S, Ye Z, Jang YY (2011) Reprogramming of EBV-immortalized B-lymphocyte cell lines into induced pluripotent stem cells. Blood 118(7):1801–1805. doi:10.1182/blood-2011-03-340620
24. Liu H, Kim Y, Sharkis S, Marchionni L, Jang YY (2011) In vivo liver regeneration potential of human induced pluripotent stem cells from diverse origins. Sci Transl Med 3(82):82ra39. doi:10.1126/scitranslmed.3002376
25. Liu H, Ye Z, Kim Y, Sharkis S, Jang YY (2010) Generation of endoderm-derived human induced pluripotent stem cells from primary hepatocytes. Hepatology 51(5):1810–1819. doi:10.1002/hep.23626
26. Choi SM, Kim Y, Shim JS, Park JT, Wang RH, Leach SD, Liu JO, Deng C, Ye Z, Jang YY (2013) Efficient drug screening and gene correction for treating liver disease using patient-specific stem cells. Hepatology 57 (6):2458–2468. doi:10.1002/hep.26237

Methods in Molecular Biology (2016) 1357: 395–401
DOI 10.1007/7651_2014_148

Published online: 18 November 2014

Induced Pluripotent Stem Cells: Generation, Characterization, and Differentiation—Methods and Protocols

Veronica Kon Graversen and Sai H. Chavala

Abstract

Reprogramming fibroblasts into induced pluripotent stem cells (iPSC) remains a promising technique for cell replacement therapy. Diverse populations of somatic cells have been examined for their reprogramming potential. Recently, ocular ciliary body epithelial cells (CECs) have been reprogrammed with high reprogramming efficiency and single transcription factor reprogramming, making them an exciting candidate for cellular reprogramming strategies.

Keywords: Ciliary body epithelial cells, Somatic cells, Induced pluripotent stem cells, Reprogramming, Transcription factors

1 Introduction

Induced pluripotent stem cells (iPSC) obtained from reprogrammed somatic cells can be differentiated into cells of all three germ layers (1, 2). Previous studies focused on fibroblasts as the main source of iPSC; however, other cell sources can be reprogrammed and have conceivable advantages for cell replacement therapy (1, 3). We recently reported efficient reprogramming methods using ciliary body endothelial cells (CECs). The ciliary body is composed of multiple ciliary processes covered by a specialized epithelium of neuroectodermal origin. We reported that CECs can be reprogrammed as a monolayer using two factors, *Oct 3/4* and *Klf4*, or with only one factor, *Oct 3/4*, preceded by sphere formation (4, 5). CECs form spheres in serum free conditions and express neural stem cell markers, similar to adult neural stem cells, and undergo a partial reprogramming process in these conditions (5–7). This partial reprogramming process can be exploited for single transcription factor iPSC reprogramming with overexpression of only *Oct 3/4* without the addition of small molecules, a characteristic shared by few known cell types (8–10). In addition, the ciliary body is a surgically accessible region, which offers an advantage for their use in regenerative ophthalmology.

2 Materials

Supplies:

1. Dissection microscope and light source.
2. 35 mm petri dishes.
3. 60 mm petri dishes.
4. 24 well plates (VWR, 73521-004).
5. 15 falcon tubes.
6. Glass pipettes (fire-polished and small borehole fire-polished).
7. Serrated curved forceps.
8. Non-serrated curved forceps.
9. Straight forceps (#5 Dumont).
10. Microdissection scissors.
11. Scalpel or micro-knife.

Solutions:

1. Regular Artificial Cerebral Spinal Fluid (ACSF) (7) consists of (*see* **Note 1**):
   (a) 124 mM NaCl.
   (b) 5 mM KCl.
   (c) 1.3 mM $MgCl_2$.
   (d) 2.4 mM $CaCl_2$.
   (e) 26 mM $NaHCO_3$.
   (f) 10 mM D-glucose.
2. HiLo ACSF: same as regular ACSF with two modifications (7):
   (a) Higher concentration of Mg^{2+}: 3.2 mM $MgCl_2$.
   (b) Lower concentration of Ca^{2+}: 0.1 mM $CaCl_2$.
3. CEC culture medium (Neurobasal (Invitrogen, 21103-049)):
   (a) Serum free with growth factors (sphere):
      - 1× B27 supplement (Invitrogen, 17504-044).
      - 2 mM L-glutamine (Invitrogen, 25030-081).
      - Basic fibroblast growth factor (20 ng/ml) (Peprotech, 100-18B).
      - Epidermal Growth Factor (20 ng/ml) (Peprotech, AF-100-15).
      - Heparin (2 μg/ml) (Sigma, H-3149).
      - 1 % penicillin/streptomycin.

   (b) Serum medium (monolayer, attached cells)—Same as (a) plus 1 % fetal bovine serum (Gibco/Invitrogen 16000-044).

4. CEC reprogramming medium (ES cell medium) (3):
   (a) Dulbecco's modified Eagle Medium (DMEM) (Sigma D-5030).
   (b) 15 % FBS.
   (c) 2 mM L-glutamine.
   (d) 100 μM nonessential amino acids.
   (e) 100 μM 2-mercaptoethanol.
   (f) Penicillin and streptomycin.
5. Dispase (VWR, 47743-724).
6. Trypsin (1.33 mg/ml) (Sigma, T-1005).
7. Hyaluronidase (0.67 mg/ml) (Sigma, H-6254).
8. Kynurenic Acid (0.2 mg/ml) (Sigma, K-3375).
9. Trypsin Inhibitor (1 mg/ml) (Roche, 10109886001).
10. Accutase (Sigma, A-6964).
11. Lentiviruses containing CMV promoter-driven human Oct 3/4, Sox2, Klf4, and c-Myc (Cellomics Technology, PLV-10030-25).
12. SNL feeder cell [a mouse fibroblast STO—(S, SIM; T, 6-thioguanine resistant; O, ouabain resistant)—cell line-derived, neomycin-resistant, leukemia inhibitory factor (LIF) producing cell line] (Cell Biolabs, CBA-316).

Tools:

1. Curved serrated forceps.
2. Curved non-serrated forceps (Dumont #7).
3. Jeweler's forceps or a Dumont #5 forceps.
4. Curved Vannas scissors.
5. Straight Vannas scissors.
6. 11 blade scalpel.

3 Methods

Begin by setting up the dissecting microscope and light source, and the autoclaved sterile dissecting instruments.

3.1 Dissection of the Adult Mouse Ciliary Body and CEC Isolation

1. Euthanize adult mice and enucleate the eyes using a curved non-serrated forceps, and place the eyes in a 35 mm petri dish with 3–5 ml of regular ACSF (*see* **Notes 2** and **3**).
2. Using the microscope, debride the external elements of the eyes by securing each eye with a curved serrated forceps, and

removing hair, blood vessels, and ocular muscles with a curved Vannas scissor.

3. Transfer the eyes into a new 35 mm petri dish containing 1–2 ml of regular ACSF.
4. Using a curved Vannas scissor poke one blade into the sclera–optic nerve junction, and attempt to bisect the eye into two halves in the sagittal plane (i.e., proceed to cut anteriorly through the middle of the cornea and back to the starting point) (*see* **Note 4**).
5. Remove the lens, vitreous and the neural retina from the eye shell with two jeweler's forceps. Move each eyecup into a new petri dish with regular ACSF.
6. Under higher magnification, use an 11 blade scalpel and a straight jeweler's forceps to cut the edge between the posterior ciliary body and RPE for the circumferential length of the eyecup. This maneuver removes the posterior segment of the eye. Then, cut the edge between the posterior iris and anterior ciliary body to remove the anterior segment of the eye. This step will result in strips of ciliary body that will then undergo enzyme treatment.
7. Transfer all of the ciliary body strips into a new 35 mm petri dish with 2 ml of room temperature Dispase for 10 min.
8. Once complete, transfer the strips into another 35 mm dish that contains 2 ml of a mixture of Kynurenic Acid, Trypsin, and Hyaluronidase and place the petri dish in an incubator at 37 °C for 10 min. Of note, Kynurenic Acid should be dissolved in HiLo ACSF, and the other two enzymes in regular ACSF. The enzyme solution should be heated in a 37 °C water bath.
9. Using the microscope, gently scrape the epithelium off the sclera using a jeweler's forceps and a curved non-serrated forceps. After enzyme treatment, the epithelial cells of the ciliary body should be easily liberated from the scleral strip.
10. Discard the scleral strips so that all that is left in the dish is the enzyme solution and the cells of interest.
11. Transfer this solution, using a fire-polished pipette, into a 15 ml centrifuge tube.
12. Aspirate and expel cell aggregates with a fire-polished pipette. Triturate approximately 30 times to break apart the cells and then place the tube in a centrifuge at 450 × *g* for 5 min.
13. Gently aspirate the supernatant off the cells.
14. Add the Trypsin inhibitor dissolved in CEC medium (1–1.5 ml) to the cells and triturate until it becomes a single cell suspension (approximately 50–60 times).
15. Centrifuge the tube at 450 × *g* for 5 min.

16. Discard the supernatant and replace with plating medium (serum-free CEC sphere medium or serum CEC monolayer growth medium) at the desired density (usually 10–20 cells/μl) in a 24-well plate. A Hemocytometer can be used to measure cell density (*see* **Note 5**).
17. Culture cells in a 37 °C CO_2 incubator.

3.2 iPSC Generation

1. For protocols involving spheres, CEC spheres, cultured for approximately 7 days, were dissociated into single cells with Accutase (Sigma) and plated at 1×10^5 cells/well on gelatin-coated six-well plates in 1 % FBS-containing CEC culture medium to promote attachment. CECs were expanded for approximately 8 days prior to reprogramming.
2. For monolayer conditions (i.e., without sphere formation), CECs were plated at 1×10^5 cells/well of gelatin-coated six-well plates in CEC growth medium, and expanded for approximately 8 days.
3. Transduction of a combination of concentrated lentiviruses containing CMV promoter-driven human *Oct 3/4*, *Sox2*, *Klf4*, and *c-Myc* were added to the cells at a multiplicity of infection of 10 with 4 μg/ml polybrene in CEC growth medium. We also successfully generated iPSC with *Oct 3/4*, *Sox2*, and *Klf4* and *Oct 3/4* and *Klf4*. We did not notice iPSC with *Oct 3/4* and *Sox2*. *Oct 3/4* alone resulted in iPSC only if sphere formation preceded reprogramming.
4. Twenty-four hours post-infection, the viral infection mixture was exchanged for fresh CEC growth medium.
5. The following day, transduced CECs were subcultured onto mitomycin C-treated SNL feeder cells in six-well plates at a split ratio of 1:4. CEC reprogramming medium (ES cell medium) without LIF containing DMEM supplemented with 15 % FBS, 2 mM L-glutamine, 100 μM nonessential amino acids, 100 μM 2-mercaptoethanol, 50 U/ml penicillin, and 50 μg/ml streptomycin was used until the characteristic iPSC morphology was noted.

4 Notes

1. Dr. van der Kooy's group reported that ACSF should be bubbled with 95 % O_2/5 % CO_2 for 15–20 min. However, we did not perform this step in our experiments. Also, the pH should be 7.35.
2. We typically perform each experiment with six mice and 12 eyes but this can vary depending on the yield required to achieve the goals of the experiment.

3. Our reprogramming studies were performed with CD1, albino mice, because in our hands we had higher sphere yields although this has not been experimentally quantified. This also has not been reported by other groups to our knowledge, and may be unique to our laboratory. The use of non-pigmented eyes also permitted us to easily detect fluorescence when performing antibody staining, which can be partially masked with pigmented CECs.
4. This step can be challenging for an individual inexperienced with this technique. One tip is to secure the eye with the serrated forceps in the non-dominant hand, and gently insert one blade of the Vannas scissor with the dominant hand into the edge of the sclera–optic nerve junction. If it is difficult to insert the blade of the scissor into the eye, poking a hole into the cornea using a 30-gauge needle can be helpful because it reduces the intraocular pressure. This allows for easier penetration of the scissor blade into the sclera–optic nerve junction.
5. Our group has found that when performing this technique in pigmented eyes, liberated pigment during the first week of cell culture can mimic bacterial contamination. We have noted extracellular pigment dispersion consistently, and have not found this to be a result of bacterial contamination with extended periods of cell culture.

Acknowledgements

SHC would like to thank and acknowledge the efforts of Dr. Derek van der Kooy and Brenda Coles who developed and shared their ciliary body dissection technique and protocol with him. SHC would also like to acknowledge Research to Prevent Blindness Career Development Award, Foundation Fighting Blindness Career Development Award, Herbst Foundation, and NEI K-08 Career Development Award. S.H.C. would like to acknowledge Research to Prevent Blindness, Foundation Fighting Blindness, National Eye Institute, and Earl and Shirley Herbst Eye Research Fund for their support.

References

1. Takahasi K, Yamanaka S (2006) Induction of pluripotent stem cells from mouse embryonic and adult fibroblast cultures by defined factors. Cell 126:663–676
2. Lowry WE, Richter L, Yachechko R, Pyle AD, Tchieu J, Sridharan R, Clark AT, Plath K (2008) Generation of human induced pluripotent stem cells from dermal fibroblasts. Proc Natl Acad Sci USA 105:2883–2888
3. Takahashi K, Okita K, Nakagawa M, Yamanaka S (2007) Induction of pluripotent stem cells from fibroblast cultures. Nat Protoc 2:3081–3089
4. Ni A, Jing Wu M, Nakanishi Y, Chavala S (2013) Facile and efficient reprogramming of ciliary body epithelial cells into induced pluripotent stem cells. Stem Cells Dev 22 (18):2543–2550

5. Ni A, Wu MJ, Chavala SH (2014) Sphere formation permits Oct 3/4 reprogramming of ciliary body epithelial cells into induced pluripotent stem cells. Stem Cells Dev, Aug 18 (Epub ahead of print)
6. Reynolds BA, Rietze RL (2005) Neural stem cells and neurospheres—re-evaluating the relationship. Nat Methods 2:333–336
7. Tropepe V, Coles BL, Chiasson BJ, Horsford DJ, Ellia AJ, McInnes RR, van der Kooy D (2000) Retinal stem cells in the adult mammalian eye. Science 287:2032–2036
8. Kim JB, Sebastiano V, Wu G et al (2009) Oct4-induced pluripotency in adult neural stem cells. Cell 136(3):411–419
9. Liu T, Zou G, Gao Y et al (2012) High efficiency of reprogramming CD34+ cells derived from human amniotic fluid into induced pluripotent stem cells with Oct4. Stem Cells Dev 21(12):2322–2332
10. Tsai SY, Bouwman BA, Ang YS et al (2011) Single transcription factor reprogramming of hair follicle dermal papilla cells to induced pluripotent stem cells. Stem Cells 29(6):964–971

Methods in Molecular Biology (2016) 1357: 403–413
DOI 10.1007/7651_2014_162

Published online: 18 December 2014

Mesoderm Differentiation from hiPS Cells

Hiroyuki Miwa and Takumi Era

Abstract

Human induced pluripotent stem (hiPS) cells are very attractive tools for modeling diseases and regenerative medicine. However, to achieve them, the efficient differentiation methods of hiPS cells into aimed cell type in vitro are necessary. Because mesoderm cells are useful in particular, we have developed the differentiation of mouse embryonic stem (mES) cells into mesoderm cells previously. In this time, these methods were improved for hiPS cells and now human mesoderm cells are able to be obtained efficiently. It is certain that the new methods are applicable to various studies and therapies.

Keywords: Human induced pluripotent stem (hiPS) cells, Mesoderm, Vascular endothelial growth factor receptor 2 (VEGFR2), Platelet-derived growth factor receptor α (PDGFRα), Bone, Cartilage

1 Introduction

Human induced pluripotent stem (hiPS) cells with pluripotency and self-renewal capability can be induced from fibroblasts or hematopoietic cells by overexpression of four factors, Oct 3/4, Sox2, Klf4, and c-Myc (1). Such hiPS cells are thus very attractive tools in terms of both basic and clinical medicine. For instance, targeted differentiation of hiPS cells into specific cells could be used in cell therapy to regenerate tissue and organ defects. In addition, inducing hiPS cells from patients and differentiating them into cells of lesional tissue could provide important cellular models of disease for basic and clinical experimentation. Furthermore, drug screening for diseases could also utilize hiPS cells differentiated from patients. However, increasing the efficiency of differentiation is paramount for realizing any or all of the proposed applications for hiPS cells. In this context, we have been investigating the differentiation of mouse embryonic stem (mES) cells and hiPS cells into mesoderm and their descendant cells, bone and cartilage, which are important in the formation of various tissues and involved in many injury and disease processes (2, 3).

For the mES cells, we are using two main methods of differentiation. One is formation of the commonly used embryoid body (EB) and involves aggregating the mES cells on low-attachment dishes and then culturing with specific factors (4). The other is culturing on OP9 cells or collagen IV-coated dishes, without any factors used during the differentiation (5). In this method, dissociated mES cells are seeded and cultured with αMEM containing fetal bovine serum (FBS). After a few days, FACS analysis showed differentiated mES cells to be positive for vascular endothelial growth factor receptor 2 (VEGFR2, FLK1) and platelet-derived growth factor receptor alpha (PDGFRα), i.e., $PDGFR\alpha^{+}VEGFR2^{+}$ (DP) cells. After another few days, the DP cells differentiated into $PDGFR\alpha^{+}VEGFR2^{-}$ (PSP) or $PDGFR\alpha^{-}VEGFR2^{+}$ (VSP) cells. The PSP is likely to be paraxial mesoderm and the VSP lateral mesoderm, because PSP cells can differentiate into cartilage and bone, while VSP differentiates into endothelial cells (6–12).

In regenerative medicine, many researchers have tried to generate mesoderm and their descendant cells from hES cells (13–16). Xu et al. (15) differentiated hESCs into fibroblast-like cells, and after EB formation followed by culture on gelatin-coated plates, these cells overexpressed human telomerase reverse transcriptase (hTERT). They also expressed CD29, CD44, CD71, and CD90 and could be differentiated into an osteocytic lineage, but not chondrocytic or adipocytic lineage. Other groups also induced hESCs into fibroblastic cells that express CD90 and CD44 and can give rise to mesoderm-descendant cells (16). Such mesodermal multipotent stem cells were also purified from in vitro ES/iPS cell culture by FACS using surface markers such as CD73 and CD105 (14). In that study, gene expression analysis isolated cells expressing surface markers such as CD44 and STRO-1, as well as other markers including DSC54, neuropilin 1, hepatocyte growth factor, forkhead box D1, and notch homolog 2 (14). The culture conditions for selectively inducing descendants of mesoderm cells from ES/iPS cells were also reported, involving the culturing of hESCs with bFGF and PDGF-AB without feeder cells and then isolating the $CD105^{+}CD24^{-}$ population by FACS (13). This population could then be differentiated into adipocytic, chondrocytic, and osteocytic lineages. Mouse iPS cell-derived EBs have also been cultured with TGF-β1 and retinoic acid (RA) to generate cells that are able to differentiate into osteoblasts (17). Similarly, hiPS cells were cultured with bFGF, PDGF-AB, and epidermal growth factor (EGF), followed by FACS isolation of the $CD105^{+}CD24^{-}$ population to generate multipotent stem cells (18).

Herein, we describe the latest method of mesoderm differentiation to generate bone and cartilage cells from hiPS cells. This was developed by combining and modifying previous methods and is expected to efficiently differentiate into mesoderm lineages (4, 5, 12, 19–21). In brief, EBs are transferred onto collagen IV-coated dishes for culturing together with various factors, including bFGF, BMP-4, and Activin A and then sorted for VEGFR2 and PDGFRα expression by FACS. The subsequent PSP contains colony-forming unit-fibroblasts (CFU-Fs) and can differentiate into bone and

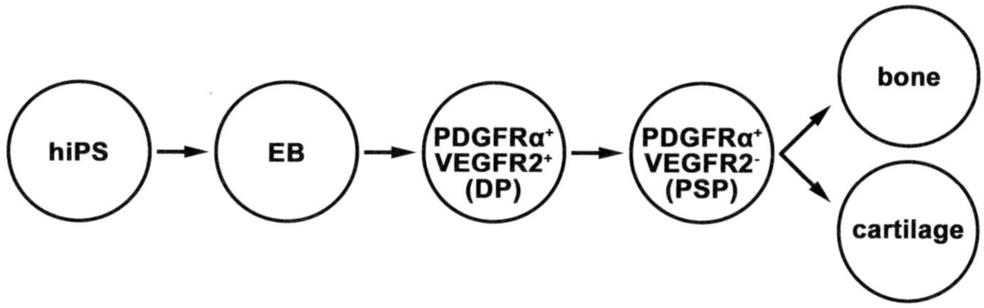

Fig. 1 Differentiation of hiPS cell into mesoderm. hiPS cells formed EBs differentiate into the three types of mesoderm cells, PDGFRα$^+$VEGFR2$^+$ population (PDGFRα and VEGFR2 double-positive population, DP), PDGFRα$^+$VEGFR2$^-$ population (PDGFRα single-positive population, PSP), and PDGFRα$^-$VEGFR2$^+$ population (VEGFR2 single-positive population, VSP). The DP is the most immature and can give rise to both the PSP and the VSP. The PSP can differentiate into bone and cartilage

cartilage cells (Fig. 1), as confirmed by CFU-F assay and staining using Alizarin Red and Alcian Blue. This method for efficiently differentiating hiPS cells into mesoderm cells should be useful for many studies and therapies in this field.

2 Materials

2.1 hiPS Cell Maintenance

1. Phosphate-buffered saline without calcium and magnesium chloride (PBS).
2. Dulbecco's Modified Eagle's Medium/Nutrient Mixture F-12 Ham (DMEM/F-12, Sigma-Aldrich). Store at 4 °C.
3. KnockOut SR (KSR, Gibco). Store at −20 °C.
4. 2-Mercaptoethanol (0.1 M, 1,000×) (2-ME).

   Add 70 μL 2-ME to 10 mL PBS and sterilized by 0.22 μm filter. Store at 4 °C.
5. L-Glutamine 200 mM (100×) (Gibco). Store at −20 °C.
6. Penicillin-streptomycin (100×) (P/S). Store at −20 °C.
7. MEM Non-Essential Amino Acid Solution (100×) (NEAA, Sigma-Aldrich). Store at 4 °C.
8. hiPS cell culture medium: DMEM/F12, 20 % KSR, 0.1 mM 2-ME, 2 mM L-glutamine, P/S, 0.1 mM NEAA. Store at 4 °C (*see* **Note 1**).
9. Fibroblast Growth Factor (basic), Human, recombinant (rhbFGF) (Wako). Store at −20 °C.
10. Trypsin 250 (Difco). Store at 4 °C.
11. Collagenase Type IV (Gibco). Store at 4 °C.
12. CTK solution: PBS, 20 % KSR, 0.25 % Trypsin, 1 mg/mL Collagenase, 1 mM $CaCl_2$. Store at −20 °C.

2.2 Formation of EB

1. hiPS cell culture medium.
2. Recombinant Human BMP-4 (R&D Systems). Store at −20 °C.
3. 100-mm Petri dish (BIO-BIK) (*see* **Note 2**).

2.3 Induction of Mesoderm Cells

1. Minimum Essential Medium Alpha Medium (αMEM, Gibco). Store at 4 °C.
2. Fetal bovine serum (FBS) (*see* **Note 3**). Store at −20 °C.
3. BMP-4.
4. Recombinant Human/Mouse/Rat Activin A (R&D Systems). Store at −20 °C.
5. bFGF.
6. Collagen Type IV Cellware 100-mm dish (BD Biosciences). Store at 4 °C.

2.4 Purification of Mesoderm Cells by FACS

1. TrypLE Express (Gibco).
2. HBSS (10×) (Gibco).
3. HBSS/BSA: HBSS, 1 % bovine serum albumin (BSA). Store at 4 °C.
4. Clear Back (human Fc receptor blocking reagent) (MBL). Store at 4 °C.
5. Biotin antihuman CD140a (PDGFRα) (BioLegend). Store at 4 °C.
6. Antihuman VEGF R2/KDR Phycoerythrin (R&D Systems). Store at 4 °C.
7. Streptavidin APC (eBioscience). Store at 4 °C.
8. 7-AAD Viability Dye (Beckman Coulter). Store at 4 °C.

2.5 Culture of Mesoderm Cells

1. αMEM.
2. FBS.
3. 2-ME.
4. P/S
5. rhPDGF-BB (R&D Systems). Store at −20 °C.
6. Mesoderm cell culture medium: αMEM, 10 % FBS, 0.1 mM 2-ME, P/S, 10 ng/mL PDGF-BB. Store at 4 °C.
7. 0.25 % Trypsin-EDTA (Gibco). Store at −20 °C.

2.6 CFU-F Assay

1. Mesoderm cell culture medium.
2. Collagen Type IV Cellware six-well plate (BD Biosciences). Store at 4 °C.
3. Leishman's eosin methylene blue solution (Merck).

2.7 Bone Cell Differentiation from Mesoderm Cells

1. Dulbecco's Modified Eagle's Medium (DMEM, Gibco). Store at 4 °C.
2. FBS.
3. P/S.
4. Dexamethasone.
5. Ascorbic acid 2-phosphate.
6. β-Glycerophosphate.
7. BMP-4.
8. 24-well culture plate (BD Biosciences).
9. Bone cell differentiation medium: DMEM, 10 % FBS, P/S, 0.1 mM dexamethasone, 50 mM ascorbic acid 2-phosphate, 10 mM β-glycerophosphate, 10 ng/mL BMP-4.
10. 4 % PFA: PBS, 4 % paraformaldehyde (PFA).
11. Alizarin Red staining solution.

    Add 0.1 g Alizarin Red S into 10 mL water.

2.8 Cartilage Cell Differentiation from Mesoderm Cells

1. αMEM.
2. FBS.
3. P/S.
4. Dexamethasone.
5. Ascorbic acid 2-phosphate.
6. Recombinant Human TGF-β3 (R&D Systems). Store at −20 °C.
7. Recombinant Human BMP-2 (R&D Systems). Store at −20 °C.
8. 24-well culture plate.
9. Cartilage cell differentiation medium: αMEM, 10 % FBS, P/S, 0.1 mM dexamethasone, 170 mM ascorbic acid 2-phosphate, 10 ng/mL TGF-β3.
10. 4 % PFA.
11. Alcian Blue staining solution.

    Add 50 mg Alcian Blue into 10 mL 0.1 M HCl.

3 Methods

3.1 In Vitro hiPS Cell Differentiation

In this method, hiPS cells are maintained on mitomycin C-treated mouse embryonic fibroblasts (MEFs) as feeder cells.

3.1.1 Formation of EB

1. Expand hiPS cells with hiPS cell culture medium and 5 ng/mL bFGF at 37 °C, 3 % CO_2 (*see* **Note 4**).

2. When confluent on a 100-mm dish, aspirate medium and wash cells twice with PBS.
3. Add 500 μL CTK solution to dissociate cells from dish.
4. Incubate at 37 °C for 5 min.
5. Remove CTK solution and add 10 mL hiPS cell culture medium.
6. Transfer cells into a 15-mL centrifuge tube and spin them down at 20 × *g* for 5 min.
7. Resuspend cells in 10 mL hiPS cell culture medium and transfer them on a gelatin-coated 100-mm dish (*see* **Note 5**).
8. Incubate at 37 °C for 30 min to remove feeder cells (*see* **Note 6**).
9. Transfer supernatant on a 100-mm Petri dish and add BMP-4 (10 ng/mL final conc.).
10. Incubate at 37 °C, 3 % CO_2 overnight.

3.1.2 Induction of Mesoderm Cells

1. Transfer EBs into a 15-mL centrifuge tube and spin them down at 20 × *g* for 5 min.
2. Resuspend EBs in 10 mL αMEM with 10 % FBS, 3 ng/mL Activin A, 10 ng/mL BMP-4, and 5 ng/mL bFGF and seed them on Collagen IV-coated dish.
3. Change medium every 2 days.
4. Culture for 5 days (*see* **Note 7**).

3.1.3 Purification of Mesoderm Cells by FACS

1. Aspirate medium and wash cells twice with PBS.
2. Add 2 mL TrypLE Express and incubate at 37 °C for 5 min.
3. Add 8 mL PBS with 10 % FBS and repeat pipetting 10–15 times.
4. Transfer cells into a 15-mL centrifuge tube and spin down them at 4 °C, 270 × *g* for 5 min.
5. Resuspend 1 × 10^7 cells in 100 μL Fc receptor blocking reagent and put them at room temperature for 5 min.
6. Add 300 μL HBSS/BSA and antibody to cell suspension and put them on ice for 20 min (*see* **Note 8**).
7. Add 400 μL HBSS/BSA and spin them down at 4 °C, 270 × *g* for 5 min.
8. Resuspend 400 μL HBSS/BSA and add 1 μL SA-APC. Put them on ice for 20 min.
9. Add 400 μL HBSS/BSA and spin them down at 4 °C, 270 × *g* for 5 min.

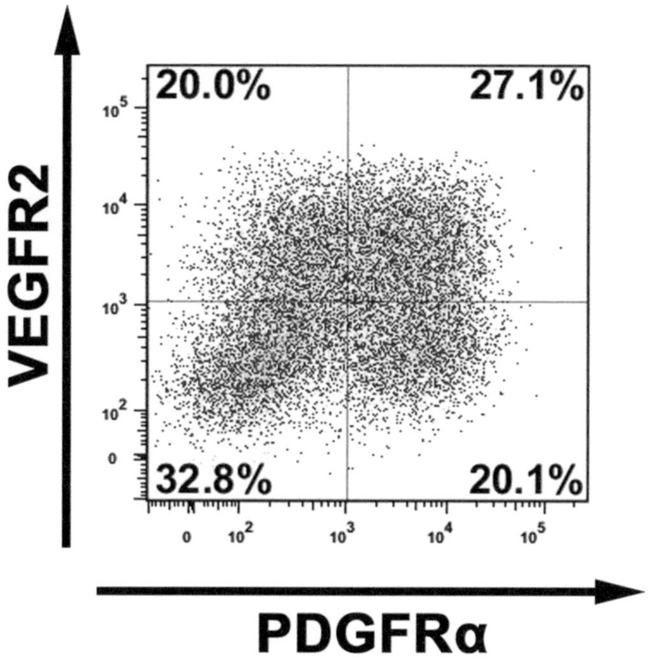

Fig. 2 Analysis of hiPS-derived mesoderm cells by FACS. hiPS cells formed EBs were cultured on Collagen IV-coated dish with αMEM containing FBS, Activin A, BMP-4, and bFGF. After 5 days of culture, cells were dissociated and analyzed with anti-PDGFRα and anti-VEGFR2 antibodies by FACS. Four populations (PDGFRα$^+$VEGFR2$^+$, DP; PDGFRα$^+$VEGFR2$^-$, PSP; PDGFRα$^-$VEGFR2$^+$, VSP; PDGFRα$^-$VEGFR2$^-$, DN) were observed in differentiated hiPS cells

10. Resuspend 800 μL HBSS/BSA and spin them down at 4 °C, 270 × *g* for 5 min.
11. Resuspend 1 mL HBSS/BSA and add 5 μL 7-AAD. Filter them into a 5-mL PS tube.
12. Sort PSP cells by FACS (Fig. 2).

3.2 Maintenance, Expansion, and Validation of Mesoderm Cells

3.2.1 Culture of Mesoderm Cells

1. Resuspend 1×10^5 hiPS-derived mesoderm cells in 10 μL mesoderm cell culture medium and put this solution on a well of Collagen IV-coated 24-well plate.
2. Incubate at 37 °C until cells attach to plate (1–3 h).
3. Add slowly 500 μL mesoderm cell culture medium into the well of plate and incubate at 37 °C, 5 % CO_2.
4. Change medium every 3 days.
5. When confluent, aspirate medium and wash cells twice with PBS.
6. Add 500 μL 0.25 % Trypsin-EDTA to dissociate cells from dish.
7. Incubate at 37 °C for 5 min.

8. Add 4.5 mL medium.
9. Transfer cells into a 15-mL centrifuge tube and spin them down at 270 × *g* for 5 min.
10. Resuspend and seed cells on new dishes (*see* **Note 9**).

3.2.2 CFU-F Assay

1. Seed $3 \times 10^{2-3}$ hiPS-derived mesoderm cells into 1 well of Collagen IV-coated six-well plate.
2. Change medium every 3 days.
3. On day 14, wash cells twice with PBS.
4. Add 1 mL Leishman's eosin methylene blue solution. Keep at 4 °C overnight.
5. Discard Leishman's eosin methylene blue solution and wash with water.
6. Dry up and count the number of colonies (*see* **Note 10**).

3.3 Differentiation into Descendants of Mesoderm Cells

Differentiation and confirmation can be performed in the same manner as described previously (14).

3.3.1 Induction of Bone Cells

1. Seed 1×10^5 hiPS-derived mesoderm cells into a well of gelatinized 24-well plate (*see* **Note 5**).
2. In the next day, change medium to 500 μL bone cell differentiation medium.
3. Change medium every 3 days.
4. The calcium deposit can be observed around on day 28.

3.3.2 Alizarin Red Staining

1. Wash cells twice with PBS.
2. Add 500 μL 4 % PFA into the well for fixture. Keep at room temperature for 10 min.
3. Wash twice with PBS.
4. Add 1 mL of Alizarin Red staining solution and keep at room temperature for 5 min.
5. Quickly wash three times with PBS.
6. Observation: calcium deposit is stained to red color.

3.3.3 Induction of Cartilage Cells

1. Resuspend 1×10^5 hiPS-derived mesoderm cells in 10 μL medium and put this solution on a well of 24-well plate.
2. Incubate at 37 °C until cells attach to plate (1–3 h).
3. Add slowly 500 μL cartilage cell differentiation medium into the well of plate.
4. One week later, change medium to new cartilage cell differentiation medium with 10 ng/mL BMP2.

5. Change medium every 3 days.
6. Analyze the cartilage generation on day 21.

3.3.4 Alcian Blue Staining

1. Wash cells twice with PBS.
2. Add 500 μL 4 % PFA into the well for fixture. Keep at room temperature for 10 min.
3. Wash twice with PBS.
4. Add 1 mL of Alcian Blue staining solution. Keep at room temperature for 30 min.
5. Quickly wash three times with PBS.
6. Observation: mucoglycoprotein is stained to blue color.

4 Notes

1. The medium is used less than 2 week as old medium affects growth rate and differentiation potency of cells.
2. Ultralow-attachment multiwell plates (Corning) can also be used for formation of EB.
3. FBS is a critical factor to induce a high rate of PDGFRα^+ cell induction. Lot no. checks of sera are highly recommended for finding appropriate serum lot. They are usually examined by the induction rate of hiPS cell-derived mesoderm cells. In general, using 20 different sera lots, the frequency of PSP cells is generated after 5 days from 10 to 20 %.
4. Avoid long-term culture. High passages (>30 passages) easily induce hiPS cells to be transformed and may lose their ability to support the in vitro hiPS cell differentiation.
5. To coat wells or dishes with gelatin before seeding cells, add 0.1 % gelatin solution into wells or dishes. Put them at room temperature for more than 5 min and aspirate gelatin solution.
6. If feeder cells remain, repeat the same progression. Or transfer supernatant into a 15-mL centrifuge tube and then put it at room temperature for 5 min. Aspirate medium and resuspend cells.
7. The proper number of days may differ from one cell line to another. At least, hiPS cells need longer differentiation than mES cells (12, 15). It is recommended to examine a rate of PSP from day 4 to 8 during induction of mesoderm cells.
8. It is recommended to check the titer of antibodies before using. The ratio in which differentiated cells can be divide into positive and negative should be decided between 1:10 and 1:1,000.
9. Differentiated cells can be maintained for more than 1 month and stored by freezing with FBS containing 10 % dimethyl sulfoxide (DMSO). When passage, split cells at 1:2–1:4.

10. A colony is defined as a cluster of more than 50 cells. It is appropriate that about 50 colonies appear per well of six-well plate.

Acknowledgment

This work was supported in part by grants from the Ministry of Health, Labour, and Welfare of Japan and Core Research for Evolutional Science and Technology (CREST), Japan Science and Technology Agency.

References

1. Takahashi K, Tanabe K, Ohnuki M, Narita M, Ichisaka T, Tomoda K, Yamanaka S (2007) Induction of pluripotent stem cells from adult human fibroblasts by defined factors. Cell 131:861–872
2. Kitagawa M, Takebe A, Ono Y, Imai T, Nakao K, Nishikawa S, Era T (2012) Phf14, a novel regulator of mesenchyme growth via platelet-derived growth factor (PDGF) receptor-α. J Biol Chem 287:27983–27996
3. Hamasaki M, Hashizume Y, Yamada Y, Katayama T, Hohjoh H, Fusaki N, Nakashima Y, Furuya H, Haga N, Takami Y, Era T (2012) Pathogenic mutation of ALK2 inhibits induced pluripotent stem cell reprogramming and maintenance: mechanisms of reprogramming and strategy for drug identification. Stem Cells 30:2437–2449
4. Keller G (2005) Embryonic stem cell differentiation: emergence of a new era in biology and medicine. Genes Dev 19:1129–1155
5. Tada S, Era T, Furusawa C, Sakurai H, Nishikawa S, Kinoshita M, Nakao K, Chiba T, Nishikawa S-I (2005) Characterization of mesendoderm: a diverging point of the definitive endoderm and mesoderm in embryonic stem cell differentiation culture. Development 132:4363–4374
6. Kabrun N, Buhring HJ, Choi K, Ullrich A, Risau W, Keller G (1997) Flk-1 expression defines a population of early embryonic hematopoietic precursors. Development 124:2039–2048
7. Nishikawa SI, Nishikawa S, Hirashima M, Matsuyoshi N, Kodama H (1998) Progressive lineage analysis by cell sorting and culture identifies FLK1 + VE-cadherin + cells at a diverging point of endothelial and hemopoietic lineages. Development 125:1747–1757
8. Yamashita J, Itoh H, Hirashima M, Ogawa M, Nishikawa S, Yurugi T, Naito M, Nakao K, Nishikawa S (2000) Flk1-positive cells derived from embryonic stem cells serve as vascular progenitors. Nature 408:92–96
9. Soriano P (1997) The PDGF alpha receptor is required for neural crest cell development and for normal patterning of the somites. Development 124:2691–2700
10. Kataoka H, Takakura N, Nishikawa S, Tsuchida K, Kodama H, Kunisada T, Risau W, Kita T, Nishikawa SI (1997) Expressions of PDGF receptor alpha, c-Kit and Flk1 genes clustering in mouse chromosome 5 define distinct subsets of nascent mesodermal cells. Dev Growth Differ 39:729–740
11. Takebe A, Era T, Okada M, Jakt LM, Kuroda Y, Nishikawa S-I (2006) Microarray analysis of PDGFR alpha + populations in ES cell differentiation culture identifies genes involved in differentiation of mesoderm and mesenchyme including ARID3b that is essential for development of embryonic mesenchymal cells. Dev Biol 293:25–37
12. Sakurai H, Era T, Jakt LM, Okada M, Nakai S, Nishikawa S, Nishikawa S-I (2006) In vitro modeling of paraxial and lateral mesoderm differentiation reveals early reversibility. Stem Cells 24:575–586
13. Lian Q, Lye E, Suan Yeo K, Khia Way Tan E, Salto-Tellez M, Liu TM, Palanisamy N, El Oakley RM, Lee EH, Lim B, Lim SK (2007) Derivation of clinically compliant MSCs from CD105+, CD24− differentiated human ESCs. Stem Cells 25:425–436
14. Barberi T, Willis LM, Socci ND, Studer L (2005) Derivation of multipotent mesenchymal precursors from human embryonic stem cells. PLoS Med 2:e161

15. Xu C, Jiang J, Sottile V, McWhir J, Lebkowski J, Carpenter MK (2004) Immortalized fibroblast-like cells derived from human embryonic stem cells support undifferentiated cell growth. Stem Cells 22:972–980
16. Stojkovic P, Lako M, Stewart R, Przyborski S, Armstrong L, Evans J, Murdoch A, Strachan T, Stojkovic M (2005) An autogeneic feeder cell system that efficiently supports growth of undifferentiated human embryonic stem cells. Stem Cells 23:306–314
17. Li F, Bronson S, Niyibizi C (2010) Derivation of murine induced pluripotent stem cells (iPS) and assessment of their differentiation toward osteogenic lineage. J Cell Biochem 109:643–652
18. Lian Q, Zhang Y, Zhang J, Zhang HK, Wu X, Zhang Y, Lam FF, Kang S, Xia JC, Lai WH, Au KW, Chow YY, Siu CW, Lee CN, Tse HF (2010) Functional mesenchymal stem cells derived from human induced pluripotent stem cells attenuate limb ischemia in mice. Circulation 121:1113–1123
19. Takashima Y, Era T, Nakao K, Kondo S, Kasuga M, Smith AG, Nishikawa S (2007) Neuroepithelial cells supply an initial transient wave of MSC differentiation. Cell 129:1377–1388
20. Yang L, Soonpaa MH, Adler ED, Roepke TK, Kattman SJ, Kennedy M, Henckaerts E, Bonham K, Abbott GW, Linden RM, Field LJ, Keller GM (2008) Human cardiovascular progenitor cells develop from a KDR+embryonic-stem-cell-derived population. Nature 453:524–528
21. Era T (2010) Mesoderm cell development from ES cells. Methods Mol Biol 636:87–103

Methods in Molecular Biology (2016) 1357: 415–421
DOI 10.1007/7651_2014_160

Published online: 18 December 2014

Enhancing Human Cardiomyocyte Differentiation from Induced Pluripotent Stem Cells with Trichostatin A

Shiang Y. Lim, Priyadharshini Sivakumaran, Duncan E. Crombie, Gregory J. Dusting, Alice Pébay, and Rodney J. Dilley

Abstract

Human induced pluripotent stem (iPS) cells are a promising source of autologous cardiomyocytes to repair and regenerate myocardium for treatment of heart disease. In this study, we describe a method for enhanced cardiomyocyte production from human iPS cells by treating embryoid bodies with a histone deacetylase inhibitor, trichostatin A (TSA), together with activin A and bone morphogenetic protein (BMP)-4. The resulting cardiomyocytes expressed cardiac-specific transcription factors and contractile proteins at both gene and protein levels. Functionally, the contractile embryoid bodies (EBs) displayed calcium cycling and were responsive to the chronotropic agents isoprenaline (0.1 μM) and carbachol (1 μM). The cardiomyocytes derived from human iPS cells may be used to engineer functional cardiac muscle tissue for studying pathophysiology of cardiac disease, for drug discovery test beds, and potentially for generation of cardiac grafts to surgically replace damaged myocardium.

Keywords: Trichostatin A, Epigenetic, Induced pluripotent stem cells, Cardiomyocyte, Differentiation, Efficiency, Cardiac tissue engineering

1 Introduction

Induced pluripotent stem (iPS) cells are an ideal source of donor cells for cardiac regenerative therapies because they have the ability to self-renew while retaining the potential to differentiate into all cell types of the body, including cardiomyocytes. Furthermore, patient-specific iPS cells offer an autologous source of cardiomyocytes for personalized therapeutic strategies to circumvent immunological issues such as rejection of transplants (1).

Cardiomyogenesis from iPS cells is a highly organized process mediated by several signaling pathways (2). Spontaneous differentiation of cardiomyocytes occurs from pluripotent cells; however, it is not very efficient, so we and others have used trichostatin A, a histone deacetylase (HDAC) inhibitor, to direct more efficient cardiomyocyte differentiation (3–6), presumably through an epigenetic mechanism. Trichostatin A belongs to the hydroxamic acid class of HDAC inhibitors and was selected because of its high potency and pan-HDAC selectivity (7).

2 Materials

2.1 Cells

1. Human iPS cell lines. iPS(foreskin)-1 and iPS(foreskin)-2 were obtained from WiCell Research Institute (Madison, WI, USA) (8). FA3 is an iPS cell line generated from patients with Friedreich's ataxia (9).
2. Human newborn foreskin fibroblasts (HFF-1) were obtained from American Type Culture Collection (ATCC, VA, USA).

2.2 Cell Culture Medium and Solutions

1. Dulbecco's phosphate-buffered saline without Ca^{2+} and Mg^{2+} (DPBS-, Invitrogen, 14190-250).
2. Dulbecco's phosphate-buffered saline with Ca^{2+} and Mg^{2+} (DPBS+, Invitrogen, 14040-182).
3. TrypLE Select enzyme (1×) (Invitrogen, 12563-011).
4. Dulbecco's Modified Eagle's Medium: Nutrient Mixture F-12 (DMEM/F-12) Glutamax (Invitrogen, 10565-042).
5. β-Mercaptoethanol (Invitrogen, 21985-023).
6. MEM nonessential amino acids (100×) (Invitrogen, 11140-050).
7. Penicillin/streptomycin (10,000 U/mL) (Invitrogen, 15140-122).
8. Human recombinant fibroblast growth factor basic protein (FGF2, Millipore, GF003).
9. Fetal bovine serum (Sigma-Aldrich, 12003C, Lot 7C0029) (*see* **Note 1**).
10. KnockOut serum replacement (KSR, Invitrogen, 10828-028).
11. Fibroblast growth medium: DMEM/F-12 Glutamax supplemented with 10 % fetal bovine serum and 50 U/mL penicillin/streptomycin.
12. Undifferentiated human iPS culture medium (iPS medium): DMEM/F-12 Glutamax supplemented with 20 % KSR, 0.1 mM nonessential amino acid, 0.1 mM β-mercaptoethanol, 50 U/mL penicillin/streptomycin, and 20 ng/mL FGF2 (*see* **Note 2**).
13. Cardiac differentiation medium: DMEM/F-12 Glutamax supplemented with 20 % fetal bovine serum, 0.1 mM nonessential amino acid, 0.1 mM β-mercaptoethanol, and 50 U/mL penicillin/streptomycin.

2.3 Chemicals

1. Mitomycin C from *Streptomyces caespitosus* (Sigma-Aldrich, M4287).
2. Gelatin from porcine skin (Sigma-Aldrich, G1890). Prepared 0.1 % gelatin in distilled water and autoclaved to solubilize the gelatin. Store at 4 °C for 3–4 months.

3. Trichostatin A (TSA, Sigma-Aldrich, T8552). Prepared aliquots of stock solution 100 μg/mL in dimethyl sulfoxide (DMSO, Sigma-Aldrich, D2650) for long-term storage at −20 °C. A final concentration of 1 ng/mL was prepared with cardiomyocyte differentiation medium before use.
4. Fibronectin from human plasma (Sigma-Aldrich, F2006). Prepared aliquots of stock solution 1 mg/mL in DPBS- for long-term storage at −20 °C. A final concentration of 10 μg/mL fibronectin was prepared with 0.1 % gelatin before use.
5. Recombinant activin A (R&D Systems, 338-AC). Prepared aliquots of stock solution 100 μg/mL in DPBS- containing 0.1 % bovine serum albumin (Sigma-Aldrich, A7030) for long-term storage at −20 °C. A final concentration of 100 ng/mL was prepared with cardiac differentiation medium before use.
6. Recombinant human bone morphogenetic protein-4 (BMP-4, PeproTech). Prepared aliquots of stock solution 20 μg/mL in DPBS- containing 0.1 % bovine serum albumin for long-term storage at −20 °C. A final concentration of 20 ng/mL was prepared with cardiac differentiation medium before use.

2.4 Plasticware

1. T-150 cm^2 tissue culture flasks (Falcon).
2. Ultra-low attachment surface flat bottom polystyrene six-well plate (Corning).
3. Center-well organ culture dishes (Falcon).
4. Glass capillary (SDR Clinical Technology), flame-pulled and sterilized with 70 % alcohol.

3 Methods

3.1 Fibroblast Expansion, Inactivation, and Feeder Layer Preparation

1. Human fibroblasts were grown in fibroblast growth medium in T-150 cm^2 flasks. Once the cells have reached 90 % confluent, the growth medium is removed by aspiration and the cells are washed twice with DPBS-. Incubate cells with 5 mL of prewarmed TrypLE solution at 37 °C for 3–5 min to detach cells.
2. Once cells have detached, add 10 mL of fibroblast growth medium to neutralize the action of the TrypLE. Fibroblasts can be divided into new T-150 cm^2 flasks at a split ratio of 1:5 for further expansion.
3. At passage 18, arrest cell division by treating 90 % confluent fibroblasts with mitomycin C at a final concentration of 10 μg/mL for 3 h at 37 °C. Following mitomycin C treatment, wash fibroblasts thoroughly with DPBS- twice and perform a cell count using trypan blue dye exclusion (*see* **Note 3**).

4. Mitomycin C-inactivated fibroblasts can be aliquoted and cryopreserved in fibroblast growth medium supplemented with 10 % DMSO and used for fibroblast feeder plate preparation upon thawing.
5. To prepare fibroblast feeder plate, mitomycin C-inactivated fibroblasts are thawed quickly and resuspended in pre-warmed fibroblast growth medium. Gently disperse fibroblasts onto the inner well of each organ culture dish at a density of 5.5×10^4 cells/cm^2. Add sterile distilled water to the outer well of each organ culture dish to maintain the humidity of the culture environment. Incubate dishes at 37 °C in 5 % CO_2 overnight and should be used within 5 days after preparation.
6. On the day of iPS cell expansion, wash new feeder plates with pre-warmed iPS medium once.

3.2 Human iPS Cell Culture and Expansion

1. To passage iPS cells, remove iPS medium by aspiration and replace with pre-warmed DPBS+. Using a flame-pulled glass capillary, dissect undifferentiated colonies of iPS cells manually into ~0.1 mm^2 pieces and detach gently from the feeder layer. Wash pieces of iPS colonies with DPBS + twice and transfer onto new feeder plates containing fresh iPS medium (*see* **Note 4**).
2. Each feeder plate should contain no more than 12 evenly spaced iPS cell pieces. Culture iPS cells at 37 °C in 5 % CO_2 and change media every 2–3 days. Passage iPS cells on a weekly basis.

3.3 Cardiomyocyte Differentiation of Human iPS Cells

1. To induce formation of embryoid bodies (EBs), dissect undifferentiated colonies of iPS cells manually into ~0.2 mm^2 pieces (Fig. 1a) and detach gently from the feeder layer. Wash pieces

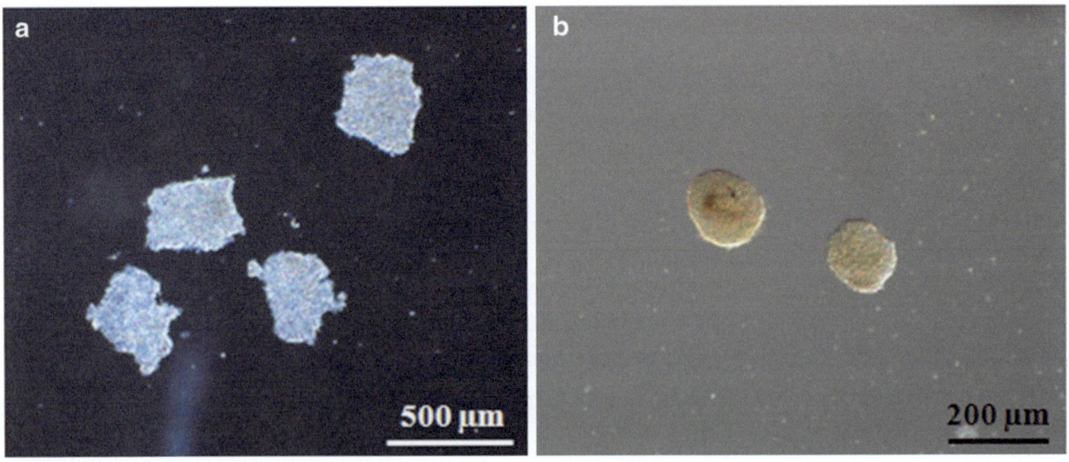

Fig. 1 (**a**) Representative phase-contrast micrograph of undifferentiated human induced pluripotent stem (iPS) cells following manual dissection. (**b**) Three-day-old embryoid bodies derived from human iPS cells in suspension

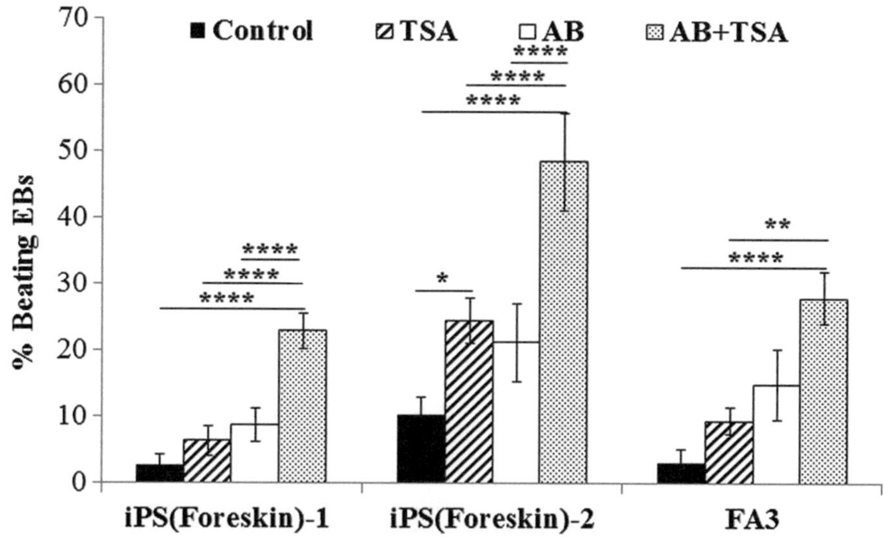

Fig. 2 Trichostatin A (TSA) enhances the cardiomyogenic effect of activin-A and BMP4 (AB) in all three human iPS cell lines tested: iPS(Foreskin)-1, iPS(Foreskin)-2 and FA3. $*p < 0.05$, $**p < 0.01$ and $****\ p < 0.0001$ by two-way ANOVA and Bonferroni *post hoc* test. Figure modified from (3)

of iPS colonies with DPBS + once and transfer to an ultra-low attachment surface six-well plate containing pre-warmed cardiac differentiation medium where they will form EBs (*see* **Note 5**) (Fig. 1b).

2. To direct cardiac differentiation during EB formation, supplement cardiac differentiation medium with 100 ng/mL activin A and 1 ng/mL TSA for 1 day (day 0–1); follow by 20 ng/mL BMP-4 and 1 ng/mL TSA for 5 days (day 1–5). Change media at day 1 and 3 (*see* **Notes 6** and 7).
3. After 6 days of EB formation, transfer EBs onto matrix-coated tissue culture plates containing pre-warmed cardiac differentiation medium. Tissue culture plates are coated with 0.1 % gelatin and 10 μg/mL fibronectin for at least 60 min at 37 °C before use (*see* **Note 8**).
4. Culture EBs in cardiac differentiation medium at 37 °C in 5 % CO_2 and change medium every 2–3 days. Beating cells appear as early as 1 day post-plating and treatment with TSA significantly increased the proportion of EBs beating (Fig. 2).
5. Cells from beating EBs can be harvested for various end-point analyses. These include:
   - Immunostaining with antibodies for cardiac progenitor markers (e.g., NKX2.5 and GATA4), cardiac contractile proteins (e.g., cardiac troponin T, cardiac troponin I, α-actinin, and myosin heavy chain), and gap junctions (connexin-43).

- Quantitative polymerase chain reaction (qPCR) to show the expression of genes encoding the cardiac-restricted transcription factors (e.g., MEF2C, GATA4, and NKX2.5) and the cardiac-specific structural and contractile proteins (e.g., ACTC1, TNNT2, TNNI3, MHL7, and MYL7).
- Electrophysiological properties of the derived cardiomyocytes with the microelectrode array recordings or patch clamping.
- Calcium transient imaging with the fluorescent Ca^{2+} indicator, Fluo-4AM (Invitrogen).

4 Notes

1. Fetal bovine serum can vary by source and lot. The batch-to-batch variability of serum may affect the rate of differentiation of human iPS cells into cardiomyocytes. Therefore, the same source and batch of serum should be used throughout the experiment to ensure consistency and reproducibility of experimental result.
2. bFGF should be added freshly to media before use. Multiple freeze/thaw cycles will result in loss of cytokine activity of bFGF and should be avoided. Therefore, aliquots of small volumes of bFGF should be prepared and stored at −20 °C.
3. Avoid excessive treatment with mitomycin C which can induce cytotoxicity and affect the quality of fibroblasts as feeder cells for human iPS cells.
4. The washing step is important to reduce the transfer of fibroblasts to new feeder plates.
5. The size of EB can affect the rate of cardiac differentiation. To ensure differentiation experiments are reproducible, EBs that are uniform in size and shape can be generated with commercially available products such as AggreWell from Stem Cell Technologies.
6. The effective time window of TSA is narrow. In a dose-response study, we have shown that the cardiomyogenic effect of TSA is most effective at 1 ng/mL in the three human iPS cell lines tested (3). At high concentration (>10 ng/mL), TSA induced cytotoxicity. We recommend a dose-response experiment should be performed to determine the optimal dose of TSA for each new pluripotent cell line. Similarly, optimization of activin A and BMP-4 growth factor concentrations may be required to achieve optimal efficiency of cardiomyogenic differentiation.
7. Duration of treatment with growth factors, activin A and BMP-4, is crucial to achieve efficient and reproducible cardiomyogenic

differentiation. Therefore, record the time when growth factors were added and change media according to the treatment duration, i.e., 24 h for activin A and 5 days for BMP-4.

8. The tissue culture plates can be coated with 0.1 % gelatin alone. We found that addition of 10 μg/mL fibronectin promotes EB attachment.

Acknowledgment

These studies were supported by grants from the National Heart Foundation and National Health and Medical Research Council of Australia (1024817; 1056589). GJD is a Principal Research Fellow of NHMRC and AP is a Career Development Fellow of NHMRC. Support is also provided by the JR and JO Wicking Trust, Friedreich's Ataxia Research Association (research grant and 2012 Keith Michael Andrus Cardiac Research Award), Tony and Gwyneth Lennon Foundation, and the Victorian State Government's Department of Innovation, Industry and Regional Development's Operational Infrastructure Support Program.

References

1. van der Heyden MA, Jonsson MK (2012) Personalized medicine and the role of induced pluripotent stem cells. Cardiovasc Res 95:395–396
2. Verma V, Purnamawati K, Manasi et al (2013) Steering signal transduction pathway towards cardiac lineage from human pluripotent stem cells: a review. Cell Signal 25:1096–1107
3. Lim SY, Sivakumaran P, Crombie DE et al (2013) Trichostatin A enhances differentiation of human induced pluripotent stem cells to cardiogenic cells for cardiac tissue engineering. Stem Cells Transl Med 2:715–725
4. Hosseinkhani M, Hasegawa K, Ono K et al (2007) Trichostatin A induces myocardial differentiation of monkey ES cells. Biochem Biophys Res Commun 356:386–391
5. Kawamura T, Ono K, Morimoto T et al (2005) Acetylation of GATA-4 is involved in the differentiation of embryonic stem cells into cardiac myocytes. J Biol Chem 280:19682–19688
6. Kaichi S, Hasegawa K, Takaya T et al (2010) Cell line-dependent differentiation of induced pluripotent stem cells into cardiomyocytes in mice. Cardiovasc Res 88:314–323
7. Vanhaecke T, Papeleu P, Elaut G et al (2004) Trichostatin A-like hydroxamate histone deacetylase inhibitors as therapeutic agents: toxicological point of view. Curr Med Chem 11:1629–1643
8. Yu J, Vodyanik MA, Smuga-Otto K et al (2007) Induced pluripotent stem cell lines derived from human somatic cells. Science 318:1917–1920
9. Liu J, Verma PJ, Evans-Galea MV et al (2011) Generation of induced pluripotent stem cell lines from Friedreich ataxia patients. Stem Cell Rev 7:703–713

Methods in Molecular Biology (2016) 1357: 423–439
DOI 10.1007/7651_2014_134

Published online: 18 November 2014

Derivation of Skeletal Myogenic Precursors from Human Pluripotent Stem Cells Using Conditional Expression of PAX7

Radbod Darabi and Rita C.R. Perlingeiro

Abstract

Cell-based therapies are considered as one of the most promising approaches for the treatment of degenerating pathologies including muscle disorders and dystrophies. Advances in the approach of reprogramming somatic cells into induced pluripotent stem (iPS) cells allow for the possibility of using the patient's own pluripotent cells to generate specific tissues for autologous transplantation. In addition, patient-specific tissue derivatives have been shown to represent valuable material for disease modeling and drug discovery. Nevertheless, directed differentiation of pluripotent stem cells into a specific lineage is not a trivial task especially in the case of skeletal myogenesis, which is generally poorly recapitulated during the in vitro differentiation of pluripotent stem cells.

Here, we describe a practical and efficient method for the derivation of skeletal myogenic precursors from differentiating human pluripotent stem cells using controlled expression of PAX7. Flow cytometry (FACS) purified myogenic precursors can be expanded exponentially and differentiated in vitro into myotubes, enabling researchers to use these cells for disease modeling as well as therapeutic purposes.

Keywords: Human embryonic stem (ES) cells, Human-induced pluripotent stem (iPS) cells, Skeletal muscle, PAX7, Myogenic precursors, Cell therapy, Muscle differentiation

1 Introduction

Embryonic stem (ES) cells are unique in terms of differentiation potential, as they are able to differentiate into all three primary germ layers and therefore potentially into all the cell types of the body. This unparalleled differentiation potential, combined with their capacity to be expanded indefinitely at the pluripotent stage, makes ES cells an ideal model to study early developmental questions since in vitro differentiation of these cells recapitulates the early stages of embryo organogenesis (1). The successful isolation of human ES cells by James Thompson in 1998 (2) and the subsequent generation of iPS cells from somatic cells (3–6) opened up an unprecedented opportunity for disease modeling and the design of patient-specific cell therapies for potential use in regenerative medicine (7, 8). Accordingly, this approach has been used extensively within the last decade to study disease pathologies as well as to investigate the therapeutic potential of produced tissue derivatives (8–15). Although many protocols have been developed for the successful generation of iPS cells (16–19), this is not the case in terms of directing the differentiation of these cells into specific cell types. This is particularly an issue for cell lineages that are not well recapitulated during the in vitro differentiation of ES/iPS cells, such as the skeletal muscle lineage.

In recent years, a few groups, including ours, have reported protocols for the derivation of skeletal myogenic precursors from hES/iPS cells (20–27). We have developed a method to efficiently generate skeletal myogenic progenitors from pluripotent ES cells through Pax3 or Pax7 induction (28–30). We have applied this method to mouse wild-type (31) and dystrophic (32) iPS cells and importantly to human ES and iPS cells (27). In each case, ES/iPS-derived myogenic progenitor cells can engraft extensively, promote improved contractility, and seed the stem cell compartment of transplanted dystrophic muscles. Here we describe a detailed method for the controlled over-expression of PAX7 from mesoderm during EB differentiation of human ES/iPS cells (27). Myogenic progenitors are purified by sorting for green fluorescent protein (GFP), which reflects PAX7 expression since ES/iPS cells are transduced with a PAX7-IRES-GFP lentiviral vector. Purified myogenic cells can be expanded exponentially in culture, while maintaining a mesenchymal/myogenic surface marker profile (such as CD90, CD44 and M-cadherin), and importantly can be differentiated into myotubes. This methodology enables researchers to derive skeletal myogenic precursors from any given human iPS cell line, which can be utilized for disease modeling, including ultrastructural and physiological analyses (33), or for in vivo regeneration studies (27). In this chapter, this protocol is described in detail, providing a step-by-step practical guide to generate inducible PAX7 human pluripotent stem cell lines, and the process involving their differentiation into skeletal myogenic progenitors and subsequent myotubes.

2 Materials

While working with human-derived cell lines or viruses, always use appropriate personal protective equipment (PPE) according to your institution's guidelines. Prepare all solutions and media using aseptic techniques under a cell culture grade laminar biosafety cabinet (class IIA). All chemicals and solvents (such as water, PBS) should be sterile and cell culture grade certified. All the cell growth media can be supplemented with Pen/Strep to prevent bacterial contamination. All the cell lines will be expanded and differentiated in a 5 % CO_2 incubator with ambient O_2 at 37 °C. For disposing waste materials, please strictly follow your institution's guidelines regarding the proper disposal of lentiviral and human-derived cell waste products.

2.1 Reagents, Medium, and Supplies Needed for Expansion of Human ES/iPS Cells

1. BD Matrigel™ hESC-qualified Matrix (354277, BD Biosciences) for coating flasks before the plating of human ES/iPS cells.
2. mTeSR™ 1 Defined, Feeder-Free Maintenance Medium for Human ES and iPS Cells (05850, Stem Cell Technologies).
3. Accumax cell detachment solution (SCR006, Millipore) for harvest and passage of the hES/iPS cells.
4. InSolution™ Y-27632 Rho-associated protein kinase inhibitor (688001, Calbiochem/Millipore).
5. PBS—phosphate-buffered saline without calcium and magnesium (10010-023, Gibco®, Life Technologies) for rinsing chelators from the culture before cell dissociation.
6. Nunc T25 and T75 cell culture-treated flasks with filter caps (Thermo Scientific).

2.2 Reagents, Medium, and Supplies Needed for Expansion and Lentivirus Production in 293T Cells

1. 293T culture medium: Dulbecco's Modified Eagle's Medium (10566-016, Gibco®, Life Technologies) supplemented with fetal bovine serum (10437-36, Gibco®, Life Technologies) to a final concentration of 10 %.
2. 0.25 % (w/v) trypsin-EDTA solution (25200-056, Gibco®, Life Technologies).
3. Plasmids for generation of doxycycline-inducible PAX7 lentiviral system (rtTA-FUGW and hPAX7-pSAM2 as viral backbones, Delta 8.9 and VSV-G as packaging and coating plasmids).
4. Lipofectamine® LTX with PLUS™ Reagent (15338-100, Life Technologies).
5. Polybrene Infection/Transfection Reagent (TR-1003-G, Millipore).
6. 60 mm cell culture dishes (BD Falcon).
7. Millex-HV Syringe Filter Unit, 0.45 μm (SLHV033RS, Millipore).

2.3 Reagents, Medium, and Supplies Needed for EB Differentiation of Human ES/iPS Cells

1. EB differentiation medium: IMDM plus Glutamax (31980-030, Gibco) containing 15 % fetal bovine serum (Gibco), 10 % horse serum (Sigma), 1 % chicken embryonic extract (US Biological), 50 μg/ml ascorbic acid (Sigma), 4.5 mM monothioglycerol, and 200 μg human holo-transferrin (2914-HT, R&D Systems).
2. 1,000× doxycycline (dox) stock solution: 0.75 mg/ml of doxycycline (D3447, Sigma) solution in sterile water for induction of PAX7.

3. InSolution™ Y-27632 Rho-associated protein kinase inhibitor (688001, Calbiochem/Millipore).
4. 60 mm Petri dishes or cell culture dishes (Nunc, Thermo) for making embryoid bodies (EBs).
5. Orbital shaker (KS-260 control, IKA) installed in the CO_2 incubator.

2.4 Reagents, Medium, and Supplies Needed for Cell Sorting Using FACS

1. 0.25 % (w/v) trypsin-EDTA solution (25200-056, Life Technologies) to disaggregate the EBs-derived cells into single cells for cell sorting.
2. FACS buffer: Sterile PBS supplemented with 2 % fetal bovine serum and 1 μg/ml of propidium iodide (PI).
3. 35 μm falcon tube with cell strainer cap (352235, Falcon/Corning) to filter the cells before sorting.

2.5 Reagents, Medium, and Supplies Needed for Expansion and Differentiation of Sorted Myogenic Progenitors

1. Expansion medium: use above-described EB differentiation medium, supplemented with 0.75 μg/ml of doxycycline (1 μl of 1,000× stock solution mentioned above per ml of complete medium) and 10 ng/ml of human recombinant basic fibroblast growth factor (FGF) (bFGF, 13256-029, Life Technologies).
2. Myotube differentiation medium: Low glucose DMEM (11885-084, Life Technologies) supplemented with 2 % horse serum (H1138-Sigma) to induce terminal differentiation of sorted myogenic precursors into myotubes.
3. Sterile 0.1 % gelatin (G9193, Sigma) solution for coating plates and flasks for cell expansion.

3 Methods

3.1 Matrigel Coating of Plates and Flasks

1. Thaw a 5 ml vial of Matrigel (hESC-qualified) at 4 °C overnight and divide in 130–150 μl aliquots. Keep these vials in a −80 °C freezer (please refer to the protein concentration and dilution factor provided on the Certificate of Analysis for each batch to determine the exact aliquot volume).
2. For coating the flasks, thaw one aliquot of Matrigel on ice and mix it with 12 ml cold DMEM/F12 medium. This solution is enough for the coating of 100 cm^2 surface area (i.e., 4 × T25 flask or 1 × T75 + 1 × T25 flask or ten wells of 6-well plates). Transfer 3 ml of the cold mixed solution into each T25 cell culture flask and tilt to cover the entire surface of the flask (for different flask sizes, use 9 ml for T75 and 1.5 ml per well of the 6-well plate accordingly).
3. Make sure the coating solution has covered the entire surface of the plate. Leave for 1 h. under the cell culture hood at room temperature to coat the flask. Unused coated flasks containing the coating solution can be wrapped in clean plastic sheets and stored in a 4 °C refrigerator to be used up to 1 week after preparation.

3.2 Thawing and Expansion of Human Pluripotent Stem Cells in mTeSR Medium

3.2.1 Thawing hES or iPS Cells

1. Warm the complete mTeSR medium in a 37 °C water bath.
2. Thaw one vial of hES or iPS cells (ideally containing about 1×10^6 cells), and mix gently with 5 ml of complete mTeSR medium for 3–5 times in a 15 ml tube.
3. Spin down the cell suspension at $200 \times g$ for 5 min at room temperature.
4. Carefully remove the supernatant using sterile Pasteur pipette.
5. Add 5 ml of warm complete mTeSR medium supplemented with 5 μl of the rho-associated, coiled-coil containing protein kinase (ROCK) inhibitor (InSolution™ Y-27632). Gently mix the medium with the cells by pipetting up and down using a 5 ml pipette for 5–8 times.
6. Remove the excess of coating solution from the T25 flask using a Pasteur pipette, and immediately (to avoid drying of the coated surface), and gently, add the cell suspension to the side or back wall of the flasks. Avoid scratching the coated surface with the pipette tip.
7. Make sure the cells have been distributed evenly on the flask surface by tilting it gently to the sides for few times. Secure the air filter cap and transfer the flask into CO_2 incubator. Grow the cells overnight.

3.2.2 Expansion of hES/iPS Cells

1. Next day, check plated cells using an inverted light microscope. For better visualization of the colonies, use low magnification (5× objective lens). The day after thawing, cells start to form small colonies. Due to the effect of the ROCK inhibitor, the cells might look elongated and have a mesenchymal/fibroblastic appearance, which is normal.
2. Remove the used medium and perform a PBS wash to remove floating dead cells. Add the same volume of fresh medium. No ROCK inhibitor supplementation is needed from now on.
3. Feed ES/iPS cells every day until appropriate confluency for passaging is achieved (70–80 %). This will depend on the cell line and proliferation rate, and it may take 3–6 days for colonies to reach maximum size. Usually ES/iPS colonies are passaged when they are large, and begin to merge, displaying a dense and bright center compared to their edge. Please refer to mTeSR-1 manual (Stem Cell Technologies) for examples of ES/iPS cell morphologies and more detailed instructions. hES/iPS cells can be split 1:4–1:6 during each passage depending on their proliferative rate.

3.2.3 Enzymatic Passaging Using Accumax

1. Before passaging the cells, make sure needed flasks and plates are coated appropriately with Matrigel, as described above.
2. Thaw a vial of Accumax at room temperature. Do not use warm water bath as Accumax will degrade in warm temperatures.

3. Use a microscope to identify and remove differentiated colonies by scraping these with a pipette tip or a Pasteur pipette.
4. Remove used medium and wash the plate once with PBS to remove any left over medium or dead cells.
5. Remove PBS and carefully add 1 ml of Accumax to each T25 flask (0.5 ml for each well of a 6-well plate or 3 ml for a T75 flask). Tilt the flasks side-to-side to ensure covering of the entire surface.
6. Transfer flask containing ES or iPS cells and Accumax to an incubator for 2 min. After this, tap the flask to see if the colonies are detached. If not, incubate for 1 more minute.
7. Break down the colonies using gentle pipetting (6–8 times) and add appropriate amount of complete mTeSR medium to inactivate the enzymes (4× volume of the used Accumax).
8. Spin down the cells at 200 × *g* for 5 min at room temperature. Resuspend in 2–4 ml of mTeSR medium and count the cells. Cells can be frozen at 1×10^6/vial (*see* **Note 1**) or passaged 1:4–1:6, as mentioned above.

3.3 Lentivirus Production

For disposing waste materials, please strictly follow your institution's guidelines regarding the proper disposal of lentiviral wastes.

3.3.1 Required Plasmids

The doxycycline (dox)-inducible PAX7 system we have developed (27) consists of two lentiviral vectors (rtTA-FUGW and hPAX7-pSAM2). Both plasmids need similar coating and packaging plasmids (VSV-G and Delta-8.9) to be transfected simultaneously in order to produce the complete virus. It is important to transfect each viral backbone vector with both packaging and coating plasmids and prepare both viruses separately in different plates. PAX7-pSAM2 vector contains an IRES-GFP sequence which allows for the identification of PAX7-induced cells by the expression of GFP. Therefore, it is very important to make sure that target hES/iPS cell lines are GFP negative before proceeding to infect them using this system. In case ES/iPS cells already express GFP, the vector can be modified to have PAX7 expression associated with another color, such as an IRES-mCherry sequence.

3.3.2 Lentiviral Production Protocol

1. Grow 239T cells in 293T culture medium (*see* Section 2.2) in a T75 flask until they reach 70–80 % confluency.
2. Aspirate the medium and rinse the flask with PBS to remove the remaining medium and floating dead cells.
3. Add 3 ml of 0.25 % trypsin to cover the entire surface of T75 flask, and incubate in CO2 incubator for 3–5 min. Check under the microscope to ensure cell dispersion. Do not agitate or tap the flask during this procedure to avoid cell clumping.

4. Add 12 ml of 293T culture medium and mix by pipetting up and down for 8–10 times to make sure the cells are separated as single cells.
5. Spin down the cells at 200 × *g* for 5 min at room temperature. Resuspend in 2–4 ml of 293T culture medium and count. The cells are ready to be seeded into 60 mm cell culture plates for transfection. This is considered day 0 of virus production.
6. At day 0, plate 1 × 10^6 293T cells per each 60 mm cell culture dish and add 5 ml of 293T culture medium. Place cells overnight in CO_2 incubator.
7. At day 1, check cells under the light microscope. The optimal confluency for transfection is around 60–70 %. If less, grow 293T cells for 1 more day. If the confluency is higher than 70 %, do not use these plates. Harvest cells, and plate them at lower density.
8. If 293T cells are at appropriate confluency, add fresh medium 1 h prior to transfection.
9. In a 15 ml polystyrene tube, add 150 μl of Opti-MEM medium and mix it with 14 μl of Lipofectamine LTX reagent.
10. In another tube, dilute 10 μg of DNA (5 μg of RTTA-FUGW or hPAX7-pSAM2 plus 2 μg of VSV-G, and 3 μg of Delta 8.9) with 150 μl of Opti-MEM medium. Then add 10 μl of PLUS reagent and mix well.
11. Add the diluted DNA into the first tube and mix well by inverting the tube for 6–8 times. Incubate this for 5 min at room temperature.
12. Add DNA-lipid complex to 293T cells by dispensing drop by drop all over the plate surface, swirling the plate gently for a few times. Incubate overnight in CO_2 incubator.
13. After 24 h, at day 2, replace used medium with fresh medium. Be careful not to detach the cells during medium exchange. At this time point, 293T cells should have reached 80–90 % confluency.
14. Viral soup can be harvested during the next 24 and 48 h (days 3 and 4). Collect the viral soup into clean tubes and filter them through a 0.45 μm syringe filter.
15. Determine the viral titer using conventional test strips to make sure the virus titer is good (such as Lenti-X GoStix, Clontech).
16. It is preferred to use fresh viruses for cell infection. However, these can be aliquoted into 1.5 ml tubes and kept at −80 °C freezers for future use. Make sure to label and aliquot each virus separately (rtTA-FUGW and hPAX7-pSAM2 lentiviral soup). Both viruses are needed for simultaneous infection to generate inducible PAX7 ES/iPS cell lines.

3.4 Infection of hES/iPS Cells with the Inducible PAX7 Lentiviral System

In order to generate dox inducible PAX7 cell lines, human ES/iPS cells have to be infected with both viruses. Spin infection is recommended, and therefore, cells to be infected need to be expanded in 6-well plates prior to infection.

1. Thaw hES/iPS cells in a T25, as described above, and grow them until ready to be passaged to 6-well plates.
2. Harvest ES/iPS cells using Accumax, as described above. After counting the number of cells, plate them in 6-well plates at different densities (5×10^4–1×10^5 cells/well), using mTeSR medium supplemented with ROCK inhibitor (*see* **Note 2**). Grow ES/iPS cells overnight.
3. The following day, check cells under the microscope, and change used medium with fresh mTeSR medium.
4. Grow ES/iPS cells until they reach up to 30–50 % confluency, which is the optimal density for viral infection.
5. Change the medium with 3 ml of fresh mTeSR medium 1 h prior to viral infection.
6. Prepare the infection master mix by adding 3 ml of fresh mTeSR medium plus 1.5 ml of PAX7-pSAM2 and 1.5 ml of rtTA-FUGW viral soups. Supplement the master mix with 6 μl of 1,000× polybrene, and mix well.
7. Replace the medium with infection master mix and spin ES/iPS cells in a plate centrifuge at 2,500 rpm for 90 min at 33 °C.
8. After spin infection, transfer the plate into a CO_2 incubator and grow cells overnight (*see* **Note 3**).
9. Check cells in the morning following the spin infection. There might be more floating dead cells due to viral infection. Rinse well the wells with PBS, and feed the cells with fresh mTeSR medium. Grow cells overnight.
10. Next day, check infected cells for the shape of colonies, confluency, and any signs of differentiation. If needed, scrape differentiated colonies and passage the cells.
11. When the colonies reach to optimal density, passage them (1:3–1:4). Remember to freeze some cells as backup at this time and label them accordingly (i.e., "iPAX7—name of the cell line, 1st infection, and passage number"). These frozen vials serve as backup in case you need them in the future.
12. Repeat the infection one more time to get more cells infected with both viruses. The goal here is to get more than 25 % of total cells infected with both viruses. In most cases, minimum two sets of spin infections are needed to get the desired levels.
13. Check the infection efficiency by performing an overnight dox induction of infected ES/iPS cells, followed by FACS analysis (described below), to make sure you have obtained the

minimum appropriate percentage of inducible PAX7-GFP ES/iPS cells. This percentage may vary depending on the cell line confluency and viral titers at the time of infection.

14. Repeat the viral infection as needed to achieve the target 25 % or more of induction rate. In our hands, if viral preparation is good, generally 2–3 sets of spin infection with fresh non-concentrated viral soup is enough for most human ES/iPS cells to get the desired rate of infection.
15. Alternatively, to improve the infection efficiency or to enrich for infected cells, concentrated viral soup can be used and/or the inducible hES/iPS cells can be purified by FACS, based on GFP expression, using low dose of dox (*see* **Note 4**).
16. Proceed to test the infection efficiency in the steps listed below.

3.5 Testing the Infection Efficiency

After infection of hES/iPS cells using both viruses, the cells have to be tested for infection efficiency. This can be done by overnight incubation with doxycycline in order to induce the expression of PAX7-IRES-GFP. Then the cells can be analyzed by FACS to determine the percentage of cells that expresses GFP (Pax7). As mentioned earlier, the objective is to get 25 % or more of the cells to express PAX7 (GFP). Before proceeding to this step, make sure to deposit frozen stock of the infected cells as backup.

1. Plate the infected cells (5×10^4 cells/well) into two wells of a 6-well plate coated with Matrigel. Grow them overnight in mTeSR medium supplemented with ROCK inhibitor.
2. In the next day, check cells under the microscope to ensure adhesion and formation of small colonies.
3. Rinse wells with PBS and add 3 ml of fresh mTeSR medium. In one of the wells, add 3 μl of 1,000 × dox stock solution to start the induction. Make sure to mix it well with the medium before adding to the well. Keep the other well without dox to serve as a non-induced GFP-negative control for the FACS analyses. Grow cells overnight.
4. The day after, harvest cells from each well using Accumax. Make sure to disaggregate the colonies into single cells (a requirement for FACS) by pipetting up and down using a 1 ml pipette tip for 12–15 times.
5. Spin down the cells and after removing the supernatant, resuspend each cell pellet with 300 μl of FACS buffer (*see* Section 2.4) using a 1 ml pipette tip. Make sure to resuspend well to make single cell suspension.
6. Filter the cells using 35 μm filter cell strainer capped falcon tubes and analyze the samples for GFP expression, using a flow cytometry system (such as BD FACSAria II Flow Cytometer, BD Biosciences).

7. Make sure to set the gate for GFP expression using the non-induced GFP-negative control cells, and then run the induced cell preparation.
8. Save the data for each sample and analyze the percentage of GFP-positive cells to determine the infection efficiency.

3.6 EB Differentiation of hES/iPS Cells

In order to differentiate iPAX7 hES/iPS cells into myogenic precursors, we have developed an embryoid body (EB) differentiation protocol (27). In this protocol, EBs are formed using a shaking method. An orbital shaker with the speed of 60 rpm placed in the CO_2 incubator is needed to perform this step of the differentiation.

1. Expand inducible PAX7 hES/iPS cells in the presence of mTeSR medium until ready to be passaged.
2. Harvest the cells using Accumax, count, and spin down.
3. Plate 1×10^6 cells/5 ml of fresh mTeSR medium supplemented with 5 μl of ROCK inhibitor in a low adherent 60 mm Petri dish.
4. Transfer the plates onto the incubator shaker with orbital shaking speed of 60 rpm and grow cells for 2 days in the CO_2 incubator (*see* **Note 5**).
5. After 2 days, check EBs under the microscope to ensure EB formation. By this time, EBs should be formed and they should look transparent and homogeneous in terms of size (27).
6. Transfer the EB containing plates under a cell culture hood and swirl the plates slowly to gather all the EBs at the center of the plate. Allow EBs to settle down for a minute.
7. Gently remove the entire medium by careful pipetting from the peripheral sides of the plate without disturbing the EBs, and replace with 5 ml of fresh EB differentiation medium supplemented with 5 μl of ROCK inhibitor.
8. Transfer the EB plates back to the shaker in the incubator, and grow them for 5 more days. Change the medium once (after 3 days) with fresh EB differentiation medium. ROCK supplementation is no longer necessary.
9. After 5 days, EBs are transferred into gelatinized flasks to be grown as monolayers.
10. Coat flasks (T25 or T75) with 0.1 % gelatin solution for 30 min in CO_2 incubator to ensure the coating of the entire surface with gelatin.
11. Aspirate the gelatin solution, and transfer intact EBs with their entire used medium into the flasks using a 5 ml pipette. Each 60 mm EB plate can be transferred into a T25 flask (or 3 in a T75). Top up with 3 ml of fresh EB medium for each T25 (or 9 ml for T75).

12. Make sure to gently shake the flasks side to side to distribute the EBs evenly all over the surface, and grow these cultures for 3 more days in the CO_2 incubator. Do not touch or move the flasks during this time to ensure the proper adherence of the EBs to the tissue culture flask surface.
13. After 3 days of culturing these cells as a monolayer, check EBs under the microscope. If the differentiation is progressing well, most of the EBs should have attached to the flasks, and cell outgrowths are emerging out of attached EBs (*see* **Note 6**).
14. This is the time point in which mesoderm is already formed (27), and cell cultures are ready to be induced for PAX7.
15. Proceed to the next step.

3.7 PAX7 Induction and Purification of Myogenic Precursors

1. After completing the above-described steps, EB-derived monolayers are ready for PAX7 induction. Therefore, remove the used medium using a Pasteur pipette, and rinse flasks with PBS to remove any dead cells or unattached EBs.
2. Start induction using fresh EB differentiation medium supplemented with dox solution. Use 1 μl of 1,000× dox stock solution/ml of EB medium, and mix well. Add 7 ml of the dox-supplemented medium for a T25 and 20 ml for a T75 flask. For cells with slow proliferation rate, supplement medium with 10 ng/ml of bFGF.
3. Grow iPAX7 ES/iPS cells for 2 days. During this induction phase, cells containing the inducible PAX7 construct start to express PAX7 and, accordingly, proliferate rapidly.
4. Two days after induction, replace half of the used medium with fresh dox-supplemented EB medium, and maintain these cultures for 2 more days (total of 4 days after starting the dox induction). If the cells are too confluent (more than 80 %), passage the cells 1:2.
5. Following 4 days of dox induction and expansion, EB-derived cell outgrowths are ready for FACS purification based on GFP, which reflects Pax7 expression (*see* **Note 7**). Remove used medium and rinse flasks once with PBS.
6. Use warm 0.25 % trypsin to harvest EB-derived cell outgrowths. Add 1 ml of trypsin to the T25 flasks (or 3 ml for a T75) and incubate these for 3 min. After 3 min, tap the flask a few times to see if EBs are detached or not. If not, incubate for 1 more minute.
7. Transfer flask(s) under the cell culture hood, and break down the detached EBs into single cell by pipetting up and down using a 2 or 5 ml pipette, while they are still in trypsin solution.
8. After disaggregating EB clumps into single cells, inactivate the trypsin using warm EB differentiation medium (use four times

of the used trypsin volume). At this time, dead cells and released DNA might form few hairy tangled clumps. Try to remove them using a 1 ml pipette tip before moving to next step.

9. Count and spin cells down for 5 min at 200 × *g*.
10. Remove the supernatant and resuspend the cell pellet in FACS buffer (*see* Section 2.4). The optimal cell density for FACS is 4–6 × 10^6/ml.
11. Filter cell preparation using a FACS cell strainer capped tube to remove any remaining clumps/debris.
12. Sort GFP^+ ($PAX7^+$) cells using a flow cytometry system (such as BD FACSAria II Flow Cytometer, BD Biosciences). As mentioned above, the PAX7-GFP-positive population should be more than 20 % of the total live cell population. Sort these cells into complete EB differentiation medium (*see* **Note 8**).
13. After cell sorting, spin them down at 200 × *g* for 5 min and plate them at the density of 5 × 10^4/cm^2 in a gelatinized T25 flask (i.e., 1.25 × 10^6 cells/T25 flask). Use 5 ml of EB differentiation medium (for T25) supplemented with 10 ng/ml of bFGF and working dilution of dox. Label the flask as P_0 post-sort.
14. Next day, check cell cultures for cell quality/survival and morphology (27). If the quality of cell sorting is good, you should see minimal cell death. Rinse once with PBS, and add 7 ml of fresh medium with dox and bFGF (for T25). At this stage, sorted myogenic precursors will look like small spindle-shaped cells similar to the morphology of mesenchymal cells.
15. Grow these cells for 2–3 more days until cells reach 80–90 % confluency (*see* **Note 9**).

3.8 Expansion and Differentiation of Myogenic Precursors

3.8.1 Expansion of $PAX7^+$-Myogenic Precursors

FACS purified $PAX7^+$ myogenic precursors are expanded in the presence of dox and bFGF ideally for up to four passages within a 2-week time period. During this timeframe, progenitor cells are harvested using trypsin every 3–4 days as they reach 80–90 % confluency. At this point, cells can be split at 1:6–1:8 or frozen down at 1 × 10^6cells/ml/cryovial for future applications (*see* **Note 10**).

1. Check expanding myogenic progenitor cell population to make sure right density (80–90 %) has been reached for passaging.
2. Remove used medium and rinse once with PBS.
3. Add 1 ml of 0.25 % trypsin per T25 flask (or 3 ml for T75).
4. Incubate for 2–3 min, as described above, and tap flask to detach cells.

5. Inactivate trypsin with complete EB medium (4× volume of used trypsin).
6. Break down clumps into a single cell suspension by pipetting for 8–10 times using 5–15 ml pipette.
7. Count and spin down at 200 × *g* for 5 min.
8. Passage cells at 1:6–1:8 into gelatin-coated flasks in EB differentiation medium supplemented with dox and bFGF for cell expansion.
9. Cells can be expanded up to four passages after initial purification. After P_4 the differentiation potential of expanding cell population decreases significantly. Therefore, it is not recommended to expand these cells beyond P_4.

3.8.2 Terminal Differentiation of Expanding Myogenic Progenitors into Myotubes

1. We have observed that cells harvested at P3 demonstrate the best differentiation potential. Therefore, for testing terminal differentiation, harvest myogenic progenitor cells at P3 and plate these in gelatinized target vessels in EB differentiation medium containing dox and bFGF for 3–4 days until 100 % confluency is reached (*see* **Note 11**), at which point culture conditions will be switched to differentiation.
2. Remove used medium and rinse flask(s) 2–3 times with PBS to remove dead cells and any trace amounts of dox.
3. Add fresh medium, which should initially consist of a 1:1 mixture of EB differentiation medium and myotube differentiation medium without any bFGF or dox supplementation. This slow transition allows the myogenic cells to gradually adapt to the low nutrient conditions of the differentiation medium.
4. After 2 days, switch the medium completely to differentiation medium. Be very gentle with cells cultures, and add medium in a very slow rate; otherwise cells might detach during the medium exchange (*see* **Note 12**).
5. Under these conditions, myotubes start to form after 3–5 days. Change the medium only if lots of dead cells are observed; otherwise there is no need for it (*see* **Note 13**).
6. After myotube formation (generally 6–8 days), cells can be fixed and stained for markers of myogenic differentiation, such as myogenin and myosin heavy chain (MHC).

4 Notes

1. For freezing, we recommend resuspending hES/iPS cells in mFreSR medium (Stem Cell Technologies). Use 1×10^6 cells/1 ml of cold freezing medium in each cryovial, and freeze these vials overnight in a −80 °C freezer, placed in an

isopropanol freezing container. Next day, frozen vials should be transferred into a liquid nitrogen tank for proper long-term preservation.

2. Since proliferation rates differ among hES/iPS cell lines, it is recommended to plate cells at different densities in order to have the wells with appropriate density for infection at the next day. Unused wells can either be frozen or discarded.

3. In case excessive cell death is observed following infection, it is recommended to replace 50 % of the infection medium (3 ml) with fresh mTeSR medium immediately after finishing the spin infection. Supplementation of the infection medium with ROCK inhibitor might also be helpful.

4. In case of low infection rate after 2–3 rounds of spin infection, one strategy is to purify this particular inducible ES/iPS cell line using a low dose of dox induction overnight. For this, expand infected hES/iPS cells in a T75 flask, and after reaching 50–60 % confluency, induce cells overnight (12 hrs) by adding 50 ng/ml of dox in mTeSR medium. This low dose of dox, in this short window (12 h), generally is not sufficient to promote cell differentiation. In the next day, harvest cells and sort for the brightest GFP-positive cells (top 50 %). Plate sorted iPax7 hES/iPS cells at the appropriate density (e.g., $5–10 \times 10^4$ cells/well of a 6-well plate) in complete mTeSR supplemented with ROCK inhibitor (do not use dox anymore). Expand sorted colonies and scrape the differentiated ones, as described above. This protocol can be repeated one more time, if necessary. This strategy will boost the inducible fraction of cell preparations significantly (up to 90 %).

5. For efficient EB generation, it is important to have the incubator fully saturated with humidity. Therefore, it is recommended to sandwich each stack of EB plates (max. of three EB plates/stack) with two plates containing sterile water. Also each morning, check the lid of each EB plate for condensation. The presence of significant condensation or water droplets at plate lid is a sign of low humidity and medium water evaporation. This might adversely affect EB growth and subsequent differentiation. In these cases, discard the culture and restart the experiments.

6. If EBs are well formed, they are bright and contain lots of live and shiny cells, and accordingly, they will attach easily after plating. In case no or very few EBs are attached, it is generally an indication of poor quality, which is likely due to cell death (in these circumstances, EBs display dark appearance). In these cases, discard the culture and restart the experiments.

7. Four days of PAX7 induction is usually enough to boost cell proliferation and have enough EB outgrows to cover most of the flask area. In case of low proliferating cell lines, cell preparations can be expanded for 2–3 more days in the presence of dox induction to provide enough cells for FACS sorting. Also bFGF supplementation might be helpful.
8. For GFP sorting, make sure to use a non-induced GFP-negative control cell population to set the sorting gates for GFP. It is also important to avoid cells with dim GFP expression. Thus, place sorting gates on the bright GFP-positive cell population. Usually GFP^+ cells are clearly well separated from the nonexpressing cells using a dot plot graph in the FACS profile (27). It is also important to check the quality of sorting (purity) and cell survival after purification is complete. To do this, separate a small sample from the sorted cell fraction, resuspend in FACS buffer containing PI, and reanalyze this by FACS. Ideally, one should observe >95 % GFP expression in purified fraction and these should be alive (negative for PI).
9. During the myogenic expansion phase, it is important to always harvest and passage the cells at 80–90 % confluency. Never allow cell cultures to reach 100 % confluency; otherwise cell quality may be affected in subsequent passages, and accordingly, their differentiation potential. The only time 100 % confluency is desired is prior to their terminal differentiation into myotubes.
10. For freezing myogenic precursors, use a mixture of 60 % complete EB differentiation medium plus 30 % FBS and 10 % of DMSO. Dox or bFGF supplementation is not necessary.
11. For the formation of multinucleated myotubes, cell density is very important. Therefore, allow cells to reach 100 % confluency before switching culture conditions to myotube differentiation; otherwise cells may not form typical multinucleated myotubes and/or detach during differentiation.
12. During the terminal differentiation phase, the cell adhesion to the plate may become loose. Therefore, be very careful during medium exchange. Always observe the corners and peripheral areas of the plate, and look for any signs of cell detachment from the flask. If sorted cells are pure, detachment is minimal. On the other hand, if sorting was not ideal and sorted population contained non-myogenic cells, plated cells might detach completely during differentiation. For this reason, it is important to check cells for GFP expression during each passage to avoid this undesirable situation. Good sorted myogenic cell cultures should express more than 95 % of PAX7-GFP during the passage times at expansion phase.

13. During the terminal differentiation into myotubes (6–8 days) and after switching medium completely to differentiation medium, usually only one medium exchange is necessary (around day 3 or 4). In case cell lines show high levels of cell death and/or cell detachment, change the medium as needed. Just be very careful to perform it very slowly to prevent further cell detaching.

References

1. Odorico JS, Kaufman DS, Thomson JA (2001) Multilineage differentiation from human embryonic stem cell lines. Stem Cells 19 (3):193–204
2. Thomson JA et al (1998) Embryonic stem cell lines derived from human blastocysts. Science 282(5391):1145–1147
3. Takahashi K et al (2007) Induction of pluripotent stem cells from fibroblast cultures. Nat Protoc 2(12):3081–3089
4. Takahashi K et al (2007) Induction of pluripotent stem cells from adult human fibroblasts by defined factors. Cell 131(5):861–872
5. Yu J et al (2007) Induced pluripotent stem cell lines derived from human somatic cells. Science 318(5858):1917–1920
6. Takahashi K, Yamanaka S (2006) Induction of pluripotent stem cells from mouse embryonic and adult fibroblast cultures by defined factors. Cell 126(4):663–676
7. Takahashi K, Yamanaka S (2013) Induced pluripotent stem cells in medicine and biology. Development 140(12):2457–2461
8. Inoue H et al (2014) iPS cells: a game changer for future medicine. EMBO J 33(5):409–417
9. Yu J, Thomson JA (2008) Pluripotent stem cell lines. Genes Dev 22(15):1987–1997
10. Park IH et al (2008) Disease-specific induced pluripotent stem cells. Cell 134(5):877–886
11. Dimos JT et al (2008) Induced pluripotent stem cells generated from patients with ALS can be differentiated into motor neurons. Science 321(5893):1218–1221
12. Ebert AD et al (2009) Induced pluripotent stem cells from a spinal muscular atrophy patient. Nature 457(7227):277–280
13. Hargus G et al (2010) Differentiated Parkinson patient-derived induced pluripotent stem cells grow in the adult rodent brain and reduce motor asymmetry in Parkinsonian rats. Proc Natl Acad Sci U S A 107(36):15921–15926
14. Lee G et al (2009) Modelling pathogenesis and treatment of familial dysautonomia using patient-specific iPSCs. Nature 461 (7262):402–406
15. Raya A et al (2009) Disease-corrected haematopoietic progenitors from Fanconi anaemia induced pluripotent stem cells. Nature 460 (7251):53–59
16. Ludwig T, Thomson A (2007) Defined, feeder-independent medium for human embryonic stem cell culture. Curr Protoc Stem Cell Biol Chapter 1:Unit 1C 2
17. Nakagawa M et al (2014) A novel efficient feeder-free culture system for the derivation of human induced pluripotent stem cells. Sci Rep 4:3594
18. Takahashi K et al (2009) Human induced pluripotent stem cells on autologous feeders. PLoS One 4(12):e8067
19. Chen G et al (2011) Chemically defined conditions for human iPSC derivation and culture. Nat Methods 8(5):424–429
20. Awaya T et al (2012) Selective development of myogenic mesenchymal cells from human embryonic and induced pluripotent stem cells. PLoS One 7(12):e51638
21. Barberi T et al (2007) Derivation of engraftable skeletal myoblasts from human embryonic stem cells. Nat Med 13(5):642–648
22. Borchin B, Chen J, Barberi T (2013) Derivation and FACS-mediated purification of PAX3+/PAX7+ skeletal muscle precursors from human pluripotent stem cells. Stem Cell Rep 1(6):620–631
23. Tedesco FS et al (2012) Transplantation of genetically corrected human iPSC-derived progenitors in mice with limb-girdle muscular dystrophy. Sci Transl Med 4(140):140ra89
24. Xu C et al (2013) A zebrafish embryo culture system defines factors that promote vertebrate myogenesis across species. Cell 155 (4):909–921
25. Zheng JK et al (2006) Skeletal myogenesis by human embryonic stem cells. Cell Res 16 (8):713–722

26. Goudenege S et al (2012) Myoblasts derived from normal hESCs and dystrophic hiPSCs efficiently fuse with existing muscle fibers following transplantation. Mol Ther 20 (11):2153–2167
27. Darabi R et al (2012) Human ES- and iPS-derived myogenic progenitors restore DYSTROPHIN and improve contractility upon transplantation in dystrophic mice. Cell Stem Cell 10(5):610–619
28. Darabi R et al (2008) Functional skeletal muscle regeneration from differentiating embryonic stem cells. Nat Med 14:134–143
29. Darabi R et al (2011) Assessment of the myogenic stem cell compartment following transplantation of Pax3/Pax7-induced embryonic stem cell-derived progenitors. Stem Cells 29:777–790
30. Filareto A, Darabi R, Perlingeiro RCR (2012) Engraftment of ES-derived myogenic progenitors in a severe mouse model of muscular dystrophy. J Stem Cell Res Ther 220:212–216
31. Darabi R et al (2011) Functional myogenic engraftment from mouse iPS cells. Stem Cell Rev Rep 7:948–957
32. Filareto A et al (2013) An ex vivo gene therapy approach to treat muscular dystrophy using inducible pluripotent stem cells. Nat Commun 4(1549):1–17
33. Skoglund G et al (2014) Physiological and ultrastructural features of human induced pluripotent and embryonic stem cell-derived skeletal myocytes in vitro. Proc Natl Acad Sci U S A 111(22):8275–8280

Methods in Molecular Biology (2016) 1357: 441–450
DOI 10.1007/7651_2014_136

Published online: 23 November 2014

Chondrogenic and Osteogenic Induction from iPS Cells

Ji-Yun Ko and Gun-Il Im

Abstract

Articular cartilage (AC) does not heal spontaneously when injured in adults. This incapacity for self-repair after damage ultimately leads to the development of osteoarthritis. In contrast, bone repairs itself without scarring. However, complete bone healing fails to occur in large defects coming from major trauma or malignant tumor resection. Cell therapy has been investigated for these musculoskeletal conditions. Induced pluripotent stem cells (iPSCs) possess the characteristics of embryonic stem cells (ESCs) in potentially unlimited proliferation while avoiding the ethical controversies. However, several issues need to be resolved before iPSCs can be considered as a potential therapeutic measure for cartilage and bone regeneration. The authors developed protocol to examine the in vitro chondrogenesis and osteogenesis from hiPSCs and in vivo cartilage and bone regeneration using animal models.

Keywords: Induced pluripotent stem cells, Chondrogenesis, Osteogenesis, Regeneration, Cell therapy

1 Introduction

Articular cartilage (AC) does not heal spontaneously when injured in adults. This incapacity for self-repair after damage has spawned extensive investigations for AC regeneration (1, 2): the most commonly probed strategy is cell-based treatments.

Embryonic stem cells (ESCs) have attracted great consideration as a cell source for regenerative medicine (3, 4) because they can potentially provide an unlimited number of progenitors cells for tissue regeneration. However, derivation of hESCs from early embryos raises ethical limitations for their use in clinic practice (5). For this dilemma, induced pluripotent stem cells (iPSCs) offer a new alleyway to avoid the controversy of using hESCs (6, 7). iPSCs function in a manner indistinguishable from ESCs by differentiating into cell types that are characteristic of the three germ layers in vitro and in vivo (8). Further, individual-specific iPSCs can be derived from patients' own cells.

Bone has properties quite different from cartilage. Bone repairs itself without scarring, which is a property uncommon in adult tissues. Most fractures heal spontaneously or with the help of surgical procedures (9). However, there are a number of clinical situations in which complete bone healing fails to occur, i.e., major trauma or malignant tumor resection resulting in critical-size bone defects (spanning > 2 cm) (10, 11). The high proliferation and differentiation capabilities of hiPSCs make them potential candidates for bone regeneration (8).

Several issues need to be resolved before iPSCs can be considered as a potential therapeutic measure for cartilage and bone regeneration. First, the persistence of differentiated phenotypes in vivo must be demonstrated. Second, simple and standardized protocols to generate "easy to grow" cell populations are necessary for the clinical application of iPSCs, so that the cells can survive in vivo transplantations and regenerate functional tissue without the risk of tumor formation. In addition, hiPSCs should demonstrate properties superior or comparable to those of MSCs in order to be regarded as a clinically viable alternative to MSCs (12, 13). To address these issues, a protocol is necessary to examine the in vitro and in vivo capacity of hiPSCs and prove their regenerative potentials for cartilage and bone regeneration.

2 Materials

2.1 Human-Induced Pluripotent Stem Cells

hiPS cell line (SC802A-1, System Biosciences In. Mountain View, CA) (*see* **Note 1**).

2.2 Medium, Buffers, and Reagents

Embryonic stem cell medium: DMEM/F12 (Welgene, Taegu, Korea) supplemented with 20 % (v/v) knockout serum replacements (KSR, Gibco BRL, Grand Island, NY, USA), 1 % antibiotics (penicillin 100 IU/ml; streptomycin 100 μg/ml; Gibco BRL), 0.1 mM nonessential amino acids (Gibco BRL), 0.1 mM β-mercaptoethanol (Sigma-Aldrich, St. Louis, MO, USA), and 4 ng/mL basic fibroblast growth factor (bFGF, R&D Systems, Minneapolis, MN, USA).

Embryoid body formation medium: ESC medium in the presence of 10^{-7} M all-trans retinoic acid (ATRA, Sigma).

Chondrogenic medium: DMEM/F-12 containing 10 % fetal bovine serum (FBS: Gibco BRL), 1 % antibiotics, 10^{-7} M dexamethasone (Sigma), 50 μM ascorbate-2-phosphate (Sigma), 50 μM L-proline (Sigma), 1 mM sodium pyruvate (Sigma), 1 % insulin-transferrin-selenium (Gibco), and 10 ng/ml TGF-β3 (R&D Systems).

Osteogenic medium: α-MEM (Welgene) containing 10 % FBS, 1 % antibiotics, 100 nM dexamethasone, 50 mM L-ascorbate-2-phosphate, and 10 mM glycerophosphate (Sigma).

Other buffers and reagents:

mTeSR1 medium (STEMCELL Technologies Inc., Vancouver, BC, Canada).

0.05 % trypsin-EDTA (Welgene).

Dulbecco's Phosphate Buffered Saline (DBPS, Welgene).

2 % alginate (Sigma).

100 mM $CaCl_2$ solution.

Dissolution solution (55 mM EDTA, 10 mM HEPES, pH 7.4).

Quant-iT™ dsDNA assay kit and Qubit fluorometer system (Invitrogen, Carlsbad, CA, USA).

Blyscan kit (Biocolor, Carrickfergus, UK).

VECTASHIELD® with 4′,6-diamidino-2-phenylindole (DAPI) mounting medium (Vector Laboratories, Burlingame, CA, USA).

Enhanced chemiluminescence (ECL) Western blot analysis detection reagent (Amersham Biosciences, Piscataway, NJ, USA).

2.3 Western Blot Antibody

Primary antibody: anti-rabbit type II collagen (COL2A1; 1:500; Abcam, Cambridge, UK), type X collagen (COL10A1; 1:500; Abcam), type I collagen (COL1A1; 1:500; Abcam), SOX-9 (1:1,000; Abcam), Runx-2 (1:500; Abcam), and bone sialoprotein (BSP; 1:500; Abcam), anti-mouse aggrecan (1:100; Abcam), and Runx-2 (1:500; Abcam).

Secondary antibody: horseradish peroxidase (HRP)-conjugated anti-rabbit IgG or anti-mouse IgG (1:2,000; Cell Signaling Technology, Beverly, MA, USA).

2.4 Immunohistochemistry (IHC) Antibody

Primary antibody: anti-rabbit NANOG 1:200 (Santa Cruz Biotechnology, Santa Cruz, CA, USA), brachyury (1:200; Abcam), COL2A1 (1:200; Abcam), COL1A1 (1:200; Abcam), SOX-9 (1:200; Abcam), Runx-2 (1:100; Abcam), and osteocalcin (1:200; Abcam), anti-mouse Oct3/4 (1:200; Santa Cruz Biotechnology), SSEA4 (1:100; R&D systems), COL10A1 (1:200; Sigma), human nuclear antigen (HN; 1:100; Chemicon Temecula, CA, USA), aggrecan (1:100; Abcam), and Runx-2 (1:100; Abcam).

Secondary antibody: Fluorescence-tagged secondary antibodies (Jackson Immunoresearch Laboratories, West Grove, PA, USA).

2.5 Staining Solution and Kit

TRACP & ALP double-stain kit (Takara Bio Inc., Tokyo, Japan).

2 % Alizarin red solution (Junsei Chemical, Tokyo, Japan).

Goldner's trichrome staining kit (Carl Roth, Karlsruhe, Germany).

3 Methods

3.1 hiPSC Culture and Differentiation into Embryoid Bodies

1. Undifferentiated hiPSCs are maintained as described previously (7, 14).
2. To prepare feeder-free hiPSCs for differentiation experiments (*see* **Note 2**), hiPSCs were passaged to Matrigel-coated polystyrene plates and cultured in the defined mTeSR1 medium (Fig. 1). A combined use of mTeSR1 medium and Matrigel-coated substrates had proven to support the feeder-independent maintenance of hiPSCs (15, 16).
3. In vitro differentiation of hiPSCs is performed using the standard embryoid body (EB) differentiation method (17) with minor modifications. For sphere formation, cells were dissociated with 0.05 % trypsin-EDTA and plated onto 6-well ultralow-attachment plates (Corning, Tewksbury, MA, USA).
4. After 2 days of sphere formation, EBs were cultured in ESC medium in the presence of 10^{-7} M all-trans retinoic acid (ATRA) for 10 days.
5. Medium is changed every other day.

3.2 Induction of In Vitro Chondrogenic Differentiation and Analysis

3.2.1 Chondrogenic Pellet Culture with hiPSCs

1. hiPSC-EBs were dissociated to a single-cell suspension by trypsinizing and then diluting to a final concentration of 5.0×10^5 cells/ml.
2. Micromass pellets were cultured as non-adherent spheres in 5 ml round tubes for 21 days.
3. The tubes were then placed in an incubator at 37 °C in a humidified 5 % CO_2 atmosphere.
4. Chondrogenic medium (CM) was changed every 3–4 days (Fig. 1).

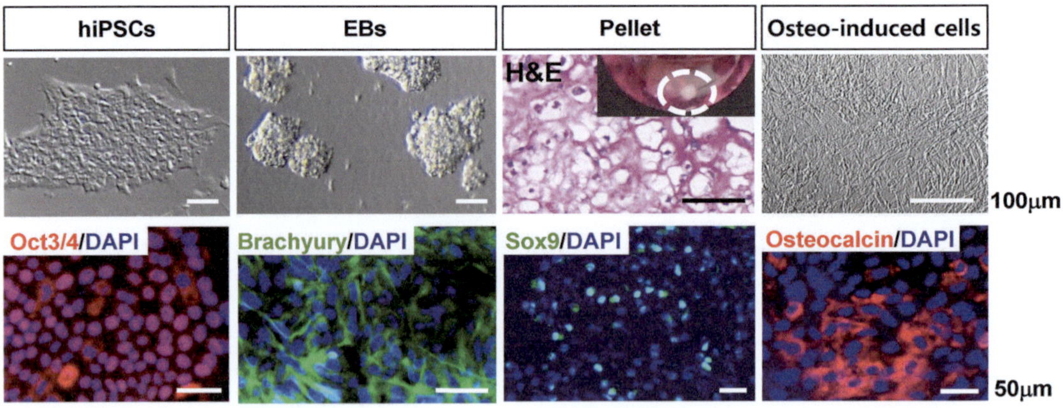

Fig. 1 Undifferentiated hiPSCs and EBs, hiPSC-chondrogenic pellets in culture tube and HE stain, and osteo-induced hiPSCs. Expression of ESC marker OCT3/4, mesodermal marker brachyury, chondrogenic marker SOX-9, and osteogenic marker osteocalcin

3.2.2 Chondrogenic Differentiation of hiPSCs in Alginate Gel

1. For constructs with alginate, dissociated hiPSCs were suspended at a density of 1.5×10^6 cells per 100 μl in 2 % alginate (Sigma).
2. Polymerization of alginate is then achieved by dropping the grafts into 100 mM $CaCl_2$ solution.
3. After instantaneous gelation, the alginate is allowed to polymerize further for a period of 8–10 min in the $CaCl_2$ solution.
4. All beads were thoroughly washed with DPBS and were cultured in a 24-well plate (Nunc, Waltham, MA, USA) under the same chondrogenic medium and conditions as in pellet culture.
5. Medium was changed every 3–4 days.

3.2.3 Biochemical Assays for DNA and GAG Quantification

1. After 21 days of in vitro culture, the alginate hydrogels were dissolved by incubating the beads for 20 min in dissolution solution.
2. The pellets were digested overnight in papain buffer at 60 °C.
3. DNA content is determined using the Quant-iT™ dsDNA assay kit and Qubit fluorometer system (Invitrogen).
4. Glycosaminoglycan (GAG) production is determined using a Blyscan kit according to the manufacturer's instructions.

3.2.4 Western Blot Analysis for Chondrogenic Protein

1. Proteins were extracted from cultures, electrophoresed, and transferred to a nitrocellulose membrane.
2. The blot is probed with anti-rabbit or anti-mouse antibody, followed by horseradish peroxidase (HRP)-conjugated anti-rabbit IgG or anti-mouse IgG.
3. This experiment is repeated in three samples, each from different individuals.

3.2.5 Immunohistochemistry (IHC)

1. The cells or sections were blocked with 5 % normal goat serum (NGS) and 0.1 % Triton X-100 in PBS at room temperature for 1 h.
2. The following primary antibodies were applied overnight at 4 °C.
3. Appropriate fluorescence-tagged secondary antibodies were used for visualization.

3.3 Induction of In Vitro Osteogenic Differentiation and Analysis

3.3.1 Osteoblast Culture of hiPSCs

1. hiPSC-EBs were dissociated to a single-cell suspension by trypsinizing and then diluting to a final concentration of 3.0×10^5 cells/ml.
2. For subsequent differentiation, single cells were plated on gelatin-coated dishes (*see* **Note 3**) and cultured with a specific induction medium (osteogenic medium [OM]).
3. The cells were incubated in OM for up to 2 weeks at 37 °C in 5 % CO_2 in a 6 cm dish at a density of 3.0×10^5 cells.
4. The medium is changed every third day.
5. The analyses were performed on day 7 and 14 to test the osteogenic differentiation of hiPSCs (Fig. 1).

3.3.2 Alkaline Phosphatase (ALP) Staining

1. ALP activity is measured using a TRACP & ALP double-stain kit followed by incubation for 7 or 14 days.
2. To fix the cultured cell samples, the culture supernatant is removed and discarded.
3. The cells were washed once with sterilized PBS and then fixed in fixation solution at room temperature for 5 min.
4. Approximately 2 ml of sterilized distilled water is added to each well to dilute the fixation solution, and the solution is then aspirated.
5. Approximately 2 ml of sterilized distilled water is again added to wash the well, and all the liquid from the well is removed and discarded.
6. For ALP staining, the substrate solution is added to a 6 cm dish onto which the cells had been fixed.
7. The amount of substrate solution is 500 μl/well. The plate is incubated at 37 °C for 45 min for the reaction.
8. The solution is then removed and discarded.
9. Subsequently, the dish is washed three times with sterilized distilled water to quench the reaction.

3.3.3 Alizarin Red Staining

1. To measure calcium deposition in the extracellular matrix, the cells were seeded in 6 cm cell culture dishes and cultured for 7 or 14 days under OM.
2. The differentiated cell cultures were washed twice with PBS and fixed in 10 % formalin for 10 min.
3. After three washes with PBS, cells were then stained with 2 % Alizarin red solution for 10 min.

3.3.4 Western Blot Analysis for Osteogenic Protein

1. Western blot analysis was performed as described previously (*see* Methods 3.2.4).

3.3.5 Immunohistochemistry (IHC)

1. IHC was performed as described previously (*see* Methods 3.2.5).

3.4 In Vivo Implantation Procedure

9-week-old male Sprague-Dawley rats are used. The animals are anesthetized with zoletil (40 mg/kg) and xylazine (10 mg/kg). The rats received daily injections of cyclosporine A (10 mg/kg, ip) to suppress immune responses.

To test the in vivo chondrogenesis, osteochondral defect model created on distal femur of immunosuppressed rats is used (12). For in vivo osteogenesis, calvarial defect model in the immunosuppressed rats is used (13).

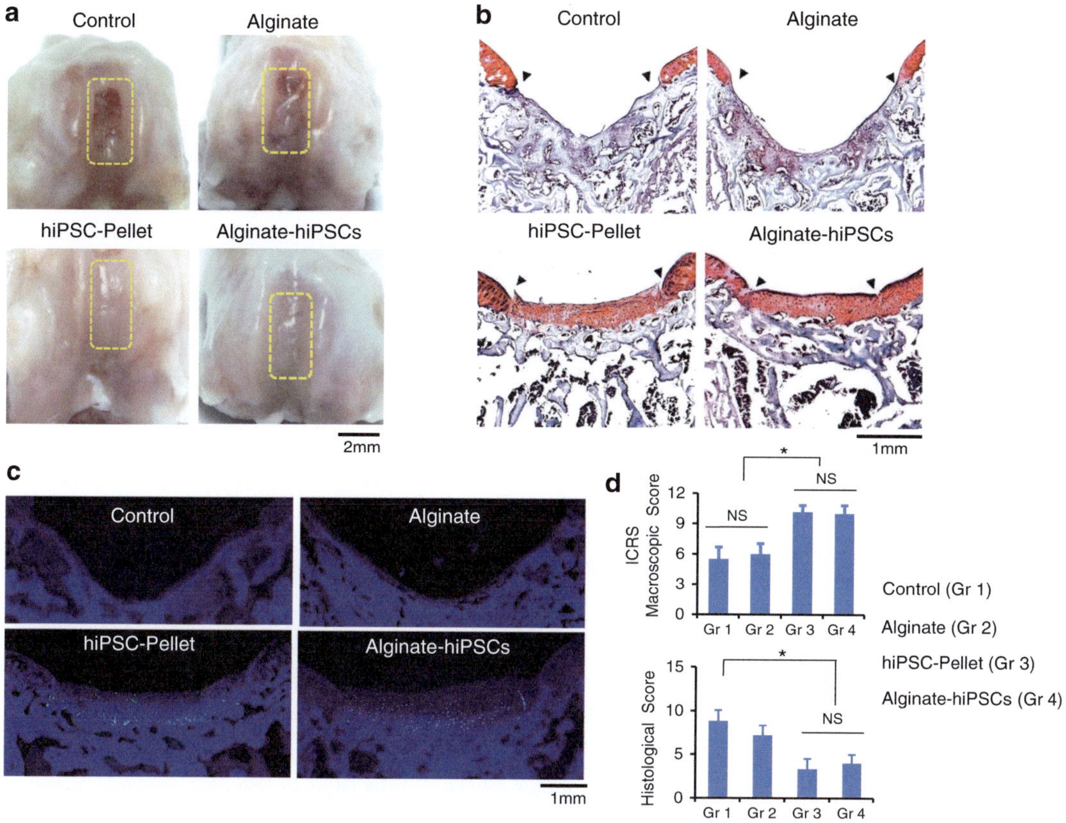

Fig. 2 In vivo repair of osteochondral defects using chondro-induced hiPSCs. Gross appearance of the defects: *yellow dots* outline the defect margin (**a**), histological finding of the defect from Safranin-O staining: *arrows* denote the margin of the defects (**b**), IHC for human antigen: positive cells are *bright green* (**c**), and ICRS macroscopic score and histological grading scale 12 weeks after implantation of hiPSCs (**d**). Bar represents mean ± SE. $N = 6$, $^{*}P < 0.05$ (reproduced from (12) with permission from Elsevier)

3.4.1 Osteochondral Defect Model for In Vivo Cartilage Regeneration

1. A 1.5 mm outer diameter trephine drill is used to create osteochondral defects (2.0 × 4.0 mm) in the trochlear groove of the femur.
2. The hiPSC-pellets or alginate-hiPSCs constructs, which were prepared in the same way as the in vitro study and cultured for 21 days, are implanted in the osteochondral defects.
3. The defects are managed using one of the following methods: no treatment (group 1), filling with alginate hydrogel only (group 2), filling with three hiPSC-pellets (group 3), and filling with alginate-hiPSCs construct (group 4).
4. After 12 weeks, the rats are sacrificed for gross and histological examinations (Fig. 2).

3.4.2 Calvarial Defect Model

1. Two full-thickness calvarial bone defects 4 mm in diameter are created without dural perforation using a surgical microdrill fitted with a trephine burr.

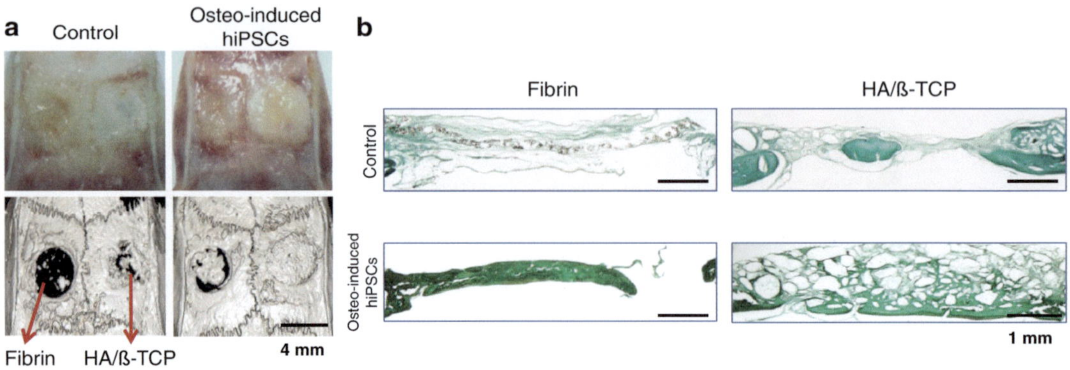

Fig. 3 In vivo healing of critical-size calvarial defect by hiPSCs after 8 weeks of implantation. (**a**) Creation of calvarial defects, gross and micro-CT findings of the defect. (**b**) Histological finding from Goldner's trichrome staining

2. The wound is thoroughly irrigated with warmed saline to remove residual bone dust.
3. Two kinds of scaffolding materials are used, fibrin glue (Tisseel® Darim, Seoul, Korea) and hydroxyapatite/β-tricalcium phosphate (HA/β-TCP, 7:3, Genoss, Suwon, Korea).
4. The harvested cells (1.5×10^6 cells) are mixed with 30 μl fibrin glue or 30 mg HA/β-TCP and placed into the calvarial defect.
5. After implantation of the cell/scaffold hybrid, the defect region is covered with polycaprolactone membrane (Genoss), and the skin is closed with staples.
6. After 8 weeks, the rats are sacrificed using carbon dioxide.
7. Calvarial bones are dissected out and underwent gross and histological analysis as well as micro computed tomography (micro-CT: NFR Polaris-G90, Nano Focus Ray, Jeonju, Korea) (Fig. 3).

3.5 Macroscopic Observation and Histology

1. Osteochondral defect model
   Regenerated cartilage is grossly examined using the International Cartilage Repair Society (ICRS) macroscopic score (18), which evaluates the degree of defect repair, integration to the border zone, and the macroscopic appearance. The histologic grading scale as described by Wakitani (19, 20) is used to evaluate the quality of the repaired tissue.
2. Calvarial defect model
   Following macroscopic examination, the calvarial bone is dissected and embedded in an Optimal Cutting Temperature (OCT) compound (aqueous embedding medium within a mold) and then frozen in a metal pan over a bath of liquid nitrogen. All frozen tissue blocks are cryosectioned to a

nominal thickness of 10 μm. Goldner's trichrome staining is performed for all specimens. The thickness of regenerated calvarial bone is obtained from the average of ten different spots in each specimen.

4 Notes

1. To avoid potential safety issues associated with the use of viruses, the hiPS cell line which was created by direct delivery of four proteins fused to a cell-penetrating peptide into human fibroblasts is used (21).
2. Undifferentiated hiPSCs are maintained on mitotically inactivated mouse embryonic fibroblast (MEF) feeder layers in embryonic stem cell medium. Medium is changed daily. For the maintenance of undifferentiated hiPSCs, cultures are passaged once every week by mechanically dissecting and transferring hiPSC colonies onto freshly prepared MEF feeders.
3. Prepare gelatin-coated dishes by adding gelatin solution (0.1 %, v/w in PBS) to culture dishes and incubating for at least 5 min in a CO_2 incubator.

References

1. Buckwalter JA, Mankin HJ (1988) Articular cartilage: degeneration and osteoarthritis, repair, regeneration, and transplantation. Instr Course Lect 47:487–504
2. Caplan AI, Goldberg VM (1999) Principles of tissue engineered regeneration of skeletal tissues. Clin Orthop Relat Res 367:S12–S16
3. Thomson JA, Itskovitz-Eldor J, Shapiro SS, Waknitz MA, Swiergiel JJ, Marshall VS et al (1998) Embryonic stem cell lines derived from human blastocysts. Science 282:1145–1147
4. Lerou PH, Daley GQ (2005) Therapeutic potential of embryonic stem cells. Blood Rev 19:321–331
5. Gaspar-Maia A, Alajem A, Meshorer E, Ramalho-Santos M (2011) Open chromatin in pluripotency and reprogramming. Nat Rev Mol Cell Biol 12:36–47
6. Takahashi K, Yamanaka S (2006) Induction of pluripotent stem cells from mouse embryonic and adult fibroblast cultures by defined factors. Cell 126:663–676
7. Takahashi K, Tanabe K, Ohnuki M, Narita M, Ichisaka T, Tomoda K et al (2007) Induction of pluripotent stem cells from adult human fibroblasts by defined factors. Cell 131:861–872
8. Itoh M, Kiuru M, Cairo MS, Christiano AM (2011) Generation of keratinocytes from normal and recessive dystrophic epidermolysis bullosa-induced pluripotent stem cells. Proc Natl Acad Sci U S A 108:8797–8802
9. Einhorn TA (2005) The science of fracture healing. J Orthop Trauma 19:S4–S6
10. Hollinger JO, Kleinschmidt JC (1990) The critical size defect as an experimental model to test bone repair materials. J Craniofac Surg 1:60–68
11. McKee MD (2006) Management of segmental bony defects: the role of osteoconductive orthobiologics. J Am Acad Orthop Surg 14: S163–S167
12. Ko JY, Kim KI, Park S, Im GI (2014) In vitro chondrogenesis and in vivo repair of osteochondral defect with human induced pluripotent stem cells. Biomaterials 35:3571–3581
13. Ko JY, Park S, Im GI (2014) Osteogenesis from human induced pluripotent stem cells: an in vitro and in vivo comparison with mesenchymal stem cells. Stem Cells Dev 23:1788–1797
14. Yu J, Vodyanik MA, Smuga-Otto K, Antosiewicz-Bourget J, Frane JL, Tian S et al (2007) Induced pluripotent stem cell lines derived from human somatic cells. Science 318:1917–1920

15. Xu C, Inokuma MS, Denham J, Golds K, Kundu P, Gold JD et al (2001) Feeder-free growth of undifferentiated human embryonic stem cells. Nat Biotechnol 19:971–974
16. Akopian V, Andrews PW, Beil S, Benvenisty N, Brehm J, Christie M et al (2010) Comparison of defined culture systems for feeder cell free propagation of human embryonic stem cells. In Vitro Cell Dev Biol Anim 46:247–258
17. Li W, Zhou H, Abujarour R, Zhu S, Young Joo J, Lin T et al (2009) Generation of human-induced pluripotent stem cells in the absence of exogenous Sox2. Stem Cells 27:2992–3000
18. van den Borne MP, Raijmakers NJ, Vanlauwe J, Victor J, de Jong SN, Bellemans J et al (2007) International Cartilage Repair Society (ICRS) and Oswestry macroscopic cartilage evaluation scores validated for use in Autologous Chondrocyte Implantation (ACI) and microfracture. Osteoarthritis Cartilage 15:1397–1402
19. Wakitani S, Goto T, Pineda SJ, Young RG, Mansour JM, Caplan AI et al (1994) Mesenchymal cell-based repair of large, full-thickness defects of articular cartilage. J Bone Joint Surg Am 76:574–592
20. Toh WS, Lee EH, Guo XM, Chan JK, Yeow CH, Choo AB et al (2010) Cartilage repair using hyaluronan hydrogel-encapsulated human embryonic stem cell-derived chondrogenic cells. Biomaterials 31:6968–6980
21. Kim D, Kim CH, Moon JI, Chung YG, Chang MY, Han BS et al (2009) Generation of human induced pluripotent stem cells by direct delivery of reprogramming proteins. Cell Stem Cell 4:472–476

Methods in Molecular Biology (2016) 1357: 451–464
DOI 10.1007/7651_2014_135

Published online: 18 November 2014

Generation of iPS Cells from Granulosa Cells

Jian Mao and Lin Liu

Abstract

Various types of somatic cells can be reprogrammed to induced pluripotent stem (iPS) cells. Somatic stem cells may generate iPS cells more efficiently than do differentiated cells. We show that granulosa cells exhibit characteristic of somatic stem cells and can be reprogrammed to iPS cells more efficiently or with few factors. Here, we describe generation of mouse and pig iPS cells from granulosa cells with high efficiency.

Keywords: Granulosa cells, Reprogramming, Induced pluripotent stem cells (iPSCs), Mouse, Pig

1 Introduction

iPS cells have been generated from many cell types, including fetal and adult fibroblasts (1, 2), hepatocytes (3), stomach cells (3), peripheral blood (4), keratinocytes (5), cord blood (6, 7), dental pulp cells (8–10), and even fully differentiated lymphocytes (T and B cells) (11–16). We show that granulosa cells with property of somatic stem cells can be reprogrammed to iPS cells with high efficiency.

It has been shown that stemness facilitates reprogramming, as shown by more efficient reprogramming of progenitor stem cells to iPS cells than of differentiated cells (17). Granulosa cells possess characteristics of multipotent stem cells (18–20). Embryonic stem cells are truly pluripotent stem cells that lack the expression of LaminA in their nuclear envelope, in addition to the expression of pluripotent genes Oct4, Nanog, and Sox2. Consistently, somatic stem cells with reduced expression of LaminA or artificial reduction of LaminA by shRNA facilitate iPS induction (21). We show that mouse and pig granulosa cells and cumulus cells from adult ovarian sections are negative for LaminA by immunocytochemistry (Fig. 1a, b). This further confirms that granulosa cells and cumulus cells exhibit stemness to some extent. Secondly, the iPS cells derived from granulosa cells may have improved safety. The standard method to achieve iPS cells is introducing Yamanaka factors into fibroblasts by retrovirus vectors (1, 2). There have been concerns about the safety of iPS cells, especially potential oncogenesis from

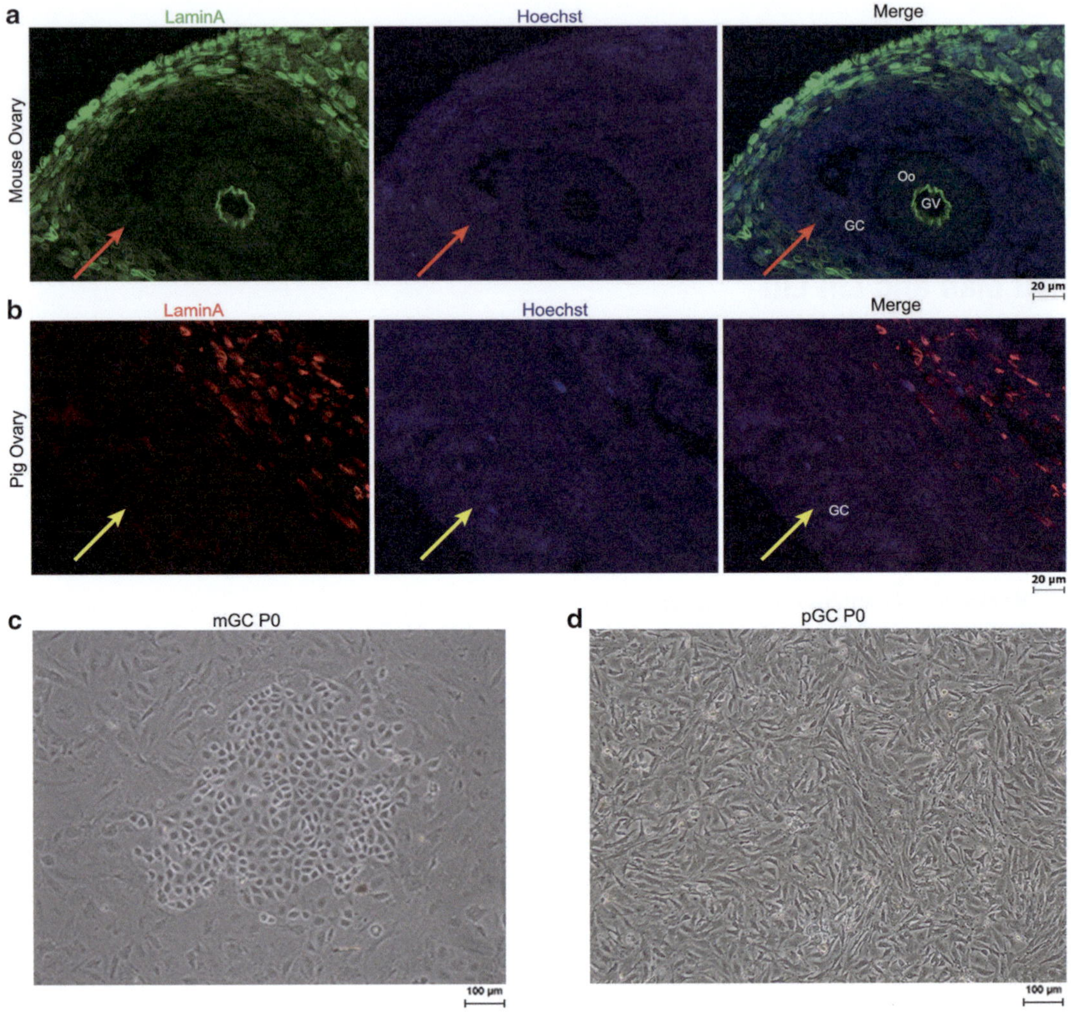

Fig. 1 Low or minimal expression levels of LaminA in granulosa cells (GCs) of growing follicles from mouse and pig ovaries. (**a**) Minimal expression of LaminA in mouse ovarian sections by immunocytochemistry using antibody against LaminA. Hoechst 33342 stained nuclei in *blue*. *Oo* oocyte, *GV* germinal vesicle; *Red arrow* indicates mouse granulosa cells. Scale bar = 20 μm. (**b**) Pig granulosa cells are negative for LaminA either. *Yellow arrow* indicates pig granulosa cells. Scale bar = 20 μm. (**c**) mGCs cultured in vitro show clonal formation like clones of stem cells. High nuclei/cytoplasm ratio can be seen. Scale bar = 100 μm. (**d**) pGCs cultured in vitro. Scale bar = 100 μm

these factors or integration of the factors. For instance, c-Myc is a well-known oncogene (22), and Klf4 also has potential carcinogenic effects (23). Reduced factors may confer more safety for iPS cells, as it has been concerns that Klf4 and c-Myc cause tumorigenesis following the transplantation of iPS cells (24). Removal of Klf4 and c-Myc would improve the safety of iPS for clinical application. Previously, the chimeric and progeny mice derived from iPS cells that are devoid of Myc transgene appear to be normal (25–27).

Notably, some precursor cells already express high levels of pluripotent factors. The advantage of expression of endogenous Yamanaka factor genes in precursors could be taken to reduce the number of reprogramming factors during iPS induction (28–31). Mouse granulosa cells express Klf4 and c-Myc endogenously, such that they can be reprogrammed into iPS cells only with two factors Oct4 and Sox2 (32). Recently, mouse iPS cells can be achieved completely by small molecules (33). Taking advantages of their stemness, granulosa cells may need fewer small molecules to be reprogrammed and/or at higher efficiency. Thirdly, granulosa cells are relatively easily accessible. Granulosa cells and cumulus cells are often by-products from in vitro fertilization (IVF) clinic or animal industry where they are removed off from oocytes and discarded. It is also worth noting that cumulus cells were the first to successfully produce clone mice, Cumulina (34). Granulosa cells also were effectively used to clone animals (35–39), including the first cloned piglets by somatic cell nuclear transfer (37), indicating that both cell types from ovarian follicles are valuable sources for successful cloning and amenable to reprogramming. Together, generation of iPS cells using granulosa cells may show great potential in future clinical applications, such as tissue engineering, regenerative medicine, derivation of patient-specific iPS cells, and cancer treatment, and in animal production and transgenesis.

2 Materials

2.1 Isolation and Culture of Granulosa Cells (See Note 1)

1. C57BL/6 female mice.
2. Pregnant mare serum gonadotropin (PMSG) (367222, Calbiochem).
3. Pasteur pipette.
4. Stereomicroscope.
5. Pig ovary (from slaughterhouse).
6. Syringe.
7. Culture medium for mouse granulosa cells: Knockout Dulbecco's modified Eagle's medium (KO-DMEM, Invitrogen), added with 20 % FBS (Hyclone), 1,000 U/ml mouse leukemia inhibitory factor (mLIF; Millipore), 0.1 mM β-mercaptoethanol (Sigma, *see* **Note 2**), 1 mM L-glutamine (Invitrogen), 0.1 mM nonessential amino acids (Sigma), and penicillin (100 U/ml)-streptomycin (100 μg/ml) (Invitrogen).
8. Culture medium for pig granulosa cells: DMEM/F12 (Invitrogen), added with 20 % FBS (Hyclone), 200 U/ml human leukemia inhibitory factor (hLIF; Millipore), 10 ng/ml basic fibroblast growth factor (bFGF, Millipore) 0.1 mM β-mercaptoethanol (Sigma), 1 mM L-glutamine (Invitrogen), 0.1 mM

nonessential amino acids (Sigma), and antibiotic-antimycotic (Invitrogen) or penicillin (100 U/ml)-streptomycin (100 μg/ml) (Invitrogen).

9. Gelatin (G-1890, Sigma).
10. Phosphate Buffered Saline (PBS) (20012, Invitrogen).
11. Trypsin-EDTA (25200, Invitrogen).

2.2 Preparation of Feeder Cells

1. Mouse embryonic fibroblasts (MEFs).
2. MEF medium: Dulbecco's modified Eagle's medium (DMEM, high glucose) (11995, Invitrogen), added with 10 % FBS (Hyclone), 1 mM L-glutamine (Invitrogen), and penicillin (100 U/ml)-streptomycin (100 μg/ml) (Invitrogen).
3. Mitomycin C (M-4287, Sigma).
4. Gelatin (G-1890, Sigma).
5. Phosphate Buffered Saline (PBS) (20012, Invitrogen).
6. Trypsin-EDTA (25200, Invitrogen).
7. Dimethyl sulfoxide (DMSO) (d-2650, Sigma).
8. Fetal bovine serum (FBS) (SH30070.03E, Hyclone).
9. Cryovials.

2.3 Induction of iPS Cells

1. Plat-E cells.
2. 293-T cells.
3. Plat-E and 293-T medium: Dulbecco's modified Eagle's medium (DMEM, high glucose) (11995, Invitrogen), added with 10 % FBS (Hyclone), 1 mM L-glutamine (Invitrogen), and penicillin (100 U/ml)-streptomycin (100 μg/ml) (Invitrogen).
4. Lipofectamine™2000 (11668-019, Invitrogen).
5. Opti-MEM (Invitrogen).
6. pMXs-mOct4/mSox2/mKlf4/mc-Myc (Addgene).
7. pMXs-pOct4/pSox2/pKlf4/pc-Myc, etc.
8. 0.45 μm filter (Millipore).
9. Polybrene (Sigma).
10. Feeder cell layers: mouse embryonic fibroblasts (MEFs).
11. Cell counting board.
12. Mouse iPSC induction medium: Knockout Dulbecco's modified Eagle's medium (KO-DMEM, Invitrogen), added with 20 % Knockout serum replacement (KSR, Invitrogen), 1,000 U/ml mouse leukemia inhibitory factor (mLIF; Millipore), 0.1 mM β-mercaptoethanol (Sigma), 1 mM L-glutamine (Invitrogen),

0.1 mM nonessential amino acids (Sigma), and penicillin (100 U/ml)-streptomycin (100 μg/ml) (Invitrogen).

13. Pig iPSC basal induction medium: Knockout Dulbecco's modified Eagle's medium (KO-DMEM, Invitrogen), added with 20 % Knockout serum replacement (KSR, Invitrogen), 200 U/ml human leukemia inhibitory factor (hLIF; Millipore), 10 ng/ml bFGF (Millipore), 0.1 mM β-mercaptoethanol (Sigma), 1 mM L-glutamine (Invitrogen), 0.1 mM nonessential amino acids (Sigma), and penicillin (100 U/ml)-streptomycin (100 μg/ml) (Invitrogen).
14. Gelatin (G-1890, Sigma).
15. Phosphate Buffered Saline (PBS) (20012, Invitrogen).
16. Trypsin-EDTA (25200, Invitrogen).
17. TrypLE (12605-010, Invitrogen).

2.4 Culture, Passage, and Cryopreservation of iPS Cells

1. Mouse iPSC culture medium: Knockout Dulbecco's modified Eagle's medium (KO-DMEM, Invitrogen), added with 20 % FBS (Hyclone), 1,000 U/ml mouse leukemia inhibitory factor (mLIF; Millipore), 0.1 mM β-mercaptoethanol (Sigma), 1 mM L-glutamine (Invitrogen), 0.1 mM nonessential amino acids (Sigma), and penicillin (100 U/ml)-streptomycin (100 μg/ml) (Invitrogen).
2. Pig iPSC basal culture medium: Knockout Dulbecco's modified Eagle's medium (KO-DMEM, Invitrogen), added with 10 % KSR (Invitrogen), 10 % FBS (Hyclone), 200 U/ml human leukemia inhibitory factor (hLIF; Millipore), 10 ng/ml bFGF (Millipore), 0.1 mM β-mercaptoethanol (Sigma), 1 mM L-glutamine (Invitrogen), 0.1 mM nonessential amino acids (Sigma), and penicillin (100 U/ml)-streptomycin (100 μg/ml) (Invitrogen).
3. Feeder cell layers: mouse embryonic fibroblasts (MEFs).
4. Gelatin (G-1890, Sigma).
5. Phosphate Buffered Saline (PBS) (20012, Invitrogen).
6. Trypsin-EDTA (25200, Invitrogen).
7. TrypLE (12605-010, Invitrogen).
8. Dimethyl sulfoxide (DMSO) (d-2650, Sigma).
9. Fetal bovine serum (FBS) (SH30070.03E, Hyclone).
10. Cryovials.

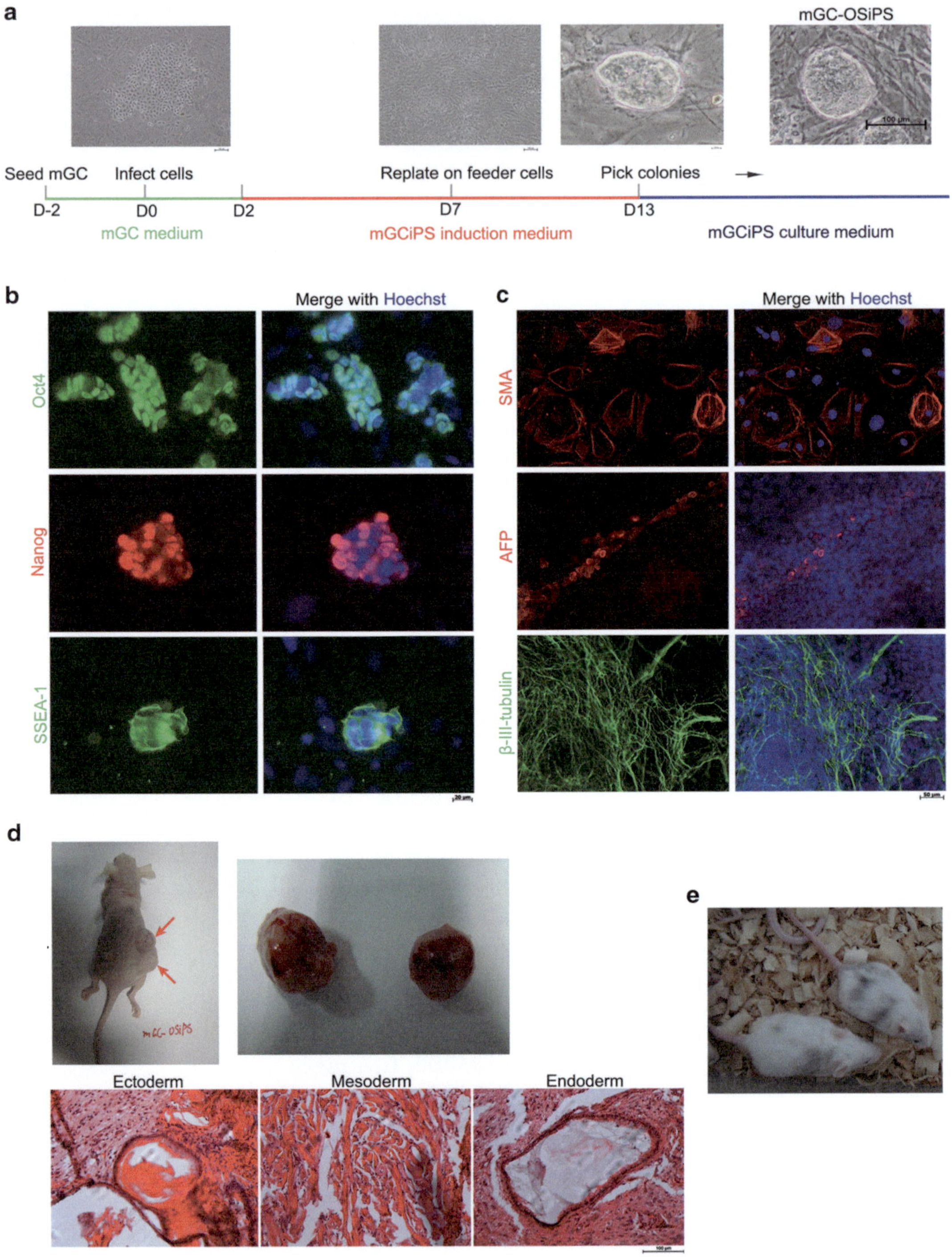

Fig. 2 Characterization of mouse iPS cells generated from mGC. (**a**) Schematic diagram of mGCiPSC generation. Scale bar = 100 μm. (**b**) mGC-OSiPS clones express strong Oct4, Nanog, and SSEA1 by immunofluorescence (scale bar = 20 μm). Nuclei stained with Hoechst 33342 (*blue*). (**c**) Differentiation in vitro of mGC-OSiPS cells by embryoid body (EB) formation. The differentiated derivatives by EB consist of

3 Methods

3.1 Generation of Mouse iPS Cells from Granulosa Cells

3.1.1 Isolation of Mouse Granulosa Cells

1. Inject 100 μl pregnant mare serum gonadotropin (PMSG) into the abdominal cavity of each C57BL/6 female mouse.
2. At 48 h after injection, sacrifice the mice and dissect the ovarian. Aspirate the granulosa cells from ovarian using a Pasteur pipette under a stereomicroscope.
3. Wash the cells in 800 μl fresh PBS and culture them in 2 ml mGC medium in 6-well plate. The medium was changed 48 h after incubation (Fig. 1c).

3.1.2 Packaging of Retroviral Vectors

1. The day before plasmid transfection, seed 5×10^6 Plat-E cells per 100 mm dish.
2. At 24 h after the seeding of Plat-E cells, pMXs-based retroviral vectors (pMXs-Oct4, Sox2, Klf4, c-Myc) were introduced into Plat-E cells using lipo-2000 transfection reagent according to the manufacturer's recommendations. 12 μg of each pMXs plasmid is transfected into one 100 mm dish.
3. Collect the viruses with 0.45 μm filter at 48 h and 72 h, respectively, after transfection. The total volume of virus suspension for each factor packaged from one 100 mm dish is 16 ml (*see* **Note 3**).

3.1.3 Induction of Mouse iPS Cells from Granulosa Cells

1. Seed ~5×10^5 freshly isolated primary granulosa cells 48 h before transduction (*see* **Note 4**).
2. Infect the granulosa cells with the pMXs-based retroviral vectors (OSKM/OSK/OS or other reprogramming factors, 2 ml virus suspension for each factor) twice in 24 h interval, each time for 12 h (*see* **Notes 5** and **6**).
3. After infection, the cells were cultured in 2 ml mouse iPSC induction medium, and the medium was changed daily.
4. At day 7, passage the cells on MEF feeders. The cells were digested by Trypsin-EDTA and counted. Seed 1×10^4 cells per 60 mm dish or ~3.3×10^3 per 6-well plate.

Fig. 2 (continued) cells representing three embryonic germ layers as shown by immunofluorescence staining using endoderm marker alpha 1-fetoprotein (AFP), mesoderm marker smooth muscle actin (SMA), and ectoderm marker β-III-tubulin (Scale bar = 50 μm). Nuclei stained with Hoechst 33342 (*blue*). (**d**) Differentiation in vivo of mGC-OSiPS cells by teratoma formation test following injection into nude mice. *Red arrows* indicate teratoma on the back of nude mice. Hematoxylin and eosin staining of teratoma tissues derived from mGC-OSiPS cells. All teratomas consist of representative derivatives of three germ layers, including epidermis (ectoderm), muscle (mesoderm), and gland epithelium (endoderm). Scale bar = 100 μm. (**e**) Chimeric mice generated from mGC-OSiPS, based on coat color

5. At day 13, pick the ES-like colonies using 200 μl tips (*see* **Note 7**). The colonies were picked to Trypsin-EDTA in 96-well plate and digested for 3–5 min and then transferred into 24-well plate which is added with mouse iPSC culture medium for culture and passage (Fig. 2a, *see* **Note 8**).

3.1.4 Culture and Passage of mGCiPS

The mGCiPS cells are cultured in mouse iPSC culture medium (Fig. 2a). The medium was changed daily, and the cells were routinely passaged every 2 days (*see* **Notes 9** and **10**).

3.1.5 Cryopreservation of mGCiPS Cells

1. Remove iPSC medium and wash mGCiPS cells once with PBS, when mGCiPS cells grow near confluence. Add 200 μl preheated 0.25 % Trypsin-EDTA into each well of 12-well plate for 1–2 min and then use 600 μl iPSC medium to block digesting.
2. Pipette mGCiPSC mass into single cell and transfer them to a sterile 1.5 ml tube.
3. Pellet the cells at 1,200 rpm for 3 min at room temperature.
4. Remove the supernatant and resuspend cells gently in 1 ml freezing medium (90 % FBS and 10 % DMSO, mixed and precooling); mix them by pipetting several times.
5. Quickly aliquot 1 ml of the cell suspension into two labeled cryovials (500 μl each) (*see* **Note 11**) and put them in a −80 °C freezer overnight, and then transfer cryovials into a liquid nitrogen tank for long-term storage.

3.1.6 Thawing of mGCiPS Cells

1. Remove the vial from liquid nitrogen and thaw cryovials by quickly warming in an incubator with 37 °C water bath (*see* **Note 12**).
2. When the ice crystals almost disappear, aseptically transfer the cell suspension into a 1.5 ml tube using a pipette filled with 500 μl of preheated iPSC culture medium to dilute DMSO.
3. Pellet the cells at 1,200 rpm for 3 min and then resuspend the pellet in fresh iPSC culture medium, plate on a new feeder well, and culture the thawed in incubator at 37 °C.
4. Change whole medium to remove floating dead cells the next day and then change iPSC culture medium daily. Cells should be ready for passaging in 2–3 days (*see* **Note 10**).

3.2 Generation of iPS Cells from Pig Granulosa Cells

3.2.1 Isolation and Culture of Pig Granulosa Cells

1. Transport pig ovaries in 37 °C PBS with 3× antibiotic-antimycotic from slaughterhouse to laboratory (*see* **Note 13**).
2. Aspirate the follicular fluid from the median size follicles using a 10 ml syringe and then transfer to 50 ml tube.
3. Pellet the cells at 1,200 rpm for 5 min and then remove the supernatants. Wash the cells with 5 ml PBS with 3× antibiotic-antimycotic.

4. Pellet the cells at 1,200 rpm for 5 min and then resuspend the pellet in fresh 3× antibiotic-antimycotic pig granulosa cells medium, and plate them on 100 mm dish with 3× antibiotic-antimycotic pGC medium (*see* **Note 14**). The granulosa cells from ten pig ovaries are plated on one 100 mm dish.
5. Change the medium with normal pGC medium after 48 h for incubation (Fig. 1d).

3.2.2 Packaging of Retroviral Vectors

1. The day before plasmid transfection, seed 7×10^6 293-T cells per 100 mm dish.
2. At 24 h after the seeding of 293-T cells, pMXs-based retroviral vectors (pMXs-Oct4, Sox2, Klf4, c-Myc) Gag-Pol and VSV-G (10:9:1) were introduced into 293-T cells using lipo-2000 transfection reagent according to the manufacturer's recommendations. A total of 12 μg plasmids are transfected into one 100 mm dish (*see* **Note 15**).
3. Collect the viruses with 0.45 μm filter at 48 h and72 h, respectively, after transfection.
4. Centrifuge the viruses suspension at 25,000 rpm, 4 °C for 90 min, discard the supernatants, and dissolve each virus (packaged from one 100 mm dish) precipitation with 800 μl H-DMEM at 4 °C overnight (*see* **Notes 3** and **16**).

3.2.3 Induction of Pig iPS Cells from Granulosa Cells

1. Seed 1×10^5 granulosa cells in a 6-well dish 24 h before transduction.
2. Infect the granulosa cells with pMXs-based retroviral vectors (OSKM or other reprogramming factors, 100 μl concentrated virus suspension for each factor) twice in 24 h interval, each time for 12 h.
3. After infection, the cells were cultured in 2 ml pig iPSC induction medium, and the medium was changed daily.
4. At day 10, passage the cells on MEF feeders. The cells were digested by TrypLE and counted. Seed 1×10^4 cells per 60 mm dish or ~3.3×10^3 per 6-well plate.
5. At day 21, pick the ES-like colonies using 200 μl tips (*see* **Note 7**). The colonies were picked to TrypLE in 96-well plate and digested for 3–5 min, and then transferred into 24-well plate which is added with pig iPSC culture medium for culture and passage (Fig. 3a, *see* **Notes 8** and **17**).

3.2.4 Culture and Passage of pGCiPS

The pGCiPS cells are cultured in pig iPSC culture medium (Fig. 3a, *see* **Note 18**). The medium was changed daily, and the cells were routinely passaged every 3–4 days (*see* **Notes 10**, **19**, and **20**).

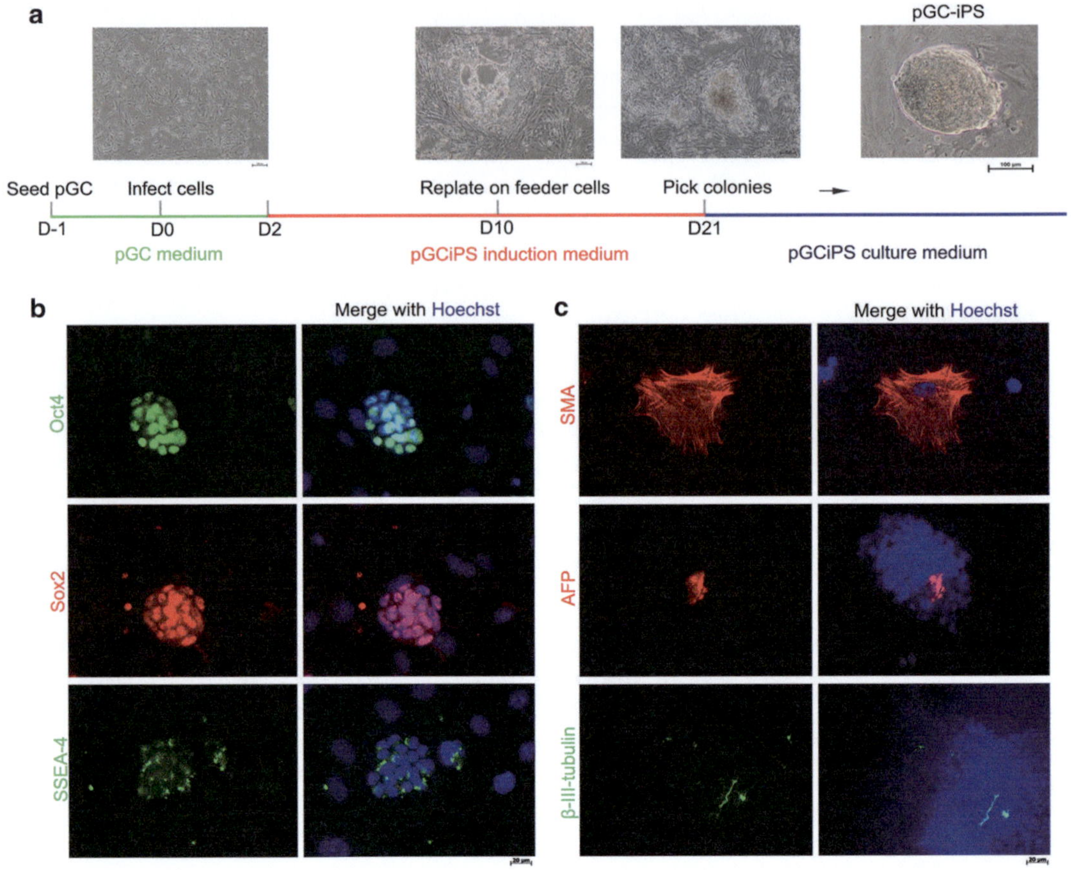

Fig. 3 Characterization of pig iPS cells generated from pGC. (**a**) Schematic diagram of pGCiPSC generation. Scale bar = 100 μm. (**b**) pGCiPS express strong Oct4, Sox2, and SSEA4 by immunofluorescence (scale bar = 20 μm). Nuclei stained with Hoechst 33342 (*blue*). (**c**) Differentiation in vitro of pGCiPS cells by embryoid body (EB) formation. The differentiated derivatives by EB consist of cells representing three embryonic germ layers as shown by immunofluorescence staining using endoderm marker alpha 1-fetoprotein (AFP), mesoderm marker smooth muscle actin (SMA), and ectoderm marker β-III-tubulin (Scale bar = 20 μm). Nuclei stained with Hoechst 33342 (*blue*)

3.2.5 Cryopreservation and Thawing of pGCiPS Cells

The methods of cryopreservation and thawing of pGCiPS cells are the same as those of mGCiPS cells (*see* **Notes 10** and **11**).

3.3 Anticipated Results

1. The stable mGC-OSiPS cells resembled typical ES cell colonies in morphology, with large nuclei and clear nucleoli and compact clonal boundaries, distinct from feeder fibroblasts (Fig. 2a). These iPS cells show pluripotency in vitro and in vivo, as evidenced by expression of multiple pluripotent stem cell markers Oct4 and Nanog in the nuclei and SSEA-1 on cell surface by immunofluorescence (Fig. 2b), differentiation into three embryonic germ layers by embryoid body formation

(Fig. 2c) and teratoma tests (Fig. 2d), as well as high efficient generation of chimeras (Fig. 2e). These data provide additional evidence in supporting the notion that reduced expression of LaminA, and stem cells can improve the reprogramming efficiency to pluripotency (21).

2. Derived pGCiPS are also morphologically similar to mES cells, with domed clonal morphology (Fig. 3a). These cells express multiple pluripotent stem cell markers Oct4 and Sox2 in the nuclei and SSEA-4 on the cell surface by immunofluorescence (Fig. 3b) and also Nanog in nuclei (data not shown). Differentiation of pGCiPS cells via EB formation yielded cells representing three embryonic germ layers indicated by tissue-specific immunofluorescence staining of SMA (cardiac muscle, mesoderm), β-III-tubulin (neurons, ectoderm), and AFP (liver, endoderm) (Fig. 3c).

4 Notes

1. All solutions and equipment in this method must be sterile, and aseptic technique should be required accordingly.
2. Be careful when using with β-mercaptoethanol, and avoid inhalation and skin contact.
3. Using fresh virus gives rise to dramatically higher infection efficiency. Virus should not be stored for longer than 1 week at 4 °C before use. Virus should never undergo more than 1 freeze/thaw cycle, as multiple freeze/thaw cycles considerably reduce infection efficiency.
4. Granulosa cells obtained from approximately 1.5 mice are seeded in one 6-well plate. Mouse granulosa cells grow much slower and their adherence to the dish requires extra time.
5. Mouse granulosa cells express Klf4 and c-Myc endogenously, and these two factors can be omitted for iPS induction.
6. Based on the basal induction medium, adding small molecules can improve reprogramming efficiency and/or replace reprogramming factors.
7. Suitable colonies should appear translucent and circular.
8. It is recommended to prepare frozen stocks of newly reprogrammed iPS cells at low passage for future use.
9. For mouse iPS cells, a 1:6 passaging ratio is usually suitable.
10. It is highly recommended that a mycoplasma test be performed when the cells grow slowly.
11. Labeling should include cell line, passage number, researcher's name, and date.

12. The necessary safety measures are required, since cryovials stored in the liquid nitrogen tank may explode unexpectedly when exposed to rapid thawing.
13. Transport time from slaughterhouse to laboratory should be as short as possible.
14. It is strongly recommended to use antibiotic-antimycotic for culture of the primary pig granulosa cells because the pig granulosa cells separated from ovaries taken from slaughterhouse are very prone to contamination.
15. Transfection for retrovirus production is performed in 293-T cells at 70 % confluence in a 100 mm dish. Avoid letting 293-T cells grow to 100 % confluence during transfection step. This can adversely affect the viral titer.
16. Retrovirus produced in VSV-G envelope can infect most mammalian cells including human cells. Production, maintenance, and infection using VSV-G pseudotype retrovirus must be performed under the containment of BL2 biosafety. When working with virus, use a designated laminar flow hood for virus-related work. Always wear protective disposable gowns and double gloves. Make sure to discard any viral waste into designated viral waste bins and rinse any disposables with bleach before disposing of them. The contents of a viral waste bin should be autoclaved in an autoclavable bag for 45 min in a standard autoclave using the sterilization program.
17. Pick up as many colonies as possible in order to achieve good colonies.
18. Small molecules can also be used to maintain the pluripotency of pGCiPS cells or convert their pluripotent state.
19. The ratio depends on cell density prior to passaging.
20. It is advisable to use TrypLE to digest pGCiPS cells.

Acknowledgments

We thank Qian Zhang, Xiaoying Ye, Kai Liu, Jihong Yuan, Jiaojiao Li, and Mengyuan Liu for helping with the experiments. We thank important funding by the China MOST National Major Basic Research Program (2011CBA01002, 2009CB941004).

References

1. Takahashi K, Yamanaka S (2006) Induction of pluripotent stem cells from mouse embryonic and adult fibroblast cultures by defined factors. Cell 126:663–676
2. Takahashi K, Tanabe K, Ohnuki M, Narita M, Ichisaka T, Tomoda K, Yamanaka S (2007) Induction of pluripotent stem cells from adult human fibroblasts by defined factors. Cell 131:861–872
3. Aoi T, Yae K, Nakagawa M, Ichisaka T, Okita K, Takahashi K, Chiba T, Yamanaka S (2008) Generation of pluripotent stem cells from adult mouse liver and stomach cells. Science 321:699–702
4. Loh YH, Agarwal S, Park IH, Urbach A, Huo H, Heffner GC, Kim K, Miller JD, Ng K, Daley GQ (2009) Generation of induced pluripotent stem cells from human blood. Blood 113:5476–5479
5. Aasen T, Raya A, Barrero MJ, Garreta E, Consiglio A, Gonzalez F, Vassena R, Bilic J, Pekarik V, Tiscornia G et al (2008) Efficient and rapid generation of induced pluripotent stem cells from human keratinocytes. Nat Biotechnol 26:1276–1284
6. Haase A, Olmer R, Schwanke K, Wunderlich S, Merkert S, Hess C, Zweigerdt R, Gruh I, Meyer J, Wagner S et al (2009) Generation of induced pluripotent stem cells from human cord blood. Cell Stem Cell 5:434–441
7. Giorgetti A, Montserrat N, Aasen T, Gonzalez F, Rodriguez-Piza I, Vassena R, Raya A, Boue S, Barrero MJ, Corbella BA et al (2009) Generation of induced pluripotent stem cells from human cord blood using OCT4 and SOX2. Cell Stem Cell 5:353–357
8. Tamaoki N, Takahashi K, Tanaka T, Ichisaka T, Aoki H, Takeda-Kawaguchi T, Iida K, Kunisada T, Shibata T, Yamanaka S et al (2010) Dental pulp cells for induced pluripotent stem cell banking. J Dent Res 89:773–778
9. Yan X, Qin H, Qu C, Tuan RS, Shi S, Huang GT (2010) iPS cells reprogrammed from human mesenchymal-like stem/progenitor cells of dental tissue origin. Stem Cells Dev 19:469–480
10. Oda Y, Yoshimura Y, Ohnishi H, Tadokoro M, Katsube Y, Sasao M, Kubo Y, Hattori K, Saito S, Horimoto K et al (2010) Induction of pluripotent stem cells from human third molar mesenchymal stromal cells. J Biol Chem 285:29270–29278
11. Brown ME, Rondon E, Rajesh D, Mack A, Lewis R, Feng X, Zitur LJ, Learish RD, Nuwaysir EF (2010) Derivation of induced pluripotent stem cells from human peripheral blood T lymphocytes. PLoS One 5:e11373
12. Loh YH, Hartung O, Li H, Guo C, Sahalie JM, Manos PD, Urbach A, Heffner GC, Grskovic M, Vigneault F et al (2010) Reprogramming of T cells from human peripheral blood. Cell Stem Cell 7:15–19
13. Seki T, Yuasa S, Oda M, Egashira T, Yae K, Kusumoto D, Nakata H, Tohyama S, Hashimoto H, Kodaira M et al (2010) Generation of induced pluripotent stem cells from human terminally differentiated circulating T cells. Cell Stem Cell 7:11–14
14. Staerk J, Dawlaty MM, Gao Q, Maetzel D, Hanna J, Sommer CA, Mostoslavsky G, Jaenisch R (2010) Reprogramming of human peripheral blood cells to induced pluripotent stem cells. Cell Stem Cell 7:20–24
15. Hanna J, Markoulaki S, Schorderet P, Carey BW, Beard C, Wernig M, Creyghton MP, Steine EJ, Cassady JP, Foreman R et al (2008) Direct reprogramming of terminally differentiated mature B lymphocytes to pluripotency. Cell 133:250–264
16. Pereira CF, Terranova R, Ryan NK, Santos J, Morris KJ, Cui W, Merkenschlager M, Fisher AG (2008) Heterokaryon-based reprogramming of human B lymphocytes for pluripotency requires Oct4 but not Sox2. PLoS Genet 4: e1000170
17. Eminli S, Foudi A, Stadtfeld M, Maherali N, Ahfeldt T, Mostoslavsky G, Hock H, Hochedlinger K (2009) Differentiation stage determines potential of hematopoietic cells for reprogramming into induced pluripotent stem cells. Nat Genet 41:968–976
18. Chronowska E (2012) Stem cell characteristics of ovarian granulosa cells - review. Ann Anim Sci 12:151–157
19. Kossowska-Tomaszczuk K, De Geyter C, De Geyter M, Martin I, Holzgreve W, Scherberich A, Zhang H (2009) The multipotency of luteinizing granulosa cells collected from mature ovarian follicles. Stem Cells 27:210–219
20. Kossowska-Tomaszczuk K, De Geyter C (2013) Cells with stem cell characteristics in somatic compartments of the ovary. Biomed Res Int 2013:310859
21. Zuo B, Yang J, Wang F, Wang L, Yin Y, Dan J, Liu N, Liu L (2012) Influences of lamin A levels on induction of pluripotent stem cells. Biol Open 1:1118–1127
22. Meyer N, Penn LZ (2008) Reflecting on 25 years with MYC. Nat Rev Cancer 8:976–990

23. Rowland BD, Peeper DS (2006) KLF4, p21 and context-dependent opposing forces in cancer. Nat Rev Cancer 6:11–23
24. Okita K, Ichisaka T, Yamanaka S (2007) Generation of germline-competent induced pluripotent stem cells. Nature 448:313–317
25. Nakagawa M, Takizawa N, Narita M, Ichisaka T, Yamanaka S (2010) Promotion of direct reprogramming by transformation-deficient Myc. Proc Natl Acad Sci U S A 107:14152–14157
26. Nakagawa M, Koyanagi M, Tanabe K, Takahashi K, Ichisaka T, Aoi T, Okita K, Mochiduki Y, Takizawa N, Yamanaka S (2008) Generation of induced pluripotent stem cells without Myc from mouse and human fibroblasts. Nat Biotechnol 26:101–106
27. Miura K, Okada Y, Aoi T, Okada A, Takahashi K, Okita K, Nakagawa M, Koyanagi M, Tanabe K, Ohnuki M et al (2009) Variation in the safety of induced pluripotent stem cell lines. Nat Biotechnol 27:743–745
28. Shi Y, Do JT, Desponts C, Hahm HS, Scholer HR, Ding S (2008) A combined chemical and genetic approach for the generation of induced pluripotent stem cells. Cell Stem Cell 2:525–528
29. Eminli S, Utikal J, Arnold K, Jaenisch R, Hochedlinger K (2008) Reprogramming of neural progenitor cells into induced pluripotent stem cells in the absence of exogenous Sox2 expression. Stem Cells 26:2467–2474
30. Kim JB, Zaehres H, Wu G, Gentile L, Ko K, Sebastiano V, Arauzo-Bravo MJ, Ruau D, Han DW, Zenke M et al (2008) Pluripotent stem cells induced from adult neural stem cells by reprogramming with two factors. Nature 454:646–650
31. Giorgetti A, Montserrat N, Rodriguez-Piza I, Azqueta C, Veiga A, Izpisua Belmonte JC (2010) Generation of induced pluripotent stem cells from human cord blood cells with only two factors: Oct4 and Sox2. Nat Protoc 5:811–820
32. Mao J, Zhang Q, Ye X, Liu K, Liu L (2014) Efficient induction of pluripotent stem cells from granulosa cells by Oct4 and Sox2. Stem Cells Dev 23:779–789
33. Hou P, Li Y, Zhang X, Liu C, Guan J, Li H, Zhao T, Ye J, Yang W, Liu K et al (2013) Pluripotent stem cells induced from mouse somatic cells by small-molecule compounds. Science 341:651–654
34. Wakayama T, Perry ACF, Zuccotti M, Johnson KR, Yanagimachi R (1998) Full-term development of mice from enucleated oocytes injected with cumulus cell nuclei. Nature 394:369–374
35. Wells DN, Misica PM, Tervit HR (1999) Production of cloned calves following nuclear transfer with cultured adult mural granulosa cells. Biol Reprod 60:996–1005
36. Park K-W, Kühholzer B, Lai L, Macháty Z, Sun Q-Y, Day BN, Prather RS (2001) Development and expression of the green fluorescent protein in porcine embryos derived from nuclear transfer of transgenic granulosa-derived cells. Anim Reprod Sci 68:111–120
37. Polejaeva IA, Chen S-H, Vaught TD, Page RL, Mullins J, Ball S, Dai Y, Boone J, Walker S, Ayares DL et al (2000) Cloned pigs produced by nuclear transfer from adult somatic cells. Nature 407:86–90
38. Gibbons J, Arat S, Rzucidlo J, Miyoshi K, Waltenburg R, Respess D, Venable A, Stice S (2002) Enhanced survivability of cloned calves derived from roscovitine-treated adult somatic cells. Biol Reprod 66:895–900
39. Keefer CL, Keyston R, Lazaris A, Bhatia B, Begin I, Bilodeau AS, Zhou FJ, Kafidi N, Wang B, Baldassarre H et al (2002) Production of cloned goats after nuclear transfer using adult somatic cells. Biol Reprod 66:199–203

Methods in Molecular Biology (2016) 1357: 465–474
DOI 10.1007/7651_2014_158

Published online: 05 December 2014

The Characteristics of Murine iPS Cells and siRNA Transfection Under Hypoxia

K. Sugimoto and Yoshihiko Hayashi

Abstract

iPS cells are attractive for the regenerative medicine. The creation of pluripotent cells from somatic cells has great potential for basic and clinical research and application. Retroviral transduction of four or three transfection factors has been shown to initiate a reprogramming process. Here, we describe the effect of transcription factors regarding the growth and differentiation of mouse iPS cells in normoxia or hypoxia. Furthermore, we introduce the function of hypoxia-inducible factors (HIFs) in mouse iPS cells in hypoxia using RT-PCR and western blotting together with HIFs knockdown techniques.

Keywords: Murine iPS cells, Transcription factors, Pluripotency, Hypoxia, HIFs

1 Introduction

Retroviral transduction of murine and human somatic cells with four transcription factors (Oct4, Sox2, Klf4, and c-Myc) initiates the conversion of infected cells into a pluripotent and embryonic stem cell-like state [1–8]. iPS cells have been recently isolated in the absence of a cariogenic c-Myc, although the transduction efficiency becomes significantly lower [9–11]. The transcriptional networks controlling stemness under hypoxia are controlled by HIFs [12], which belong to the key regulators of cell reaction such as erythropoiesis, apoptosis, and proliferation [13] to the lack of cell oxygen. Environmental oxygen tension regulates the HIF expression [14]. It is important and meaningful to understand how hypoxia may regulate pluripotency in iPS cells through HIFs [15–18].

Thus, we have investigated the effects of four or three transcription factors for the growth and differentiation of murine iPS cells under normoxia or hypoxia. Furthermore, the function of HIFs in murine iPS cells under hypoxia has been examined in respect to the morphology and expression of transcription factors [19] by using siRNAs, RT-PCR, and Western blotting for these purposes.

2 Materials

Prepare all solution using ultrapure water (prepared by purifying deionized water to attain a sensitivity of 18 MΩ at room temperature) and analytical grade reagents.

2.1 Cell Culture

1. Dish coating reagent (ReproCoat®, ReproCell, Japan) for on-feeder culture. Store it at room temperature.
2. Murine iPS cell: Four transcription factors inducing (iPS-MEF-Ng-20D-17) and three transcription factors inducing (iPS-MEF-Ng-178B-5) purchased from Riken CELL BANK (Japan).
3. Feeder cell: mouse embryonic fibroblasts (MEFs) (ReproCell).
4. Medium: DMEM supplemented with 0.1 mM NEAA, 1,000 U/ml mouse LIF, 15 % FBS, 1 % penicillin/streptomycin, and 0.1 mM 2-mercaptoethanol. Store at 4 °C.
5. Multigas incubator: the regulatory ability of O_2 concentration (2–80.0 %) and the accuracy of O_2 concentration (±0.5 %).

2.2 siRNA Application for Knockdown

1. siRNA for HIF-1α: Mm_Hif1α_4FlexiTube siRNA (QIAGEN, Germany).
2. siRNA HIF-2α: Mm_Epas1 (Hif2α)_5FlexiTube siRNA (QIAGEN).
3. siRNA for HIF-3α: Mm_Hif3α_5FlexiTube siRNA (QIAGEN).
4. 15 μl of HiPerfect transfection reagent (QIAGEN).
5. siRNA for control: AllStars Negative Controls (QIAGEN) siRNA (no homology to any known mammalian gene).

2.3 Real-Time PCR Analysis

1. PBS(−) (Wako, Japan).
2. Trizol® reagent (Life Technologies, USA).
3. Nanog primer: F, 5′-AGGGTCTGCTACTGAGATGCTCTG-3′, and R, 5′-CAACCACTGGTTTTTCTGCCACCG-3′.
4. Oct4 primer: F, 5′-CTGTAGGGAGG GCTTCGGGCACTT-3′, and R, 5′-CTGAGGGCCAGGCAGGAGCACGA G-3′.
5. Sox2 primer: F, 5′-GGCAGCTACAGCATGATGCAGGAGC-3′, and R, 5′-CTGGTCATGGAGTTGTACTGCAGG-3′.
6. Glyceraldehyde phosphate dehydrogenase (GAPDH): F, 5′-GCA CAGTCAAGGCCGAGAAT-3′, and R, 5′-GCCTTCTCCATGG TGGTGAA-3′.
7. SuperScript™ First-Strand Synthesis System (Life Technologies).

2.4 Western Blotting

1. Lysis buffer: 1 M HEPES, 10 % NP-40, 1 mM NaF, 1 mM Na_3VO_4, 0.5 M EDTA, 100 % protease inhibitor cocktail.
2. 2× sodium dodecyl sulfate (SDS)-polyacrylamide gel electrophoresis (PAGE) sample buffer (nacalai tesque, Japan).
3. SDS-PAGE. Gels: NuPAGE® Bis-Tris Mini Gels (Life Technologies)
4. Prestained Protein Markers (nacalai tesque).
5. iBlot® Transfer Stack (Life Technologies, USA).
6. iBlot® Gel Transfer Device (Life Technologies).
7. Anti-mouse-Nanog antibody (ReproCELL, Japan).
8. Anti-mouse-Oct4 antibody (Cell Signaling Technology, USA).
9. Anti-mouse-Sox2 antibody (Cell Signaling Technology).
10. Horseradish peroxidase (HRP)-conjugated secondary antibodies (iBlot® Western Detection Kits, Life Technologies).
11. Supplied reagents (iBlot® Chemiluminescent Kit, Life Technologies).

3 Methods

Carry out all procedures at room temperature unless otherwise specified.

3.1 Cell Culture

1. One hour before cell adjustment, pour ReproCoat® into a dish for cell attachment to culture dish and leave it at rest for 37 °C, 30 min.
2. Feed and coculture murine iPS cells on MEFs in the maintaining medium (*see* **Notes 1** and **2**)
3. After one passage, seed murine iPS cells in a 60-mm culture dish at a density of 1.0×10^5 cell/dish (*see* **Note 3**).
4. Culture iPS cells in a humidified atmosphere of 20 % or 5 % oxygen at 37 °C (*see* **Note 4**).
5. In addition to the morphologic observation, count the cell number at 3, 5, and 7 days using a hemocytometer under a phase contrast microscope. Culture both four and three factors transduced iPS cells under 5 % oxygen and observe a significantly rapid cell growth compared to under 20 % oxygen (Fig. 1).

3.2 siRNA Application for Knockdown

1. Use three transcription factors inducing murine iPS cells for siRNA experiments on a 60-mm culture dish at 1.0×10^5 cells/dish under 5 % oxygen for one passage.
2. Culture iPS cells under 5 % oxygen for 48 h.

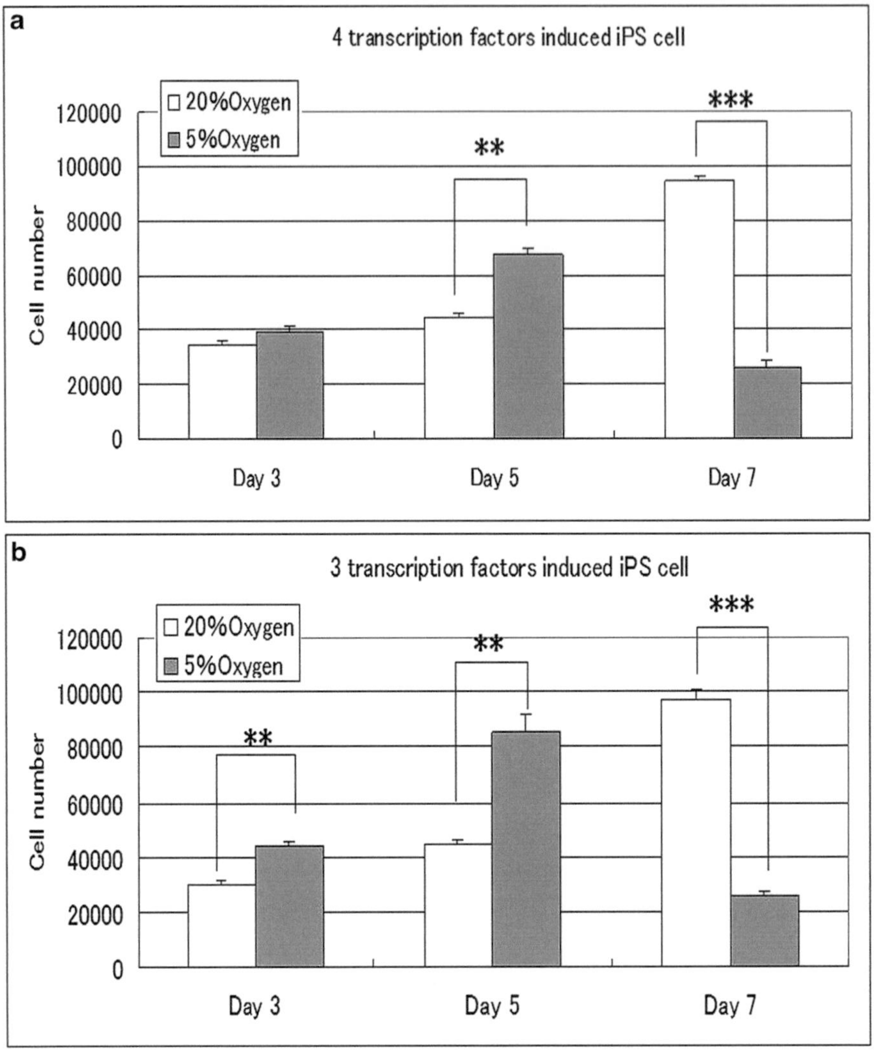

Fig. 1 The number of mouse iPS cells in 5 % and 20 % oxygen on days 3, 5, and 7 (data from triplicate samples. $^{**}P < 0.01$, $^{***}P < 0.001$). (**a**) iPS cells transduced using 4 transcription factors. (**b**) iPS cells transduced using 3 transcription factors (reproduced from [19] with permission from Wiley)

3. For each transfection, add 50 nM siRNAs {Product Names: Mm_Hif1α_4FlexiTube siRNA, Mm_Epas1 (Hif2α)_5FlexiTube siRNA, Mm_Hif3α_5FlexiTube siRNA} (QIAGEN, Germany) to each dish and mix with 15 μl of HiPerfect transfection reagent (QIAGEN) into 100 μl of DMEM.
4. Incubate a sample for 5–10 min to make a transfection complex.
5. Add complex into a cell in a dropwise manner.
6. Turn a plate calmly and equalize the distribution of the transfection complex.

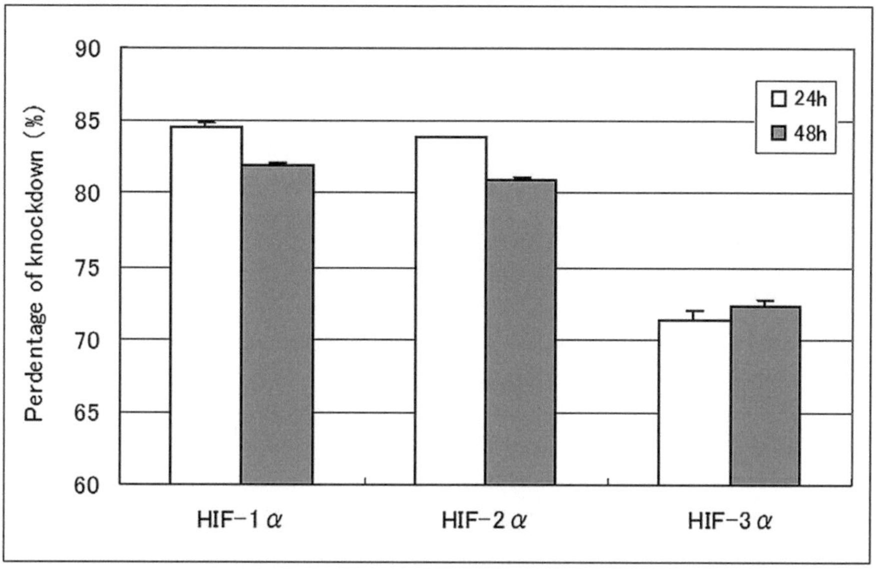

Fig. 2 Knockdown efficiency of HIFs in three transcription factors induced mouse iPS cells at 24 and 48 h (data from duplicate samples) (reproduced from [19] with permission from Wiley)

7. At 48 h after transfection, harvest cells and prepare extracts for mRNA and protein analyses.
8. Calculate knockdown efficiency of siRNA into iPS cells in comparison with mRNA expression of control groups without transfection procedure. Following siRNA and quantitative RT-PCR, confirm the knockdown of HIFs at 24 and 48 h, respectively, compared to transfection control siRNA (Fig. 2).
9. Observe iPS cell morphology and colony formation after silencing HIFs by siRNA (Fig. 3).
10. Using real-time PCR, observe a reduction on the expression of pluripotency marker genes (Nanog, Oct4, and Sox2) when silenced HIFs compared to the control group (Fig. 4).

3.3 Real-Time PCR Analysis

1. Rinse cells twice with PBS(−) and lyse by adding the Trizol® reagent (Life Technologies, USA).
2. Add 200 μl of chloroform in 1 ml of Trizol.
3. Take a vortex and leave at rest for 2–3 min.
4. Centrifuge at 4 °C in 3,000–7,000 × *g* force for 5 min.
5. Pour a supernatant into the Eppendorf tube (*see* **Note 5**).
6. Add 500 μl of isopropyl alcohol and mix using a vortex.
7. Centrifuge at 4 °C in 10,000 × *g* force for 10 min.
8. Remove the supernatants and transfer it into 500 μl of 70–75 % ethanol.

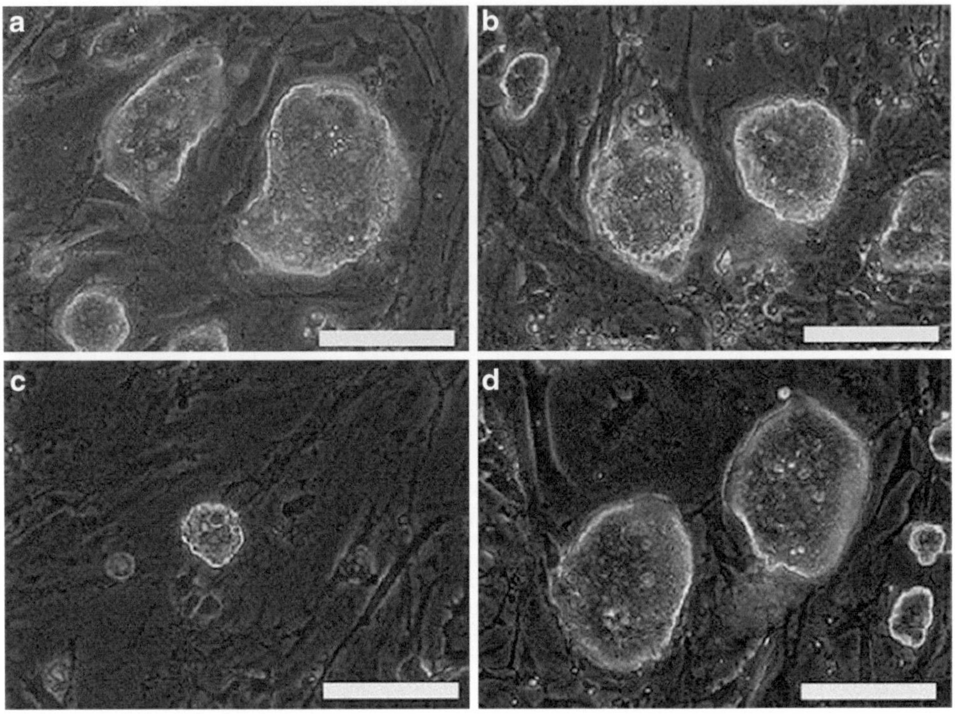

Fig. 3 Representative phase contrast photographs of colony morphology 48 h after HIFs knockdown in three transcription factors induced mouse iPS cells under 5 % oxygen. (**a**) Control group. (**b**) HIF-1α knockdown group. (**c**) HIF-2α knockdown group. (**d**) HIF-3α knockdown group. Scale bar = 100 μm (reproduced from [19] with permission from Wiley)

9. Centrifuge at 4 °C in 10,000 × *g* force for 5 min and throw away the supernatant.
10. Dissolve a pellet in diethylpyrocarbonate (DEPC) and measure total RNA quantity by using a spectrophotometer.
11. Synthesize the first-strand cDNA from the total RNA following the manufacturer's instructions using the SuperScript™ First-Strand Synthesis System for RT-PCR (Life Technologies).
12. For the PCR reaction, add the Brilliant SYBR Green QPCR Master Mix® (Life Technologies) containing the components necessary to carry out PCR amplifications, including Taq DNA polymerase (Nippon gene, Japan) and SYBR Green I® (Agilent Technologies, USA) as the reporter fluorescent dye, ROX as the reference dye, and cDNA as the PCR template to a 0.2 μl PCR tube.
13. Amplify cDNA under the following conditions with a real-time PCR system (Mx3000P™, Agilent Technologies): 94 °C for 5 min (denaturation), followed by 35 cycles at 94 °C for 45 s, 60 °C for 1 min, and 72 °C for 1 min (*see* **Note 6**).

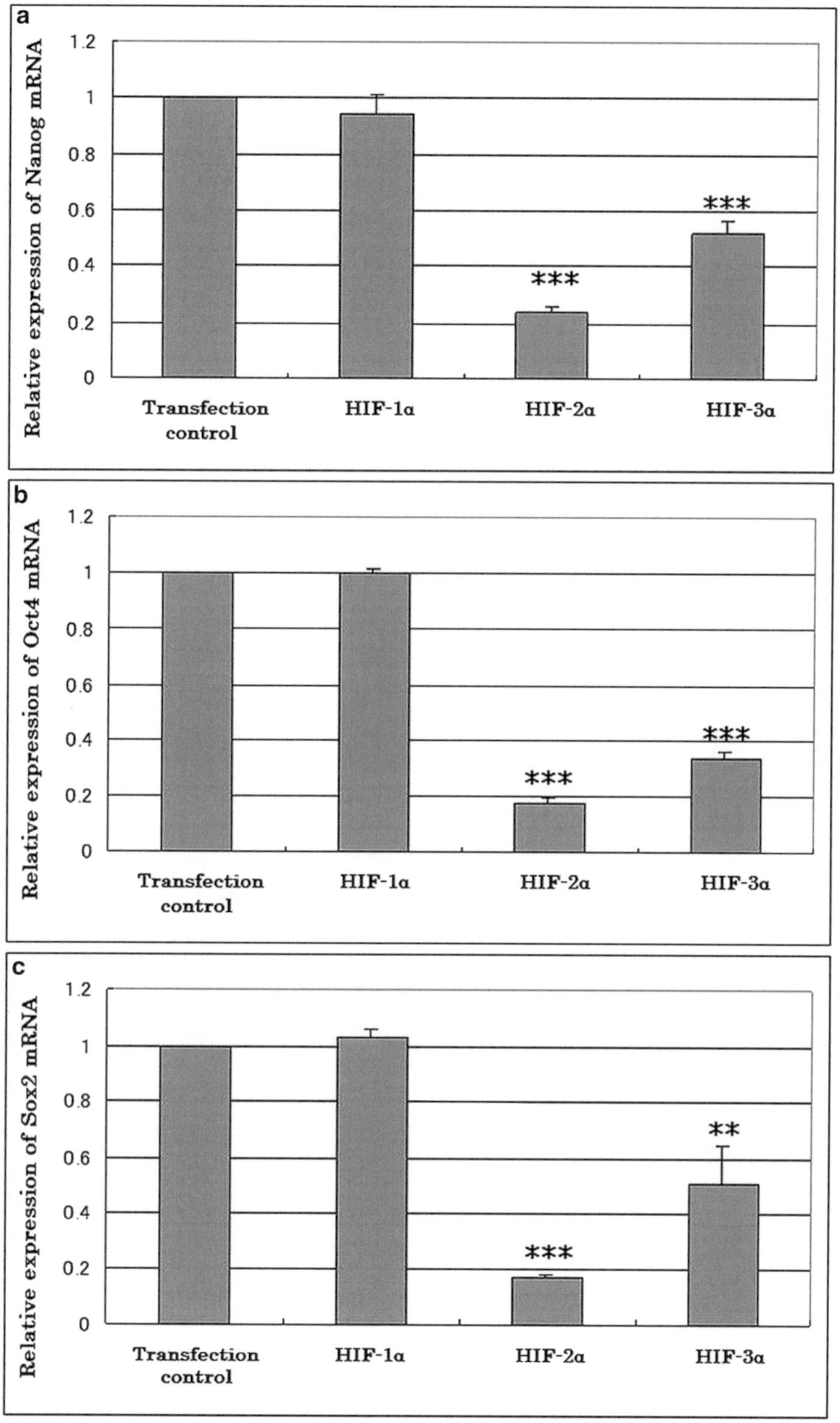

Fig. 4 Pluripotency marker mRNAs expression 48 h after HIFs knockdown in three transcription factors induced mouse iPS cells under 5 % oxygen (data from triplicate samples. $^{**}P < 0.01$, $^{***}P < 0.001$). (**a**) mRNA expression of Nanog. (**b**) mRNA expression of Oct4. (**c**) mRNA expression of Sox2 (reproduced from [19] with permission from Wiley)

3.4 Western Blotting

1. Seed murine iPS cells in 100-mm culture dishes at a density of 1×10^6 cells and carry out siRNA experiments in 5 % oxygen condition.
2. After the transfection of siRNA by the above-mentioned protocols, culture murine iPS cells in 5 % oxygen.
3. After 48 h of cell culture, retrieve cells with a rubber scrapper (Thermo Fisher Scientific, USA).
4. Lyse cells in each group in a lysis buffer (1 M HEPES, 10 % NP-40, 1 mM NaF, 1 mM Na_3VO_4, 0.5 M EDTA, 100 % protease inhibitor cocktail, deionized water).
5. Sonicate with a cell disrupter for 1 min in ice-cold water.
6. After centrifugation of the lysate at $70 \times g$ force for 10 min at 4 °C, subject supernatants to a Western blot analysis.
7. Determine protein concentrations in each group using the micro-Lowry method.
8. Denature 30 μl of sample in 2× SDS-PAGE sample buffer and separate by SDS-PAGE (*see* **Note 7**).
9. Transfer gels using the iBlot® Transfer Stack (Life Technologies, USA) on the iBlot® Gel Transfer Device (Life Technologies).
10. Use recommended P3 program (20 V for 7 min) for all transfers.
11. Detect specific antibody bindings with a HRP-conjugated secondary antibodies (iBlot® Western Detection Kits, Life Technologies) and visualize using the supplied reagents (iBlot® Chemiluminescent Kit, Life Technologies) (*see* **Note 8**).
12. Quantify band density of each group by a densitometric analysis using the Scion Image software program.
13. Calculate the ratio of the densitometric value of the experimental group to the value of the control group.

4 Notes

1. Pour the ReproCoat® over the surface of the dish, discard the excess, and allow the residue to dry before seeding feeder cells.
2. Discard the feeder cells prepared before over 4 days to maintain pluripotency of iPS cells.
3. Never seed the iPS cells in the density of under 5.0×10^4 cells/dish because of the occurrence of mitosis and proliferation.
4. The characteristics of the colony of iPS cells are an obscure border and a high density of cells inside the colony which consists of a nearly circular polygon.

5. Pay attention to not suck in the portion of protein together with a supernatant.
6. The mRNA level of each gene relative to that of GAPDH is calculated using the "Comparative Quantification" method with the Aligent® kit.
7. When added Prestained Protein Markers in electrophoretic gel, pay attention to diffuse it into other lanes.
8. Anti-mouse-Nanog antibody diluted to 1:500, anti-mouse-Oct4 antibody diluted to 1:500, and anti-mouse-Sox2 antibody diluted to 1:500 are used as primary antibodies. The specific antibody binding detected with HRP-conjugated secondary antibodies is visualized using supplied reagents.

Acknowledgements

This work was supported by Grants-in-Aid for Scientific Research (B) from Japan Society for The Promotion of Science, Contract Grant Number 22390360 to Y.H.

References

1. Hochedlinger K, Jaenisch R (2003) Nuclear transplantation, embryonic stem cells, and the potential for cell therapy. N Engl J Med 349:275–286
2. Rideout WM 3rd, Hochedlinger K, Kyba M, Daley GQ, Jaenisch R (2002) Correction of a genetic defect by nuclear transplantation and combined cell and gene therapy. Cell 109:17–27
3. Maherali N, Sridharan R, Xie W, Utikal J, Eminli S, Arnold K, Stadtfeld M, Yachechko R, Tchieu J, Jaenisch R, Plath K, Hochedlinger K (2007) Directly reprogrammed fibroblasts show global epigenetic remodeling and widespread tissue contribution. Cell Stem Cell 1:55–70
4. Meissner A, Wernig M, Jaenisch R (2007) Direct reprogramming of genetically unmodified fibroblasts into pluripotent stem cell. Nat Biotechnol 25:1177–1181
5. Okita K, Ichisaka T, Yamanaka S (2007) Generation of germline-competent induced pluripotent stem cell. Nature 448:313–317
6. Takahashi K, Tanabe K, Ohnuki M, Narita M, Ichisaka T, Tomoda K, Yamanaka S (2007) Induction of pluripotent stem cells from adult human fibroblasts by defined factors. Cell 131:861–872
7. Takahashi K, Yamanaka S (2006) Induction of pluripotent stem cells from mouse embryonic and adult fibroblast cultures by defined factors. Cell 126:663–676
8. Wernig M, Meissner A, Foreman R, Brambrink T, Ku M, Hochedlinger K, Bernstein BE, Jaenisch R (2007) In vitro reprogramming of fibroblasts into a pluripotent ES-cell-like state. Nature 448:318–324
9. Yu J, Vodyanik MA, Smuga-Otto K, Antoseiwicz-Bourget J, Frane JL, Tian S, Nie J, Josdottir GA, Ruotti V, Stewart R, Slukvin II, Thomson JA (2007) Induced pluripotent stem cell lines derived from human somatic cells. Science 318:1917–1920
10. Nakagawa M, Koyanagi M, Tanabe K, Takahashi K, Ichisaka T, Aoi T, Okita K, Mochiduki Y, Takizawa N, Yamanaka S (2008) Generation of induced pluripotent stem cells without Myc from mouse and human fibroblasts. Nat Biotechnol 26:101–106
11. Wernig M, Meissner A, Cassady JP, Jaenisch R (2008) C-Myc is dispensable for direct reprogramming of mouse fibroblasts. Cell Stem Cell 2:10–12
12. Bertout JA, Patel SA, Simon MC (2008) The impact of O_2 availability on human cancer. Nat Rev Cancer 8:967–975
13. Semenza GL (2000) HIF-1: mediator of physiological responses to hypoxia. J Appl Physiol 88:1474–1480

14. Mohyeldin A, Garzon-Muvdi T, Quinones-Hinojosa A (2010) Oxygen in stem cell biology: a critical component of the stem cell niche. Cell Stem Cell 7:150–161
15. Wenger RH (2002) Cellular adaptation to hypoxia: O_2-sensing protein hydroxylases, hypoxia-inducible transcription factors, and O_2-regulated gene expression. FASEB J 16:1151–1162
16. Semenza GL, Wang GL (1992) A nuclear factor induced by hypoxia via *de novo* protein synthesis binds to the human erythropoietin gene enhancer at a site required for transcriptional activation. Mol Cell Biol 12:721–732
17. Ivan M, Kondo K, Yang H, Kim W, Valiando J, Ohh M, Salic A, Asara JM, Lane WS, Kaelin WG (2001) HIFα targeted for VHL-mediated destruction by proline hydroxylation: implications for O_2 sensing. Science 292:464–468
18. Masson N, Willam C, Maxwell PH, Pugh CW, Ratcliffe PJ (2001) Independent function of two destruction domains in hypoxia-inducible factor-α chains activated by prolyl hydroxylation. EMBO J 20:5197–5206
19. Sugimoto K, Yoshizawa Y, Yamada S, Igawa K, Hayashi Y (2013) Effects of hypoxia on pluripotency in murine iPS cells. Microsc Res Tech 76:1084–1092

Methods in Molecular Biology (2016) 1357: 475–483
DOI 10.1007/7651_2014_138

Published online: 20 November 2014

Hepatic Differentiation from Murine and Human iPS Cells Using Nanofiber Scaffolds

Taiji Yamazoe, Nobuaki Shiraki, and Shoen Kume

Abstract

The induced pluripotent stem (iPS) cells of murine and human are capable to differentiate into any cell type of the body through recapitulating normal development, similarly as the embryonic stem (ES) cells. Lines of evidence support that both ES cells and iPS cells are induced to differentiate in vitro by sequential treatment of humoral cues such as growth factors and chemicals, combined with the use of certain microenvironments including extracellular matrices and scaffolds.

Here, we describe the procedure to potentiate hepatic lineage cells differentiation from murine and human iPS cells, using growth factor cocktails and nanofiber scaffolds. Nanofiber scaffolds have a three-dimensional surface mimicking the fine structures of the basement membrane in vivo, allow the iPS cells to differentiate into the definitive endoderm and mature hepatocyte-like cells more efficiently than the two-dimensional conventional culture plates.

Keywords: Hepatic differentiation, Microenvironment, Extracellular matrices, Nanofiber scaffolds

1 Introduction

The iPS cells and ES cells have the ability to differentiate into any cell type of our body through mimicking normal developmental processes (1, 2). Therefore, these stem cells can serve as an attractive cell source for a large number of cells needed in biomedical research and regenerative therapies.

There are two majorly considerable conditions to culture ES and iPS cells, one is humoral cues in the culture medium and the other is the components of extracellular matrices and scaffolds.

Based on lines of evidence in developmental biology, hepatic differentiation from stem cells has been established (3, 4). This utilized not only the growth factors that are indispensable for liver organogenesis in vivo but also small chemicals that are theoretically expected to evoke intracellular signaling pathways. Activin A, for instance, is a ligand of the TGF-β superfamily and is used to induce endoderm differentiation, and hepatocyte growth factor is used to differentiate cells to adopt differentiation into the hepatic lineages (3, 5).

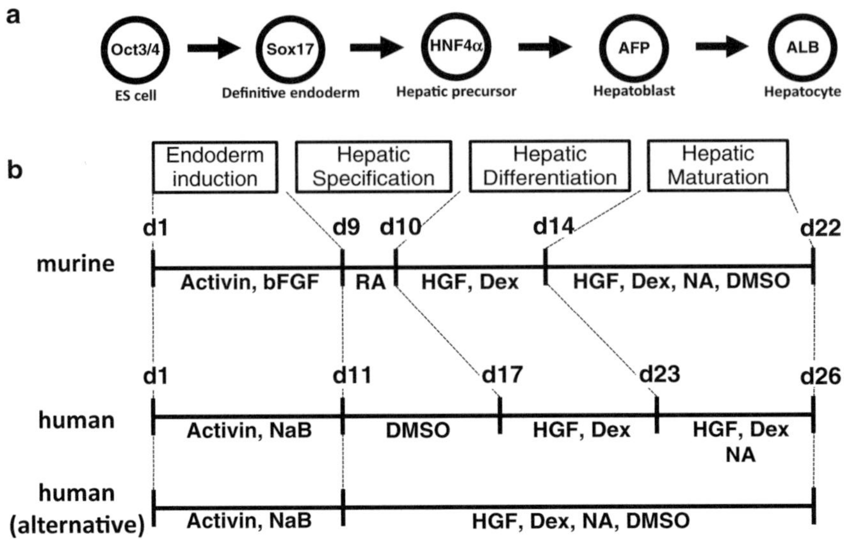

Fig. 1 Scheme for hepatic differentiation program using nanofiber matrices. (**a**) Developmental time course shows a line of differentiating cell profile recapitulating normal developmental process and exhibiting specific marker genes. (**b**) Schedule of medium change shows sequential treatment of specific differentiation cues for each differentiation time windows. *bFGF* basic fibroblast growth factor, *HGF* hepatocyte growth factor, *Dex* dexamethasone, *NA* nicotinamide, *DMSO* dimethyl sulfoxide, *NaB* sodium butylate

Another important factor is the microenvironment including extracellular matrices and scaffolds. We previously reported that culturing ES/iPS cells on a mesonephric cell line, M15, in the presence of specific growth factors, resulted in an efficient induction of endoderm-derived tissues, such as the liver, pancreas, and intestine (6–9). We showed that M15 cells provide basement membrane components, including lama5, on which ES cells could differentiate into regional-specific lineages of the definitive endoderm (10, 11). We then developed an efficient differentiation procedure using synthetic nanofiber matrices for hepatic lineage cells and pancreatic beta cells (12, 13). The nanofiber matrices show a highly integrated three-dimensional structure that resembles the basement membrane, and provide appropriate guidance cues to modulate cell behavior (14). Here, we demonstrate the nanofiber-based procedure for hepatic differentiation from murine and human iPS or ES cells. This procedure including sequential treatment of growth factors and chemicals to induce endoderm and hepatic lineage cells for 22 days in murine and for 26 days in human (Fig. 1).

2 Materials

2.1 Murine iPS Cell Differentiation

1. Culture Plate (96-well plate, Corning Costar Ultra-Web Synthetic Polyamine Surface, 3873XX1).

2. PBS (*see* **Note 1**).
3. 0.25 % Trypsin-EDTA (Invitrogen, 25200-072).
4. Mouse endoderm differentiation basal medium (store at 4 °C)

DMEM (Invitrogen, 11995-075, high Glucose)	500 mL
AlbuMAX II (Invitrogen, 11021-029) (*see* **Note 2**)	6 mL
Insulin-Transferrin-Selenium-G (Invitrogen, 41400-045)	5 mL
Penicillin/streptomycin	
(P/S) (Nacalai Tesque, 26252-94)	5 mL
L-Glutamine (Nacalai Tesque, 16948-04)	5 mL
MEM nonessential amino acids solution	
(NEAA) (Invitrogen, 11140-050)	5 mL
0.1 M 2-mercaptoethanol (*see* **Note 3**)	500 μL

5. Supplements for mouse endoderm differentiation medium

   Activin A (R&D, 338-AC): Stock solution at 10 μg/mL in 0.1 % (w/v) BSA/PBS. Aliquot into 100 μL and store at −80 °C. Once thawed, keep at 4 °C. Add to mouse endoderm differentiation basal medium at a final concentration of 10 ng/mL.

   bFGF (Peprotech, 100-18B-2): Stock solution at 5 μg/mL in 0.1 % (w/v) BSA/ PBS. Aliquot into 100 μL and store at −80 °C. Once thawed, keep at 4 °C. Add to mouse endoderm differentiation basal medium at a final concentration of 5 ng/mL.

6. Mouse hepatic specification basal medium (store at 4 °C)

RPMI (Invitrogen, 11875-093)	500 mL
B27 supplement (Invitrogen, 17504-044)	10 mL
P/S (Nacalai Tesque, 26252-94)	5 mL
L-Glutamine (Nacalai Tesque, 16948-04)	5 mL
NEAA (Invitrogen, 11140-050)	5 mL
0.1 M 2-mercaptoethanol (*see* **Note 3**)	500 μL

7. Supplements for mouse hepatic specification medium

   Stemolecule™ All-Trans Retinoic Acid (ATRA; Stemgent, #130-095-571): Stock solution at 10 mM in DMSO (Sigma, D2650). Aliquot into 100 μL and store at −80 °C. Once thawed, keep at 4 °C with protection from light. Add to mouse hepatic specification basal medium at a final concentration of 10^{-6} M.

8. Mouse hepatic differentiation and maturation basal medium (store at 4 °C)

DMEM (Invitrogen, 11995-075, high Glucose)	500 mL
KSR (Invitrogen, 10828-028)	58 mL
P/S (Nacalai Tesque, #26252-94)	5.8 mL
L-Glutamine (Nacalai Tesque, #16948-04)	5.8 mL
NEAA (Invitrogen, 11140-050)	5.8 mL
0.1 M 2-mercaptoethanol (*see* **Note 3**)	580 μL

9. Supplements for mouse hepatic differentiation medium

   Dexamethasone (Dex) (Sigma, D8893): Stock solution at 1 mM in EtOH. Aliquot into 100 μL and store at −80 °C. Once thawed, keep at 4 °C. Add to mouse hepatic differentiation and maturation basal medium at a final concentration of 1 μM.

   HGF (Peprotech, 100-39): Stock solution at 10 μg/mL in 0.1 % (w/v) BSA/PBS. Aliquot into 100 μL and store at −80 °C. Once thawed, keep at 4 °C. Add to mouse hepatic differentiation and maturation basal medium at a final concentration of 10 ng/mL.

10. Supplements for mouse hepatic maturation medium

    Nicotinamide (Sigma, N0636-100G): Stock solution at 1 M in PBS. Aliquot into 5 mL and store at −20 °C. Once thawed, keep at 4 °C. Add to mouse hepatic differentiation and maturation basal medium at a final concentration of 1 mM.
    Dimethyl Sulfoxide (DMSO) Hybri-Max (Sigma, D2650): Add to mouse hepatic differentiation and maturation basal medium at a final concentration of 1 % (v/v).

11. Mouse iPS plating medium

DMEM (Invitrogen, 11995-075)	500 mL
FBS (Hyclone)	58 mL
P/S (Nacalai Tesque, 26252-94)	5.8 mL
L-Glutamine (Nacalai Tesque, 16948-04)	5.8 mL
NEAA (Invitrogen, 11140-050)	5.8 mL
0.1 M 2-mercaptoethanol (*see* **Note 3**)	580 μL

2.2 Human iPS Cell Differentiation

1. Culture Plate (96-well plate, Corning Costar Ultra-Web Synthetic Polyamine Surface, 3873XX1).
2. PBS (*see* **Note 1**).
3. 0.25 % Trypsin-EDTA (Invitrogen, 25200-072).
4. Matrigel (BD, 354234) (*see* **Note 4**).

5. Y27632 (Wako, 253-00513) (*see* **Note 5**).
6. Human endoderm differentiation basal Medium (store at 4 °C)

RPMI (Invitrogen, 11875-093)	500 mL
B27 supplement (Invitrogen, 17504-044)	10 mL
P/S (Nacalai Tesque, 26252-94)	5 mL
L-Glutamine (Nacalai Tesque, 16948-04)	5 mL
NEAA (Invitrogen, 11140-050)	5 mL
0.1 M 2-mercaptoethanol (*see* **Note 3**)	500 μL

7. Supplements for human endoderm differentiation medium

   Activin A (R&D, 338-AC): Stock solution at 100 μg/mL in 0.1 %(w/v) BSA/ PBS. Aliquot into 50 μL and store at −80 °C. Once thawed, keep at 4 °C. Add to human endoderm differentiation basal medium at a final concentration of 100 ng/mL. Sodium butyrate (Sigma, B5887-250): Stock solution at 1 M in PBS. Aliquot into 50 μL and store at −20 °C. Once thawed, keep at 4 °C. Add to human endoderm differentiation basal medium at a final concentration of 100 μM.
8. Human hepatic specification basal medium (store at 4 °C)

KnockOut DMEM/F12 (Invitrogen, 12660-012)	500 mL
KSR (Invitrogen, 10828-028)	125 mL
P/S (Nacalai Tesque, 26252-94)	6.5 mL
L-Glutamine (Nacalai Tesque, 16948-04)	6.5 mL
NEAA (Invitrogen, 11140-050)	6.5 mL
0.1 M 2-mercaptoethanol (*see* **Note 3**)	650 μL

9. Supplements for human hepatic specification medium

   Dimethyl Sulfoxide (DMSO) Hybri-Max (Sigma, D2650). Add to human hepatic specification basal medium at a final concentration of 1 % (v/v).
10. Human hepatic differentiation and maturation basal medium (store at 4 °C)

DMEM (Invitrogen, 11995-075, high Glucose)	500 mL
KSR (Invitrogen, 10828-028)	58 mL
P/S (Nacalai Tesque, 26252-94)	5.8 mL
L-Glutamine (Nacalai Tesque, 16948-04)	5.8 mL
NEAA (Invitrogen, 11140-050)	5.8 mL
0.1 M 2-mercaptoethanol (*see* **Note 3**)	580 μL

11. Supplements for human hepatic differentiation medium

    Dexamethasone (Dex) (Sigma, D8893): Stock solution at 1 mM in EtOH. Aliquot into 100 μL and store at −80 °C. Once thawed, keep at 4 °C. Add to human hepatic differentiation and maturation basal medium at a final concentration of 1 μM.

    HGF (Peprotech, 100-39): Stock solution at 10 μg/mL in 0.1 % (w/v) BSA/PBS. Aliquot into 100 μL and store at −80 °C. Once thawed, keep at 4 °C. Add to human hepatic differentiation and maturation basal medium at a final concentration of 10 ng/mL.

12. Supplements for human hepatic maturation medium

    Nicotinamide (Sigma, N0636-100G): Stock solution at 1 M in PBS. Aliquot into 5 mL and store at −20 °C. Once thawed, keep at 4 °C. Add to human hepatic differentiation and maturation basal medium at a final concentration of 0.5 mM.

    Dimethyl Sulfoxide (DMSO) Hybri-Max (Sigma, #D2650): Only for 2 step method, add to human hepatic differentiation and maturation basal medium at a final concentration of 0.5 % (v/v).

13. Trypsin stop medium

DMEM (Invitrogen, 11995-075)	500 mL
FBS (Hyclone)	58 mL
P/S (Nacalai Tesque, 26252-94)	5.8 mL
L-Glutamine (Nacalai Tesque, 16948-04)	5.8 mL
NEAA (Invitrogen, 11140-050)	5.8 mL
0.1 M 2-mercaptoethanol (*see* **Note 3**)	580 μL

14. Human iPS cell plating medium

    Use appropriate maintenance medium for your ES cells or iPS cells treated with Y27632 at a final concentration of 10 μM.

3 Methods

Carry out all procedure in clean bench and keep the cells in CO_2 incubator with 37 °C, 90 % humidity and 5 % CO_2.

All medium and solution should be use at room temperature or warmed to 37 °C. DO NOT use cold medium at a refrigerator.

Take 200 μL of every differentiation medium for one well of 96-well nanofiber plates.

3.1 Murine iPS Cell: Plating

1. Remove iPS maintenance medium from the cells.
2. Wash with PBS.

3. Add 1 mL 0.25 % trypsin-EDTA solution to the culture dish. Let stand for 5 min at 37 °C and confirm under microscope for detachment of cells.
4. Disperse the cells into a single-cell suspension by pipetting with a P1000 pipet.
5. Add 4 mL mouse iPS plating medium and collect the cells by centrifugation at 180 × *g* for 5 min.
6. Resuspend the pellet with mouse iPS plating medium, cell count, and adjust to a final cell density of 2.5×10^4 cells/mL.
7. Plate 200 μL ES cell suspension into each well of 96-well synthetic nanofiber plate.
8. Incubate at 37 °C under 5 % CO_2 overnight.

3.2 Murine iPS Cell: Differentiation

1. Change medium with fresh mouse endoderm differentiation medium supplemented with both activin and bFGF on day 1, 3, 5, and 7 (*see* **Note 6**).
2. Change medium with fresh mouse hepatic specification medium supplemented with ATRA on day 9, and culture for 24 h.
3. Change medium with fresh mouse hepatic differentiation medium supplemented with both Dex and HGF on day 10 and 12.
4. Change medium with fresh mouse hepatic maturation medium supplemented with all of Dex, HGF, nicotinamide, and DMSO on day 14 and 16. By changing medium every 2 days it is capable to extend culture.

3.3 Human iPS Cell: Preconditioning

1. In daily changing fresh medium, treat human ES cells or iPS cells with 10 μM Y27632 (ROCK inhibitor) 1 day before plating.

3.4 Human iPS Cell: Plate Preparation

1. Add 50 μL of ten times diluted Matrigel stock solution (final 20 times dilution) onto each well of nanofiber 96-well plate and incubate for more than 3 h at 37 °C under 5 % CO_2.

3.5 Human iPS Cell: Plating

1. Remove medium from the cells.
2. Wash with PBS.
3. Add 1 mL 0.25 % trypsin-EDTA solution to the culture dish. Let stand for 5 min at 37 °C and confirm under microscope for detachment of cells.
4. Disperse the cells into a single-cell suspension by pipetting with a P1000 pipet.
5. Add 4 mL Trypsin stop Medium and collect the cells by centrifugation at 4 °C, 180 × *g* for 5 min.

6. Resuspend the pellet with appropriate ES (iPS) maintenance medium, cell count, and adjust to a final cell density of 5.0×10^5 cells/mL. Add Y27632 into cell suspension to adjust final concentration to 10 μM.
7. Plate 200 μL ES cell suspension into each well of 96-well synthetic nanofiber plate.
8. Incubate at 37 °C under 5 % CO_2 overnight.

3.6 Human iPS Cell: Differentiation

Change medium every 2 days in the indicated period with specified medium.

1. Change medium with fresh human endoderm differentiation medium supplemented with both activin and sodium butyrate on day 1, 3, 5, 7, and 9 (*see* **Note 6**).
2. Change medium with fresh human hepatic specification medium supplemented with DMSO on day 11, 13, and 15.
3. Change medium with fresh human hepatic differentiation medium supplemented with both Dex and HGF on day 17, 19, and 21.
4. Change medium with fresh human hepatic maturation medium supplemented with all of Dex, HGF, nicotinamide on day 23 and 25. By changing medium every 2 days it is capable to extend culture.

3.7 Human iPS Cell: Differentiation (2 Step Methods, Alternative for Responsive Cell Line)

1. Change medium with fresh human endoderm differentiation medium supplemented with both activin and sodium butyrate on day 1, 3, 5, 7, and 9.
2. Change medium with fresh human hepatic differentiation medium supplemented with all of Dex, HGF, nicotinamide, and DMSO every 2 days from day 11 to 25. By changing medium every 2 days it is capable to extend culture.

4 Notes

1. Dissolve three tablets of PBS (SIGMA, #P4417-100TAB) in 600 mL ultrapure water, autoclave, and store at room temperature.
2. Dissolve 25 g AlbuMAX II in 125 mL ultrapure water with stirring. Sterilize them with filtration (Millipore, SCGPS05RE). Aliquot into 2 mL and store at −20 °C.
3. Dilute 2-mercaptoethanol (Sigma, M7522) to 0.1 M with PBS (i.e., 2-mercaptoethanol 100 μL/PBS 14.1 mL). Store at 4 °C and use within 1 month.

4. Dilute 5 mL Matrigel with 5 mL DMEM (Invitrogen, 11995-075, high Glucose). Aliquot into 100 μL and store at −20 °C. Dilute 10 times with DMEM before use.
5. Dissolve 5 mg Y27632 in 1.5 mL sterile ultrapure water to make 10 mM stock solution. Aliquot into 50 μL and store at −80 °C.
6. While changing medium for the first time, as cells are attached on the nanofiber surface weakly, dispense the fresh medium gently, not blowing them up.

References

1. Takahashi K, Tanabe K, Ohnuki M, Narita M, Ichisaka T, Tomoda K et al (2007) Induction of pluripotent stem cells from adult human fibroblasts by defined factors. Cell 131:861–872
2. Takahashi K, Yamanaka S (2006) Induction of pluripotent stem cells from mouse embryonic and adult fibroblast cultures by defined factors. Cell 126:663–676
3. Basma H, Soto-Gutiérrez A, Yannam GR, Liu L, Ito R, Yamamoto T et al (2009) Differentiation and transplantation of human embryonic stem cell-derived hepatocytes. Gastroenterology 136:990–999
4. Si-Tayeb K, Noto FK, Nagaoka M, Li J, Battle MA, Duris C et al (2010) Highly efficient generation of human hepatocyte-like cells from induced pluripotent stem cells. Hepatology 51:297–305
5. D'Amour KA, Agulnick AD, Eliazer S, Kelly OG, Kroon E, Baetge EE (2005) Efficient differentiation of human embryonic stem cells to definitive endoderm. Nat Biotechnol 23:1534–1541
6. Shiraki N, Yoshida T, Araki K, Umezawa A, Higuchi Y, Goto H et al (2008) Guided differentiation of embryonic stem cells into Pdx1-expressing regional-specific definitive endoderm. Stem Cells 26:874–885
7. Shiraki N, Umeda K, Sakashita N, Takeya M, Kume K, Kume S (2008) Differentiation of mouse and human embryonic stem cells into hepatic lineages. Genes Cells 13:731–746
8. Umeda K, Suzuki K, Yamazoe T, Shiraki N, Higuchi Y, Tokieda K et al (2012) Albumin gene targeting in human embryonic stem cells and induced pluripotent stem cells with helper-dependent adenoviral vector to monitor hepatic differentiation. Stem Cell Res 10:179–194
9. Ogaki S, Shiraki N, Kume K, Kume S (2013) Wnt and Notch signals guide embryonic stem cell differentiation into the intestinal lineages. Stem Cells 31:1086–1096
10. Higuchi Y, Shiraki N, Yamane K, Qin Z, Mochitate K, Araki K et al (2010) Synthesized basement membranes direct the differentiation of mouse embryonic stem cells into pancreatic lineages. J Cell Sci 123:2733–2742
11. Shiraki N, Yamazoe T, Qin Z, Ohgomori K, Mochitate K, Kume K et al (2011) Efficient differentiation of embryonic stem cells into hepatic cells in vitro using a feeder-free basement membrane substratum. PLoS One 6: e24228
12. Yamazoe T, Shiraki N, Toyoda M, Kiyokawa N, Okita H, Miyagawa Y et al (2013) A synthetic nanofibrillar matrix promotes in vitro hepatic differentiation of embryonic stem cells and induced pluripotent stem cells. J Cell Sci 126:5391–5399
13. Sakano D, Shiraki N, Kikawa K, Yamazoe T, Kataoka M, Umeda K et al (2014) VMAT2 identified as a regulator of late-stage β-cell differentiation. Nat Chem Biol 10:141–148
14. Nur-E-Kamal A, Ahmed I, Kamal J, Schindler M, Meiners S (2006) Three-dimensional nanofibrillar surfaces promote self-renewal in mouse embryonic stem cells. Stem Cells 24:426–433

Methods in Molecular Biology (2016) 1357: 485–486
DOI 10.1007/7651_2015_308

Published online: 18 November 2015

Erratum to: Generation of iPS Cells from Granulosa Cells

Jian Mao and Lin Liu

Erratum to: Methods in Molecular Biology
DOI: 10.1007/7651_2015_135

In a previous version of this chapter an incorrect figure caption was given for Figs. 2 and 3. The correct figure caption is shown below.

Fig. 2 Characterization of mouse iPS cells generated from mGC. (**a**) Schematic diagram of mGCiPSC (mGC-OSiPSCs) generation by Oct4 and Sox2. Scale bar = 100 μm. (**b**) mGC-OSiPS clones express strong Oct4, Nanog, and SSEA1 by immunofluorescence (scale bar = 20 μm). Nuclei stained with Hoechst 33342 (*blue*). (**c**) Differentiation in vitro of mGC-OSiPS cells by embryoid body (EB) formation. The differentiated derivatives by EB consist of cells representing three embryonic germ layers as shown by immunofluorescence staining using endoderm marker alpha 1-fetoprotein (AFP), mesoderm marker smooth muscle actin (SMA), and ectoderm marker β-III-tubulin (scale bar = 50 μm). Nuclei stained with Hoechst 33342 (*blue*). (**d**) Differentiation in vivo of mGC-OSiPS cells by teratoma formation test following injection into nude mice. *Red arrows* indicate teratoma on the back of nude mice. Hematoxylin and eosin staining of teratoma tissues derived from mGC-OSiPS cells. All teratomas consist of representative derivatives of three germ layers, including epidermis (ectoderm), muscle (mesoderm), and gland epithelium (endoderm). Scale bar = 100 μm. (**e**) Chimeric mice generated from mGC-OSiPS, based on coat color

The online version of the original chapter can be found at
DOI 10.1007/7651_2015_135

Fig. 3 Characterization of pig iPS cells generated from pGC. (**a**) Schematic diagram of pGCiPSC (R7F-piPSCs) generation by 7F (Oct4, Sox2, Klf4, c-Myc, Nanog, hTert, and p53i) and cultured under R2i (PD0325901 and SB431542) medium. Scale bar = 100 μm. (**b**) pGCiPS express strong Oct4, Sox2, and SSEA4 by immunofluorescence (scale bar = 20 μm). Nuclei stained with Hoechst 33342 (*blue*). (**c**) Differentiation in vitro of pGCiPS cells by embryoid body (EB) formation. The differentiated derivatives by EB consist of cells representing three embryonic germ layers as shown by immunofluorescence staining using endoderm marker alpha 1-fetoprotein (AFP), mesoderm marker smooth muscle actin (SMA), and ectoderm marker β-III-tubulin (scale bar = 20 μm). Nuclei stained with Hoechst 33342 (*blue*)

Methods in Molecular Biology (2016) 1357: 487–496
DOI 10.1007/978-1-4939-3055-5

INDEX

K

L

M

O

P

MIX
Papier aus verantwortungsvollen Quellen
Paper from responsible sources
FSC® C105338

If you have any concerns about our products,
you can contact us on
ProductSafety@springernature.com

In case Publisher is established outside the EU,
the EU authorized representative is:
Springer Nature Customer Service Center GmbH
Europaplatz 3, 69115 Heidelberg, Germany

Printed by Libri Plureos GmbH
in Hamburg, Germany